Marginal Basin Geology

Volcanic and associated sedimentary
and tectonic processes in modern and ancient
marginal basins

Marginal Basin Geology

Volcanic and associated sedimentary
and tectonic processes in modern and ancient
marginal basins

edited by
B. P. Kokelaar & M. F. Howells
School of Environmental Sciences, Ulster
Polytechnic, Co. Antrim, and Wales Geological
Survey Unit, British Geological Survey,
Aberystwyth

1984

Published for
The Geological Society
by Blackwell Scientific Publications
Oxford London Edinburgh
Boston Palo Alto Melbourne

Published by
Blackwell Scientific Publications
Osney Mead, Oxford OX2 0EL
8 John Street, London WC1N 2ES
23 Ainslie Place, Edinburgh EH3 6AJ
52 Beacon Street, Boston, Massachusetts 02108, USA
706 Cowper Street, Palo Alto, California 94301, USA
107 Barry Street, Carlton, Victoria 3053, Australia

First published 1984

DISTRIBUTORS

USA and Canada
   Blackwell Scientific Publications Inc.
   PO Box 50009, Palo Alto
   California 94303
Australia
   Blackwell Scientific Book Distributors
   31 Advantage Road, Highett
   Victoria 3190

British Library Cataloguing in Publication Data

Marginal basin geology : volcanic and associated
   sedimentary and tectonic processes in modern and
   ancient marginal basins. (Geological Society
   special publications, ISSN 0305-8719; No. 16)
   1. Submarine geology
   I. Kokelaar, B.P.   II. Howells, M.F.
   III. Series
   551.46'08   QE39

   ISBN 0-632-01073-8

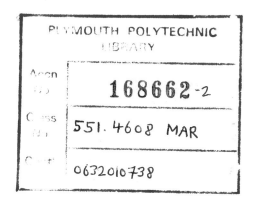
Typeset by Spire Print Services Ltd,
Salisbury, Wilts
Printed in Great Britain by
Alden Press Ltd, Oxford

Contents

Preface

Relatively small semi-isolated oceanic basins, spatially associated with active or inactive volcanic-arc and trench systems, are common features at the margins of major oceans. Many such marginal basins clearly show evidence of development by crustal extension during subduction of oceanic lithosphere, commonly in a position above the descending slab. The extension may occur before the development of an arc or be initiated in a fore-arc, intra-arc or back-arc position. Some basins, however, are not directly related to subduction but develop due to irregular and complex motions within a system of small plates, caused by movements of major plates outside the system. Other basins occur which are trapped fragments of major ocean lithosphere, for example where an oceanic transform boundary has changed to a trench with active subduction.

Basins developed within continental crust at destructive plate margins are also widely recognized but are less common. These ensialic marginal basins may show slight to marked crustal thinning, with or without complete rifting and subsequent emplacement of oceanic lithosphere.

The apparent paradox of the occurrence of extensional tectonism in destructive plate margin environments has been considered widely and was discussed in some detail at a meeting of the Royal Society in 1980 (*Phil. Trans. R. Soc. London*, **A300**, 1981). Various models have been proposed to account for the major features of marginal basins, and the results from deep-sea drilling and geophysical surveys have been of paramount importance in their formulation.

The determination of plate tectonic environments and processes is more difficult in older rocks. The spatial relationships and tectonic settings become increasingly obscured by deformation, metamorphism and erosion in long-lived orogenic belts, and by such modification in successive orogenies. Commonly, in studies of Lower Palaeozoic rocks for example, the original environment has to be reconstructed after it has been fragmented, with many of the fragments well removed from their initial positions, some lost and perhaps with a large proportion of the remainder being obscured by younger strata. In addition, the rock chemistry may be substantially changed. However, despite these factors, most types of extant marginal basins have been recognized in the geological record, generally as variously deformed tectonic slices, with numerous examples from the Cenozoic and Mesozoic, and increasing numbers from the Palaeozoic. Also, it has become apparent that many ophiolite complexes originated in marginal basin settings.

Because of the wealth of data available from modern basins and the wide recognition of ancient analogues, it seemed timely in 1982 to convene a meeting to consider marginal basins with emphasis on rock types, associations and processes. The aim was to establish a dialogue between researchers studying modern and ancient environments wherein the data from the modern basins would provide a framework for more precise determination of the origin of the ancient rocks, and also detailed studies from the geological record would facilitate improved understanding of the widely spaced deep-sea samples and geophysical profiles. The meeting held on the 9th and 10th September 1982 at Keele University, was organized by the Volcanic Studies Group of the Geological Society. This volume consists largely of papers presented at the meeting and includes a modified version of a field guide prepared for the pre-conference field discussion meeting (1–8 September 1982). Although by no means exhaustive, the volume reflects the state of research into marginal basin geology and emphasizes the need to integrate observations from modern and ancient basins. Hopefully, in the diverse topics and approaches, the interested reader will find new perspectives and stimulation for future research.

ACKNOWLEDGMENTS: We are indebted to Professor G. Kelling and Dr R. A. Roach of the Department of Geology at Keele University for housing the conference so successfully. Financial support for the conference was provided by the Geological Society and the Royal Society. We particularly thank Professor B. W. Langlands of the School of Environmental Sciences at Ulster Polytechnic for his continuous encouragement and support in many ways during both the preparation of the meeting and the completion of the volume. We also thank the Director of the British Geological Survey for his sustained interest and support. The very considerable assistance throughout from Drs R. A. Roach, R. E. Bevins and A. J. Reedman is greatly appreciated. We especially acknowledge the exceptional commitments of Pat

Andrews and to her we are most grateful. Lastly, we thank the numerous referees and advisors for their interest and time, and the authors for their forbearance.

B. PETER KOKELAAR, School of Environmental Sciences, Ulster Polytechnic, Shore Road, Newtownabbey, Co. Antrim BT 37 0QB, U.K. Present address: Department of Environmental Studies, University of Ulster, Jordanstown, Newtonabbey, Co. Antrim BT37 0QB, U.K.

MALCOLM F. HOWELLS, Wales Geological Survey Unit, British Geological Survey, Bryn Eithyn Hall, Llanfarian, Aberystwyth, Dyfed SY23 4BY, U.K.

PROCESSES

Submarine volcaniclastic rocks

R. V. Fisher

SUMMARY: The type, relative abundance and stratigraphical relationships of volcanic rocks that comprise island volcanoes are a function of (i) depth of extrusion beneath water, (ii) magma composition, and (iii) lava–water interactions. The water depth at which explosions can occur is called the pressure compensation level (PCL) and is variable. Explosive eruptions that occur above the PCL and below sealevel can give rise to abundant hydroclastic and pyroclastic debris. Below the PCL, clastic material cannot form explosively; it forms from lava by thermal shock. The volcaniclastic products are widely dispersed in basins adjacent to extrusion sources by three principal kinds of marine transport processes. These are slides, sediment gravity flows and suspension fallout. Volcaniclastic debris can be derived in subaqueous and subaerial-to-subaqueous environments (i) directly from eruptions, (ii) from remobilization of juvenile volcaniclastics, or (iii) from epiclastic material which initially develops above sealevel.

Sediment gravity flows (fluids driven by sediment motion) exhibit the phenomenon of *flow transformation*. This term is used here for the process by which (i) sediment gravity flow behaviour changes from turbulent to laminar, or vice versa, within the body of a flow, (ii) flows separate into laminar and turbulent parts by gravity, and (iii) flows separate by turbulent mixing with ambient fluid into turbulent and laminar parts. Dominant kinds of subaqueous volcaniclastic sediment gravity flows are debris flows, hot or cold pyroclastic flows and turbidites. Fine grained material can be thrown into suspension locally during flow transformations or underwater eruptions, but thin, regionally distributed subaqueous fallout tephra is mostly derived from siliceous Plinian eruptions.

Volcaniclastic rocks in marine sequences occur in many kinds of sedimentary environments and tectonic settings (Table 1). Some are the result of eruptions on land which deliver fallout ash, lava and pyroclastic flows into water. Others are derived from underwater eruptions which extrude lava flows and volcanic fragments of various kinds, or are pyroclastic, hydroclastic or epiclastic materials reworked from land or remobilized under water. This review is mostly concerned with subaqueous sediment gravity flow deposits, because of their abundance in marine environments adjacent to volcanic regions.

The site of most voluminous volcanism is at divergent plate boundaries where basaltic sheet flows, pillow basalts and smaller amounts of pillow breccias and hyaloclastites are formed. An estimated 4–6 km^3 of such material is added each year to the Earth's crust at mid-ocean ridges (Nakamura 1974). The second important site of volcanism is at convergent plate boundaries. Basaltic and andesitic island arcs develop at converging oceanic plates; converging oceanic and continental plates give rise to dominantly andesitic to rhyolitic volcanic chains on the edges of continents. Perhaps the most varied and complex environments are the different kinds of back-arc settings. A third important site of oceanic volcanism is represented by intra-plate seamounts and ocean islands. Sub-marine volcaniclastic deposits are most abundant near island arcs and ocean islands because large volumes of volcaniclastic material are transported from land into the sea during eruptions and during subsequent erosion. Volcanogenic sedimentation and tectonics, an area of considerable research, is reviewed by Mitchell & Reading (1978).

Underwater eruptions

Whether underwater eruptions are effusive or explosive is determined by (i) the depth (pressure) of the water column (Fig. 1), (ii) the composition of the magma, especially amounts of volatiles, and (iii) the extent of interaction between magma and water. Subsequent transport depends upon slope, which in turn partly depends upon the growth rate of the volcano and the initial lava-to-clastic ratio of extruded products. The manner of transport also depends upon whether or not eruption columns can develop. When vapour pressure in the magma exceeds water pressure, vesiculation commences. As depth decreases, vesicles become more abundant (Moore 1965, 1970; Moore & Schilling 1973) and, at a shallow level, explosions caused by exsolution of magmatic volatiles can occur. This depth, which is variable, is here called the pressure compensation level (PCL) and 'volatile fragmentation depth' by

5

TABLE 1. Simplified partitioning of environment, kind of extrusion, transport and emplacement processes and some typical pyroclastic deposits

Environment	Eruptions	Transport	Emplacement	Deposits
		Directly from eruptions		
Subaqueous (marine, lacustrine, sub-ice)	Effusive (shallow or deep)	Lava flows	Congealing flows	Massive flows; pillow lavas; hyaloclastites.
	Explosive (shallow)	Dilute suspension in water	Suspension fallout	Thin, well sorted, normally graded beds. May be reworked by bottom currents. Local occurrence near source.
		Pyroclastic flows	Suspension fallout (dilute suspensions from tops of flows).	Thin, fine grained well sorted, normally graded beds. May be reworked by bottom currents. Rest on turbidites.
			Turbulent flows from tops of mass flows.	Thin sequence of fairly well sorted beds, may be cross-bedded. Tops may be reworked by bottom currents. Rest on pyroclastic flows.
			Laminar mass flows.	Thick, poorly sorted, poorly bedded, non-welded. Mixing with water can result in lahars. Tops may be reworked by bottom currents. May contain rip-ups. Bases erosive to non-erosive.

(Flows also develop from slump and flow of water-logged pyroclastic debris on subaqueous slopes of volcanoes to give turbidites and lahars.)

Subaerial eruptions with subaqueous deposition

— Effusive —— Lava flows from land into water. —— Congealing flows, pillows, broken pillows. Explosive disruption. —— Massive flows. Broken pillow complexes and hyaloclastites. Explosive disruption may produce littoral cones.

— Explosive —— Turbulent suspensions in air. —— Fallout on water to bottom. —— Thin, well sorted, normally graded beds. Sharp bases, bioturbated tops. May be in deep sea 100s of km from source.

Pyroclastic flows from land into water (may be destroyed by explosive disruption upon entering water).

—— Dilute suspension fallout from tops of flows. —— Thin, well sorted, normally graded. Partly derived from air fall and surge from land which does not enter water. Rest on turbidites. May be reworked by bottom currents.

—— Turbulent flows from tops of mass flows. —— Thin sequence of fairly well sorted beds, may be cross-bedded. Rest on pyroclastic flows. Tops may be reworked by bottom currents.

—— Laminar mass flows. —— Thick, poorly sorted, poorly bedded. May be welded to base.

Subaerial

— Effusive —— Flows —— Congealing flows —— Massive forms, sometimes mostly rubble. No pillows.

— Explosive —— Ballistic; turbulent suspension in air. —— Fallout from air —— Thin to thick, well sorted; may show normally graded bedding.

Pyroclastic flows.

—— Fallout. Derived from tops of flows. —— Thin, well sorted beds.

—— Turbulent suspensions (pyroclastic surge.) —— Thin sequences, fairly well sorted, commonly well bedded, may be cross-bedded.

—— Laminar mass flows. —— Thick to thin, poorly sorted, massive to poorly bedded, welded to non-welded.

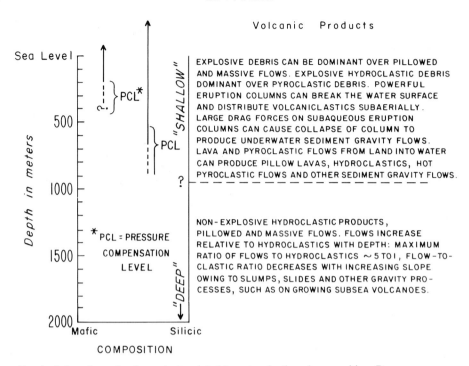

FIG. 1. Submarine volcanic products related to water depth and composition. Pressure compensation levels (PCL) are not specifically known and depend upon magmatic gas pressures, volumes and expansion rates relative to pressure exerted by the water column.

(Fisher & Schmincke 1984). It might exceed 1000 m for silicic and volatile-rich mafic alkalic magmas, but apparently is less than 500 m for most mafic basaltic magmas (McBirney 1963), well within the range of marine shelves that fringe continents and many islands, and of the tops of some submerged islands. Geological evidence suggests that the PCL for explosive alkali basaltic volcanism may be 500–1000 m (Staudigel & Schmincke 1981), but it is less than 200 m below water level for most basaltic eruptions (Moore & Fiske 1969; Jones 1970; Fleet & McKelvey 1978; Allen 1980; Furnes *et al.* 1980).

In addition to pyroclastic fragmentation caused by explosive expansion of volatiles from within magma, there are hydroclastic processes, i.e. fragmentation caused by contact of magma and water (Fisher & Schmincke 1984). Hydroclastic processes can be non-explosive or explosive. Explosive processes can only take place above the PCL, but non-explosive fragmentation can occur at all depths.

Non-explosive hydroclastic processes include granulation and thermal spalling due to stresses set up in magma undergoing rapid quenching (Rittmann 1962; Carlisle 1963; Honnorez

1972), kinetic processes of falling, slumping and breaking of detached pillows on steep slopes, rupturing of growing pillows (Rittmann 1962) and implosions of evacuated pillows where water pressure exceeds internal pillow pressures (Moore 1975).

Explosive hydroclastic processes occur when (i) pore water in rocks is rapidly vaporized by an underlying magma body, or when (ii) water becomes mixed with magma beneath the surface, or under a lava at the surface. Mixing processes are complex (e.g. see Bennett 1972; Colgate & Sigurgeirsson 1973; Wohletz 1980; Kokelaar 1983). Pillowed lava flows and hyaloclastites, either mafic or silicic in composition, are prime evidence of subaqueous environments. In areas where fossils or other evidence (e.g. turbidites, carbonates, and facies associations such as shelf, slope or submarine fan) are absent, pillow lavas are commonly the only evidence of a subaqueous environment.

Mafic pillowed and sheet lava are the two chief forms of lava flows at mid-ocean ridges, as directly observed by submersibles and inferred from drilling through the ocean crust (Bellaiche *et al.* 1974; Needham & Francheteau 1974; Ballard & Moore 1977; Lonsdale 1977; Ballard *et al.* 1979). Mafic volcaniclastic debris is commonly overlooked in deep-sea environments because rock recovery from drill cores averages about 20% and gives a bias in abundance toward massive lava. Robinson *et al.* (1980), however, obtained up to 20% hyaloclastite in Cretaceous crust on the Bermuda Rise from three drill site cores with an average recovery of 70%.

Subaqueous mafic pillow flows and hyaloclastites are most abundant but silicic pillows and hyaloclastites have recently been described (Bevins & Roach 1979; Furnes *et al.* 1980; De Rosen-Spence *et al.* 1980). According to Dimroth *et al.* (1979), in comparison to subaqueous basic lavas, subaqueous silicic flows are thicker and less extensive, lava pods and lobes are similar but larger, quickly chilled rinds on pods and lobes are thicker, and vesicles are larger.

It is generally thought that massive silica-rich lava flow sheets cannot form in submarine environments. However, thick, massive and extensive (up to 2000 km²) dacite-andesite and rhyodacite lava flows (Cas 1978) are interbedded and conformable with marine Palaeozoic flysch deposits in Australia. Cas (1978) suggests that the lava flows were erupted in water depths great enough to prevent escape of most volatiles. The flows remained mobile because volatiles were retained, and the rate of viscosity increase from cooling was considerably less than the rate of emplacement.

Transport processes and deposits

There are four principal kinds of clastic (volcanic or non-volcanic) deposits on the sea floor. These are, from proximal to distal locations, slump and slide (olistostromes), mass flow, turbidite and pelagic rain (suspension fallout) (Fig. 2). In volcanic as well as in non-volcanic regions, initial debris can be epiclastic, but in volcanic regions there also can be subaqueous lava flow, pyroclastic flow and suspension fallout deposits derived directly from subaqueous or subaerial eruptions (Tables 1 and 2). Moreover, pyroclastic material deposited on steep volcano slopes or shelf margins can be later remobilized as slides and transformed to submarine lahars, turbidity currents and suspension fallout (Fig. 3). The presence of primary and reworked pyroclastic and hydroclastic debris indicates penecontemporaneous volcanism (and tectonism) whereas epiclastic volcanic debris may not.

Flow transformations

Middleton & Hampton (1973, 1976) distinguish between (i) fluid gravity flow in which fluid is moved by gravity and entrains or drives the sediment parallel to the bed (e.g. rivers, ocean currents), and (ii) sediment gravity flow in which movement is by gravity and the sediment motion moves the interstitial fluid (e.g. grain flow, debris flow) (Fig. 4). Lowe (1979) and Nardin *et al.* (1979) have classified sediment gravity flows based upon flow rheology and particle support mechanism.

Sediment gravity flows exhibit the phenomenon of *flow transformation* (Fisher 1983). This term refers to changes within a flow from laminar to turbulent, or vice versa, in turn related chiefly to particle concentration, thickness of the flow and flow velocity (slope). Transformations from slides or slumps to flows (e.g. Naylor 1980) could be considered in another class of transformations but are not discussed here. Stanley *et al.* (1978) use the word 'transforma-

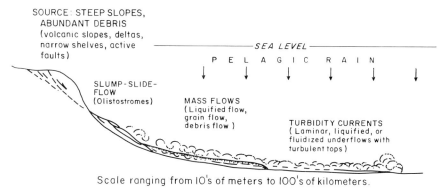

Fig. 2. Conditions for initiation of flows and spectrum of main subaqueous transport processes in non-volcanic regions, but which also occur on subaqueous slopes of volcanoes (see Fig. 3).

TABLE 2. *Simplified list of particles, processes, environments and rocks in subaqueous settings*

Origin of particles	Processes of emplacement[b]	Kinds of deposits	Sites of origin[b]	Sites of deposition[c]
Hydroclastic[a]	Slides	Olistostromes	Volcanic	Beach
Pyroclastic	Laminar flow	Debris flows	eruption	Delta
Epiclastic	Turbulent flow	Turbidites	Volcano	Shelf
	Suspension fallout	Suspension fallout	slopes	Slope
		Pyroclastic flows	Sedimentary	Fan
			aprons	Plain

[a] Can be of explosive or non-explosive origin.
[b] Subaerial or subaqueous.
[c] Sediments can be supplied directly from an eruption or can be remobilized from steep slopes such as delta fronts, continental slopes, etc. Lava flows can occur in proximal environments.

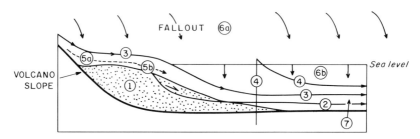

FIG. 3. Sources for subaqueous volcanic debris, and transport processes. This figure combines conditions given in Fig. 2 with volcanics derived directly from an active volcano. (1) Clastic wedge. Hydroclastic, pyroclastic and epiclastic debris. Dominance of particle type determined by type of volcanic activity and periods of inactivity. (2) Slide–debris flow–turbidity current transitions. (3) Pyroclastic flow from land into water. Transitions to subaqueous lahars and turbidity currents. (4) Pyroclastic flows and suspension fallout from powerful shallow underwater eruptions. Column may or may not breach water surface. (5) Lava flows from land into water; (5a) massive flows; (5b) pillows and hyaloclastites. (6) Suspension fallout; (6a) from subaerial eruptions into water; (6b) from underwater eruptions or flow transformations. (7) Deep eruptions below PCL; flows and pillow lavas.

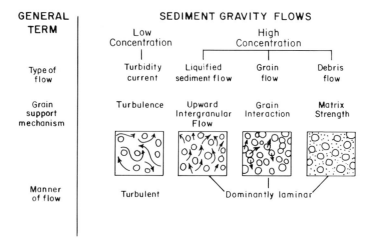

FIG. 4. Classification of sediment gravity flows. Modified from Middleton & Hampton (1973, 1976).

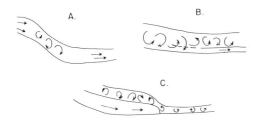

FIG. 5. Schematic representation of (a) body transformations; (b) gravity transformations; and (c) surface transformations. More than one kind of transformation can occur in a single flow and, following separation of flows, transformations can occur again.

tion' but they describe 'transitions' of depositional features, from mass flow to turbidity current deposits, rather than the transformation processes as is done here.

At least three kinds of flow transformations appear to be important in subaqueous environments (Fig. 5):

(i) *Body transformations*, where flow behaviour changes from turbulent to laminar, or vice versa, depending upon the Reynolds and Bingham numbers (see below) within the body of the flow, without significant additions or losses of water.

(ii) *Gravity transformations*, in which flows are initially turbulent but become gravitationally segregated to form a high-concentration laminar underflow with an overriding lower concentration turbulent flow that can move independently.

(iii) *Surface transformations*, where flow separations arise due to the drag effects of water on the top, or nose, of a flow, or due to incorporation of fluid at a hydraulic jump or beneath the nose of a flow.

The concept of flow transformation focuses upon the mass dynamics of flow which chiefly determine sedimentary textures, structures and sequential relationships of massive and laminated units. Various combinations of these three kinds of flow transformations evolving from one another are explained in Fig. 5. Also, fine grained sediment may be thrown into suspension during transformation to produce suspension fallout. *Fluidization transformations* develop by elutriation of fine particles by upward moving fluids from a dense phase bed to produce a turbulent dilute phase above the bed. This type of transformation is best known in gas–solids systems and is believed to produce ash cloud surges from the upper parts of hot pyroclastic flows (Sparks 1976; Fisher 1979; Sheridan 1979; Wilson 1980; Moore & Sisson 1981), and possibly from base surges (Wohletz & Sheridan 1979). Fluidized water-saturated mass flows may occur in subaqueous settings (Middleton & Hampton 1973, 1976; Carter 1975), but the products of elutriation have not been described or perhaps not recognized in the subaqueous realm. As discussed below, in the section on subaqueous pyroclastic flows, gas–solids suspensions might also occur beneath water.

The phenomenon of body transformations is suggested by the experiments and speculations of Kuenen (1952) and Morgenstern (1967) who proposed that slumps or debris flows can change to turbidity currents, with no change in water content, when their velocity is great enough to produce internal turbulence. This can be related to Reynolds and Bingham numbers. Middleton (1970) speculated that changes, from laminar to turbulent and back to laminar flow can occur in large subaqueous grain flows. Fisher's (1971) description of the Parnell Grit, New Zealand, where there are large shale rip-ups in a massive 9 m thick, poorly sorted bed with an inversely graded base, suggests that transformation from turbulent to laminar flow occurred but does not demonstrate a preceding laminar to turbulent transformation.

Gravity transformations are shown experimentally to occur in high-concentration turbidity currents with the development of a 'quick bed' (Middleton 1967). The quick bed is a laminar flowing grain flow and/or fluidized sediment flow (Gonzalez-Bonorino & Middleton 1976) which forms at the base of an initially turbulent high-concentration turbidity current (solids concentration ≥ 30% by volume) after passage of the head. Lowe (1982) describes gravity transformations and resulting deposits in high-concentration turbidity currents. He concludes that individual flows become gravitationally segregated, with low-density flows evolving from high-density flows as grain concentration increases toward the base of a flow. Gravity transformations also occur subaerially where initially turbulent pyroclastic flows develop into low-concentration, turbulent pyroclastic surges flowing above high-concentration, laminar block-and-ash flows (Fisher *et al*. 1980; Fisher & Heiken 1982). The phenomenon has also been invoked to explain features of the blast deposits derived from the 18 May 1980 eruption of Mount St Helens (Moore & Sisson 1981). The pyroclastic surge may undergo continued gravity transformation to develop a secondary block-and-ash flow.

Hampton (1972) reviews various subaqueous transformations and shows how surface transformations can occur whereby sediment stripped from the nose and surface of a subaqueous debris flow becomes turbulently mixed with water to form a turbidity current that continues beyond the debris flow. Another kind of surface transformation is described by van Andel & Komar (1969) and Komar (1971). They suggest that at the base of slopes, rapid flows slow down and may go from supercritical conditions (Froude No. > 1) to subcritical (Froude No. < 1). Turbulence develops at this transformation which is known as a hydraulic jump, thereby incorporating ambient fluid into the flow and reducing its density. A further type is described by Allen (1971) who concludes that the frontal parts of slumps may be transformed into turbidity currents by the inflow and mixing of water at their bottom through 'tunnels' that occur along their fronts. The formation of lahars from hot pyroclastic flows that flow into water (e.g. Macdonald 1972) could be classed as a surface transformation.

The manner by which particles are supported in the final stages of flow determines in large part the textures and structures of a bed. The manner of support is a direct function of the particle concentration and the amount of fine grained cohesive sediment mixed with pore-water, which in turn greatly affect whether or not laminar to turbulent or turbulent to laminar transformations take place.

The following sections deal mainly with subaqueous debris flows and pyroclastic flows, and their flow transformations. The large and important topic of olistostromes is neglected. They are commonly regarded as bouldery mudstones that originate from submarine slides and slumps (e.g. McBride 1978), although Naylor (1981) cautions that the term should be used descriptively for bouldery mudstones without reference to origin.

Debris flows and lahars

Lahars are debris flows composed of volcanic materials. They may form deposits in subaerial or subaqueous environments, and can originate in several ways (Table 3). Lahars can develop differently from non-volcanic debris flows, but flow and depositional processes are essentially similar. They are sediment gravity flows in which the large fragments (sand, gravel, boulders) are supported mainly by the yield strength of the matrix (Fig. 4). Yield strength in many debris flows is mainly a function of water content and the type and amount of clay which together form the pore fluid (Hampton 1975). Debris flows are modified Bingham plastics (Johnson 1970) in which internal shear stress (τ) must overcome matrix cohesion, a frictional resistance, and internal viscous forces, in order to flow (Middleton & Hampton 1973). Debris flows have a non-flowing central region (rigid plug) where internal shear stress is less than the yield strength, τy. The competence of a debris flow (D) is defined by Hampton (1975) as the diameter of the largest clast that it can carry ($D = 8.8\ \tau y\ g^{-1}$) ($\rho_{\text{clast}} - \rho_{\text{matrix}}$). Much current research on debris flow deals with identifying

TABLE 3. *Origin of subaqueous lahars*

I. Direct and immediate results of eruption
    1. Movement of lahars directly from land into water. Subaerial lahars formed by:
        (a) Eruption through crater lake, snow or ice.
        (b) Heavy rain during eruption.
        (c) Flow of hot pyroclastic material into rivers or onto snow or ice.
    2. Hot pyroclastic flow that moves directly from land into water and becomes mixed with the water.
II. Indirectly related to eruptions
    1. Triggering by earthquakes or sudden liquefaction of water-soaked, pyroclastic or hydroclastic debris previously deposited in shelf environments, or on steep submerged volcano slopes.
    2. Reworking of subaerially deposited pyroclastic debris which enters water from land. Subaerial origin can include:
        (a) Triggering of water-soaked debris by earthquake.
        (b) Bursting and rapid drainage of crater lakes.
        (c) Dewatering of large avalanches originating from collapse of volcano side.
III. Not related to contemporaneous volcano activity
    1. Epiclastic or hydrothermally altered material which becomes water-soaked and moves from land into water.
    2. Epiclastic material on steep coasts and frontal parts of deltas and shelfs which becomes mobilized during slumps and slides.

---

environments of deposition (especially marine), as aids in constructing facies models and determining tectonic environments (e.g. Stanley *et al*. 1978; Leitch & Cawood 1980; Nemec *et al*. 1980; Swarbrick & Naylor 1980; Gloppen & Steel 1981).

The plastic behaviour of debris flows is the result of high concentration of particles relative to trapped water, and the abundance of fine compared with coarse grained particles. A relatively small percentage ($<$ 10%) of clay-size fragments has a surprisingly large effect on inhibiting turbulence (Hampton 1972). High concentration, irrespective of particle sizes, inhibits turbulence, results in high bulk density and large apparent viscosity values, and contributes to high strength properties. Such flows can support large particles despite laminar flow, and come to rest ('freeze') with steep fronts when internal shear stresses decrease below the threshold strength on low slopes. Moreover, they can flow over soft erodible materials without significantly disturbing the underlying bed. Possibly, they can flow beneath water as far as 500 km, over slopes as low as 0.1° (Embley 1976).

Although debris flows move in laminar fashion (Johnson 1970), turbulence might develop within the body of a flow without mixing with ambient water (a body transformation). The onset of turbulence in Bingham plastics is a function of the ratio of the Reynolds number (Re = $Ud\rho/\mu$) to the Bingham number (B = $\tau_c d/\mu U$) where U = average velocity, d = thickness of the flow, ρ = density, μ = dynamic viscosity and τ_c = strength of the plastic. Figure 6 shows that turbulence occurs at Re $\geqslant$ 1000 B. This is equivalent to $\rho U^2/\tau_c \geqslant$ 1000 (i.e. Re/B $\geqslant$ 1000), which Hiscott & Middleton (1979) call the Hampton number. Hiscott & Middleton (1979) give an excellent review of the criteria for turbulence, calculations for strength of debris flows, and depositional processes of mass flows (also see Lowe 1976, 1979, 1982). Turbulence, therefore, can develop within the body of a debris flow, despite high concentration, if it is large and velocity is high enough (see also Middleton 1970). If turbulence develops, erosion and mixing with bottom materials and ambient water may occur. This in turn can modify the behaviour of the fluid. Entrainment of clays would tend to reduce turbulence (Hampton 1972), but entrainment of water could lower the relative solids concentration which would favour development of turbulence. With enough water, a transformation from

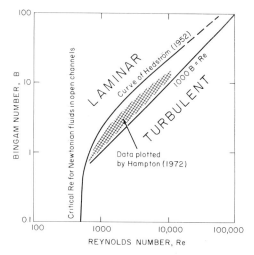

FIG. 6. Relationship of Bingham to Reynolds number for development of turbulence in a Bingham plastic. (After Hiscott & Middleton 1979.)

laminar to turbulent flow could be irreversible, although gravity transformations could occur later.

Plugs are easily deformable masses of debris that form within the interiors of debris flows where shear stresses fall below the strength of the fluid, and within which there is little internal shearing motion. The plug rides on an underflow of mobile debris believed to be laminar (Johnson 1970, p. 500; Hampton 1972). Movement in the underflow occurs because stresses are high enough (a function of velocity and weight). As velocity decreases, stresses decrease, and the lower part of the flow progressively 'freezes' as the yield strength is reached. Thus, the underflow progressively becomes part of the plug. Freezing locks in the textures, structures and fabrics inherited by the plug from upstream before the internal yield strength of the flow is reached.

Characteristic features of deposits from subaerial and subaqueous high-concentration flows include poor or no internal bedding, matrix support of large fragments, poor sorting, and inverse grading in a narrow zone at their base. Inverse grading has been recorded and discussed by many authors. For example in lahars by Schmincke (1967), in subaqueous debris flows and other kinds of sedimentary gravity flows by Fisher & Mattinson (1968), Hiscott & Middleton (1979) and Naylor (1980), and in subaerial pyroclastic flows (in which water or abundant clay is not important in the transport or depositional process) by

Sparks (1976). Elongate particles in these deposits are commonly oriented roughly parallel to the depositional surface or are imbricated. Such orientation is most strongly developed near the base. Basal contacts are commonly non-erosive, and thick, coarse grained debris flows may overlie fine grained, easily erodible material in sharp contact. Large particles may be impressed upon fine grained sediments due to loading. These features occur in subaqueous volcanic debris flows as well as in subaerial counterparts (Fisher 1971), and they are used to recognize debris flows in present-day marine basins (e.g. Carey & Sigurdsson 1980; Sigurdsson *et al.* 1980) and within the geological record (e.g. Cook *et al.* 1972; Cas 1979; Lewis 1976; Hiscott & Middleton 1979; Swarbrick & Naylor 1980; Leitch & Cawood 1980). Subaqueous debris flows appear to differ from subaerial debris flows in several respects, including the ratio of maximum particle size (MPS) to bed thickness, textures, degree of grading and imbrication, and association with other facies (Fig. 7). The bed thickness of subaqueous debris flows is commonly 3–10 times greater than the maximum fragment size, whereas in subaerial deposits, bed thickness is only 2–4 times greater (Nemec *et al.* 1980; Gloppen & Steel 1981). Many subaerial debris flows are abruptly overlain by a thinner, finer grained and better sorted mantle of massive conglomerate or laminated and cross-bedded granule sandstone, interpreted to have been winnowed by water flow following deposition of the debris flow. Commonly the subaqueous deposits are capped by massive or rippled fine grained sandstone which is gradational with the underlying conglomerate. The sandstone is apparently emplaced at the same time as the underlying material rather than the result of later reworking. Subaqueous debris flows can be expected to interfinger with lacustrine or marine sediments.

A subaqueous volcanic debris flow in the Lesser Antilles (Carey & Sigurdsson 1980) is referred to as a sediment gravity flow or pyroclastic gravity flow by Sigurdsson *et al.* (1980), because the structures and textures are similar to subaerial debris flows. The deposit originated from pyroclastic flows which entered the sea and became mixed with water, as indicated by incorporated pelagic clay. The clay content increases with distance from the source, thereby indicating continued mixing during transport. Thus it appears that the flow was turbulent during most of its transport but was transformed (body transformation) to laminar flow in the last stages of movement. There appears to have been little difference between the flow processes of the sediment gravity flow described by Carey & Sigurdsson (1980) and those of the massive division of subaqueous pyroclastic flows described by Fiske & Matsuda (1964) and Bond (1973) (see below).

Large amounts of subaqueous volcaniclastic debris were deposited as debris flows within an Archaean volcaniclastic sequence in the Noranda region, Quebec (Tasse *et al.* 1978). In proximal sections, a significant portion of debris flows (type A beds, Fig. 8) are massive or inversely graded, with rare parallel laminations. Distally, the sequence has a higher percentage of normally graded beds and beds with parallel laminations. The lateral changes can be attributed to surface transformations in developing turbidity currents. Their facies variations in terms of mean bed thicknesses and primary grading features are diagrammatically illustrated in Fig. 8.

Subaqueous pyroclastic flows

Subaqueous pyroclastic flows can develop from pyroclastic flows that move into water from land (e.g. 1902 Mt Pelée: Lacroix 1904). Also, they are postulated to form from entirely underwater eruptions (Fiske 1963; Fiske *et al.* 1963; Fiske & Matsuda 1964; Kokelaar *et al.*, this volume). Many workers do not believe that hot pyroclastic flows can be deposited under water, although some authors report welded tuff from marine sequences (see Table 4). Moreover, Sparks *et al.* (1980b) present theoretical arguments suggesting that welding can occur beneath water. Wright & Mutti (1981) discuss many of the problems encountered in interpreting subaqueous pyroclastic flows.

A   B

FIG. 7. Comparison of depositional features of (a) subaerial debris flows, and (b) subaqueous debris flows. (After Nemec *et al.* 1980.)

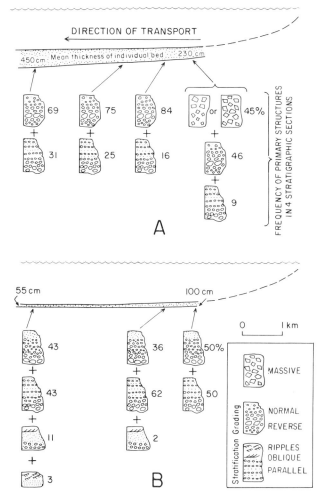

FIG. 8. Lateral variations of primary structures and mean bed thickness in Archaean subaqueous pyroclastic flow deposits, Canada. (a) Mass flow deposits. (b) Turbidite deposits as interpreted by Tasse *et al.* (1978). (From Lajoie 1979.)

Non-welded deposits

Evidence for subaqueous deposition in ancient deposits is determined indirectly on stratigraphical grounds, using sedimentary sequence associations (facies models) or interbedding with pillow lavas or fossiliferous sediments (Fiske & Matsuda 1964; Howells *et al.* 1973; Yamada 1973; Niem 1977). Subaqueous pyroclastic flow deposits consist of lithic fragments, crystals, glass shards and pumice in variable proportions and sizes. Shale rip-ups and blocks up to several metres in diameter derived from underlying beds may be incorporated near the base of a sequence (Yamada 1973; Niem 1977).

Shattered crystals are common in subaqueous

pyroclastic flows (Fiske & Matsuda 1964; Fernandez 1969; Yamada 1973; Niem 1977; Fig. 9), as are glassy fragments produced by sudden quenching. Fine perlitic cracks are characteristic of essential, non-vesicular, vitric clasts (Kato *et al.* 1971).

Non-welded subaqueous pyroclastic flow deposits characteristically are massive to poorly bedded and poorly sorted in their lower part (lower division), and are thinly bedded in their upper part (upper division) (Fig. 10). The lower division can form 50% or more of the total sequence (Fiske & Matsuda 1964; Bond 1973; Niem 1977). The two-division sequence is interpreted in terms of a waning, initially voluminous underwater eruption (Fiske & Mat-

TABLE 4. Subaqueous pyroclastic flow occurrences (from Fisher & Schmincke 1984)

Environment	Age	Location	Welding and temperature	Remarks	Reference
Marine	1883	Krakatau, Java	Non-welded.	Generated destructive tsunamis.	Self & Rampino 1981
	28,000 yr BP	Grenada Basin, Lesser Antilles	Non-welded. Resemble debris flow deposits.	Derived from subaerial hot pyroclastic flows on Dominica.	Carey & Sigurdsson 1980; Sigurdsson et al. 1980
	Late Quaternary	Subaqueous west flank, Dominica, Lesser Antilles	Deposits not cored.	Deposits traceable to subaerial flows (block and ash; welded ignimbrite).	Sparks et al. 1980a
	Middle Miocene	Santa Cruz Island, Calif., U.S.A.	Non-welded. Pumice-rich. Massive beds, some with pinkish oxidized tops.	Within a marine sequence.	Fisher & Charleton 1976
	Miocene	Tokiwa Formation, Japan	Non-welded. Pumice-rich.	Interbedded conformably with fossil-rich beds.	Fiske & Matsuda 1964
	Oligocene	Is. of Rhodes, Greece	Non-welded.	Deep water, marine.	Wright & Mutti 1981
	Oligocene–Eocene	Mt Rainier Natl Park, Wash., U.S.A.	Non-welded. Massive lower to bedded upper divisions.	Marine shelf deposition inferred.	Fiske 1963; Fiske et al. 1963
	Early Oligocene–Late Eocene	Philippines	Non-welded. Pyroclastic turbidites.	Interbedded with marine limestone.	Garrison et al. 1979
	Paleogene to Cretaceous	Philippines	Non-welded to welded (?) tuff.	Within a marine sequence.	Fernandez 1969
	Lower Mesozoic	Sierra Nevada, California, U.S.A.	Metamorphosed (shards not preserved).	Ponded in shallow marine calderas.	Busby-Spera 1981a,b
	Permian	Alaska, U.S.A.	Non-welded. Massive lower to bedded upper divisions.	Structures and sequence similar to Tokiwa Formation, Japan.	Bond 1973
	Mississippian	Oklahoma and Arkansas, U.S.A.	Non-welded. Turbidite structures.	Interbedded within marine flysch sequence.	Niem 1977
	Ordovician	Ireland	Welded.	Associated with shallow-water deltaic sediments.	Stanton 1960; Dewey 1963
	Ordovician	Wales, United Kingdom	Welded.	In marine sequence. Traceable to subaerial pyroclastic flows.	Francis & Howells 1973; Howells et al. 1979; Howells & Leveridge 1980
	Lower Ordovician	S. Wales, U.K.	Welded.	Within marine sequence.	Lowman & Bloxam 1981
	Archaean	Rouyn-Noranda, Canada	Non-welded.	Structures and sequence similar to Tokiwa Formation, Japan.	Dimroth & Demarcke 1978; Tasse et al. 1978
Lacustrine	Plio–Pleistocene	Japan	Non-welded. Resemble turbidites.	Interbedded with lacustrine rocks. Flow from land into water.	Yamada 1973
	Mio–Pliocene	Japan	Non-welded. Thermoremanent magnetism suggests emplacement at 500°C	Massive units resembling debris flows.	Kato et al. 1971; Yamazaki et al. 1973

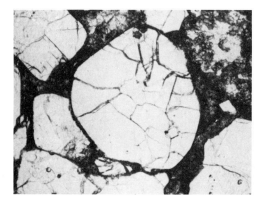

FIG. 9. Shattered quartz crystals from a subaqueous pyroclastic flow deposit, Eugenia Formation, Baja, California (J. Hickey, pers. comm. 1982).

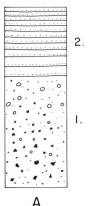

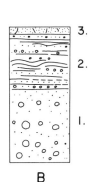

FIG. 10. Examples of non-welded subaqueous pyroclastic flow deposits. (a) Example interpreted to have developed from waning eruption by Fiske & Matsuda (1964); massive division (1) overlain by doubly graded sequence (2). (b) Obispo Tuff, interpreted to have developed from passage of one flow or several pulses of flow (Fisher 1977). Massive division (1), overlain by (2) sequence of planar to wavy beds or beds with low-angle cross-beds, fine to medium grained tuff to lapilli tuff layers, and (3) capped by a thin, very fine grained bioturbated tuff of probable suspension fallout origin. Interval (2) may reflect several pulses of flow, possibly representing retrogressive slumping (Morgenstern 1967). Pumice lapilli commonly decrease in size from bottom to top indicating that the tuff was water logged prior to flow, thereby supporting the idea that flow was remobilized pyroclastic debris from an unstable subaqueous slope.

suda 1964). Yamada (1973, this volume), however, recognizes five divisions similar to a Bouma sequence (Bouma 1962). A 2-fold sequence capped by a thin, very fine grained, bioturbated pelagic layer occurs in the Obispo Tuff, California (Fig. 10b), where basal contacts are generally sharp on undisturbed strata, although flute marks, grooves and load casts occur locally. Where bottom contacts are plane and sharp, missing underlying strata may indicate erosion (Yamada 1973).

The lower division of a subaqueous pyroclastic flow unit is a single, coarse grained bed commonly lacking internal structures or laminations (Fiske & Matsuda 1964; Bond 1973), although some beds have incipient cross-laminations outlined by a crude orientation of coarse fragments (Kato *et al.* 1971; Yamazaki *et al.* 1973). Larger and more dense fragments generally occur near the base of the massive divisions, with crystals, glass shards and pumice becoming more abundant toward the top.

Dense, lithic fragments may be inversely graded near the base, as in debris flows (Yamada 1973).

The upper division of non-welded subaqueous pyroclastic flow deposits is commonly composed of many thin, fine to coarse grained ash beds. In some sequences, each bed may be normally graded, and the entire sequence of beds becomes finer grained upward. This is called a doubly graded sequence (Fiske & Matsuda 1964). Not all upper divisions are doubly graded (Niem 1977).

Double grading is interpreted by Fiske & Matsuda (1964) to signify contemporaneous waning subaqueous volcanism with deposition from thin turbidity flows following deposition of the massive bed. Deposits without double grading, such as those described by Niem (1977) (see also Fig. 10b), are taken to signify sloughing from oversteepened pyroclastic slopes on the edge of a volcano where transformations from slides to sediment gravity flows could occur.

Welded deposits

Most controversial are welded pyroclastic flow deposits interpreted to be subaqueous on stratigraphic grounds. The best known example is from the Ordovician of Snowdonia in Wales. There, a welded tuff within the Capel Curig Formation is interpreted to be the submarine correlative of subaerial welded tuff (Francis & Howells 1973; Howells et al. 1973; Howells et al. 1979; Kokelaar et al., this volume a,b). Francis & Howells (1973) suggest that subaerially produced pyroclastic flows entered the sea and remained hot enough to weld after subaqueous emplacement. The subaerial welded tuff consists of an uninterrupted sequence of two or more flow units of greater combined thickness (70 m) than correlative subaqueous units (< 50 m). The base of the subaerial tuff is planar, contains fiamme and overlies a conglomerate with a reddened top, and welding is restricted to the middle of the sheet. The subaqueous pyroclastic flow deposits are welded to their base, and are interbedded with marine siltstones containing brachiopods. Basal contacts of the subaqueous welded tuffs tend to be irregular with zones resembling large-scale load and flame structures. These flow deposits are massive in the lower and central portions and grade imperceptibly upward through ill-defined, faintly bedded zones to evenly bedded tuff which locally has broad, low-angle cross-bedding suggestive of marine deposition. Their tops are in sharp contact with overlying sandstone with similar broad cross-bedded structures. Kokelaar (1982) accounts for the large load-structure protrusions at the base of the subaqueous welded tuffs by fluidization of subjacent wet sediments following emplacement of a hot pyroclastic flow.

Origin of subaqueous pyroclastic flows

Three principle ways in which subaqueous pyroclastic flows can be generated are given in Fig. 11. Massive lower divisions with finer grained bedded tops can develop in each situation. Mixing with water by flow transformation to form sediment gravity flows and turbidity currents can also occur in each case. The occurrence of subaqueous welded tuff in Wales correlative with subaerial welded tuff (Francis & Howells 1973) suggests that hot pyroclastic flows that enter water from land (case A) may do so without mixing with water and retain

enough heat to weld. As shown by Sparks et al. (1980b), welding is theoretically possible. This would be likely to occur where changes in slope are insufficient to cause laminar to turbulent body transformations. Such transformations are more likely to occur where sudden steepening of slope could cause turbulence and mixing, thereby producing a cool sediment gravity flow (Sigurdsson et al. 1980). Also as shown for case A (Fig. 11), turbulent ash cloud surges, which can develop above dense flows on land, can flow over the water (Anderson & Flett 1903). Additionally, surface transformations on the underwater flow could produce turbidity currents. In each instance, thin, graded beds with small total volume can develop by fallout at the top of flow units.

For case B (Fig. 11), Fiske & Matsuda (1964) envisage an eruption column emerging with great force from an underwater vent. Large amounts of material are ejected into water, fall back to the sea floor, and form a

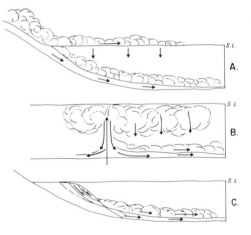

Fig. 11. Means by which subaqueous pyroclastic flows can develop. (a) Hot pyroclastic flow from land into water (Francis & Howells 1973; Yamada 1973; Sigurdsson et al. 1980). Density separation of flow and ash cloud surge at air–water interface; surface transformation to form thin turbidite sequence. Flow may weld if not mixed with water, or be transformed to debris flow if mixed. (b) Pyroclastic flow develops from column collapse (Fiske 1963; Fiske & Matsuda 1964); waning eruption may produce overlying laminated doubly graded sequence. Suspended material can form thin fallout layers and floating pumice can form pumice rafts. (c) Sediment gravity pyroclastic flows develop from slumping of unstable slopes composed of pyroclastic debris.

dense water-rich debris flow deposited as a thick, massive layer. In the closing stages of eruption, small turbidity currents entrain pumice lapilli and crystals of plagioclase and quartz which are deposited as thinly graded beds above the massive beds.

Ash from a submarine eruption cloud that breaks the water surface, or that has spread widely away from the vent within the water body, continues to settle after the eruption ends, but in decreasing amounts and becoming finer grained. Thus, there is an insufficient supply to maintain the continuous flow that deposited the thick structureless part of the pyroclastic flow. Instead, intermittent thin turbidity currents follow one another in close succession. As the particle-rich column continues to settle, successively smaller amounts of finer grained and less-dense ash fall out, and the density currents become less frequent and finally end. Thus, the total volume of the upper division is considerably greater than that developed at the top of a single flow entering water from land (case A), although distally Bouma sequences may dominate in case A (Yamada 1973, this volume).

A vertical eruption column formed entirely beneath water, such as that proposed by Fiske & Matsuda (1964), may not develop into hot pyroclastic flows upon collapse because turbulent mixing with water along its margin would be enhanced by drag effects in water which would cool the materials. Possibly, however, a voluminous 'boiling-over' type of eruption without a vertical column (Fisher & Schmincke 1984) could be extruded at rates great enough to produce a flow protected from the water by a carapace of steam and retain enough heat to become welded. As yet, however, no subaqueous vents for welded tuffs have been positively identfied, although one has been postulated by Kokelaar *et al.* (this volume a, b) to explain a welded tuff sequence on Ramsey Island in SW Wales.

Slumping of unstable material (case C) composed exclusively of pyroclastic material is probably common on submerged slopes of volcanoes, but has rarely been documented. It is to be expected, however, that flow transformations from subaqueous lahars to turbidity currents can develop sequences similar to those produced directly from volcanic eruptions. A thick (*c.* 15 m) coarse grained pyroclastic flow within the Obispo Tuff, California is interpreted to have formed by cold water mobilization of pyroclastic material off steep slopes (Fisher 1977).

Subaqueous pyroclastic flows interbedded with lacustrine sediments of Onikobe Caldera, Japan (Yamada 1973), closely resemble turbidite sequences. The flows may have originated from subaerial eruptions at the margin of the caldera, but there is little evidence to indicate whether hot pyroclastic flows entered the water from land, or if slumping of unstable slopes on the margin of the lake initiated the flows.

Sparks *et al.* (1980b) have proposed on theoretical grounds that if a hot pyroclastic flow manages to pass through the air–water interface, subaqueous welding is not only possible, but is enhanced. Hydrostatic pressure increases by 1 bar for every 10 m water depth, and if water is 'drawn' into the hot pyroclastic flow it can be absorbed into glass shards. At increased pressures, water content of glass can be higher, resulting in lower viscosities which favour welding at lower temperatures (Smith 1960; Riehle 1973). Calculations of the viscosity of glass shards as a function of water depth suggest that the viscosity of glass decreases substantially with depth despite cooling of the flow (Sparks *et al.* 1980b). For subaerial pyroclastic flows at 800°C, the viscosity of rhyolitic glass is about 2×10^{11} poise, but, at 755°C under 500 m of water, this is reduced by dissolution of water to an estimated 2×10^{9} poise. Thus glass under the latter conditions would compact much faster than in a subaerial flow under the same uniaxial stress. A major difficulty with the theory, however, is the slow diffusion rate of water into glass. In addition, there is a question of how water could be drawn into a flow and into contact with glass participles. Turbulent mixing is the dominant process by which water can gain access to a flow, but this would also tend to cool or explosively destroy the flow (Walker 1979). A rapid increase of pressure, from loading by the water column, would inhibit volatile escape and is a more likely cause of welding. Water surrounding the hot flows would create a thin vapour carapace around the flow during cooling (Francis & Howells 1973; Yamazaki *et al.* 1973), provided that turbulent mixing does not occur at the surface of the flow.

Submarine ash layers

The initial depositional process of marine fallout ash is by the settling of particles introduced into water from the air or from underwater flows or eruptions. The process therefore depends upon the size, shape and density of fragments, and the vertical and lateral distribution is influenced by outside factors such as wind and water current velocities and directions. For more complete reviews of this subject, the reader is referred to Kennett (1981) and Fisher & Schmincke (1984). Data from widespread submarine fallout layers are useful in many ways. For example, (i) in correlating over long distances between sequences deposited in different environments (Ninkovich & Heezen 1965, 1967; Huang *et al.* 1973; Watkins *et al.* 1978; Keller *et al.* 1978; Hahn *et al.* 1979; Thunell *et al.* 1979; Ninkovich 1979), (ii) in solution of problems of magma genesis and evolution (Scheidegger *et al.* 1978, 1980), (iii) in assessments of volcanic cycles (Hein *et al.* 1978), (iv) in derminations of production rates of tephra and eruption column heights (Shaw *et al.* 1974; Huang *et al.* 1973, 1975, 1979; Watkins & Huang 1977; Ninkovich *et al.* 1978), (v) in estimations of the duration of eruptions (Ledbetter & Sparks 1979), (vi) as aids in determining rates of plate movements and arc polarity (Ninkovich & Donn 1976; Sigurdsson *et al.* 1980), (vii) in determination of prevailing wind and water current directions (Eaton 1964; Nayudu 1964; Ninkovich & Shackleton 1975; Hampton *et al.* 1979), (viii) in making inferences about climatic changes (Kennett & Thunell 1975, 1977) and (ix) in solving problems concerning diagenesis (Coombs 1954; Fisher & Schmincke 1984).

Widespread marine fallout layers can occur in back-arc and fore-arc regions, depending mainly on wind directions, and occur in all marine sedimentary environments. In the Lesser Antilles, marine fallout layers are the dominant volcaniclastic type in front of the arc (Sigurdsson *et al.* 1980), but in the Cordilleran region of western North America they are most abundant behind the Cascades volcanic arc, presently as well as in ancient deposits (Eaton 1964; Dickinson 1974).

Island volcanoes

Irrespective of magma composition and marginal or intra-plate setting, volcanoes in the sea can be considered to evolve through two overlapping stages (Fig. 12). Stage A develops beneath the PCL and stage B above the PCL (L. Ayres, personal communication). As suggested by previous discussions, volcanic island foundations (below the PCL) can be composed of massive and pillowed lava flows with lesser amounts of hyaloclastites (up to 20%). As volcanoes grow and slopes steepen in stage A, talus development and slides cause mechanical fragmentation and movement of debris into deeper water, resulting in a decrease in the lava-to-clastic ratio outward from the source. This is observed on sea-mounts (Stanley & Taylor 1977; Lonsdale & Spiess 1979). Above the PCL, however, there is a wider range of transport processes, kinds of fragments and consequent lateral facies changes in rock types.

Ayres (pers. comm.) infers a shield shape for the foundations of island volcanoes in stage A, with steeper slopes developing in stage B. In Fig. 12, lava-to-clastic ratio is estimated at up to 5:1 in early growth stages and in flatter summit regions of cones, decreasing perhaps to 2:1 or less away from the source as slopes steepen, to entirely clastic on surrounding abyssal plains. In addition to lava flows, the intrusion of sills may be important in the growth of an island volcano. On La Palma, Canary Islands, for example, Staudigal & Schmincke (1981) report 2000 m of sills which contributed significantly to the elevation of the volcano.

Rocks of stage B_1, above the PCL but below water level, can be dominantly explosive hydroclastic debris. At DSDP Site 253 (Ninetyeast Ridge, Indian Ocean) for example, there is at least 388 m of bedded Middle Eocene basaltic

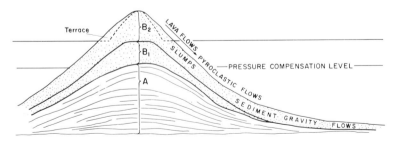

FIG. 12. Generalized model of an island volcano. Stage A, foundation of island volcano beneath pressure compensation level (PCL); lava (L) > hydroclastics (H); may contain a significant volume of sills. Stage B_1, beneath sealevel and above PCL; L < H. Stage B_2, subaerial part; pyroclastics (P) > H > L if andesitic, but L > P if basaltic shield. Clastic apron enlarges rapidly in stage B_2 and sediment gravity flows and fallout deposits increase in flanks and surrounding basin.

hydroclastic debris, believed to have been erupted in about 150 m of water (Fleet & McKelvey 1978).

In stage B$_2$, lava flows of basaltic islands, such as Hawaii, cap the hydroclastic material as the vents become subaerial (Moore & Fiske 1969). Along the fringes of such islands, explosion debris, pillows and pillow breccias can be generated where lavas flow into the sea (Moore 1975; Peterson 1976; Fisher 1968). Erosion commences to produce epiclastic volaniclastic debris which increases in volume as the volcano grows in size. However, at volcanoes where pyroclastic material is abundant relative to lava flows, such as andesitic stratocones, emerged slopes in stage B$_2$ are steeper than shield volcanoes, and pyroclastic debris is delivered to the sea as lahars, pyroclastic flows and fallout ash directly from eruptions, or by remobilization of loose pyroclastic deposits. The volumes of pyroclastic material stored in clastic wedges on submerged slopes of andesite volcanoes is consequently greater than for shield volcanoes, and so is the volume of debris deliverered to adjacent basins as sediment gravity flows and suspension fallout. The rate and volume of debris supplied during periods of volcanic activity and inactivity is discussed by Kuenzi *et al.* (1979) and Vessel & Davies (1981). The rate and volume of supply from fringing terraces to adjacent slopes and basins can also be modified by eustatic sealevel changes. During high stands of sealevel, more sediment can be stored in the clastic wedges and less reaches the basins than during low stands, when sedimentation rate and volume of sediment increases in the basins (Klein *et al.* 1979). Primary and reworked volcanic products become more diverse as an island emerges, because erosion is contemporaneous with eruptions, and sediment aprons with unstable slopes can grow more rapidly than before emergence. Reefs can form, and in areas of long-continued volcanism and erosion, fringing shelves become the sites for development of deltas and fan-deltas with deep-sea fans and abyssal turbidite plains farther away (Lonsdale 1975).

Facies of volcanic rocks reflect distance to source. Dykes, sills, plutons, lava flows, coarse grained pyroclastic rocks and their reworked equivalents, occur at or near the source. Farther away, lava flows and intrusive rocks disappear and volcaniclastic sediments become finer grained (Dickinson 1968; Mitchell 1970). Fallout tuffs occur within both of these near- and intermediate-distance regions, but may be dispersed globally and occur at great distances from source volcanoes. These relationships apply to both subaerial and submarine environments.

It is also true that (i) volcanic rocks are emplaced in the order in which they are extruded, and (ii) the eroded products of an inactive volcano tend to be emplaced in reverse order of extrusion. Erosion can extend down to the source plutons (Dickinson & Rich 1972; Ingersoll 1978; Mansfield 1979). This simplified scheme (Fig. 13) is only a basic framework, however, because erosion of a volcanic edifice can alternate with periods of active growth, adjacent volcanoes can supply material to basins at different times, distant volcanoes can supply ash during any part of the constructive or destructive cycles, constant settling of pelagic and hemipelagic material from marine waters can occur, and sediments from land masses and islands can be supplied from opposite sides of a basin. Further complications arise, for example, where moving plates carry source areas away from the depositional site (diverging plates), toward the depositional site (converging plates), or laterally along transform boundaries.

Irrespective of lateral facies changes and despite the many complexities, it is possible to

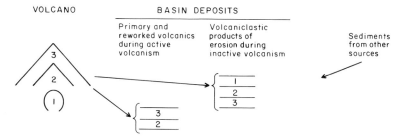

FIG. 13. Simplified stratigraphic scheme for deposition of volcaniclasitc debris from active and inactive volcanoes in adjacent basins. (1) plutonic sources and epiclastic products, (2) and (3) volcanic products in sequence of emplacement from active volcanoes. Erosion of the volcanic pile produces epiclastic debris in inverse order.

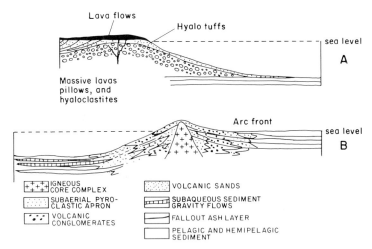

Lava flows
Hyalo tuffs
sea level
A
Massive lavas
pillows, and
hyaloclastites

Arc front
sea level
B

IGNEOUS
CORE COMPLEX

SUBAERIAL PYRO-
CLASTIC APRON

VOLCANIC
CONGLOMERATES

VOLCANIC SANDS

SUBAQUEOUS SEDIMENT
GRAVITY FLOWS

FALLOUT ASH LAYER

PELAGIC AND HEMIPELAGIC
SEDIMENT

FIG. 14. Facies model of island volcanoes. (a) Newly emergent intra-plate basaltic shield volcano (from Moore & Fiske 1969). (b) Island-arc andesitic stratocone (from Sigurdsson *et al.* 1980).

gain considerable information, about many of the aspects discussed above, from vertical sequences (in outcrop or drill-hole samples), if the origin of clasts can be determined (e.g. pyroclastic, hydroclastic, epiclastic) and depositional processes can be interpreted from rock textures and structures (see Schmincke & von Rad 1979).

Work in the Lesser Antilles has led to a facies model of a calc-alkaline island-arc volcano (Fig. 14) (Carey & Sigurdsson 1978; Sigurdsson *et al.* 1980; Sparks *et al.* 1980a; see also Carey & Sigurdsson, this volume). In the Lesser Antilles, dominant kinds of sediment transport processes and deposit types are highly asymmetric. Volcaniclastic sediments form 34% of the total cored sediment behind the arc and 4% of the total in front of the arc, a proportion which does not significantly vary within 300 km of the arc. Sediment gravity flows of various kinds form

98% of the volcaniclastic sediments in the back-arc basin, and ash-fall layers form 60% of the volcaniclastic sediments in the fore-arc region. This asymmetry is attributed to wind directions, arc slopes (steeper toward the back-arc) and ocean current directions.

Research in the Lesser Antilles documents sedimentary asymmetry associated with arc polarity. Sedimentary asymmetry, however, depends upon arc geometry (facing direction, trend direction of the arc) relative to wind and ocean current directions. Therefore, volcaniclastic facies as related to arc polarity will differ in different arcs.

ACKNOWLEDGMENTS: I thank James Hickey (UCSB), Douglas Coombs (Otago), Hans-Ulrich Schmincke (Bochum) and Cathy Busby-Spera (Princeton) for critical, penetrating and extremely helpful reviews of this manuscript.

References

ALLEN, C. C. 1980. Icelandic subglacial volcanism: thermal and physical studies *J. Geol.* **88**, 108–17.

ALLEN, J. R. L. 1971. Mixing at turbidity current heads, and its geological implications. *J. sediment. Petrol.* **41**, 97–113.

ANDERSON, T. & FLETT, J. S. 1903. Report on the eruption of the Soufriere in St Vincent in 1902 and on a visit to Montagne Pelee in Martinique, Part I. *Philos. Trans. R. Soc. London.* **A200**, 353–553.

BALLARD, R. D. & MOORE, J. G. 1977. *Photographic Atlas of the Mid-Atlantic Ridge.* Springer, New York.

——, HOLCOMBE, R. T. & VAN ANDEL, T. H. 1979. The Galapagos Rift at 86°W. III. Sheet flows, collapse pits, and lava lakes of the rift valley. *J. geophys. Res.* **84**, 5407–22.

BELLAICHE, G., CHEMINEE, J. L., FRANCHETEAU, J., HEKINIAN, R., LEPICHON, X., NEEDHAM, H. D. & BALLARD, R. D. 1974. Rift Valley's inner floor: First submersible study. *Nature, Lond.* **250**, 558–60.

BENNETT, F. D. 1972. Shallow submarine volcanism. *J. geophys. Res.* **77**, 5755–60.

BEVINS, R. E. & ROACH, R. A. 1979. Pillow lava and isolated pillow breccia of rhyodacitic composition from the Fishguard Volcanic Group, Lower Ordovician, SW Wales, United Kingdom. *J. Geol.* **87**, 193–201.

BOND, G. C. 1973. A late Paleozoic volcanic arc in the eastern Alaska Range, Alaska. *J. Geol.* **81**, 557–75.

BOUMA, A. H. 1962. *Sedimentology of Some Flysch Deposits; A Graphic Approach to Facies Interpretation.* Elsevier, Amsterdam, Netherlands.

BUSBY-SPERA, C. 1981a. Silicic ash-flow tuffs interbedded with submarine andesitic and sedimentary rocks in lower Mesozoic roof pendants, Sierra Nevada, California (Abst.). *Bull. geol. Soc. Am.* **13**, 47.

——, 1981b. Early Mesozoic submarine epicontinental calderas in the southern Sierra Nevada, California. *Eos*, **62**, 1061.

CAREY, S. N. & SIGURDSSON, H. 1978. Deep-sea evidence for distribution of tephra from the mixed magma eruption of the Soufriere on St. Vincent, 1902: ash turbidites and air fall. *Geology*, **6**, 271–4.

—— & SIGURDSSON, H. 1980. The Roseau Ash: deep-sea tephra deposits from a major eruption on Dominica, Lesser Antilles Arc. *J. Volcan. geoth. Res. Amsterdam*, **7**, 67–86.

CARLISLE, D. 1983. Pillow breccias and their aquagene tuffs, Quadra Island, British Columbia. *J. Geol.* **71**, 48–71.

CARTER, R. M. 1975. A discussion and classification of subaqueous mass-transport with particular application to grain-flow, slurry-flow, and fluxoturbidites. *Earth Sci. Rev.* **11**, 145–77.

CAS, R. 1978. Silicic lavas in Paleozoic flyschlike deposits in New South Wales, Australia: Behaviour of deep subaqueous silicic flows. *Bull. geol. Soc. Am.* **89**, 1708–14.

—— 1979. Mass-flow arenites from a Paleozoic interarc basin, New South Wales, Australia: mode and environment of emplacement. *J. sediment. Petrol.* **49**, 29–44.

COLGATE, S. A. & SIGURGEIRSSON, T. 1973. Dynamic mixing of water and lava. *Nature, Lond.* **244**, 552–5.

COOK, H. E., MCDANIEL, P. N., MOUNTJOY, E. W. & PRAY, L. C. 1972. Allochthonous carbonate debris flows at Devonian bank ('reef') margins, Alberta, Canada. *Bull. Can. Pet. Geol.* **20**, 439–97.

COOMBS, D. S. 1954. The nature and alteration of some Triassic sediments from Southland, New Zealand. *Trans. R. Soc. N.Z.* **82**, 65–109.

DE ROSEN-SPENCE, A. F., PROVOST, G., DIMROTH, E., GOCHNAUER, K. & OWEN, V. 1980. Archean subaqueous felsic flows, Rouyn-Noranda, Quebec, Canada, and their Quaternary equivalents. *Precambrian Res.* **12**, 43–77.

DEWEY, J. F. 1963. The Lower Paleozoic stratigraphy of central Murrisk County, Mayo, Ireland, and the evolution of the South Mayo Trough. *J. geol. Soc. London*, **119**, 313–43.

DICKINSON, W. R. 1968. Sedimentation of volcaniclastic strata of the Pliocene Koroimavua Group

in northwest Viti Levu, Fiji. *Am. J. Sci.* **266**, 440–53.

—— 1974. Sedimentation within and beside ancient and modern magmatic arcs. *In*: DOTT JR, R. H. & SHAVER, R. H. (eds) *Modern and Ancient Geosynclinal Sedimentation.* Spec. Pub. Soc. Econ. Paleont. Miner. **19**, 230–9.

—— & RICH, E. L. 1972. Petrographic intervals and petrofacies in the Great Valley Sequence, Sacramento Valley, California. *Bull. Geol. Soc. Am.* **83**, 3007–24.

DIMROTH, E. & DEMARCKE, J. 1978. Petrography and mechanism of eruption of the Archean Dalembert tuff, Rouyn-Noranda, Quebec, Canada. *Can. J. Earth Sci.* **15**, 1712–23.

——, COUSINEAU, P., LEDUC, M., SANSCHAGRIN, Y. & PROVOST, G. 1979. Flow mechanisms of Archean subaqueous basalt and rhyolite flows. *In*: *Current Res. Part A, Geol Surv. Canada, Paper 79-1A*, 207–11.

EATON, G. P. 1964. Windborne volcanic ash: A possible index to polar wandering, *J. Geol.* **72**, 1–35.

EMBLEY, R. W. 1976. New evidence for occurrence of debris flow deposits in the deep sea. *Geology*, **4**, 371–4.

FERNANDEZ, H. E. 1969. Notes on the submarine ash flow tuff in Siargao Island, Surigao del Norte (Philippines). *Philipp. Geol.* **23**, 29–36.

FISHER, R. V. 1968. Puu Hou littoral cones, Hawaii. *Geol. Rdsch.* **57**, 837–64.

—— 1971. Features of coarse-grained, high-concentration fluids and their deposits. *J. sediment. Petrol.* **41**, 916–27.

—— 1977. Geologic guide to subaqueous volcanic rocks in the Nipomo, Pismo Beach and Avila Beach areas. *Penrose Conf. geol. Soc. Am. The Geology of Subaqueous Volcanic Rocks.* 1–29.

—— 1979. Models for pyroclastic surges and pyroclastic flows. *J. Volcan. geoth. Res. Amsterdam*, **6**, 305–18.

—— 1983. Flow transformations in sediment gravity flows. *Geology*, **11**, 273–4.

—— & CHARLETON, D. W. 1976. Mid-Miocene Blanca Formation, Santa Cruz Island, California. *In*: HOWELL, D. G. (ed.) *Aspects of the Geologic History of the California Continental Borderland.* Misc. Pub. Pacific Section Am. Assoc. Petrol. Geol. **24**, 228–40.

—— & HEIKEN, G. H. 1982. Mt Pelée, Martinique: May 8 and 20 pyroclastic flows and surges. *J. Volcan. geoth. Res. Amsterdam* **13**, 339–71.

—— & MATTINSON, J. M. 1968. Wheeler Gorge turbidite—conglomerate series; inverse grading. *J. sediment. Petrol.* **38**, 1013–23.

—— & SCHMINCKE, H.-U. 1984. *Pyroclastic Rocks.* Springer, Heidelberg.

——, SMITH, A. L. & ROOBOL, M. J. 1980. Destruction of St Pierre, Martinique by ash cloud surges, May 8 and 20, 1902. *Geology*, **8**, 472–6.

FISKE, R. S. 1963. Subaqueous pyroclastic flows in the Ohanapecosh Formation, Washington. *Bull. geol. Soc. Am.* **74**, 391–406.

—— & MATSUDA, T. 1964. Submarine equivalents of

ash flows in the Tokiwa Formation, Japan. *Am. J. Sci.* **262**, 76–106.

——, HOPSON, C. A. & WATERS, A. C. 1963. Geology of Mount Rainier National Park, Washington. *Prof. Pap. U.S. geol. Surv.* **444**, 1–93.

FLEET, A. J. & McKELVEY, B. C. 1978. Eocene explosive submarine volcanism, Ninetyeast Ridge, Indian Ocean. *Mar. Geol.* **26**, 73–97.

FRANCIS, E. H. & HOWELLS, M. F. 1973. Transgressive welded ash-flow tuffs among the Ordovician sediments of NE Snowdonia, North Wales. *J. geol. Soc. London*, **129**, 621–41.

FURNES, H., FRIDLEIFSSON, I. B. & ATKINS, F. B. 1980. Subglacial volcanics—on the formation of acid hyaloclastites. *J. Volcan. geoth. Res. Amsterdam*, **8**, 95–110.

GARRISON, R. E., ESPIRITU, E., HORAN, L. J. & MACK, L. E. 1979. Petrology, sedimentology, and diagenesis of hemipelagic limestone and tuffaceous turbidites in the Aksitero Formation, Central Luzon, Philippines. *Prof. Pap. U.S. geol. Surv.* 1112.

GLOPPEN, T. G. & STEEL, R. J. 1981. The deposits, internal structure and geometry in six alluvial fan–fan delta bodies (Devonian-Norway)—A study in the significance of bedding sequence in conglomerates. *Spec. Publ. Soc. econ. Paleontol. Mineral., Tulsa*, **31**, 49–69.

GONZALEZ-BONORINO, G. & MIDDLETON, G. V. 1976. A Devonian submarine fan in western Argentina. *J. sediment. Petrol.* **46**, 56–69.

HAHN, G. A., ROSE JR, W. I. & MEYERS, T. 1979. Geochemical correlation of genetically related rhyolitic ash-flow and air-fall ashes, central and western Guatemala and the equatorial Pacific. *Spec. Pap. geol. Soc. Am.* **180**, 101–12.

HAMPTON, M. A. 1972. The role of subaqueous debris flow in generating turbidity currents. *J. sediment. Petrol.* **42**, 775–93.

—— 1975. Competence of fine-grained debris flows. *J. sediment. Petrol.* **45**, 834–44.

——, BOUMA, A. H. & FROST, T. P. 1979. Volcanic ash in surficial sediments of the Kodiak shelf—an indicator of sediment dispersal patterns. *Mar. Geol.* **29**, 347–56.

HEIN, J. R., SCHOLL, D. W. & MILLER, J. 1978. Episodes of Aleutian Ridge explosive volcanism. *Science*, **199**, 137–41.

HISCOTT, R. N. & MIDDLETON, G. V. 1979. Depositional mechanics of thick bedded sandstones at the base of a submarine slope, Tourelle Formation (Lower Ordovician), Quebec, Canada. *Spec. Publ. Soc. econ. Paleontol. Mineral., Tulsa*, **27**, 307–26.

HONNOREZ, J. 1972. La Palagonitisation: l'alteration sous-marine du verre volcanique basique de Palagonia (Sicile). *Vukan-institut Immanuel Friedlaender* No. 9, *Birkhauser Verlag Basel und Stuttgart*.

HOWELLS, M. F. & LEVERIDGE, B. E. 1980. The Capel Curig Volcanic Formation. *Rep. Inst. Geol. Sci. Lond.* **80**(6), 1–23.

——, —— & EVANS, C. D. R. 1973. Ordovician ash-flow tuffs in eastern Snowdonia. *Rep. Inst. Geol. Sci. Lond.* **73**(3), 1–33.

——, ——, ADDISON, R., EVANS, C. D. R. & NUTT, M. J. C. 1979. The Capel Curig Volcanic Formation, Snowdonia, North Wales; variations in ash-flow tuffs related to emplacement environment. *In*: HARRIS, A. L., HOLLAND, C. H. & LEAKE, B. L (eds) *The Caledonides of the British Isles–Reviewed. Spec. Publ. geol. Soc. London*, **8**, 611–8. Blackwell Scientific Publications, Oxford.

HUANG, T. C., WATKINS, N. D. & SHAW, D. M. 1975. Atmospherically transported volcanic glass in deep-sea sediments: volcanism in sub-antarctic latitudes of the south Pacific during late Pleistocene time. *Bull. geol. Soc. Am.* **86**, 1305–15.

——, CAREY, S., SIGURDSSON, H. & DAVIS, A. 1979. Correlations and contrasts of deep-sea ash deposits from the Lesser Antilles in the Western Equatorial Atlantic and Eastern Caribbean at latitude 14°N. *Eos*, **59**, 1119.

WATKINS, N. D., SHAW, D. M. & KENNETT, J. P. 1973. Atmospherically transported volcanic dust in South Pacific deep sea sedimentary cores at distances over 3000 km from the eruptive source. *Earth planet. Sci. Lett.* **20**, 119–24.

INGERSOLL, R. V. 1978. Petrofacies and petrologic evolution of the Late Cretaceous forearc basin, northern and central California. *J. Geol.* **86**, 335–52.

JOHNSON, A. M. 1970. *Physical Processes in Geology*. Freeman, Cooper & Co., San Francisco.

JONES, J. G. 1970 Intraglacial volcanoes of the Laugarvatn region, southwest Iceland, II. *J. Geol.* **78**, 127–40.

KATO, I., MUROI, I., YAMAZAKI, T. & ABE, M. 1971. Subaqueous pyroclastic flow deposits in the upper Donzurubo Formation, Nijo-san district, Osaka, Japan. *J. Geol. Soc. Japan*, **77**, 193–206.

KELLER, J., RYAN, W. B. F., NINKOVICH, D. & ALTHERR, R. 1978. Explosive volcanic activity in the Mediterranean over the past 2,000,000 yr as recorded in deep-sea sediments. *Bull. geol. Soc. Am.* **89**, 591–604.

KENNETT, J. P. 1981. Marine tephrochronolgy. *In*: EMILIANI, C. (ed.) *The Oceanic Lithosphere*, Vol. 7, John Wiley & Sons, Inc. Pp. 1373–436.

—— & THUNELL, R. C. 1975. Global increase in Quaternary volcanism. *Science*, **187**, 497–503.

—— & —— 1977. On explosive Cenozoic volcanism and climatic implications. *Science*, **196**, 1231–4.

KLEIN, G. DEV., OKADA, H. & MITSUI, K. 1979. Slope sediments in small basins associated with a Neogene active margin, western Hokkaido Island, Japan. *Spec. Publ. Soc. econ. Paleontol. Mineral., Tulsa*, **27**, 359–74.

KOKELAAR, B. P. 1982. Fluidization of wet sediments during the emplacement and cooling of various igneous bodies. *J. geol. Soc. London*, **139**, 21–33.

—— 1983. The mechanism of Surtseyan Volcanism. *J. geol. Soc. London*, **140**, 939–44.

KOMAR, P. D. 1971. Hydraulic jumps in turbidity currents. *Bull. geol. Soc. Am.* **82**, 1477–88.

KUENEN, PH. H. 1952. Estimated size of the Grand Banks turbidity current. *Am. J. Sci.* **250**, 874–84.

KUENZI, W. D., HORST, O. H. & McGEHEE, R. V. 1979. Effect of volcanic activity on fluvial-deltaic sedimentation on a modern arc-trench gap, southwestern Guatemala. *Bull. geol. Soc. Am.* **90**, 827–38.

LACROIX, A. 1904. *La Montagne Pelée et Ses Eruptions.* Masson, Paris.

LAJOIE, J. 1979. Facies models 15. Volcaniclastic rocks. *Geosci. Can.* **6**, 129–39.

LEDBETTER, M. T. & SPARKS, R. S. J. 1979. Duration of large-magnitude explosive eruptions deduced from graded bedding in deep-sea ash layers. *Geology*, **7**, 240–4.

LEITCH, E. C. & CAWOOD, P. A. 1980. Olistoliths and debris flow deposits at ancient consuming plate margins: an eastern Australian example. *J. sediment. Petrol.* **25**, 5–22.

LEWIS, D. W. 1976. Subaqueous debris flows of early Pleistocene age at Motunau, North Canterbury, New Zealand. *N.Z. Jl Geol. Geophys.* **19**, 535–67.

LONSDALE, P. F. 1975. Sedimentation and tectonic modification of the Samoan archipelagic apron. *Bull. Am. Ass. Petrol. Geol.* **59**, 780–98.

—— 1977. Abyssal pahoehoe with lava coils at the Galapagos Rift. *Geology*, **5**, 147–52.

—— & SPIESS, F. N. 1979. A pair of young cratered volcanoes on the east Pacific rise. *J. Geol.* **87**, 157–73.

LOWE, D. R. 1976. Grain flow and grain flow deposits. *J. sediment. Petrol.* **46**, 188–99.

—— 1979. Sediment gravity flows: their classification and some problems of application to natural flows and deposits. *Spec. Publ. Soc. econ. Paleontol. Mineral., Tulsa*, **27**, 75–82.

—— 1982. Sediment gravity flows. II. Depositional models with special reference to the deposits of high-density turbidity currents. *J. sediment. Petrol.* **52**, 279–97.

LOWMAN, R. D. W. & BLOXAM, T. W. 1981. The petrology of the lower Paleozoic Fishguard Volcanic Group and associated rocks E of Fishguard, N Pembrokeshire (Dyfed), South Wales. *J. geol. Soc. London*, **138**, 47–68.

MACDONALD, G. A. 1972. *Volcanoes.* Prentice-Hall, New Jersey.

MANSFIELD, C. F. 1979. Upper Mesozoic subsea fan deposits in the southern Diablo Range, California: Record of the Sierra Nevada magmatic arc. *Bull. geol. Soc. Am.* **90**, 1025–46.

McBIRNEY, A. R. 1963. Factors governing the nature of submarine volcanism. *Bull. Volcanol.* **26**, 455–69.

McBRIDE, E. F. 1978. Olistostrome in the Tesnus Formation (Mississippian–Pennsylvanian), Payne Hills, Marathon region, Texas. *Bull. Geol. Soc. Am.* **89**, 1550–8.

MIDDLETON, G. V. 1967. Experiments on density and turbidity currents. III. Deposition of sediment. *Can. J. Earth Sci.* **4**, 475–505.

—— 1970. Experimental studies related to problems of flysch sedimentation. *Spec. Pap. geol. Ass. Can.* **7**, 253–72.

—— & HAMPTON, M. A. 1973. Sediment gravity flows: Mechanics of flow and deposition. *In*: MIDDLETON, G. V. & BOUMA, A. H. (eds) *Turbidites and Deep-water Sedimentation.* Short Course, Soc. Econ. Paleont. Mineral. Pacific Section, May 1973, pp. 1–38.

—— & —— 1976. Subaqueous sediment transport and deposition by sediment gravity flows. *In*: STANLEY, D. J. & SWIFT, D. J. P. (eds) *Marine Sediment Transport and Environmental Management.* John Wiley & Sons, pp. 197–218.

MITCHELL, A. H. G. 1970. Facies of an early Miocene volcanic arc, Malekula Island, New Hebrides. *Sedimentology*, **14**, 201–43.

—— & READING, H. G. 1978. Sedimentation and tectonics. *In*: READING, H. G. (ed.) *Sedimentary Environments and Facies*, **14**, 439–76. Blackwell Scientific Publications, Oxford.

MOORE, J. G. 1965. Petrology of deep-sea basalt near Hawaii. *Am. J. Sci.* **263**, 40–52.

—— 1970. Water content of basalt erupted on the ocean floor. *Contrib. Mineral. Petrol.* **28**, 272–9.

—— 1975. Mechanism of formation of pillows. *Am. Sci.* **63**, 269–77.

—— & FISKE, R. S. 1969. Volcanic substructure inferred from dredge samples and ocean-bottom photographs, Hawaii. *Bull. geol. Soc. Am.* **80**, 1191–202.

—— & SCHILLING, J.-G. 1973. Vesicles, water, and sulfur in Reykjanes Ridge basalts. *Contrib. Mineral. Petrol.* **41**, 105–18.

—— & SISSON, T. W. 1981. Deposits and effects of the May 18 pyroclastic surge. *In*: LIPMAN, P. W. & MULLINEAUX, D. R. (eds) *The 1980 Eruptions of Mount St Helens, Washington. Prof. Pap. U.S. geol. Surv.* **1250**, 421–38.

MORGENSTERN, N. R. 1967. Submarine slumping and the initiation of turbidity currents. *In*: RICHARDS, A. F. (ed.) *Marine Geotechnique*, pp. 189–220.

NAKAMURA, K. 1974. Preliminary estimate of global volcanic production rate. *In*: FURUMOTO, S., MINIKAMI, T. & YUHARA, K. (eds) *The Utilization of Volcano Energy.* The Proceedings of U.S.–Japan Cooperative Science Seminar, Sandia Laboratories, Albuquerque, New Mexico. Pp. 273–84.

NARDIN, T. R., HEIN, F. J., GORSLINE, D. S. & EDWARDS, B. D. 1979. A review of mass movement processes, sediment and acoustic characteristics, and contrasts in slope and base-of-slope systems *versus* canyon–fan–basin floor systems. *Spec. Publ. Soc. econ. Paleontol. Mineral., Tulsa*, **27**, 61–73.

NAYLOR, M. A. 1980. The origin of inverse grading in muddy debris flow deposits—a review. *J. sediment. Petrol.* **50**, 1111–6.

—— 1981. Debris flow (olistostromes) and slumping on a distal passive continental margin: the Palombini limestone-shale sequence of the northern Apennines. *Sedimentology*, **28**, 837–52.

NAYUDU, Y. R. 1964. Volcanic ash deposits in the Gulf of Alaska and problems of correlation of deep-sea ash deposits. *Mar. Geol.* **1**, 194–212.

NEEDHAM, J. D. & FRANCHETEAU, J. 1974. Some characteristics of the rift valley in the Atlantic Ocean near 36°48′N. *Earth planet. Sci. Lett.* **22**, 29–43.

NEMEC, W., POREBSKI, S. J. & STEEL, R. J. 1980.

Texture and structure of resedimented conglomerates: examples from Ksiaz Formation (Famenian–Tournasian), southwestern Poland. *Sedimentology*, 27, 519–38.

NIEM, A. R. 1977. Mississippian pyroclastic flow and ash-fall deposits in the deep-marine Ouachita flysch basin, Oklahoma and Arkansas. *Bull. geol. Soc. Am.* **88**, 49–61.

NINKOVICH, D. 1979. Distribution, age and chemical composition of tephra layers in deep-sea sediments off western Indonesia. *J. Volcan. geoth. Res. Amsterdam*, **5**, 67–86.

—— & DONN, W. L. 1976. Explosive Cenozoic volcanism and climatic implications. *Science*, **194**, 899–906.

—— & HEEZEN, B. C. 1965. Santorini tephra. *In: Submarine Geology and Geophysics* (Colston Papers No. 17), **27**, 413–53. Butterworths Scientific Publications, London.

—— & —— 1967. Physical and chemical properties of volcanic glass shards from Pozzolana ash, Thera Island, and from upper and lower ash layers in eastern Mediterranean deep-sea cores. *Nature, Lond.* **213**, 582–4.

—— & SHACKLETON, N. J. 1975. Distribution, stratigraphic position and age of ash layer 'L', in the Panama Basin region. *Earth planet. Sci. Lett.* **27**, 20–34.

——, SPARKS, R. S. J. & LEDBETTER, M. T. 1978. The exceptional magnitude and intensity of the Toba eruption, Sumatra: An example of the use of deep-sea tephra layers as a geological tool. *Bull. Volcanol.* **41–3**, 286–98.

PETERSON, D. W. 1976. Processes of volcanic island growth, Kilauea Volcano, Hawaii, 1969–1973. *In:* FERRAN, O. G. (ed.) *Proceedings of the Symposium 'Andean and Antarctic Volcanology Problems'*. Spec. Series Internat. Assoc. of Volcanology and Chemistry of the Earth's Interior, 172–89.

RIEHLE, J. R. 1973. Calculated compaction profiles of rhyolitic ash-flow tuffs. *Bull. geol. Soc. Am.* **84**, 2193–216.

RITTMANN, A. 1962. *Volcanoes and Their Activity*. John Wiley and Sons, New York.

ROBINSON, P. T., FLOWER, M. F. J., SWANSON, D. A. & STAUDIGEL, H. 1980. Lithology and eruptive stratigraphy of Cretaceous oceanic crust, western Atlantic ocean. *Init. Rep. Deep Sea drill. Proj.* **51–3**, 1535–55.

SCHEIDEGGER, K. F., JEZEK, P. A. & NINKOVICH, D. 1978. Chemical and optical studies of glass shards in Pleistocene and Pliocene ash layers from DSDP Site 192, northwest Pacific Ocean. *J. Volcan. geoth. Res. Amsterdam*, **4**, 99–116.

——, CORLISS, J. B., JEZEK, P. A. & NINKOVICH, D. 1980. Composition of deep-sea ash layers derived from north Pacific volcanic arcs: variations in time and space. *J. Volcan. geoth. Res. Amsterdam*, **7**, 107–37.

SCHMINCKE, H.-U. 1967. Graded lahars in the type section of the Ellensburg Formation, south-central Washington. *J. sediment. Petrol.* **37**, 438–48.

—— & VON RAD, U. 1979. Neogene evolution of Canary Island volcanism inferred from ash layers and volcaniclastic standstones of DSDP Site 397 (Leg 47A). *Init. Rep. Deep Sea drill. Proj.* **47**(1), 703–25.

SELF, S. & RAMPINO, M. R. 1981. The 1883 eruption of Krakatau. *Nature, Lond.* **294**, 699–704.

SHAW, D. M., WATKINS, N. D. & HUANG, T. C. 1974. Atmospherically transported volcanic dust in deep sea sediments: theoretical consideration. *J. geophys. Res.* **79**, 3087–97.

SHERIDAN, M. F. 1979. Emplacement of pyroclastic flows: a review. *Spec. Pap. geol. Soc. Am.* **180**, 125–36.

SIGURDSSON, H., SPARKS, R. S. J., CAREY, S. N. & HUANG, T. C. 1980. Volcanogenic sedimentation in the Lesser Antilles Arc. *J. Geol.* **88**, 523–40.

SMITH, R. L. 1960. Ash flows. *Bull. geol. Soc. Am.* **71**, 795–842.

SPARKS, R. S. J. 1976. Grain size variations in ignimbrites and implications for the transport of pyroclastic flows. *Sedimentology*, 23, 147–88.

——, SIGURDSSON, H. & CAREY, S. 1980a. The entrance of pyroclastic flows into the sea. I. Oceanographic and geologic evidence evidence from Dominica. Lesser Antilles. *J. Volcan. geoth. Res. Amsterdam*, **7**, 87–96.

——, —— & —— 1980b. The entrance of pyroclastic flows into the sea. II. Theoretical considerations on subaqueous emplacement and welding. *J. Volcan. geoth. Res. Amsterdam*, **7**, 97–105.

STANLEY, D. J. & TAYLOR, P. T. 1977. Sediment transport down a seamount flank by a combined current and gravity process. *Mar. Geol.* **23**, 77–88.

——, PALMER, H. D. & DILL, R. F. 1978. Coarse sediment transport by mass flow and turbidity current processes and downslope transformations in Annot Sandstone Canyon–Fan Valley systems. *In:* STANLEY, D. J. & KELLING, G. (eds) *Sedimentation in Submarine Canyons, Fans and Trenches.* Dowden, Hutchinson & Ross, Stroudsburg, Penn. 85–115.

STANTON, W. I. 1960. The lower Paleozoic rocks of south-west Murrisk, Ireland. *J. geol. Soc. London*, **116**, 269–96.

STAUDIGEL, H. & SCHMINCKE, H.-U. 1981. Structural evolution of a seamount: evidence from the uplifted intraplate seamount on the island of La Palma, Canary Islands. *Eos*, **62**, 1075.

SWARBRICK, R. E. & NAYLOR, M. A. 1980. The Kathikas melange, SW Cyprus, Cretaceous submarine debris flows. *Sedimentology*, **26**, 63–78.

TASSE, N., LAJOIE, J. & DIMROTH, E. 1978. The anatomy and interpretation of an Archean volcaniclastic sequence, Noranda region, Quebec. *Can. J. Earth Sci.* **15**, 874–88.

THUNELL, R., FEDERMAN, A., SPARKS, S. & WILLAMS, D. 1979. The age, origin and volcanological significance of the Y-5 ash layer in the Mediterranean. *Quat. Res.* **12**, 241–53.

VAN ANDEL, TJ. H. & KOMAR, P. D. 1969. Ponded sediments of the Mid-Atlantic Ridge between 22° and 23° North latitude. *Bull. geol. Soc. Am.* **80**, 1163–90.

VESSEL, R. K. & DAVIES, D. K. 1981. Nonmarine sedimentation in an active forre arc basin. *Spec. Publ. Soc. econ. Paleontol. Mineral., Tulsa*, **31**, 31–45.

WALKER, G. P. L. 1979. A volcanic ash generated by explosions where ignimbrite entered the sea. *Nature, Lond.* **281**, 642–6.

WATKINS, N. D. & HUANG, T. C. 1977. Tephras in abyssal sediments east of the North Island, New Zealand: Chronology, paleowind velocity, and paleoexplosivity. *N.Z. Jl Geol. Geophys.* **20**, 179–98.

——, SPARKS, R. S. J., SIGURDSSON, H., HUANG, T. C., FEDERMAN, A., CAREY, S. & NINKOVICH, D. 1978. Volume and extent of the Minoan tephra from Santorini Volcano: new evidence from deep-sea sediment cores. *Nature, Lond.* **271**, 122–6.

WILSON, C. J. N. 1980. The role of fluidization in the emplacement of pyroclastic flows: An experimental approach. *J. Volcan. geoth. Res. Amsterdam*, **8**, 231–49.

WOHLETZ, K. H. 1980. Explosive Hydromagmatic Volcanism. Thesis, Ph.D. Arizona State University (unpub.).

—— & SHERIDAN, M. F. 1979. A model of pyroclastic surge. *Spec. Pap. geol. Soc. Am.* **180**, 177–94.

WRIGHT, J. V. & MUTTI, E. 1981. The Dali Ash, Island of Rhodes, Greece: a problem in interpreting submarine volcanigenic sediments. *Bull. Volcanol.* **44–2**, 153–67.

YAMADA, E. 1973. Subaqueous pumice flow deposits in the Onikobe Caldera, Miyagi Prefecture, Japan. *J. geol. Soc. Japan*, **79**, 589–97.

YAMAZAKI, T., KATO, I., MUROI, I. & ABE, M. 1973. Textural analysis and flow mechanism of the Donzurubo subaqueous pyroclastic flow deposits. *Bull. Volcanol.* **37**, 231–44.

RICHARD V. FISHER, Department of Geological Sciences, University of California, Santa Barbara, CA 93106, U.S.A.

Subaqueous pyroclastic flows: their development and their deposits

E. Yamada

SUMMARY: Subaqueous pyroclastic flows are generated by eruptions in shallow water or close to a shore. Their flow mechanism is similar to that of their subaerial counterparts, except that they incorporate steam and not air. Near the source area, their deposits are non-sorted and resemble subaerial flow deposits. In some cases evidence of high temperature is recognized. Within a short distance (less than a few km) they are sorted and five subdivisions can be determined. In upward succession these are: (i) massive graded pumice tuff; (ii) parallel-laminated pumice tuff; (iii) parallel-laminated sandy tuff; (iv) parallel-laminated fine tuff; and (v) massive very fine tuff. With increasing distance from the source, and laterally from the main direction of the flow, the lower divisions gradually die out. Deposition of the upper three divisions is greatly influenced by sedimentary processes. The thickness and constitution of each division vary according to the volume and nature of the associated eruption.

Subaerial pyroclastic flow deposits have been studied intensively, and some flows have been observed and recorded in historic times. The deposits are abundant in continental areas and island arcs. However, there are many pyroclastic flow deposits which are intercalated with marine or lacustrine sediments, and these commonly show lithological facies different to those of subaerial flows, though their constituents are similar. In this paper such deposits are called subaqueous pyroclastic flow deposits. They are especially common in marginal and intra-arc basins. The Green Tuff (Neogene), in the Japanese islands, is characterized by such subaqueous pyroclastic flow deposits, which have been altered and commonly show a greenish tint.

In this paper the constituents, vertical and lateral variations of facies, and the eruption and deposition environments of subaqueous pyroclastic flow deposits are briefly reviewed. Emphasis is placed on those deposits which the author has studied. The processes of their formation are discussed on the basis of this data.

Constituents of subaqueous pyroclastic flow deposits

The constituents of subaqueous pyroclastic flow deposits are subdivided into essential, accessory, and accidental. The essential particles are the pyroclastic fragments of juvenile magma directly related to the eruption. They comprise blocks and shreds of pumice, glass shards and dust, and crystal fragments. Accretionary lapilli may also occur. In pyroclastic flow deposits of the Upper Donzurubo Formation (Kato *et al.*

1971), slightly vesiculated perlite fragments are also considered as essential particles. Pumice is usually white, with abundant capillary-tubed vesicles, but grey, denser pumice (scoria) and grey and white banded pumice also occur. These essential particles range from andesite to rhyolite in composition and include one or more of the following crystals; plagioclase, quartz, K-feldspar, hornblende, pyroxene, and biotite.

Accessory particles are those derived from the cognate volcanic edifice from which the pyroclastic flow was erupted. They include fragments of rhyolite, dacite or andesite lava and scoria. Occasionally fragments of contrasting composition are included. These fragments may be torn from the vent or incorporated on the flanks of the pile during transport of the pyroclastic flow.

Accidental particles are either of the basement rocks of the volcanic edifice, plucked from the wall of the vent, or sediments scoured from the substrate during flow beyond the edifice. In some deposits, blocks of incompletely lithified silt, up to several metres in diameter, are included.

Thus the constituent particles of the subaqueous pyroclastic flow deposits considered here are like those in subaerial pyroclastic flow deposits. This suggests a similar mode of eruption. The relative proportions and sizes of the constituent particles vary in different deposits. Moreover they show marked vertical and lateral variations in a single deposit.

Fiske (1963) distinguished three types of subaqueous pyroclastic flow deposits. Deposits of the first type originate from fairly homogeneous domes, spines or lava flows which were erupted into water and fragmented by steam-blast explosions. They mainly contain only one or two kinds of lithic fragments. Deposits of the second type result from subaqueous phreatomagmatic eruptions and therefore contain a variety of lithic fragments and variable amounts of pumice. The third type of deposit is produced by underwater eruptions of rapidly vesiculating magma and therefore contains abundant pumice and glass shards. In this paper, only the third type is considered, but the eruption environment may be subaerial or subaqueous.

Vertical sequences of facies

Most subaqueous pyroclastic flow deposits have a characteristic vertical sequence of facies (Fig. 1). However, there are several exceptions. For example, the vertical sequence of the subaqueous pyroclastic flow deposits in the Upper Donzurubo Formation (Yamazaki *et al.* 1973) is similar to that of subaerial deposits. This similarity may be due to the very shallow depth of water (several tens of metres) and the short distance of travel under water (less than 3 km) which did not allow sorting to occur.

The vertical sequence of subaqueous pyroclastic flow deposits can commonly be divided into five, from the bottom (i) to the top (v): (i)

massive graded pumice tuff; (ii) parallel-laminated pumice tuff; (iii) parallel-laminated sandy pumice tuff; (iv) parallel-laminated fine tuff; and (v) massive very fine tuff. The lower two divisions correspond to the massive part of the flow described by Fiske & Matsuda (1964), and the parallel-laminated sandy pumice tuff and the parallel-laminated fine tuff divisions correspond to the thin-bedded part. In places the massive very fine tuff division at the top is missing, probably due to removal by currents. The vertical sequence shows double grading (Fiske & Matsuda 1964).

Subaqueous pyroclastic flows commonly scour the underlying sediments and the boundary with the underlying strata is sharply defined. However, the base of the Garth Tuff of the Capel Curig Volcanic Formation (Francis & Howells 1973) is very irregular with large-scale flame-like structures, highly transgressive protrusions and detached bodies (see below).

The massive graded division lacks bedding although in places accidental silt clasts, scoured from below the flow during transport, occur in crude beds. This division is comparatively rich in accidental and accessory fragments; in some deposits accessory fragments form up to 60% of total volume. Through most of the division the accessory lithic fragments show normal size grading, and close to the base they are commonly inversely graded. White, capillary-

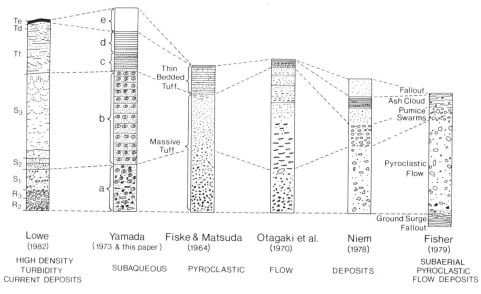

FIG. 1. Comparison of the vertical facies sequences of subaqueous pyroclastic flow deposits. The deposits of high-density turbidity currents and of subaerial pyroclastic flow deposits are shown for comparison. No relative vertical scale is implied.

tubular pumice clasts show inverse size grading through the division. The grading characters are controlled mainly by the settling velocity of each fragment in the subaqueous pyroclastic flow; the grey pumice and the banded pumice show grading characters different to that of the white pumice. In the ash-flow tuffs of the Capel Curig Volcanic Formation (Francis & Howells 1973) this division is welded.

Above the massive division, the parallel-laminated pumice tuff division has ill-defined bedding at its base which becomes gradually more distinct upwards. Each lamina is crudely graded according to the density and size differences of the particles. The maximum size of pumice clasts in this division gradually increases upwards from the base, and reaches its peak near the top. Commonly the maximum size of pumice clasts decreases abruptly from this level to the contact with the overlying parallel-laminated sandy pumice tuff division.

The parallel-laminated sandy pumice tuff division consists mainly of sand-sized pumice shreds with lithic and crystal fragments. The lamination, defined by density and size differences of particles, exists throughout the division. Each lamina is thinner than that below.

The parallel-laminated fine tuff division consists mainly of very fine white tuff which grades upwards from the underlying parallel-laminated sandy pumice tuff division. Its laminae are defined by concentrations of fine, dark-coloured crystal fragments and size grading of glass shards. Cross-lamination and accretionary lapilli occur locally in this division, and in the parallel-laminated sandy pumice tuff division. The massive very fine tuff division grades upwards from the underlying parallel-laminated fine tuff division.

This vertical sequence is characteristic of subaqueous pyroclastic flow deposits and is similar to that determined in deposits from high-density turbidity currents by Lowe (1982) (Fig. 1).

Lateral change of facies

A model of proximal to distal facies changes in subaqueous pyroclastic flow deposits is shown in Fig. 2. The facies within a few km of the eruption centre is little known, probably because it is very difficult to trace a flow to its source. In the Onikobe caldera (Yamada 1973), the subaqueous pumice flow deposits about 2 km from the source include cobble- to pebble-sized lithic fragments through most of the sequence. In places, accessory blocks, 1–2 m in diameter, are included and some are radially jointed due to rapid cooling *in situ* (Fig. 3). Several irregular thin flow units occur. Dark fine laminae, slightly inclined to each other and defined by concentrations of mafic crystals, are locally developed. Grading in each unit is mostly indistinct or absent. In the Green Tuff of Japan, thick subaqueous pumice flow deposits are common. These are massive and show almost no stratification, though pseudo-eutaxitic texture due to diagenetic flattening of argillized pumice clasts is locally developed. Similar deposits have been reported in the Delta River sequence (Bond 1973). These may also represent a proximal facies of subaqueous pyroclastic flow deposits.

Between 2 and 4 km from its source, the subaqueous pumice flow deposits of the Onikobe caldera gradually assume a vertical sequence similar to those shown in Fig. 1. In this transition zone, between the proximal and distal facies, several thin subaqueous pumice flow units can be distinguished in the upper part of the main eruption unit. Each thin flow unit shows scour-and-fill structure at its base and comprises massive, coarse pumice tuff in the

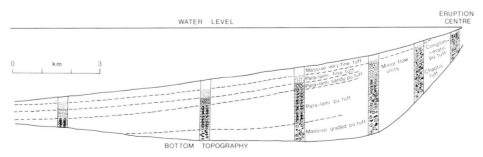

FIG. 2. Schematic cross-section of subaqueous pyroclastic flow deposits, showing the proximal to distal changes of facies. The vertical scale is greatly exaggerated. (Para-lami = parallel-laminated; Pu = pumice.)

FIG. 3. Closely spaced radial joints in an andesite block embedded in the proximal facies (about 2 km from the source) of a subaqueous pumice flow deposit in the Onikobe caldera, Japan. The scale length is 1 m.

erates interdigitate with the upper part of the pumice flow deposits.

In the Wadaira Tuff D (Fiske & Matsuda 1964), the characteristic vertical sequence changes away from the source and towards the edges of the main course of the flow channel, with the lower massive graded and the parallel-laminated pumice tuff divisions gradually disappearing within 1–2 km. In other subaqueous pyroclastic flow deposits, the characteristic vertical sequence continues for 20–30 km with little change—e.g. in the subaqueous pyroclastic flow deposits in the Vati Group (Wright and Mutti 1981) and in the Hatton Tuff (Niem 1977).

In the Beavers Bend Tuff (Niem 1977) the thickness decreases away from the source, from 12 to 3 m, over a distance of 30 km. In the Wadaira Tuff D the thickness decreases more sharply from the source, 60–10 m, over a distance of 1.7 km, although this is probably marginal to the main flow channel. In contrast, the tuff in the Vati Group (Wright & Mutti 1981) shows little thickness variation (7.5 m) in more than 20 km, and in the main ash-flow tuffs of the Capel Curig Volcanic Formation (Howells & Leveridge 1980), little thickness variation occurs. Here a transition from subaerial welded tuff to subaqueous welded tuff facies has been determined.

lower part; parallel laminations are gradually developed upwards. Further upwards in the minor flow unit, cross-stratification is locally developed and parallel-laminated fine tuff beds lie at the top. Locally, in the upper part of the deposit, drifted pumice blocks, up to tens of cm in diameter, are embedded and eddies which formed behind the blocks have truncated the parallel laminae. In places, accretionary lapilli, 0.5 to 3 cm are scattered in the upper sandy and fine tuffs. Parallel laminations do occur in this zone but most are slightly inclined, bevelled or irregularly wavy.

The components of the subaqueous pyroclastic flow deposits beyond 4 km from the source are similar to those described in the previous section. In the subaqueous pumice flow deposit of the Onikobe caldera, the maximum diameter of lithic fragments gradually decreases from 6 cm at about 4 km to 0.3 cm at about 6 km away from the source. A similar decrease in the diameter of lithic fragments away from the source is described by Otagaki *et al.* (1970) in the Shakanai area. A decrease in the size of crystals and pumice clasts with distance from the source has been recognized in the Hatton and the Beavers Bend Tuffs by Niem (1977). In places beyond 4 km from the source the lower portions of the characteristic vertical sequence are absent, probably because of emplacement on topographic highs. Locally in the marginal parts of the Onikobe caldera, talus conglom-

Environments of eruption and emplacement

By geological mapping of pyroclastic flows and the study of the associated strata, various environments of eruption and emplacement for individual deposits have been interpreted. The Wadaira tuff member (Fiske & Matsuda 1964) is considered to have been erupted under seawater and emplaced in open sea 400–500 m deep. It is considered to have travelled between 12 and 18 km from its source. The subaqueous pyroclastic flow deposits in the Shakanai area (Otagaki *et al.* 1970) are postulated to have been erupted and deposited in similar environments to the Wadaira tuffs, although they are less extensive.

Fiske (1963) interpreted the pyroclastic flow deposits of the Ohanapecosh Formation as having been erupted under shallow water or partly above water, and deposited in a shallow sea or lake. The main ash-flow tuffs of the Capel Curig Volcanic Formation have been interpreted (Howells & Leveridge 1980), from the associated sediments, to have been erupted subaerially, following which they transgressed a transitional zone into littoral and offshore submarine environments. They travelled more than 35 km, of which about 10 km was under seawater.

The subaqueous pyroclastic flow deposits in the Onikobe caldera (Yamada 1973) were probably erupted just above the water level and were deposited in a lake 200–500 m deep. The flows travelled a maximum of 6–8 km and were stopped or deflected by the caldera wall. The Hatton and the Beavers Bend Tuffs (Niem 1977) were erupted in shallow water and deposited in a deep-water basin at bathyal depth.

Discussion of the processes of formation of subaqueous pyroclastic flow deposits

Eruption

The subaqueous pyroclastic flow deposits so far reported are considered to have been erupted on land or in shallow water. Subaerial pyroclastic eruptions generate subaqueous pyroclastic flows when subaerial pyroclastic flows transgress into the sea or deep and extensive lakes. The eruption mode of these subaqueous pyroclastic flows is not affected by the aqueous environment. Shallow subaqueous eruptions may, however, be more violent than subaerial eruptions because the fragmented magma may rapidly exchange heat with the water which explosively turns to steam (see Self & Sparks 1978). Deep subaqueous eruptions are not likely to produce the type of subaqueous pyroclastic flow deposits considered in this paper, as the high ambient pressure will inhibit the exsolution of vapour from the magma. McBirney (1963) suggests that explosive vesiculation will not normally take place at depths greater than 500 m. Exceptionally, when the water content of rhyolitic magma approaches 3%, explosive eruptions may occur at depths approaching 2 km. However, McBirney (1971) questioned whether siliceous magma can vesiculate significantly at depths of more than a few tens of metres.

Generation

Eruption columns, from which pyroclastic flows are generated, probably reach higher levels in a subaerial and a shallow subaqueous environment (less than a few tens of metres) than in deeper water. This is because in the latter environment the vesiculation is less explosive and the viscous and inertial resistances of water are great. However, particles rise and fall more slowly in water than in air, and as a result their sorting is better. The generation of subaqueous pyroclastic flows from subaqueous eruption columns may occur in a manner similar to that described by Sparks *et al.* (1978) for subaerial pyroclastic flows. However, in moderate depths of water the lower eruption column and the slower settling velocity of particles will reduce the initial momentum of the pyroclastic flow to much below that of flows generated from collapse of subaerial columns.

When a small subaerial pyroclastic flow transgresses into water, its outer surface will be cooled suddenly and the flow will be impeded because of the higher density and viscosity of water than air. But the main body of the flow will continue along the bottom of the lake or the sea as a high-density turbidity current. The associated light fine-ash cloud may extend over the water surface for some distance, as witnessed during the 1902 eruption of Mt Pelée (Anderson & Flett 1903, quoted in Macdonald 1972). The interstitial medium of the high-density turbidity current will be the ambient water. However, when a large amount of pyroclastic material flows rapidly and continuously into water, the steam generated at its interface with water is incorporated and facilitates flow with a vapour medium for some distances from the shore. Deposition in the littoral zone may result in phreatic explosions (Walker 1980).

Flow and deposition

Normally, the subaqueous pyroclastic flow will be a high-density current continuously supplied by particles from an eruption column. Near the source the flow will be turbulent and deposition will take place from the high-shear boundary layer at the bottom, as occurs in subaerial pyroclastic flows (Fisher 1966). Therefore, the deposits near the source area tend to be non-graded and non-sorted. However, during transport along the sea or lake floor, sorting will gradually take place and the flow will become density and size graded, vertically as well as laterally.

The massive graded division will be deposited rapidly by freezing of the basal traction carpet and by suspension sedimentation from the high-density turbidity current. The overlying parallel-laminated pumice tuff division results from traction carpet layers of the successive small flows (Lowe 1982). The parallel-laminated sandy pumice tuff and fine tuff divisions will be deposited slowly from less dense turbidity currents, which separated from and lagged behind the denser preceding currents.

The massive fine tuff division will be deposited slowly from suspended fine-ash clouds.

This simplified model of flow and deposition of subaqueous pyroclastic flow deposits will be modified, especially in the upper divisions, by ocean or other currents and by the complexity of pyroclastic eruptions.

Post-depositional cooling

After deposition of subaerial ash-flow layers, welding, devitrification and vapour-phase crystallization commonly take place. In subaqueous ash-flow deposits such post-depositional processes are rarely reported. However, in the Capel Curig Volcanic Formation (Francis & Howells 1973), and on Siargao Island (Fernandez 1969), subaqueous welded ash-flow deposits have been reported. In the Green Tuff (Neogene) in the Japanese islands, welded tuffs are in places intricately associated

FIG. 4. Specimen from the basal part of a subaqueous welded pyroclastic flow tuff in the marine Shirahama Group (late Miocene), Izu Peninsula, Japan. Welding of pumice shreds intensifies upwards. The scale is in cm.

with massive pumice flow tuffs. This association may be the proximal facies of subaqueous pumice flow deposits, although typically the Green Tuff comprises distal facies subaqueous pyroclastic flow deposits intercalated with marine fossiliferous siltstone and sandstone beds. In the marine Shirahama Group (Neogene), on the Izu Peninsula of Japan, a welded pumice flow tuff occurs with the welding gradually becoming more pronounced upwards to about 30 cm above the base, and above it becomes less distinct. In the intensely welded part, the pumice lapilli are like black (although a few are reddish-brown) obsidian (Fig. 4), but consist of highly compressed bundles of clay minerals. In the less-welded tuffs, the pumice lapilli are buff-yellow and bubble wall texture is still discernible. The matrix is sandy and contains no glass shards. Columnar cooling joints and closely jointed accessory blocks in the subaqueous welded tuffs indicate that they were at high temperature when deposited. Kato et al. (1971) considered that the Donzurubo subaqueous pyroclastic flow deposits were emplaced at a high temperature because the remanent magnetism in embedded essential rock fragments showed a consistent orientation. In considering the temperature of deposition, the transgressive base of the subaqueously welded tuff of the Capel Curig Volcanic Formation was interpreted to have resulted from heating of the underlying water logged sediments which became fluidized and displaced by steam (Kokelaar 1982). Thus subaqueous pyroclastic flow deposits may retain high temperatures, especially in the areas close to the source or the shore where the flow plunged into water. The increased solubility of water in silicic glass with increased pressure may favour welding under water (Sparks et al. 1980).

Conclusions

Subaqueous pyroclastic flow deposits have a characteristic vertical sequence (Fig. 1), and their facies change laterally (Fig. 2). The vertical sequence is generated mainly by sorting of pyroclastic materials during flow. These characteristic facies sequences only vary slightly in different deposits. The proximal facies are greatly influenced by the eruption mode and history. The distal facies and the upper part of the deposits are affected by ocean or other currents prevalent in the area of emplacement.

ACKNOWLEDGMENTS: An early form of this paper was read at the Penrose Conference on Subaqueous Volcanism at Santa Barbara, California, in November 1977. The author profited from free discussions with participants at the conference, especially with Professor Richard V. Fisher, and also expresses hearty thanks to the anonymous reviewers of the manuscript, whose valuable suggestions greatly improved this paper.

References

BOND, G. C. 1973. A late paleozoic volcanic arc in the eastern Alaska Range, Alaska. *J. Geol.* **81**, 557–75.

FERNANDEZ, H. E. 1969. Notes on the submarine ash flow tuff in Siargao Island, Surigao del Norte. *Philipp. Geol.* **23**, 29–36.

FISHER, R. V. 1966. Mechanism of deposition from pyroclastic flow *Am. J. Sci.* **264**, 350–63.

—— 1979. Models for pyroclastic surges and pyroclastic flows. *J. Volcanol. geoth. Res. Amsterdam*, **6**, 305–18.

FISKE, R. S. 1963. Subaqueous pyroclastic flows in the Ohanapecosh formation, Washington. *Bull geol. Soc. Am.* **74**, 391–406.

—— & MATSUDA, T. 1964. Submarine equivalents of ash flows in the Tokiwa Formation, Japan. *Am. J. Sci.* **262**, 76–106.

FRANCIS, E. H. & HOWELLS, M. F. 1973. Transgressive welded ash-flow tuffs among the Ordovician sediments of NE Snowdonia, N. Wales. *J. geol. Soc. London*, **129**, 621–41.

HOWELLS, M. F. & LEVERIDGE, B. E. 1980. The Capel Curig Volcanic Formation. *Rep. Inst. geol. Sci. London.* **80** (6), 25 pp.

KATO, I., MUROI, I., YAMAZAKI, T. & ABE, M. 1971. Subaqueous pyroclastic flow deposits in the Upper Donzurubo Formation, Nijo-san district, Osaka, Japan. *J. geol. Soc. Japan*, **77**, 193–206.

KOKELAAR, B. P. 1982. Fluidization of wet sediments during the emplacement and cooling of various igneous bodies *J. geol. Soc. London*, **139**, 21–33.

LOWE, D. R. 1982. Sediment gravity flows. II. Depositional models with special reference to the deposits of high-density turbidity currents. *J. sediment. Petrol.* **52**, 279–97.

MACDONALD, G. A. 1972. *Volcanoes.* Prentice-Hall, Englewood Cliffs, New Jersey.

McBIRNEY, A. R. 1963. Factors governing the nature of submarine volcanism. *Bull. Volcanol.* **26**, 455–69.

—— 1971. Oceanic volcanism: A review. *Rev. Geophys. Space Phys.* **9**, 523–56.

NIEM, A. R. 1977. Mississippian pyroclastic flow and ash-fall deposits in the deep-marine Ouachita flysch basin. Oklahoma and Arkansas. *Bull. geol. Soc. Am.* **88**, 49–61.

OTAGAKI, T., ABE, Y., TSUKADA, Y., KIMURA, A., OSADA, T. & FUJIOKA, H. 1970. Geology and ore deposit of the Shakanai Mine (4). On the occurrence of ore deposit and pyroclastic rocks near the ore bodies. *Mining Geology*, **20**, 315–27. (in Japanese).

SELF, S. & SPARKS, R. S. J. 1978. Characteristics of widespread pyroclastic deposits formed by the interaction of silicic magma and water. *Bull. Volcanol.* **41**, 196–212.

SPARKS, R. S. J., SIGURDSSON, H. & CAREY, S. N. 1980. The entrance of pyroclastic flows into the sea. II. Theoretical considerations on subaqueous emplacement and welding. *J. Volcanol. geoth. Res. Amsterdam*, **7**, 97–105.

——, WILSON, L. & HULME, G. 1978. Theoretical modelling of the generation, movement, and emplacement of pyroclastic flows by column collapse. *J. geophys. Res.* **83**, 1727–39.

WALKER, G. P. L. 1979. A volcanic ash generated by explosions where ignimbrite entered the sea. *Nature, Lond.* **281**, 642–6.

WRIGHT, J. V. & MUTTI, E. 1981. The Dali ash, island of Rhodes, Greece: a problem in interpreting submarine volcanogenic sediments. *Bull. Volcanol.* **44**, 153–67.

YAMADA, E. 1973. Subaqueous pumice flow deposits in the Onikobe Caldera, Miyagi Prefecture, Japan. *J. geol. Soc. Japan*, **79**, 585–97.

YAMAZAKI, T., KATO, I., MUROI, I & ABE, M. 1973. Textural analysis and flow mechanism of the Donzurubo subaqueous pyroclastic flow deposits. *Bull. Volcanol.* **37**, 231–44.

EIZO YAMADA, Geological Survey of Japan, 1-1-3 Higashi, Yatabe, Ibaraki, 305 Japan.

A model of volcanogenic sedimentation in marginal basins

S. Carey & H. Sigurdsson

SUMMARY: Marginal basins adjacent to oceanic island arcs receive volcaniclastic debris from the bordering volcanic arc, back-arc spreading centre, and, to a lesser degree, the remnant arc. The volcanic arc is volumetrically the most important source, with abundant volcaniclastics being produced by explosive subaerial and/or subaqueous eruptions and secondary erosion of the arc complex. Transport of material to the deeper parts of marginal basins occurs by passive settling through the water column or by a variety of sediment gravity flows generated by primary eruptions and secondary remobilization processes. Deposition of arc-derived material at the base of the arc flank produces thick volcaniclastic aprons and results in a marked asymmetry in the accumulation of volcanogenic sediment within basins. The apron facies are complex and not analogous to typical deep-sea fan systems because of the influence of arc volcanism on the nature and delivery of sediment to the deep sea. Back-arc spreading centres supply a small volume of deep-water hyaloclastites and hydrothermally derived volcanogenic sediments which are incorporated into the basal parts of sedimentary sequences. The dispersal and deposition of volcanogenic sediments are profoundly influenced by geological processes associated with marginal basin evolution such as back-arc spreading, faulting, and island-arc splitting.

Marginal basins, as defined by Karig (1971a), are semi-isolated basins or a series of basins lying behind the volcanic chains of island-arc systems. Lawver & Hawkins (1978) expanded the definition to include small areas of oceanic lithosphere between two island arcs, arcs and continental masses or between two continental fragments.

These basins are the sites of complex sedimentation, usually with several genetically distinct yet interactive sediment sources. However, piston coring, DSDP drilling, and studies of uplifted marginal basin sequences all demonstrate the volumetric importance and diagnostic potential of the arc-derived volcaniclastic debris. This review therefore deals primarily with the generation, transport and deposition of both primary and secondary volcaniclastic material derived from the island arc.

Marine geophysical evidence indicates that the formation and evolution of marginal basins is coupled directly to the tectonics of lithospheric plate subduction (Karig 1971a). Because volcanogenic sedimentation within marginal basins is fundamentally related both in time and space to this tectonism, we begin with a brief discussion of the origin and evolution of the basin environment.

Origin and evolution of the marginal basin environment

It is now widely accepted that most marginal basins have formed by extension somewhat analogous to the creation of new oceanic crust at mid-ocean ridge spreading centres. This interpretation is supported by basin morphology (Karig 1970, 1971a; Karig et al. 1978; Bibee et al. 1980), the mapping of linear magnetic anomalies (Watts & Weissel 1975; Louden 1976), recovery of fresh basaltic rocks within basins (Karig 1971b; Hawkins 1977; Tarney et al. 1977) and seismic refraction profiles which indicate an oceanic-type basinal crust. Extensional basins have been classified as either active or inactive by Karig (1971a). Active basins are presently undergoing extension by the formation of new oceanic crust, e.g. the back-arc areas of the Tonga, Kermadec, Mariana, Bonin, New Hebrides, and Scotia island-arc systems. Inactive basins were formed by extension at some time in the past but presently show no evidence of active spreading. Inactive basins are further subdivided into two groups based on crustal heat flow (Karig 1971a). Inactive basins with high heat flow include the Seas of Japan and Okhotsk and the Parece Vela Basin, whereas inactive basins with normal heat flow include the South Fiji and Aleutian basins. In this classification, heat flow is assumed to be indicative of the relative age of a basin with high heat flow implying a more youthful basin. These generalizations apply to many marginal basins, particularly those of the Western Pacific, but there are exceptions. For example, there is convincing evidence that the Bering Sea may represent an old piece of oceanic crust trapped by the initiation of subduction along the Aleutian island arc (Cooper et al. 1976) and Uyeda & Ben-Avraham (1972) have suggested a similar origin for the western Philippine Sea.

Karig (1971b) proposed a tectonic model for the development of marginal basins in the Philippine Sea which begins with lithospheric plate subduction and the creation of a trench-volcanic arc system with active, typically calc-alkaline volcanism (Fig. 1a). Extension, developed as a result of the subduction process, causes the volcanic arc to split along zones of weakness, such as magma conduits and zones of active intrusion, and steep normal faults define the boundaries of the resulting rift (Fig. 1b). New oceanic crust is then produced along an axial high and the remnant arc migrates away from the active volcanic arc (Fig. 1c). As the basin grows, the flanks of the axial high subside in a manner similar to the subsidence of newly formed oceanic crust at mid-ocean ridge spreading centres (Fig. 1d). A new cycle of basin formation begins when extension in the relatively mature back-arc basin ceases and the volcanic arc rifts again. Bibee *et al.* (1980) proposed a model to explain the evolution of discrete marginal basins and the often complex magnetic anomaly patterns (Fig. 2). They point out that the major difference between back-arc and mid-ocean ridge spreading is that the driving mechanism in the back-arc area is more localized and closely tied to the volcanic arc-trench system. Symmetrical spreading in the back-arc causes the spreading centre to migrate continually away from the volcanic arc and out of the tensional stress regime (Fig. 2c). Eventually, a new cycle of basin formation begins when the spreading centre reverts to the site of major tensional stress, either within the basin or back at the volcanic arc (Fig. 2d).

The nature of the back-arc spreading process and whether it is truly analogous to mid-ocean ridge spreading or represents a type unique to the back-arc environment has been a point of controversy. Early estimates of basin spreading rates suggested a lack of correlation between spreading rate and axial high morphology (Karig 1971b). However, recent studies, for example in the Mariana Trough, do show a correlation between spreading rate (slow) and axial high morphology which is similar to that on typical mid-ocean ridges (Bibee *et al.* 1980; Hussong & Fryer 1980; Hussong & Uyeda 1981).

Several important features which are relevant to the processes of volcanogenic sedimentation emerge from all these studies. Marginal basins are dynamic tectonic and sedimentary environments where new oceanic crust is formed in a manner analogous to the production of new crust at slow-spreading mid-ocean ridges. Typically, the new crust is morphologically complex, consisting of a series of ridges and troughs, some *en échelon* and cut at high angles by numerous faults. These features affect sedimentation by providing localized sediment traps and large-scale barriers to sediment dispersal (Karig & Wageman 1975; Karig *et al.* 1978; Bibee *et al.* 1980). Of particular importance to facies

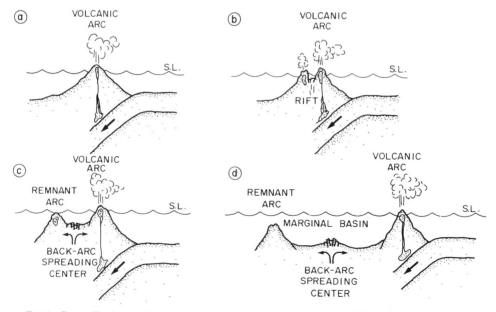

FIG 1. Generalized tectonic evolution of a marginal basin after Karig (1971a,b).

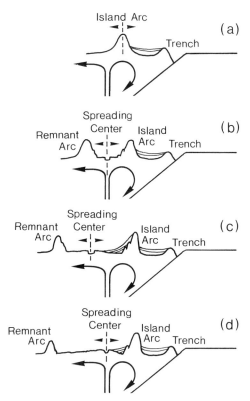

FIG. 2. Model to explain rift-jumping in back-arc basins from Bibee *et al.* (1980). Basin spreading results in the migration of the spreading centre away from the tensional forces developed by subduction (b,c). The change in stress field eventually causes the spreading centre to be re-established back towards the island arc (d).

variations of volcanogenic sediments is the long-term tectonic evolution of marginal basins. The growth and subsidence of marginal basin crust leads to lateral and vertical facies changes as the locus of sedimentation continually moves with respect to different sources (Karig & Moore 1975).

The model of volcanogenic sedimentation presented in this paper is based on the tectonic model depicted in Fig. 1. This will be used as the long-term (1–10 Ma) tectonic framework within which volcanogenic sedimentation takes place. Admittedly not all marginal basins are subject to such an evolution; for example, basins created by the trapping of old oceanic crust would remain tectonically stable compared with extensional basins. Nevertheless, as a generalization, crustal extension is regarded as a common feature of marginal basin development.

Sources and transport mechanisms of volcanogenic sediment

There are two major sources of volcanogenic sediments supplied to marginal basins adjacent to oceanic island arcs:

(i) the bordering volcanic arc; and
(ii) the back-arc spreading area (Fig. 3).

The former is divided into the subaerial and subaqueous parts in recognition of the role that external water plays in the generation of volcaniclastics by magma fragmentation and erosion of pre-existing deposits. The remnant arc may also supply some epiclastic debris but this

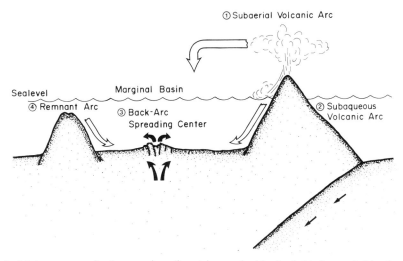

FIG. 3. Major sources of volcanogenic sediment in marginal basins behind oceanic island arcs.

is considered to be a relatively minor contribution (Karig & Moore 1975).

The great water depth of back-arc spreading systems inhibits the fragmentation of freshly erupted magma by internal vesiculation or phreatomagmatic explosions (McBirney 1971). Hyaloclastites can be produced, however, by spalling and granulation as hot lava interacts with cool seawater. These deposits typically consist of sand- and silt-size glassy basaltic fragments. Unlike fragments produced by internal vesiculation of dissolved gases, hyaloclastite fragments exhibit a blocky, angular morphology bounded by fracture surfaces (Heiken 1974). The occurrence of deep-water hyaloclastites at oceanic spreading centres was shown on DSDP Leg 46 at 23°N, 160 km E of the Mid-Atlantic Ridge (Schmincke *et al.* 1978; Dick *et al.* 1976).

Deep-water hyaloclastites are generally restricted to the axial zone of the spreading centre during formation and can only be dispersed significant distances by bottom currents. At site 396B (DSDP Leg 46) the occurrence of graded bedding and scouring features indicates some type of current activity but it is not possible to determine its extent. Because of their limited dispersal, deep-water hyaloclastite sequences are likely to be found at the base of marginal basin sedimentary sequences, where they are incorporated during the spreading process.

An additional source of volcanogenic sediments from back-arc spreading centres, albeit indirect, is from hydrothermal systems created by the circulation of seawater into newly formed oceanic crust (Fig. 4). At mid-ocean ridge spreading centres three different types of hydrothermal deposits have been identified:

(i) Fe-sulphide ore deposits (Francheteau *et al.* 1979);
(ii) Fe- and Mn-rich sediment (Bostrom & Peterson 1965; and
(iii) Mn-rich crusts (Scott *et al.* 1974).

It has been suggested that all three deposits are created by precipitation from various mixtures of a hot, acidic, reducing solution which has circulated through and leached the basaltic pile, with a cooler end-member solution approximating to seawater in composition (Edmond *et al.* 1979). Of these three main types only the Fe- and Mn-rich sediment has so far been documented at back-arc spreading centres (Bonatti *et al.* 1979; Cronan *et al.* this volume).

In the Philippine Sea, Fe- and Mn-rich sediment was found in the basal sections of sites 291, 294 and 295 with at least 58 m of ferruginous clay overlying basaltic basement at site 295 (Bonatti *et al.* 1979). Major and minor element geochemistry of this material demonstrated a striking similarity to hydrothermal sediments found on normal mid-ocean ridge spreading centres. Because the hydrothermal deposits are formed and distributed about the spreading axis, they are found mainly at the base of marginal basin sedimentary sequences, like the deep-water hyaloclastites.

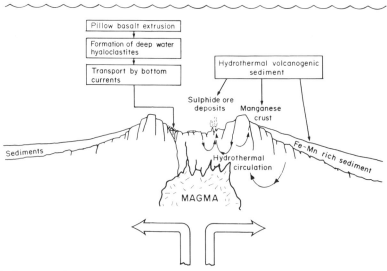

FIG. 4. Sources and transport mechanisms of volcanogenic sediments from back-arc spreading centres.

TABLE 1. *Volume % of fragmental material in island arcs, continental volcanic arcs, ocean basins, and oceanic islands from Garcia (1978)*

Area	Volume % fragmental	Magma type[a]
Island arcs		
Aleutians-Alaska	95	CA
Bonin-Izu	85	CA-Th
Indonesia	99	CA
Grecian	83	Alk
Lesser Antilles	95	CA-Alk
Marianas	90	Th
Continental arcs		
Andes	97	CA
Cascades	90	CA
Central America	99	CA
Ocean basins		
Pacific	3	Th
Atlantic	6	Th
Indian	3	Th
Oceanic islands		
Iceland	29	Th
Hawaii	1–2	Th-Alk
Canary Islands	20	Alk

[a]Alk = alkalic, CA = calc-alkaline, Th = tholeiitic.

In contrast to the back-arc spreading centre, the bordering volcanic arc is a more diverse and complex source of volcanogenic sediment. Explosive volcanism is a characteristic and ubiquitous feature of volcanic arcs. Fragmentation of volatile-rich magmas by internal vesiculation and the interaction with external water can produce enormous quantities of volcaniclastic debris during the course of a single eruption. The efficiency of this process can be appreciated by comparing the volumes of fragmental material in island and continental arcs with oceanic islands and basins (Table 1). The majority of island and continental arcs have in excess of 90% by volume of fragmental material, compared to less than 40% for oceanic islands and less than 10% for oceanic basins.

The principal source of volcaniclastic material in the subaerial regions of volcanic arcs is highly explosive eruptions from composite volcanoes and calderas. These eruptions vary according to magma volume, magma composition, volatile content, degree of crystallinity, eruption rate and vent morphology.

Erupted material consists largely of three components:

(i) vesiculated magma fragments;
(ii) euhedral and fragmented crystals; and
(iii) lithics.

Vesiculated magma fragments can range in grain size from pumiceous blocks to micron-size glass shards (Heiken 1974).

Volcaniclastics produced by subaerial arc volcanism can be transported to marginal basins by:

(i) direct fallout from the atmosphere with subsequent passive settling through the water column;
(ii) direct entrance of pyroclastic flows or lahars into the sea (e.g. Francis & Howells 1973; Yamazaki *et al.* 1973; Carey & Sigurdsson 1980; Sparks *et al.* 1980a, 1980b; Self & Rampino 1981); and
(iii) secondary reworking of pyroclastic deposits by fluvial and aeolian processes.

In many instances the transport of material may involve several episodes of storage and remobilization in different environments (Fig. 5). (For a review of the production and transport of volcaniclastics see Fisher (this volume).)

Superimposed upon the primary production and transport of volcaniclastics from subaerial arc volcanism is the secondary erosion of old pyroclastics and lava to form volcanogenic epiclastic sediment. Unlike the episodic influx of 'primary' volcaniclastics derived directly from subaerial eruptions, the supply of epiclastic material is somewhat more continuous and more a function of specific erosion processes.

In the subaqueous volcanic arc, eruptions may occur over a wide range of water depths (perhaps several km) and involve magmas ranging in composition from basalt to rhyolite. The subaqueous production of volcaniclastic material can occur either by non-explosive means (granulation, spalling, and brecciation of subaqueous lava flows), or by explosive processes (phreatic or phreatomagmatic eruptions) (Honnorez & Kirst 1975). External water is a major factor in determining the nature of eruptions. Depending on the depth, it can act either to inhibit fragmentation of magma by suppressing internal vesiculation due to high hydrostatic pressure, or to stimulate fragmentation by explosive expansion to a gas.

Explosive eruptions will be more frequent in the shallow areas of the volcanic arc, although the production of volcaniclastic debris by non-explosive processes can still occur. Evidence, particularly from Iceland (Allen 1980; Furnes *et al.* 1980), suggests depths of 200–300 m for the transition from non-explosive to explosive behaviour. A broad spectrum of eruption types is to be expected depending upon magma composition, water depth, and the relative impor-

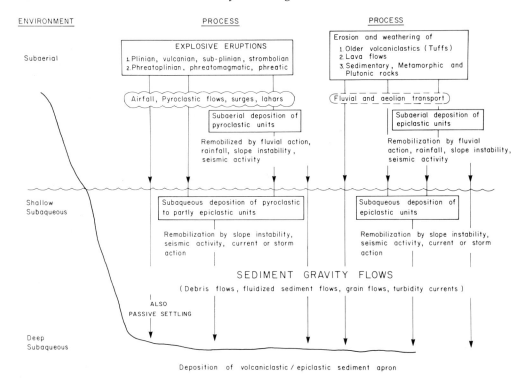

FIG. 5. Summary of volcanogenic sediment production and transport processes from the subaerial volcanic arc.

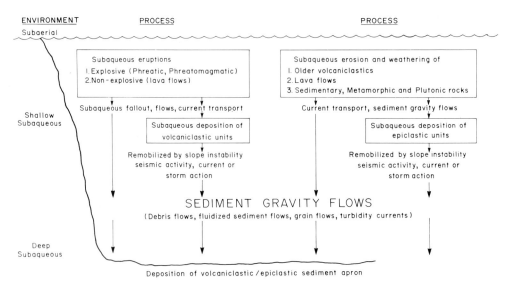

FIG. 6. Summary of volcanogenic sediment production and transport processes from the subaqueous volcanic arc.

tance of juvenile *v.* non-juvenile water. Figure 6 is an attempt to summarize the major processes of sediment production and transport in the subaqueous arc environment.

Models of volcanogenic sedimentation in marginal basins

Volcanogenic sedimentation in marginal basins occurs within a tectonically dynamic environment of plate subduction and back-arc spreading. The sources and transport of sediment have been briefly discussed and these aspects are now considered within a volcano-tectonic framework for a typical oceanic island arc. Clearly this is a somewhat simplistic case, as basins commonly have other volumetrically important sediment inputs, such as of terrigenes, which dilute the volcanogenic components. However, this dilution does not affect many of the basic processes of volcanogenic sediment formation and transport. Thus some aspects of the model can be extrapolated to basins with more complex configurations and heterogeneous sediment sources.

The adopted model of marginal basin evolution in an oceanic island-arc setting is from Karig (1971a,b) (Fig. 1). With this model two different data sets have been used to construct an integrated volcanogenic sedimentation model for marginal basins:

(i) results from DSDP sites located in back-arc basins of the Western Pacific and;
(ii) results from piston cores collected in the Grenada back-arc basin of the Lesser Antilles island arc.

Each data set has it own limitations, but in combination they provide a more complete picture of sedimentation. The DSDP results allow changes in sedimentation over long periods of time (1–10 Ma) to be examined and related to the major volcano-tectonic processes, such as splitting of active arcs, back-arc spreading, episodes of island-arc volcanism and hydrothermal circulation. In contrast, although piston cores are not very useful for determining long-term changes in sedimentary styles, they can, when collected in large numbers, provide the correlation control for determining lateral facies changes over shorter periods of time (up to 1 Ma). In particular, they provide information on the nature of growth processes on the volumetrically important volcaniclastic apron which develops at the base of the island-arc flank.

Previous models

Several models have been proposed for sedimentation in marginal basins (e.g. Karig & Moore 1975; Klein 1975b, 1977). These deal with some aspects of volcanogenic sedimentation but do not emphasize details of sediment production and transport. Furthermore, volcanological interpretations have not been fully incorporated into these models.

In the model of Karig & Moore (1975), sedimentation is discussed within the tectonic framework of an actively spreading marginal basin isolated from terrigenous sources. Four sediment sources are proposed:

(i) volcaniclastic debris from the volcanic arc;
(ii) montmorillonitic clays produced by the weathering of mafic volcanics;
(iii) biogenic components; and
(iv) wind-blown dust.

The volcaniclastic contribution of the active volcanic arc is volumetrically the most important (Karig & Moore 1975) and this results in a marked asymmetry in sediment accumulation within basins. The thickest sedimentary sequences are found in the extensive volcaniclastic aprons at the base of the active-arc flank and also, to a lesser extent, at the base of the remnant arc. As in larger ocean basins, the minimum sediment accumulation occurs over the active spreading ridge.

Karig & Moore (1975) illustrate the changes in the distribution of sedimentary facies during basin evolution (Fig. 7). In the early stages of rifting, the small inter-arc basin is flooded by coarse volcaniclastic material derived from volcanic activity along the arc (Fig. 7a). With continued spreading the basin widens to a point where the arc-derived volcaniclastics can no longer be dispersed basin-wide. In its place a brown montmorillonitic clay, rich in volcanic glass and phenocrysts, accumulates towards the remnant-arc side of the basin. Closer to the remnant arc, the clay grades into biogenic calcareous ooze (Fig. 7b). The volcaniclastic apron at the base of the volcanic arc flank progrades into the basin by processes somewhat analogous to submarine fan growth. If the basin continues to spread it is possible that biogenic calcareous ooze will be deposited directly on the new crust created at the back-arc spreading centre. (Fig. 7c).

Karig & Moore (1975) further emphasize three factors which can affect facies associations in the basin:

(i) cessation of basin spreading;
(ii) cessation of arc volcanism; and
(iii) resurgence of arc volcanism.

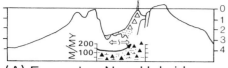

(A) Example : New Hebrides

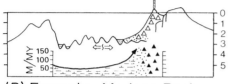

(B) Example : Mariana Basin

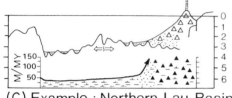

(C) Example : Northern Lau Basin

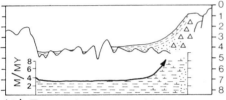

(D) Example : Parece Vela Basin

FIG. 7. Tectonic and sedimentary evolution of a marginal basin, from Karig & Moore (1975). At the base of each example the sedimentation rate and distribution of surficial sedimentary facies (see text) is shown: (a) Initial stage; volcaniclastics cover floor of entire basin. (b) Early differentiation of facies. (c) Advanced differentiation of facies; nannofossil ooze has developed near remnant arc. (d) Following cessation of spreading and volcanism; most of the basin has subsided below the CCD.

Cessation of basin spreading would allow the volcaniclastic apron to prograde across the spreading centre and the montmorillonite-rich and calcareous ooze facies. If spreading continues while volcanism ceases, then the volcaniclastic apron in turn will be buried by nannofossil and radiolarian oozes while the basin floor is above the CCD (carbonate compensation depth), and by brown clays when the floor is below the CCD (Fig. 7d). In contrast to the

montmorillonite-rich clay, this brown clay contains less volcanic glass, feldspar, quartz and montmorillonite, reflecting reduced input from arc volcanism. Finally, if a resurgence of volcanism occurs, then a new volcaniclastic apron will be generated which will cover the facies deposited during the lull in volcanic activity. This model is important because it emphasizes the importance of the volcaniclastic apron and its formation by volcanism in the subaerial and subaqueous arc. The apron may be one of the more diagnostic features of marginal basin sedimentation.

Klein (1975b) proposed a model based on Leg 30 DSDP cores from the South Fiji, Hebrides, and Coral Sea marginal basins. Seven sedimentary facies were discriminated for the three areas, which were then interpreted within a tectonic framework for marginal basin evolution (Fig. 8). This model differs from that of Karig & Moore (1975) in many respects. In Klein's tectonic scheme, basin initiation occurs within oceanic crust even though there is evidence that the South Fiji Basin, for instance, probably formed by splitting of a pre-existing frontal arc (Karig 1970). Another important difference, in part due to the differences in tectonic models, is that Klein's model does not suggest asymmetry in sediment accumulation resulting from the deposition of an arc-derived

1 Plate Splitting; Sediment Collapse.

4 Spreading Ceases; Pelagic Sedimentation.

2 Early Spreading; High Slope Gradients Coarse-Grained Turbidity Currents & Debris Flow.

5 Resurgent Tectonic Stage; Deposition of Turbidite Clastic Wedge.

KEY

▦ Sandy Turbidites

▥ Pelagic Carbonate Ooze & Chalk

▮ Pelagic Clay

■ Clayey Turbidites

▨ Coarse Grained Turbidites Debris Flow CGL.& SS.

◿ Ocean Crust

3 Late Spreading; Low Slope Gradients, Sandy Turbidity Currents.

FIG. 8. Sedimentary evolution of SW Pacific marginal basins based on Leg 30 DSDP cores, from Klein (1975a,b).

volcaniclastic apron. However, the importance of these aprons has been clearly demonstrated by Karig (1975). Klein (1977) later argued that the lack of asymmetry within the SW Pacific basins was due to the complex curvature of island-arc sources. He also indicated that clastic sediment accumulation in basins can be symmetrical if terrigenous and volcaniclastic aprons occur on opposite sides of the basin, as is the case for the Sea of Japan. The interfingering of different clastic aprons is well illustrated in the Shikoku Basin where at least three different aprons are derived from the Japanese islands, the Palau–Kyushu ridge and the Iwo Jima ridge (White *et al*. 1980).

Furthermore, in Klein's model the cessation of basin spreading and arc volcanism occur at the same time. Although obviously related to the subduction process in some way, there would appear to be no *a priori* reason for back-arc spreading to be synchronous with arc volcanism. There are several examples, such as the Japan Sea, of inactive marginal basins which are bordered by presently active volcanic arcs. Karig & Moore (1975) favour a more independent association of the two processes, which can result in various combinations of frontal arc volcanic cycles superimposed upon a non-steady basin spreading. Evidence for alternately intensified arc volcanism and back-arc spreading comes from the recent results of Legs 59 and 60 of the DSDP programme. Scott & Kroenke (1981) show that initial periods of back-arc spreading are coincident with periods of low volcanic activity along the active arc.

DSDP results from Western Pacific marginal basins

After the initial work by Karig & Moore (1975) and Klein (1975a,b, 1977), two additional legs of the DSDP project were completed which provided new data for refining models of volcanogenic sedimentation in marginal basins. Legs 59 and 60 in the Mariana arc area are a transect across an active intra-oceanic volcanic arc, back-arc basin, remnant arc, inactive back-arc basin, and additional remnant arc (Kroenke *et al*. 1980; Hussong *et al*. 1981). These features correspond, from E to W, to the Mariana Ridge, Mariana Trough, West Mariana Ridge, Parece Vela Basin and the Palau–Kyushu Ridge (Fig. 9).

The results of Legs 59 and 60 (Scott & Kroenke 1981; Hussong & Uyeda 1981) support the original working model of Karig (1971a,b) for the tectonic evolution of this area. From W to E, successive marginal basins

and remnant arcs become younger towards the presently active Mariana island arc. Each basin originated by the splitting of an active volcanic arc, followed by back-arc spreading. This has resulted in an extensive eastward migration of the active zone of lithospheric subduction. Whereas each of the remnant arcs only preserves the record of volcanism associated with the spreading of one basin, the eastward migrating subduction area contains a complete volcanic record of all basin-forming cycles (Hussong & Uyeda 1981).

The sites drilled on Legs 59 and 60 are ideal for studying the long-term (several to tens of millions of years) changes in sedimentary facies which occur in marginal basins, because they are located in a series of basins which are at different stages of development. By comparing sites which are similarly located with respect to major basinal features, such as the active or remnant arc, but are found in basins of different ages, it is possible to relate sedimentary facies transitions to the tectonic and volcanic processes which control basin development. Of particular importance is the volcaniclastic apron at the base of the island-arc flank. All data have been obtained from Kroenke *et al*. (1980) and Hussong *et al*. (1981).

Site 455, 54 km E of the Mariana Trough spreading axis, lies within the volcaniclastic apron from the Mariana island arc (Fig. 9). Drilling had to be abandoned but the sequence recovered illustrates the nature of the volcaniclastic wedge. A basal unit of vitric mudstone, chalk and tuff is overlain by 73 m of volcanic sand and gravel consisting of angular basaltic glass and pumice. Although no sedimentary structures were preserved, this coarse debris is interpreted as having been emplaced by high-concentration sediment gravity flows originating from the arc. This sequence is overlain by vitric mud and nannofossil ooze with abundant interbeds of sandy vitric ash. The volcaniclastic material reflects derivation from shallow water and possibly subaerial explosive eruptions.

Site 456 was drilled slightly to the W, on a topographic high (Fig. 9), and basaltic basement was encountered at 134 m sub-bottom depth. This basement and the overlying sediment show striking evidence of hydrothermal alteration. The uppermost basalt, rich in chlorite and epidote, is metamorphosed to greenschist facies. Pyrite occurs in both the basalts and the overlying sediments, and recrystallized clay, quartz, and wairakite in some of the sediments suggest alteration temperatures in excess of 200°C (Natland & Hekinian 1981). The lower part of the sedimentary sequence (456:

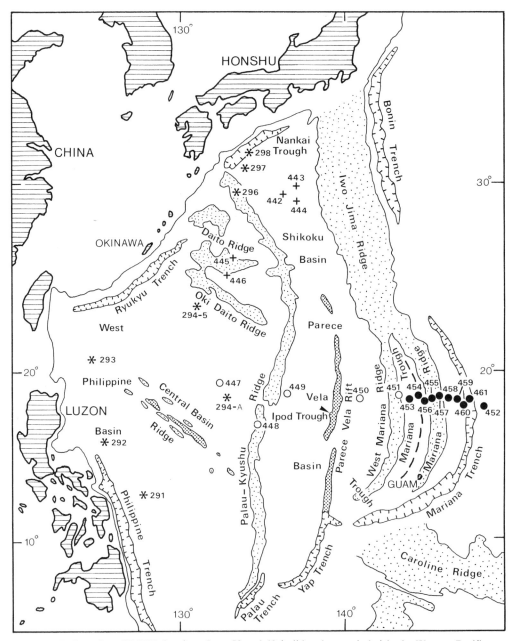

FIG. 9. Location of DSDP sites from Legs 59 and 60 (solid and open circles) in the Western Pacific (Hussong *et al.* 1981).

56–120 m) was deposited rapidly when the site was close to the back-arc spreading centre and mainly under the influence of arc-derived volcaniclastic turbidites. Subsequent uplift isolated the site from turbidites, but ash-fall layers continued to accumulate with nannofossil and siliceous oozes. This site has thus been continu-

ously influenced by arc-derived volcaniclastics, although the dominant style of deposition changed with time due to vertical tectonic movements.

Site 450, located W of the West Mariana Ridge in the Parece Vela Basin (Fig. 9), demonstrates how deposition close to the volcanic arc changes with time. The sequence records major changes in sedimentation which can be related to spreading in the Parece Vela Basin, volcanism on the West Mariana Ridge and splitting of the active arc. Basalts, at a sub-bottom depth of 332 m, show evidence of intrusion into unconsolidated sediment and true basement was not reached. As at site 456, the basalt–sediment interface shows evidence of hydrothermal alteration. Tuffaceous sediment above the basalt shows reddish-brown alteration and a thin (1 cm) yellowish-white band occurs directly at the contact. The succeeding 257 m comprises fine vitric tuffs, vitric tuffs, and tuffaceous volcaniclastic conglomerates. Sedimentary structures include parallel and ripple laminations, cross-bedding, load casts, and both normal and inverse grading, indicating emplacement by sediment gravity flows and possible reworking by bottom currents.

This volcaniclastic sequence was derived during a period of middle Miocene volcanism along the West Mariana Ridge, and as this activity waned the supply of material to the volcaniclastic wedge was reduced. The waning is marked by the overlying 57 m of interbedded pelagic clay and dark ash. Within this unit a transition occurs at about 37 m sub-bottom depth, below which point the sediment contains calcareous fossils and above which they are absent. This transition may represent subsidence of the site area to below the CCD during the waning of volcanism along the West Mariana Ridge. The uppermost 26 m, of ash-poor pelagic clay, was deposited when volcanism along the ridge had ceased and the arc had begun to split, forming the Mariana Trough.

In summary, the sequence at site 450 reflects:

(i)   initial creation of back-arc crust;
(ii) influx of volcaniclastic debris from the active arc producing an apron at the base of the arc's slope as the basin widens by back-arc spreading;
(iii) waning of arc volcanism allowing a pelagic facies to become interbedded with arc volcaniclastics; and
(iv) cessation of arc volcanism and back-arc subsidence as ash- and carbonate-poor pelagic clay is deposited.

Other DSDP sites are useful for tracing the sedimentary changes at the remnant-arc side of evolving basins. Site 453, located about 10 km E of the eastern edge of the West Mariana Ridge, and site 449, located E of the Palau–Kyushu Ridge (Fig. 9), show that the arc-derived volcaniclastic influx generally decreases with time in these areas and is replaced by dominantly pelagic sedimentation. This coincides with the widening of basins by back-arc spreading and the cessation of arc volcanism as splitting of the arc initiates a new cycle of basin formation. Furthermore, these sites illustrate that by comparison with the volcanic arc the remnant arc is not an important source of volcaniclastic debris. In general these observations support the original model of Karig & Moore (1975). Unfortunately the DSDP results only allow the broadest of generalizations to be made regarding volcanogenic sedimentation. Each basin exhibits unique characteristics which are related to the nature and orientation of major sediment sources (Klein 1977). Nevertheless, the similarity of sedimentary sequences and facies transitions in sites from different basins points out the fundamental control of sedimentation by cyclic volcano-tectonic processes.

Volcanogenic sediments from the Grenada Basin, Lesser Antilles

The back-arc area of the Lesser Antilles is occupied by the N–S elongated Grenada Basin which lies between the barely emergent Aves Ridge to the W and the volcanic arc to the E (Fig. 10). The basin contains 4.2 km of sediment resting on slightly thickened crust of oceanic affinity (Boynton *et al.* 1979). The depth to igneous crust, great thickness of sediment and fast sub-Moho velocity all suggest that the basin is in excess of 40 Ma old and may have been created by back-arc spreading (Kearey 1974; Boynton *et al.* 1979).

The Aves Ridge is a N–S trending structural high separating the Grenada Basin to the E from the more extensive Venezuela Basin to the W. Emergence of the ridge above sealevel only occurs at tiny Aves Island where calcareous and arenaceous sediments are exposed and surrounded by coral reefs (Kearey 1974). Several hypotheses for the origin of the Aves Ridge have been presented (Edgar *et al.* 1971; Malfait & Dinkleman 1972; Kearney 1974). Both geophysical and geochemical observations suggest that the ridge represents a remnant arc displaced from the present volcanic arc by back-arc spreading (Grenada Basin), and as such can be compared to the back-arc ridges of the Mariana and Tonga–Kermadec arc systems of the Western Pacific (Karig 1970, 1971a,b).

Some of the twenty-nine piston cores recovered from the Grenada Basin (Fig. 10) have been described by Carey & Sigurdsson (1978, 1980) and Sigurdsson *et al.* (1980). Volcanogenic sediments in these cores have been

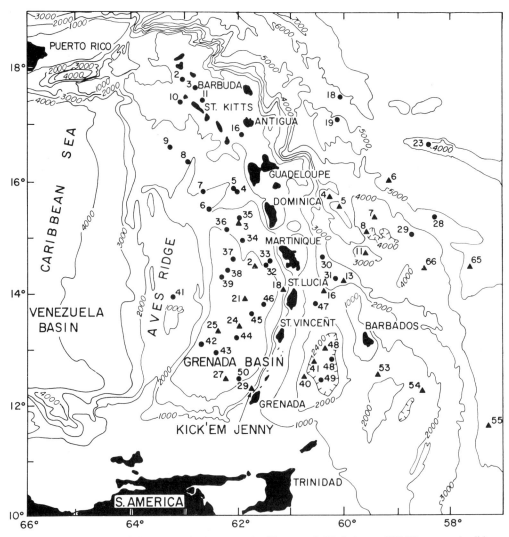

Fig. 10. Location of piston cores in the Lesser Antilles area. Solid circles are EN-20 cores and solid triangles are GS7605 cores.

classified using a combination of lithological (Gretchin *et al*. 1978) and genetic terminology. The term pyroclastic is restricted to deposits formed directly by explosive eruptions, such as ash-fall, pyroclastic flows and lahars. Commonly, the primary character of the pyroclastics is difficult to determine; features which are useful for discrimination include homogeneity of volcanic components, particularly volcanic glass, degree of particle rounding, and the amount and nature of non-volcanogenic components. However, because the erosion products of young pyroclastics can resemble certain facies of primary pyroclastics, such classifications must be viewed with caution.

The associated genetic terminology, while concerned to some extent with volcanological processes, is more directed towards the mechanisms of emplacement. Except for direct ash-fall deposits, most of the volcanogenic sediments have been emplaced by some type of sediment gravity flow. A classification was therefore based on the idealized sedimentary structures of sediment gravity flows presented by Middleton & Hampton (1973, 1976).

A chronostratigraphic framework for the cores was established using a combination of foraminifera biostratigraphy (Ericson & Wollin 1968; Broecker & van Donk 1970; Imbrie & Kipp 1971; Prell 1976) and the correlation of radiocarbon-dated subaerial pyroclastic deposits with marine pyroclastic deposits.

The stratigraphy of the cores is presented in Figs. 11–13. The basin contains a wide variety of thick (up to 4.5 m), coarse grained (up to 6.5 cm diameter pumices) pyroclastic and epiclastic gravity flow deposits interbedded with a predominantly grey-green silty hemipelagic clay. Near the top of some cores a transition from the grey-green hemipelagic clay to a thin sequence of brown pelagic clay is observed. This transition has been noted in other equatorial Atlantic cores and has been interpreted as marking the boundary between the Holocene and the last glacial period (Wisconsinan) (Damuth 1977). As a consequence of the rapid influx of volcaniclastics from the Lesser Antilles island arc, the only biostratigraphic boundary penetrated in the basin was the Z–Y (12,500 yr BP). Volcanogenic turbidites are most abundant in the Grenada Basin although debris flows are the thickest and coarsest grained. Rarely are all of the Bouma divisions present in a single turbidite; more commonly only one or two of the divisions occur. Of all the turbidites examined, 76% include division A (massive, graded), 41%

division B (parallel laminae), 9% division C (ripple, wavy, or convoluted laminations), 6% division D (upper parallel laminae) and 33% division E (inter-turbidite). As the sediment cores were not X-rayed, some of the upper division may have escaped detection.

TABLE 2. *Bouma sequences of volcaniclastic turbidites in the Grenada Basin*

Bouma divisions[a]	% of total turbidites
A (alone)	40
A, B	22
A, E	12
B (alone)	6
A, B, E	5
B, E	5
B, C	2
B, D, E	2
A, B, C, E	2
C (alone)	1
B, D	1
D, E	1
B, C, E	1

[a]Bouma (1962)

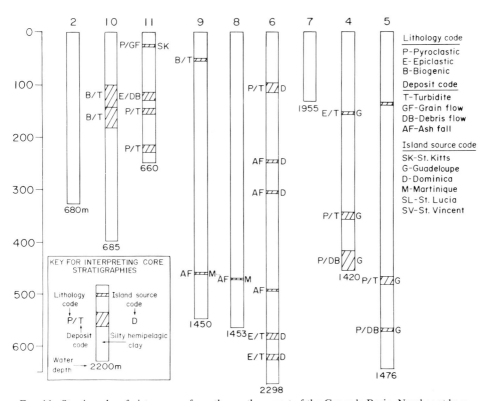

FIG. 11. Stratigraphy of piston cores from the northern part of the Grenada Basin. Number at base of each core refers to water depth (m) at core site.

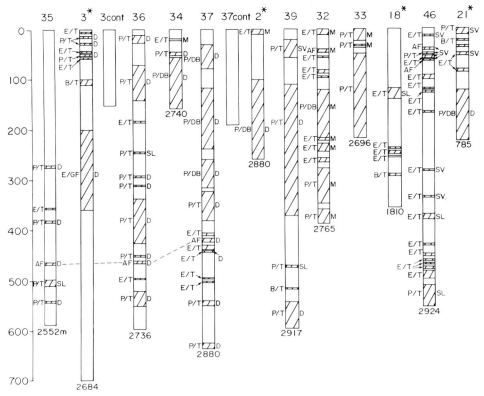

FIG. 12. Stratigraphy of piston cores from the central part of the Grenada Basin. Number at base of each core refers to water depth (m) at core site. *GS7605 cores.

The most common turbidites consist of division A alone and a combination of A and B (Table 2). Numerous other combinations exist (Table 2) but are generally less abundant. The turbidites are lithologically quite variable, consisting of homogenous pumice and glass shards, angular crystals (mostly plagioclase, hypersthene, hornblende, minor augite and Fe–Ti oxides), and lithics. Others contain well-rounded crystals, lithics, and reworked biogenic debris. The former are likely to be associated with a single eruption and the latter with erosion of pre-existing pyroclastic deposits or lava flows.

Typically the pyroclastic and epiclastic debris-flow deposits have a chaotic appearance, with pumices up to 6.5 cm in diameter set in a fine matrix consisting of crystals, lithics, glass shards, biogenic fragments and clay. Locally the turbidites are closely associated with debris-flow deposits. In core EN20-44 (Figs 10 and 13) a debris-flow deposit, with pebble-sized pumices and clay clasts at its base, is directly overlain by a homogenous E Bouma division. This may be an example of turbidite generation

during debris-flow transport (Hampton 1972) or flow transformation (see Fisher, this volume).

One of the debris-flow deposits has been correlated with the subaerial Roseau pyroclastic-flow deposits on the island of Dominica (Carey & Sigurdsson 1980). It was probably formed by pyroclastic flows entering the sea along the W coast of Dominica, although temporary staging of pyroclastic material in shallow water followed by remobilization cannot be ruled out. In some cores, turbidites containing material of Roseau composition overlie the debris-flow deposits and are probably associated with the post-eruptive erosion of the subaerial pyroclastic flows by the Roseau river.

The stratigraphy of the Grenada Basin cores indicates that hemipelagic deposition has been interrupted periodically by the sudden influx of pyroclastic and epiclastic debris transported by a variety of sediment gravity flows. The pyroclastic material has, for the most part, been derived from subaerial eruptions on the islands of Guadeloupe, Dominica, Martinique, St Lucia, and St Vincent. Epiclastic deposits are

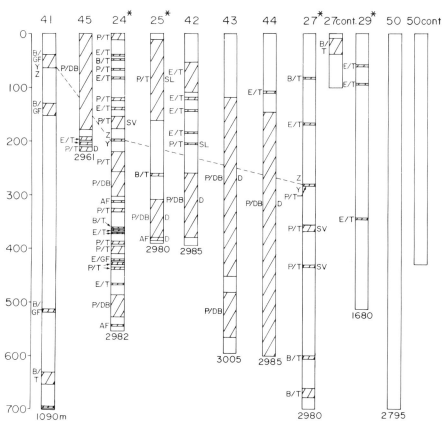

FIG. 13. Stratigraphy of piston cores from the southern part of the Grenada Basin. Number at base of each core refers to water depth (m) at core site. *GS7605 cores.

commonly mixtures of material from more than one island source, indicating a complex sedimentological history.

With the exception of some of the pelagic and hemipelagic deposits in the northern cores, most of the sediments in the deeper part of the Grenada Basin belong to a volcaniclastic apron facies as defined by Karig & Moore (1975). These authors suggested that the volcaniclastic apron facies may be similar to a submarine fan complex but that 'depositional processes on the apron are not well known and may be different from those operating on other submarine fans'. Klein (1975a,b) compared the facies sequence of DSDP site 286 in the Hebrides Basin with ancient submarine fan deposits (Mutti & Ricci-Lucchi 1972) and modern submarine fan sub-environments (Normark 1970; Hein 1973). However, as already stated, the DSDP sites are not particularly useful for evaluating laterally and vertically complex depositional systems. In contrast the large number of cores from the Grenada Basin provides a good opportunity to

study the growth of the volcaniclastic apron and to compare it to submarine fan complexes. Even in this case the number of cores available for study relative to the size of the basin is admittedly small.

Recent models of submarine fan complexes subdivide the depositional system into a series of sub-environments (Normark 1974, 1978; Walker 1978). Three major subdivisions have been recognized, based on detailed bathymetric surveying of modern deep-sea fans. These are an upper fan characterized by a single, deep, leveed channel which is usually connected to a landward submarine canyon system, a middle fan constructed by the coalescence of migrating suprafan lobe deposits, and a smooth lower fan which laterally grades into a basin plain. From data of both modern and ancient fans, Walker (1978) incorporated five facies of deep-water clastic rocks into a model of submarine fan deposition (Fig. 14). Coarse grained debris-flows and disorganized-bed conglomerates are restricted to the upper fan within the main

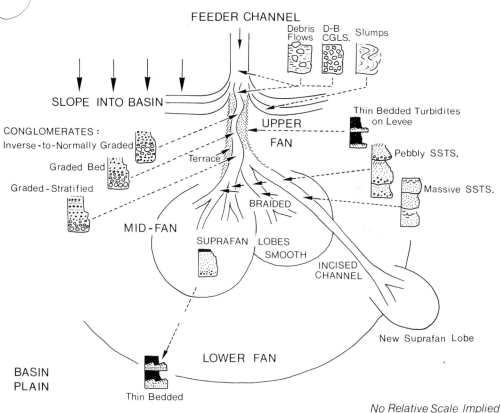

FIG. 14. Model of submarine-fan deposition, relating facies, fan morphology, and depositional environment, from Walker (1978). D-B indicates disorganized-bed conglomerates.

feeder channel. Downslope within the main channel these deposits are progressively replaced by inverse-to-normal graded, graded, and graded stratified conglomerates. Outside the main channel, thin-bedded turbidites are deposited on the leveed slopes. In the mid-fan area channellized massive and pebbly sandstones coalesce into lenticular bodies. In this area migration of braided channels tends to erode any finer material although finer grained turbidites may be deposited within the channels. The lower fan area is characterized by hemipelagic and pelagic sediments interbedded with the 'classical' turbidites.

These proximal to distal facies transitions are diagnostic features of submarine fan deposition. The volcaniclastic deposits in the Grenada Basin were examined to see whether such a model is applicable to the marginal basin volcaniclastic apron. Although only thin sequences can be compared, there does not appear to be any systematic relationship between facies type and location in the basin (Figs 10–13). For example, the thickest and coarsest grained

debris-flow deposit is found in cores farthest from the source island and some cores more proximal to island sources contain thin-bedded turbidites interbedded with hemipelagic clay. A comparison of Bouma sequences in turbidites from proximal (to arc) and distal cores also shows no systematic trends. Furthermore, electron microprobe analysis of volcanic glass, shows that volcanogenic sediment from different islands is intercalated in many areas (Fig. 15).

Based on these observations, it is proposed that the Grenada Basin volcaniclastic apron is not analogous to a simple or compound submarine fan depositional system. It is also inferred that this is likely to be the case for other marginal basin volcaniclastic aprons. Unlike the single source, radially distributed submarine fan, the volcaniclastic apron is better interpreted as an elongate coalescence of multiple volcaniclastic depositional systems which show few systematic facies transitions of gravity flow deposits. The fundamental difference between the fan and apron systems is the nature

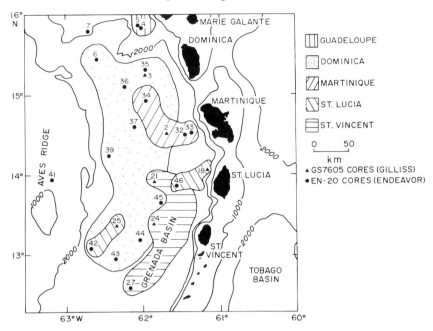

FIG. 15. Correlation of volcanogenic sediment gravity flows in the Grenada Basin based on electron microprobe analyses of glass shards.

and delivery of sediment to deep water in each case. Deep-sea fans usually develop from a single submarine canyon, through which sediment from a continental shelf or subaerial fluvial source is passed. In contrast, the apron receives sediment from multiple sources; volcanic islands arranged along a curvilinear trend. More importantly, sediment is not necessarily supplied to the apron via well-defined submarine canyons. Unlike most deep-sea fans the apron is subject to massive influxes of sediment derived from island-arc volcanism. This sediment will be delivered to deep water at positions which are more strongly governed by the location of the eruption and its associated ashfall, pyroclastic flows and lahars. Because arc volcanism results in both more random sediment dispersal and commonly enormous influxes of volcaniclastics, it is not surprising that systematic facies transitions are rare. Cas *et al.* (1981) have arrived at a similar interpretation for the lower Devonian Kowmung volcaniclastics of the northern Lachan fold belt, New South Wales, where deposits were derived from multiple sources and no remnant canyon or channels have yet been discovered.

Volcanogenic sedimentation model

A combination of the DSDP results from the Western Pacific, the stratigraphy of cores from

the Grenada Basin, and the tectonic model of marginal basin evolution can be used to arrive at an integrated volcanogenic sedimentation model for back-arc areas of oceanic island arcs. The model is deliberately simplistic so as not to limit its applicability to more complex situations. Details of specific processes such as sediment production and transport have already been discussed and will not be elaborated in the model summary. Many of the conclusions are similar to those of Karig & Moore (1975) and also to some aspects of Klein (1975a, 1977), but the model concentrates more on the nature and growth of the volcaniclastic apron.

Stage 1: Initial rifting and development of inter-arc basin

During the early stages of marginal basin development, rifting of an island arc along a zone of weakness and subsequent back-arc spreading creates a narrow trough with steep, fault-scarped margins.

The steep, unstable margins are conducive to the generation of mass-gravity flows (slumps, slides, debris flows and turbidity currents) which inundate most of the basin floor (Fig. 16). These mass flows incorporate both epiclastic and pyroclastic debris although if this period corresponds to a lull in volcanic activity along the arc (Scott & Kroenke 1981), then the deposits will be dominantly epiclastic. Large

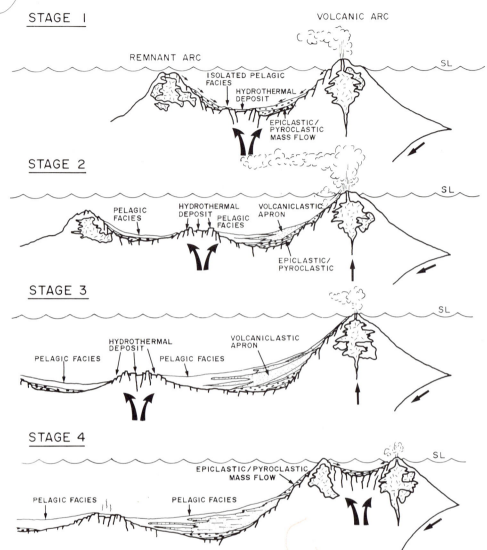

FIG. 16. Distribution of volcanogenic sediments in an evolving marginal basin. Stage 1. Early rifting and high influx of volcaniclastics. Stage 2. Basin widening by back-arc spreading, and active island-arc volcanism. Thick volcaniclastic apron developed at base of active-arc flank. Stage 3. Basin maturity, waning of arc volcanism allows a transgression of pelagic deposition over volcaniclastic apron. Stage 4. Basin inactivity, back-arc spreading ceases and splitting of volcanic arc initiates a new cycle of basin formation.

tectonic displacements during this stage can expose deep-seated parts of the original arc in the vicinity of the remnant arc, as shown by DSDP site 453 (Fig. 9). Hydrothermally derived volcanogenic sediments may be deposited in association with the clastic deposits about back-arc spreading centres and on nearby flanks. Elsewhere the rapid influx of sediment at this stage is not likely to favour the production of Fe- and Mn-rich sediments.

If basin spreading results in physical barriers to clastic sediment dispersal, then a pelagic facies will be deposited, especially on the basin side near the remnant arc. A low intensity of arc volcanism at this time would also allow the pelagic facies to extend near to the base of the volcanic arc. Because basin subsidence at this stage is slight, deposition of the pelagic facies will be above the CCD and may therefore be calcareous. Dispersed volcanic ash, clays derived from the weathering of volcanic rocks, and aeolian dust will be the other major components of the pelagic facies.

State 2: Back-arc spreading and island-arc volcanism

Continued back-arc spreading results in basin widening, subsidence, and cooling as the newly formed crust moves away from the ridge axis. The steep basin margins are modified to smoother topography by the deposition of mass-flow deposits. At this point an increase in arc volcanism leads to the development of a volcaniclastic apron at the base of the arc (Fig. 16). This apron includes three important components:

(i) primary volcaniclastic influx from subaerial and/or subaqueous arc eruptions;
(ii) an epiclastic influx from erosion of the subaerial and/or subaqueous arc complex; and
(iii) a continuous accumulation of pelagic biogenic and aeolian debris.

The continuous deposition of pelagic components and the intermittent deposition of epiclastic debris is periodically interrupted by massive influxes of volcaniclastics from subaerial and/or subaqueous eruptions. Transport of volcaniclastics will be by debris flows, grain flows, turbidity currents and passive settling through the water column (ash-fall) (Figs 5 and 6). These different sedimentation processes result in a complex array of superimposed facies with few systematic proximal-to-distal transitions such as those observed in submarine fans. Influx from multiple sources arranged in a curvilinear trend (island arc) produces a volcaniclastic apron which is elongate parallel to the trend of the arc and wedge-shaped in profile. In some cases, palaeocurrent evidence in apron deposits indicates transport parallel to the volcanic arc.

The combination of the three different components results in an overall high sedimentation rate for the apron complex. In the Grenada Basin this is roughly 0.2 km Ma^{-1} which compares to 0.1–0.3 km Ma^{-1}, reported by Karig & Moore (1975). Because the most rapidly accumulating component is directly linked to eruptions, the growth of the apron reflects the tempo of island-arc volcanism. Increased arc volcanism will lead to a progradation of the apron into areas previously dominated by pelagic sedimentation and decreased activity will be marked by the interfingering and eventual replacement of the apron facies by pelagic sedimentation alone. For the Mariana Trough, Sample & Karig (1982) have shown that the

growth rate of the volcaniclastic apron appears to be positively correlated with the variation in subduction rate along the Mariana arc.

Various patterns can also be envisaged if the role of back-arc spreading is taken into account. Cessation of basin spreading during a period of continuous arc volcanism would allow the apron to prograde towards the basin interior and replace pelagic deposition, whereas if spreading was faster than the propagation of the apron then increasing areas of the basin would receive pelagic deposits. The dominant pelagic deposition during this stage will be, depending on the depth of the CCD and productivity of the local water, calcareous or siliceous ooze or brown clay with abundant dispersed volcanic ash. As in stage 1, hydrothermally derived sediments will be produced at the back-arc spreading centre and incorporated into the base of the sedimentary sequence as the crust migrates away from the central axial high.

Stage 3: Basin maturity

In later stages of basin development back-arc spreading begins to draw to a close (Fig. 16). The cessation may be due to the migration of the spreading centre away from the region of tensional stress induced by the subduction process (Bibee *et al.* 1980). Whatever the cause, the next step in the tectonic development will be displacement of the rift to a position closer to the volcanic arc, or splitting of the arc itself to initiate another cycle of basin formation, or inactivity. On cessation of basin spreading, volcanism in the arc may either continue or begin to wane. If it continues, then volcaniclastic deposition may prograde into the basin. If volcanism wanes, as seems to be indicated by many of the DSDP sites in the Western Pacific, then pelagic sedimentation will prevail. At this point the pelagic deposits will become ash-poor and are likely to be brown clays as a result of subsidence below the CCD.

Stage 4: Old age

After a new cycle of basin formation is initiated the older basin becomes inactive (Fig. 16). Influx of epiclastic debris from the splitting of the volcanic arc may occur, but for the most part pelagic sedimentation will now be the dominant process. Dispersed ash from the arc at the leading edge of the new basin may be transported to the inactive basin but the abundance of this material should decrease as the inactive basin is moved progressively further away from the source by back-arc spreading in the new basin.

ACKNOWLEDGMENTS: This paper benefited greatly from discussions with and reviews by Dick Fisher and Malcolm Howells. One of the authors (S.C.) would also like to express appreciation to the Geological Society and the conveners of the Marginal Basin Conference for providing financial assistance for travel to England. This work was supported by NSF grants OCE-75-21197 and OCE-77-25789.

References

ALLEN, C. C. 1980. Icelandic subglacial volcanism: thermal and physical studies. *J. Geol.* **88**, 108–17.

BIBEE, L. D., SCHOR, G. G. & LU, R. S. 1980. Inter-arc spreading in the Mariana Trough. *Mar. Geol.* **35**, 183–97.

BONATTI, E., KOLLA, V., MOORE, W. S. & STERN, C. 1979. Metallogenesis in marginal basins: Fe-rich basal deposits from the Philippine Sea. *Mar. Geol.* **32**, 21–37.

BOSTROM, K. & PETERSON, M. N. A. 1965. Precipitates from hydrothermal exhalation on the East Pacific Rise. *Econ. Geol.* **61**, 1258–65.

BOUMA, A. H. 1962. *Sedimentology of Some Flysch deposits: A Graphic Approach to Facies Interpretations.* Elsevier, Amsterdam.

BOYNTON, C. H., WESTBROOK, G. K., BOTT, M. H. P. & LONG, R. E. 1979. A seismic refraction investigation of crustal structure beneath the Lesser Antilles island-arc. *Geophys. J. R. astr. Soc.* **58**, 371–93.

BROECKER, W. S. & VAN DONK, J. 1970. Insolation changes, ice volumes, and the ^{18}O record in deep-sea cores. *Rev. Geophys. Space Phys.* **8**, 169–98.

CAREY, S. & SIGURDSSON, H. 1978. Deep-sea evidence for distribution of tephra from the mixed magma eruption of the Soufriere of St. Vincent, 1902: ash turbidites and air fall. *Geology*, **6**, 271–74.

—— & SIGURDSSON, H. 1980. The Roseau Ash: deep-sea tephra deposits from a major eruption on Dominica, Lesser Antilles Arc. *J. Volcan. geoth. Res. Amsterdam*, **7**, 67–86.

CAS, R. A., MCA. POWELL, C., FERGUSSON, C. L., JONES, J. G., ROOTS, W. D. & FERGUSSON, J. 1981. The Lower Devonian Kowmung Volcaniclastics: a deep-water succession of mass-flow origin, northeastern Lachlan Fold Belt, N.S.W. *J. geol. Soc. Aust.* **28**, 271–88.

COOPER, A. K., MARLOW, S. W. & SCHOLL, D. W. 1976. Mesozoic magnetic lineations in the Bering Sea marginal basin. *J. geophys. Res.* **81**, 1916–34.

DAMUTH, J. E. 1977. Late Quaternary sedimentation in the western equatorial Atlantic. *Bull. Geol. Soc. Am.* **88**, 695–710.

DICK, H., HEIRTZLER, J., DMITRIEV, L. *et al.* 1976. Glass-rich basaltic sand and gravel within the oceanic crust at 22°N. *Nature, Lond.* **262**, 768–70.

EDGAR, N., EWING, J. & HENNION, J. 1971. Seismic refraction and reflection in Caribbean Sea. *Bull. Am. Ass. Petrol. Geol.* **55**, 833–70.

EDMOND, J. M., MEASURES, M., MANGUM, B., GRANT, B., SCLATER, F. R., COLLIER, A. &

HUDSON, A. 1979. On the formation of metal-rich deposits at ridge crests. *Earth planet. Sci. Lett.* **46**, 19–30.

ERICSON, D. B. & WOLLIN, G. 1968. Pleistocene climates and chronology in deep-sea sediments. *Science*, **162**, 1227–34.

FRANCHETEAU, J., NEEDHAM, H. D., CHOUKROUNE, P. *et al.* 1979. Massive deep-sea sulphide ore deposits discovered by submersible on the East Pacific Rise: Project RITA, 21°N. *Nature, Lond.* **277**, 522–8.

FRANCIS, E. H. & HOWELLS, M. F. 1973. Transgressive welded ash-flow tuffs among the Ordovician sediments of NE Snowdonia, North Wales, *J. geol. Soc. London*, **129**, 621–41.

FURNES, H., FRIDLEIFSSON, I. B. & ATKINS, F. B. 1980. Subglacial volcanics—on the formation of acid hyaloclastites. *J. Volcan. geoth. Res. Amsterdam*, **8**, 95–110.

GARCIA, M. O. 1978. Criteria for the identification of ancient volcanic arcs. *Earth Sci. Rev.* **14**, 147–65.

GRECHIN, V. I., NIEM, A. R., MAHOOD, R. O., ALEXANDROVA, V. A. & SAKHAROV, B. A. 1981. Neogene tuffs, ashes, and volcanic breccias from offshore California and Baja California, DSDP Leg 63: sedimentation and diagenesis. *In:* YEATS, R. S., HAG, B. U. *et al.* (eds) *Init. Rep. Deep Sea drill. Proj.* **63**, 631–57. U.S. Government Printing Office, Washington, D.C.

HAMPTON, M. A. 1972. The role of subaqueous debris flow in generating turbidity currents. *J. sediment. Petrol.* **42**, 775–93.

HAWKINS, J. W. 1977. Petrologic and geochemical characteristics of marginal basin basalts. *In:* TALWANI, M. & PITMAN, W. C. (eds) *Island Arcs, Deep Sea Trenches, and Back-Arc Basins*, 355–66. American Geophysical Union, Washington, D.C.

HEIKEN, G. H. 1974. An atlas of volcanic ash. *Smithson. Contr. Earth Sci.* **12**.

HEIN, J. R. 1973. Increasing rates of movement with time between California and the Pacific plate: from Delgas submarine fan source areas. *J. geophys. Res.* **75**, 239–54.

HONNOREZ, J. & KIRST, P. 1975. Submarine basaltic volcanism: morphometric parameters for discriminating hyaloclastites from hyalotuffs. *Bull. Volcanol.* **39**, 1–25.

HUSSONG, D. M. & FRYER, D. B. 1980. Tectonic evolution of the marginal basins behind the Mariana Arc. *Abst. Prog. geol. Soc. Am.* **12–3**, 112.

—— & UYEDA, S. 1981. Tectonics in the Mariana Arc: results of recent studies, including DSDP leg 60. *Oceanol. Acta*, 203–12.

——, —— *et al.* (eds) 1981. *Init. Rep. Deep Sea drill. Proj.* **60**, U.S. Government Printing Office, Washington, D.C.

IMBRIE, J. & KIPP, N. G. 1971. A new micropaleontological method for quantitative paleoclimatology: application to a late Pleistocene Caribbean core. *In*: TUREKIAN, K. K. (ed.) *The Late Cenozoic Glacial Ages*, 71–181. Yale University Press, New Haven.

KARIG, D. E. 1970. Ridges and basins of the Tonga-Kermadec Island arc system. *J. geophys. Res.* **75**, 239–55.

—— 1971a. Origin and development of marginal basins in the Western Pacific. *J. geophys. Res.* **76**, 2542–61.

—— 1971b. Structural history of the Mariana Island Arc System. *Bull. geol. Soc. Am.* **82**, 323–44.

—— 1975. Basin genesis in the Philippine Sea. *In*: KARIG, D. E. & INGLE, J. C. JR. *et al.* (eds) *Init. Rep. Deep Sea drill. Proj.* **31**, 857–80. U.S. Government Printing Office, Washington, D.C.

—— & MOORE, G. F. 1975. Tectonically controlled sedimentation in marginal basins. *Earth planet. Sci. Lett.* **26**, 233–8.

—— & WAGEMAN, J. M. 1975. Structure and sediment distribution in the northwest corner of the west Philippine Basin. *In*: KARIG, D. E. & INGLE, J. C. JR. *et al.* (eds) *Init. Rep. Deep Sea drill. Proj.* **31**, 615–20. U.S. Government Printing Office, Washington, D.C.

——, ANDERSON, R. N. & BIBEE, L. D. 1978. Characteristics of back arc spreading in the Mariana Trough. *J. geophys. Res.* **83**, 1213–26.

KEAREY, P. 1974. Gravity and seismic reflection investigations into the crustal structure of the Aves Ridge, eastern Caribbean. *Geophys. J. R. astr. Soc.* **38**, 435–48.

KLEIN, G. 1975a. Depositional facies of leg 30 DSDP sediment cores. *In*: ANDREWS, J. E. & PACKHAM, G. *et al.* (eds) *Init. Rep. Deep Sea drill. Proj.* **30**, 423–42. U.S. Government Printing Office, Washington, D.C.

—— 1975b. Sedimentary tectonics in Southwest Pacific marginal basins based on leg 30 Deep Sea Drilling Project cores from the South Fiji, Hebrides, and Coral Sea Basin. *Bull. geol. Soc. Am.* **86**, 1012–8.

—— 1977. Subaqueous gravity flow sedimentation in the marginal basins of the western Pacific Ocean. *Abst. Prog. geol. Soc. Am.* **9**, 1052.

KROENKE, L. & SCOTT, R. *et al.* (eds) 1980. *Init. Rep. Deep Sea drill. Proj.* **59**. U.S. Government Printing Office, Washington, D.C.

LAWVER, L. A. & HAWKINS, J. W. 1978. Diffuse magnetic anomalies in marginal basins: their possible tectonic and petrologic significance. *Tectonophysics*, **45**, 323–39.

LOUDEN, K. 1976. Magnetic anomalies in the West Philippine basin. *Mongr. Am. geophys. Un.* **19**, 253–67. Int. Woolard Symp., Washington, D.C.

MALFAIT, B. & DINKELMAN, M. 1972. Circum-Caribbean tectonic and igneous activity and the evolution of the Caribbean plate. *Bull. geol. Soc. Am.* **83**, 251–72.

MCBIRNEY, A. R. 1971. Oceanic volcanism: a review. *Rev. geophys. Space Phys.* **9**, 523–56.

MIDDLETON, G. & HAMPTON, M. 1973. Sediment gravity flows: mechanics of flow and deposition. *In*: *Turbidites and Deep-Water Sedimentation*, 1–38. Soc. econ. Paleontol. Mineral. short course.

—— & HAMPTON, M. 1976. Subaqueous sediment transport and deposition by sediment gravity flows. *In*: STANLEY, D. & SWIFT, D. (eds) *Marine Sediment Transport and Environmental Management*, 197–220. Wiley, New York.

MUTTI, E. & RICCI-LUCCHI, F. 1972. Le torbiditi dell'Appenino settentrionale: introduzione all'abalsi di facies. *Mem. geol. Soc. Ital.* **11**, 161–99.

NATLAND, J. H. & HEKINIAN, R. 1981. Hydrothermal alteration of basalts and sediments at DSDP site 456, Mariana Trough. *In*: HUSSONG, D. M. & UYEDA, S. *et al.* (eds) *Init. Rep. Deep Sea drill. Proj.* **60**, 759–68. U.S. Government Printing Office, Washington, D.C.

NORMARK, W. R. 1970. Growth patterns of deep-sea fans. *Bull. Am. Ass. Petrol. Geol.* **54**, 2170–95.

—— 1974. Submarine canyons and fan valleys: factors affecting growth patterns of deep-sea fans. *In*: *Modern and Ancient Geosynclinal Sedimentation*. Spec. Publ. Soc. econ. Paleont. Mineral., Tulsa, **19**, 56–68.

—— 1978. Fan valleys, channels and depositional lobes on modern submarine fans: character for recognition of sandy turbidite environments. *Bull. Am. Ass. Petrol. Geol.* **62**, 912–31.

PRELL, W. L. 1976. Late Pleistocene faunal, sedimentary, and temperature history of the Columbia Basin, Caribbean Sea. *In*: CLINE, R. M. & HAYS, J. D. (eds) *Investigation of Late Quaternary Paleoceanography and Paleoclimatology*. Mem. geol. Soc. Am. **145**, 201–20.

SAMPLE, J. C. & KARIG, D. E. 1982. A volcanic production rate for the Mariana Island Arc. *J. Volcan. geoth. Res. Amsterdam*, **13**, 73–82.

SCHMINCKE, H. U., ROBINSON, P. T., OHNMACHT, W. & FLOWER, M. 1978. Basaltic hyaloclastites from hole 396B, DSDP Leg 46. *In*: DMITRIEV, L. & HEIRTZLER, J. L. *et al.* (eds) *Init. Rep. Deep Sea drill. Proj.* **46**, 341–56. U.S. Government Printing Office, Washington, D.C.

SCOTT, M. R., SCOTT, R. B., RONA, P. A., BUTLER, L. W. & NALWALK, A. J. 1974. Rapidly accumulating manganese deposit from the median valley of the Mid-Atlantic Ridge. *Geophys. Res. Lett.* **1**, 355–8.

SCOTT, R. B. & KROENKE, L. 1981. Periodicity of remnant arcs and back-arc basins of the South Philippine Sea. *Oceanol. Acta*, 193–202.

SELF, S. & RAMPINO, M. 1981. The 1883 eruption of Krakatau. *Nature, Lond.* **294**, 699–704.

SIGURDSSON. H., SPARKS, R. S. J., CAREY, S. & HUANG, T. C. 1980. Volcanogenic sedimentation in the Lesser Antilles Arc. *J. Geol. Chicago*, **88**, 523–40.

SPARKS, R. S. J., SIGURDSSON, H. & CAREY, S. N. 1980a. The entrance of pyroclastic flows into the sea, I. Oceanographic and geologic evidence from Dominica, Lesser Antilles. *J. Volcan. geoth. Res. Amsterdam* **7**, 87–96.

——, ——, ——, 1980b. The entrance of pyroclastic flows into the sea, II. Theoretical considerations on subaqueous emplacement and welding. *J. Volcan. geoth. Res. Amsterdam*, **7**, 97–106.

TARNEY, J., SAUNDERS, A. D. & WEAVER, S. 1977. Geochemistry of volcanic rocks from the island arcs and marginal basins of the Scotia Arc Region. *In*: TALWANI, M. & PITMAN, W. C. (eds) *Island Arcs, Deep Sea Trenches, and Back-Arc Basins*, 367–78. American Geophysical Union, Washington, D.C.

UYEDA, S. & BEN-AVRAHAM, Z. 1972. Origin and development of the Philippine Sea. *Nature (Phys. Sci.)* **240**, 176–8.

WALKER, R. G. 1978. Deep-water sandstone facies and ancient submarine fans: models for exploration for stratigraphic traps. *Bull. Am. Ass. Petrol. Geol.* **62**, 932–66.

WATTS, A. B. & WEISSEL, J. K. 1975. Tectonic history of the Shikoku marginal basin. *Earth planet. Sci. Lett.* **25**, 239–50.

WHITE, S. M. *et al*. 1980. Sediment synthesis: DSDP Leg 58, Philippine Sea. *In*: DE VRIES KLEIN, G. & KOBAYASHI, K. *et al*. (eds) *Init. Rep. Deep Sea drill. Proj.* **58**, 963–1014. U.S. Government Printing Office, Washington, D.C.

YAMAZAKI, T., KATO, I., MUROI, I. & ABE, M. 1973. Textural analysis and flow mechanism of the Donzurubo subaqueous pyroclastic flow deposits. *Bull. Volcanol.* **37**, 231–44.

STEVEN CAREY & HARALDUR SIGURDSSON, Graduate School of Oceanography, University of Rhode Island, Kingston, RI 02881, U.S.A.

Geochemical characteristics of basaltic volcanism within back-arc basins

A. D. Saunders & J. Tarney

SUMMARY: Back-arc basins are formed by extensional processes similar to those occurring at mid-ocean ridges. However, whereas the magmas erupted along the major ocean ridges are predominantly LIL element-, Ta- and Nb-depleted N-type MORB, many back-arc basins are floored by basalts transitional between N-type MORB and island arc or even calc-alkaline basalts (viz. enrichment of LIL elements (K, Rb, Ba, Th) relative to HFS elements (Nb, Ta, Zr, Hf, Ti)). On a broad scale, it is possible to relate basalt composition, tectonic setting of the basin, and maturity of the adjacent subduction zone. Thus, the Parece Vela Basin, formed during the earliest stages of the Mariana subduction system, is floored by basalts indistinguishable from N-type MORB, whereas the later Mariana Trough is erupting N-type MORB *and* basalts with calc-alkaline characteristics, commonly in close spatial proximity. The calc-alkaline component is best developed in narrow, ensialic basins such as Bransfield Strait, where the extension is adjacent to mature, continent-based magmatic arcs. This range of compositions, from N-type MORB to calc-alkaline basalt, can be satisfactorily explained only by invoking chemical variations in the composition of the mantle material supplying the back-arc basin crust. Two major processes may be suggested: (i) selective contamination of the mantle wedge by LIL-enriched hydrous fluids, perhaps together with sediments, derived from the descending, dehydrating oceanic lithosphere; and (ii) repeated melt (and incompatible element) extraction during basalt genesis. The former process will enrich the mantle source of back-arc basalts with LIL elements; the latter will deplete the source in all incompatible elements, but the net effect of both processes is to increase the LIL/HFS element ratio of the source regions. Consequently, as the subduction zone matures, the LIL/HFS element ratio of successive back-arc basalts will be expected to increase, from initial N-type MORB 'background' values, to ratios more typical of island-arc basalts. The model has implications for mantle dynamics in back-arc regions, because transfer of material from the subducted slab may destabilize the overlying mantle, potentially leading to diapiric uprise when tectonic conditions permit extension.

There now exists a considerable amount of geochemical and geophysical data with which it is possible to review the character and petrogenesis of basalts erupted within back-arc basins. The processes of basin formation, particularly crustal accretion, may not differ greatly from those occurring in the major ocean basins, but the intimate association between back-arc basins and contemporaneous island arcs implies that subducted components may be involved in the genesis of back-arc basalts. In addition to influencing the chemistry of back-arc basalts, slab-derived material may also play an important role in the thermal and dynamic evolution of the underlying mantle, e.g. in the initiation of buoyant mantle diapirs. This paper reviews the geochemistry of back-arc basalts, considers various processes which could account for the observed characteristics, and discusses their influence on mantle evolution and dynamics in arc–back-arc regions.

Back-arc basins: a working definition

Marginal basins are loosely defined as 'semi-isolated basins or series of basins lying behind the volcanic chains of island-arc systems' (Karig 1971). Back-arc basins represent but one major category of marginal basins. Almost thirty marginal basins have now been recognized (Weissel 1981), the majority occurring in the Western Pacific, and it is convenient to group them into four main types:

(i) Those formed by back-arc extension (back-arc basins *sensuo stricto*).
(ii) Those formed by predominantly transform-style tectonics at zones of oblique plate convergence (e.g. the Gulf of California).
(iii) Those formed by sea-floor spreading unrelated to specific zones of plate convergence (e.g. the Woodlark Basin).
(iv) Those formed by entrapment of older

oceanic lithosphere (e.g. the Aleutian Basin).

Extensional back-arc basins are formed by the accretion of new oceanic crust between diverging 'active' and 'remnant' island-arc fragments (Karig 1970, 1971), and post-date the age of the associated subduction zone. In most examples of back-arc extension, the associated island arc is built on oceanic crust, in which case the basin is said to be *intra-oceanic* or ensimatic (Table 1). In at least two examples (the South Sandwich Islands and the Mariana Islands), the active arc apparently sits on crust formed during the latest period of back-arc extension (Barker 1972; Hussong *et al*. 1981). Rarely, the volcanic arc may be rooted on a sliver of continental crust separated from the continental margin by an *ensialic* back-arc basin. Such basins may be floored by attenuated continental crust, particularly during the earliest stages of formation (e.g. the proto-Gulf of California: Karig & Jensky 1972), but in this account we are concerned with ensialic basins floored by mafic (essentially oceanic) crust (e.g. Bransfield Strait).

Back-arc extension is commonly episodic, with successive arc-basin pairs being accreted on to the plate margin to form festoons of active and remnant arcs separated by active or inactive back-arc basins. Most back-arc basins are relatively short-lived, rarely remaining active for longer than 15 Ma, because of the effects of 'roll-back' at the adjacent subduction zone (Molnar & Atwater 1978). 'Roll-back' or 'trench-suction' effectively causes the active arc to migrate trenchwards, with the result that the axis of back-arc spreading becomes increasingly remote from the subduction zone and, by inference, from the locus of mantle diapirism and the main heat source. This is consistent with the observed thermal and topographic data, which demonstrate that basin heat flow and basement elevation gradually diminish with age (Watanabe *et al*. 1977). In addition, remnant-arc fragments subside to form submarine ridges (e.g. Kyushu–Palau Ridge).

All back-arc basins for which reasonably comprehensive geochemical data are currently available are listed in Table 1. The Gulf of California is included because it superficially resembles most 'conventional' back-arc basins, even though its extension has resulted from transform-style tectonics (Larson 1972; Moore & Buffington 1968), and it is probably not presently underlain by a subducted slab (Dickinson & Snyder 1979).

Volcanism in back-arc basins

Active back-arc basins have high heat flow, and patterns of sediment distribution and magnetic lineations consistent with an extensional origin (Karig 1970, 1971). It is reasonable to assume therefore that the uppermost crustal structure in back-arc basins is formed by mechanisms broadly similar to those occurring in the major ocean basins, i.e. mantle fusion beneath an axial spreading centre, creation of sub-ridge magma reservoirs, dyke injection, and eruption of basaltic pillow and sheet lavas. However, few back-arc basins have such well-defined magnetic anomaly patterns as normal ocean basins. It has been suggested that this reflects a lack of coherent spreading, with the possibility of numerous small spreading axes scattered over a diffuse zone in the centre of the basin (Karig 1970, 1971; Lawver *et al*. 1976; Lawver & Hawkins 1978). The diffuse-spreading model is particularly attractive for regions such as the Lau Basin (Lawver *et al*. 1976), although Weissel (1977) proposed a more organized, normal plate-like evolution, maintaining that the observed magnetic patterns have been complicated by spreading-centre jumps. Conversely, some back-arc basins, such as that in the E Scotia Sea behind the South Sandwich volcanic arc, have well-developed lineations and a clearly defined spreading axis (Barker 1972; Barker & Hill 1981).

Poorly defined magnetic signatures may also result from the low geomagnetic latitude of equatorial marginal basins, the irregular basement topography of many back-arc basins, and the thick sedimentary carapace covering early-formed basin crust. The latter will be particularly important if high sedimentation rates occur at the spreading axis, when Layer 2 may be emplaced as sills rather than as flows. This is demonstrably the case in the central Gulf of California, where sedimentation rates exceed 1000 m Ma^{-1}, and where sills predominate over flows (Saunders *et al*. 1982a). Such high rates are more likely to occur during the initial stages of back-arc rifting, when juvenile basins will be subjected to sediment influx from the immediately adjacent arc or continental margin.

The available data imply that the structure of back-arc basin crust is broadly similar to that of oceanic crust, and that either can be represented by ophiolite complexes (Hawkins 1977). It is interesting to note that one of the few truly autochthonous ophiolites, the Mesozoic 'rocas verdes' mafic complex of

TABLE 1. *Selected characteristics of representative back-arc basins*

Basin	Age (E: early; L: late; M: middle)	Contemporaneous island arc	Spreading half rate (mm yr^{-1})	Predominant basalt geochemistry	References
Intra-oceanic					
Mariana Trough	L. Mioc–Rec	Mariana Is.	16–22	N-type MORB + LIL-enriched (island-arc) basalts	1–5
Lau Basin	L. Mioc–Rec	Tonga Is.	~38	N-type MORB+ LIL-enriched (island-arc) basalts	1, 6–8
Shikoku Basin	L. Olig–E. Mioc	Iwo–Jima Ridge	~20–30 }	N-type MORB	1, 9–11
Parece Vela Basin	L. Olig–E. Mioc	W Mariana Ridge	~20–30 }		1, 10, 12
E Scotia Sea	M. Plio–Rec	S Sandwich Is.	25–35	N-type MORB + LIL-enriched (island-arc) basalts	13–15
Ensialic					
Bransfield Strait	L. Plio–Rec.	[S Shetland Is.]	~22	Transitional N-type MORB/calc-alkaline basalts	16–18
Rocas Verdes	L. Jur–E. Cret	[Patagonian Batholith]	—}	N-type MORB and	19, 20
Gulf of California	M. Plio–Rec	[Baja Peninsula]	30}	LIL-enriched basalts	21, 22

References: 1: Weissel (1981); 2: Hussong & Uyeda (1981); 3: Hart *et al.* (1972); 4: Fryer *et al.* (1981); 5: Wood *et al.* (1981); 6: Gill (1976); 7: Hawkins (1976); 8: Weissel (1977); 9: Watts & Weissel (1975); 10: Mrozowski & Hayes (1979); 11: Marsh *et al.* (1980); 12: Mattey *et al.* (1980); 13: Barker (1972); 14: Barker & Hill (1981); 15: Saunders & Tarney (1979); 16: Barker & Griffiths (1972); 17: Roach (1978); 18: Weaver *et al.* (1979); 19: Dalziel *et al.* (1974) and Dalziel (1981); 20: Saunders *et al.* (1979); 21: Larson (1972); 22: Saunders *et al.* (1982a, 1982c).
Island arcs listed in square brackets may not have been extant during marginal basin formation.

southern Chile, is a fossil marginal basin, and preserves the normal gabbro-sheeted dyke–pillow lava sequence (Dalziel *et al.* 1974; Dalziel 1981). It is the disputed origin of many ophiolites, as ocean crust of back-arc crust, that has in part prompted detailed geochemical studies of the possible diagnostic characters of basalts from the two environments (e.g. Pearce 1975; Pearce *et al.* 1981; Saunders *et al.* 1980b; Pearce *et al.*, this volume).

Geochemistry of back-arc basin basalts

General considerations

Most samples of igneous rocks recovered from the floors of back-arc basins are basaltic in composition, usually olivine- (more rarely nepheline- or quartz-normative tholeiites, and are virtually indistinguishable from ocean ridge tholeiites on the basis of their mineralogy and major-element chemistry (e.g. Hart *et al.* 1972; Hawkins 1976, 1977; Saunders & Tarney 1979; Wood *et al.* 1980, 1981; Mattey *et al.* 1980; Marsh *et al.* 1980). More evolved rocks do occur in ensialic basins, either plagiogranites as in the 'rocas verdes' marginal basin ophiolite (Saunders *et al.* 1979), or subaerially exposed basalt to rhyodacite lavas of calc-alkaline affinity as on axial islands in Bransfield Strait (Weaver *et al.* 1979), but they have not been recovered from ensimatic basins. Trace element and isotopic data may, however, be used to characterize the mantle source regions of back-arc basalts.

Analytical requirements and data selection

Multi-element diagrams are increasingly used to highlight geochemical variations between igneous rocks from different regions. However, these diagrams require a comprehensive selection of precise trace-element determinations to be made on the same sample, a requirement only rarely met by existing data in published studies. In the pattern described in the following account there are, unfortunately, some omissions of certain critical elements. Two elements, Th and Ta, are of particular importance, Th because it is one of the few large ion lithophile (LIL) elements (Fig. 1) which is relatively immobile during alteration of basalt by seawater, and Ta because it can be analysed to very low levels while retaining high precision (unlike Nb). Both Th and Ta can be determined with high precision by neutron activation analysis.

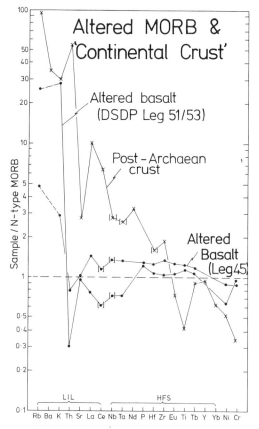

FIG. 1. Selected altered ocean ridge basalts and post-Archaean crustal average, normalized against fresh N-type ocean ridge basalt (see Appendix 1). DSDP Leg 51/53: highly altered basalt (LOI ~ 5.8%) from Cretaceous Atlantic Ocean crust (sample 417A-27-1, 118–120 cm) (Joron *et al.* 1979); Leg 45: slightly altered basalt (LOI ~ 1.9%) from upper Miocene Atlantic Ocean crust (sample 395A-37-1, 136–141 cm) (Bougault *et al.* 1978). Post-Archaean crustal average values from Taylor (1964), with Nb and Ta values taken from average granodioritic compositions (unpublished data). The dashed line represents normalized N-type MORB.

Seawater alteration of basalt

The degree to which LIL element enrichment is due to secondary alteration or primary magmatic processes is of considerable significance to the study of back-arc and ocean ridge basalts, because the main differences in chemistry are based on relative abundances of LIL elements. It has long been recognized that low-temperature interaction between seawater and basalt causes enrichment of K, Rb, Sr, [87]Sr, U and Cs (Hart *et al.* 1974; Mitchell & Aumento 1977) and, under extreme conditions, even the light rare earth elements (LREE) (Ludden & Thompson 1979; Terrell *et al.* 1979). Most basalts recovered by drilling have, because of their exposure to porewaters, undergone some degree of alteration. Higher grades of hydrothermal alteration (greenschist to amphibolite grade metamorphism) may lead to leaching of some of these elements (Saunders *et al.* 1979; Stern & Elthon 1979).

To demonstrate the effects of low-grade alteration, data for two basalts, one slightly altered and the other more severely altered, are plotted in Fig. 1. In this diagram, the data have been normalized against a representative mid-ocean ridge tholeiite (N-type MORB: see below), thus facilitating direct comparison between altered and fresh basalt. The following observations may be made: (i) K and Rb are strongly enriched in both samples (no data are available for Ba) and Sr less so (although this is a function of the absolute primary abundance of Sr); (ii) the high field strength (HFS) elements (Nb to Yb on Fig. 1) are essentially immobile (cf. Pearce & Cann 1973); and (iii) Th is also apparently immobile. Indeed, Th abundances in the altered basalts are less than in the normalizing basalt. This is not unexpected as many fresh basalts from the East Pacific Rise and Mid-Atlantic Ridge have much lower Th contents (< 0.1 ppm: see below) than average MORB.

It is clear (Fig. 1) that the use of the LIL elements K, Rb, Sr and possibly Ba in evaluating primary magmatic LIL element distribution is open to uncertainty when studying even slightly altered submarine basalts, particularly deeper drilled samples. This uncertainty is compounded in meta-igneous suites affected by higher grade alteration and metamorphism. Nonetheless, we consider that Th is less mobile, at least during low-grade alteration.

Trace element and isotope chemistry of back-arc basin basalts: a broad perspective

In this section the evidence which demonstrates that many back-arc basin basalts have a chemistry transitional between ocean ridge tholeiites and island-arc basalts (cf. Gill 1976; Saunders & Tarney 1979) is considered. The extent to which island-arc magma characteristics (in particular the enrichment in LIL elements relative to HFS elements) are developed in back-arc basalts is variable, even within individual basins; both LIL-enriched and LIL-depleted varieties can occur in close proximity.

It can be shown that, at least on a general scale, the degree of LIL-enrichment in back-arc basalts is dependent on the maturity of the adjacent subduction zone, rather than simply on the evolutionary state of the basin and contemporaneous arc. To illustrate such features, it is convenient to make direct geochemical comparisons with mid-ocean ridge tholeiites.

The predominant basalt type erupted along the ocean ridge system has been termed a 'normal' or N-type MORB (Sun *et al.* 1979), and contains very low abundancies of the highly incompatible elements Rb, Ba, K, Th, La, Ce, Nb and Ta (Kay *et al.* 1970; Hart 1971; Tarney *et al.* 1980). It is generally accepted that the low abundances of these elements, relative to other, less-incompatible species (e.g. Zr, Hf, heavy rare earth elements (HREE)) implies derivation of parental magmas from a source which has undergone previous episodes of melt extraction. In addition, low [87]Sr/[86]Sr (0.7025–0.7028) and high [143]Nd/[144]Nd (0.5131–0.5133) ratios suggest that this extraction began early in Earth history (Gast 1968; O'Nions *et al.* 1977).

Trace-element data for a representative N-type MORB are compiled in Appendix 1. These data are not intended to be typical of all N-type MORB, but do serve to compare with data from back-arc basalts. This has been done by normalizing back-arc basalt analyses to N-type MORB (Figs. 2, 4–6). The form of the normalized data curves is of particular interest, rather than merely the absolute abundance levels.

Data for a basalt from the Parece Vela Basin, an inactive back-arc basin to the W of the Mariana Trough system, are plotted on Fig. 2a. The slight deviation from the N-type MORB pattern probably results from secondary alteration; fresh basaltic glass recovered from the basin floor retains the low K_2O levels typical of N-type MORB, whereas crystalline interiors of basalt fragments have higher K_2O levels (Mattey *et al.* 1980). Note also the low abundance of Th in this sample. The Parece Vela Basin, and its northerly continuation, the Shikoku Basin, were active during mid-Oligocene to early Miocene time, separating the contemporary island arc, the West Mariana Ridge, from the remnant arc, the Palau–Kyushu Ridge, by as much as 1000 km (Karig 1971; Mrozowski & Hayes 1979). It could be suggested that the absence of LIL-enriched magmas may be a function of the width or maturity of the basin. However, it should be noted that the analysed samples were recovered by drilling from crust formed when the basin was only about 250 km

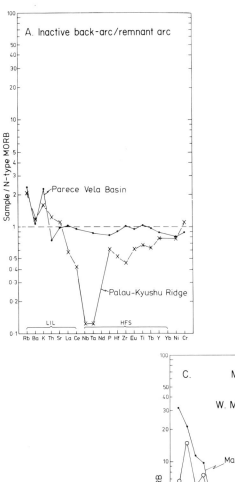

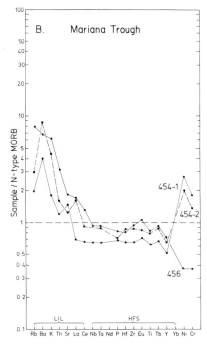

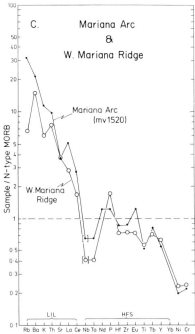

FIG. 2. Normalized multi-element plots of basalts and basaltic andesites from the Mariana arc–back-arc system. (a) Basalts from the Palau–Kyushu remnant arc (448A-36-4, 130–133 cm) and Parece Vela Basin (451-45-1, 20–32 cm) (Wood *et al.* 1980); (b) basalts from the Mariana Trough back-arc basin (454-1: 454A-5-4, 15–18 cm; 452-2: 454A-16-1, 111–114 cm; 456: 456A-13-1, 23–26 cm) (Wood *et al.* 1981); (c) basalts and basaltic andesites from the West Mariana Ridge remnant arc (451-46-1, 24–30 cm) (Wood *et al.* 1980) and Mariana Arc (MV 1520) (Wood *et al.* 1981).

(Site 499) and 400 km (Site 450) wide (Kroenke *et al.* 1980).

Basalts with N-type MORB geochemistry have also been recovered from the narrower Mariana Trough and Lau Basins (Gill 1976; Hawkins 1976; Wood *et al.* 1981), but here they are much less abundant than LIL element enriched varieties. Fresh basalts dredged from the axial zone of the Mariana Trough have higher contents of K, Rb, Ba and Sr, slightly higher $^{87}Sr/^{86}Sr$ ratios (0.7028–0.7030), and higher La_n/Sm_n ratios than N-type MORB (Hart *et al.* 1972; Fryer *et al.* 1981); they appear to be transitional towards island-arc tholeiites. Tholeiitic basalts showing strong LIL element enrichment similar to calc-alkaline basalts were recovered at DSDP Site 454 in the Mariana Trough (Fig. 2b). Some of this enrichment (particularly in Rb, K and Ba) is doubtless due to secondary alteration, but the high Th levels are probably primary (Wood *et al.* 1981). Of particular interest are basalts from Site 454 showing a range from N-type MORB to calc-alkaline basalts (Fig. 2b), implying a close spatial and temporal association of two distinct magma types.

Dredged basaltic glasses from the Mariana Trough are enriched in volatile species, particularly H_2O, relative to N-type MORB (Fig. 3), which is consistent with the high vesicularity of some Mariana Trough basalts (Garcia *et al.* 1979).

Trace-element data for lavas from the island arcs and remnant arcs of the Mariana system indicate that the degree of LIL element enrichment has increased in successive episodes of arc magmatism over the last 35 Ma. The earliest arc, preserved as the Palau–Kyushu Ridge, is composed of volcanic rocks belonging to the island-arc tholeiite series (Mattey *et al.* 1980). The degree of LIL element enrichment is minimal (Fig. 2a), although the abundances of HFS elements, particularly Nb and Ta, are significantly lower than in most N-type MORB. Basalts and andesites from the later arcs, the West Mariana Ridge and Mariana Arc, have trace-element compositions typical of calc-alkaline suites: strong LIL element enrichment and low Ta and Nb abundances (Fig. 2c; Dixon & Batiza 1979; Wood *et al.* 1981). Superficially it would appear that as the subduction zone matures, the magmas erupted in successive arcs and associated back-arc basins become increasingly LIL element enriched.

Available geochemical data for basalts from the back-arc basin in the E Scotia Sea are consistent with this observation. Samples recovered from the presently active spreading centre have

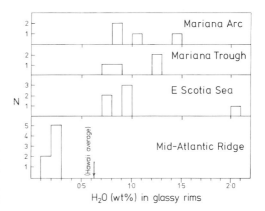

FIG. 3. Frequency histograms of H_2O contents of abyssal basaltic glasses recovered from back-arc basins and mid-ocean ridges. Data sources: Mariana Trough and Arc, Garcia *et al.* (1979); E Scotia Sea, Muenow *et al.* (1980); mid-ocean ridges and Hawaiian average, Delaney *et al.* (1978).

a range of compositions similar to basalts from the Mariana Trough, which concurs with the suggestion that the present spreading episode is but the latest of a series of basin openings which formed the central Scotia Sea (Barker 1972; Barker & Hill 1981). Rapid sea-floor spreading in the E Scotia Sea began approximately 8 Ma ago, and favours accretion on the trenchward side (Barker 1972). The adjacent island arc, the South Sandwich Islands, is erupting magmas of the island-arc tholeiite series and, like the Mariana Islands, appears to be resting on crust generated during the present episode of back-arc extension.

Two (D20 and D23) of four dredge hauls from the spreading axis in the E Scotia Sea comprise sparsely vesicular basalts which are chemically similar to N-type MORB (Fig. 4), although there is a slight tendency towards enrichment in LIL elements, including Th. The greatest enrichment in LIL elements relative to HFS elements is observed in the basalts from dredges 22 and 24, which are also exceptionally vesicular. D24.14 (Fig. 4), for example, is an unaltered quartz-normative basalt depleted in HFS elements relative to N-type MORB, but enriched in Rb, Ba, and K. This chemical similarity to island arc basalts is further supported by moderately high $^{87}Sr/^{86}Sr$ ratios (0.7031–0.7034 in dredge 24 basalts) and high magmatic water contents (at least 2.04% in D24.11: Fig. 3; Muenow *et al.* 1980). Basalts from dredges 20 and 23 also have higher magmatic water contents (Fig. 3) and $^{87}Sr/^{86}Sr$

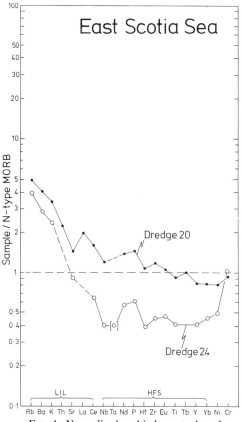

FIG. 4. Normalized multi-element plot of basalts from the E Scotia Sea back-arc basin. Dredge 20 (sample D20.35) and dredge 24 (D24.14) data from Saunders & Tarney (1979) and unpublished data.

isotope ratios (0.7028–0.7030) than N-type MORB.

It now remains to consider briefly the geochemistry of lavas erupted within back-arc basins formed along continental margins. Unlike the intra-oceanic basins of the Western Pacific, ensialic back-arc basins (floored by mafic crust) are rare: only five examples are known (Bransfield Strait, Sea of Japan, Andaman Sea, Gulf of California and the Mesozoic 'rocas verdes' basin of S Chile) and of these, the Gulf of California and the Andaman Sea possess tectonic characteristics atypical of back-arc basins in general. The scarcity of ensialic basins may be due to the inherent difficulty of splitting the continental lithosphere, or simply because the present motion vectors and configurations of the major plates are not conducive to basin formation (Molnar & Atwater 1978). Nonetheless, it would appear that 'unusual' tectonic conditions must prevail before ensialic basin

formation can commence. The Andaman Sea developed at an obliquely convergent plate margin, and it has been suggested that the Sea of Japan resulted from the subduction of part of the Kula Ridge (Uyeda & Miyashiro 1974).

In the case of Bransfield Strait, the onset of collision of the Aluk Ridge with the northern Antarctic Peninsula led to the cessation of calc-alkaline magmatism ('arc switch-off') during Tertiary times (Herron & Tucholke 1976; Saunders *et al.* 1982b). At present, only a small remnant of spreading axis is preserved in Drake Passage, to the NW of the South Shetland Islands. A pronounced slowing of spreading rate at this centre between 6 and 4 Ma ago appears to have been the cause of rifting and extension behind the South Shetland Islands (Barker & Griffiths 1972; Barker & Burrell 1977), perhaps because of an increase in the roll-back vector of the subducted slab. Extension behind the South Shetland Islands began less than 2 Ma ago, and has formed the 60 km wide Bransfield Strait; volcanicity is recorded on the axial Bridgeman and Deception Islands (Weaver *et al.* 1979).

The tectonic and volcanic setting of the Gulf of California is similar to that of Bransfield Strait, except that convergence between the Pacific and North American Plates is highly oblique. The resulting separation of the Baja Peninsula from the continent of Mexico is therefore dominated by transform-style faulting, linking short spreading centres (Larson 1972). As with the northern Antarctic Peninsula, however, encroachment of young Pacific crust appears to have caused arc switch-off before the basin began to open (Dickinson & Snyder 1979). Ridge–trench collision occurred approximately 28 Ma ago, with the resultant triple point migrating southwards along the plate boundary (Atwater & Molnar 1973). A consequence of this boundary configuration is that no subducted oceanic lithosphere has underlain the Baja region since 10–15 Ma ago, although the Gulf began to open only 5 Ma ago (Dickinson & Snyder 1979).

Both Bransfield Strait and the Gulf of California occur within regions with a long history of calc-alkaline magmatism (far longer than, for example, the Mariana or South Sandwich systems), although the chemistry of their eruptive products is quite different. Lavas from Deception Island, on the axis of Bransfield Strait, display a wide range of compositions, from basalt to rhyodacite, and belong to the sub-alkaline series of Irvine & Barager (1971) (Weaver *et al.* 1979). The basalts lie within or close to Kuno's (1966) high-alumina basalt (calc-alkaline) series, and do not show strong iron-enrichment. Bridgeman Island lavas have a more restricted composition, being quartz-normative basalts and basaltic andesites. Absolute abundances of LIL elements are significantly higher than in N-type MORB and indeed most ensimatic back-arc basalts (Fig. 5), and more closely resemble those recorded in equivalent rocks from calc-alkaline island arcs (e.g. West Mariana Ridge: Fig. 2c). Bransfield Strait lavas also have very low abundances of Nb (Fig. 5) and their affinities with island-arc suites is further supported by their high $^{87}Sr/^{86}Sr$ ratios (0.7035–0.7039: Weaver *et al.* 1979). For comparison, two calc-alkaline basalts from the South Shetlands and the eastern Antarctic Peninsula are included in Fig. 5. Although not contemporaneous with the Bransfield Strait activity, these samples demonstrate the strong LIL-enrichment found in continental

calc-alkaline magmas; the characteristic low Ta and Nb levels are also seen.

Samples recovered from the Gulf of California show a more restricted range of compositions; with the exception of a basalt-andesite sequence from Tortuga Island, a small seamount in the central part of the Gulf, they are mostly tholeiitic basalts. Basalts recovered from the mouth of the Gulf are typical of N-type MORB erupted elsewhere along the East Pacific Rise, showing strong LREE depletion and having very low abundances of Rb, K, Ba, Th, Nb and Ta (Fig. 6) (Lopez *et al.* 1978; Saunders *et al.* 1982a). Dolerites and basalts from the Guaymas Basin, in the central part of the Gulf, also show mineralogy and major-element chemistry typical of N-type MORB, but they are enriched in K, Rb, Ba, Sr, LREE and Th (Fig. 6) relative to basalts from the Gulf mouth region. Some of this enrichment, particularly in

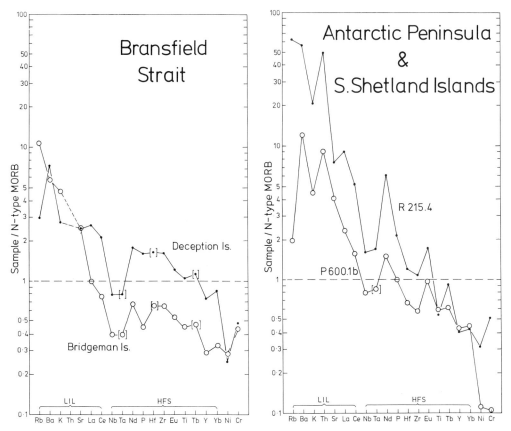

FIG. 5. Normalized multi-element plot of basalts from Bransfield Strait, South Shetland Islands, and the Antarctic Peninsula. Deception Island (sample B.138.2) and Bridgeman Island (sample P.640.lb) data from Weaver *et al.* (1979); South Shetland Islands (sample P.600.lb: Tertiary basalt, Fildes Peninsula, King George Island) and Antarctic Peninsula (sample R.215.4: Mesozoic basaltic andesite, Jason Peninsula, Graham Land) from unpublished data.

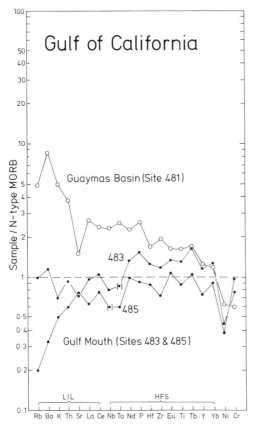

FIG. 6. Normalized multi-element plot of basalts recovered during Legs 64 and 65 of the Deep Sea Drilling Project from the Gulf of California. Data sources: Site 483 (65-483-15-2, 1–6 cm) and Site 485 (65-485A-35-5, 70–72 cm) from Saunders (1983); Site 481 (64-481-15-3, 27–29 cm) from Saunders *et al.* (1982c). New Ba values for Site 483 (4 ppm) and 485 (14 ppm) samples supercede previously published data.

(Verma 1982), similar to N-type MORB, although andesites from Tortuga Island have higher ratios (up to 0.7036: Batiza *et al.* 1979); again, the calc-alkaline 'signature' in the central Gulf basalts is much less obvious than it is for Bransfield Strait. This would concur with the absence of a subducted slab beneath the Gulf region, and the suggestion that the area is now underlain by Pacific Ocean-type mantle (Saunders *et al.* 1982a).

Generalized comparisons of the chemistry of basalts from different back-arc basins are contentious because of the wide range of compositions within even the same basin. Variation in absolute abundances of incompatible elements, rather than relative differences between LIL and HFS element contents, may be due to various degrees of fractional crystallization or partial melting. To reduce this effect, selected analyses of back-arc basalts (in general the most LIL-enriched examples from each basin) have

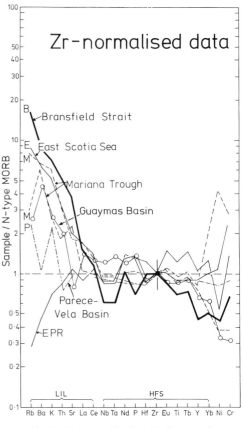

FIG. 7. Zr-normalized multi-element plots of back-arc basalts from various basins, taken from Figs 1, 2, 4–6.

Rb, K and Ba, is due to secondary processes, but the high Sr abundances are considered to be primary (Saunders *et al.* 1982a,c). This suggestion is supported by the high Sr contents seen in subaerially erupted lavas from Tortuga Island (Batiza *et al.* 1979).

The degree of LIL element enrichment in the Guaymas Basin samples is much less than that in basalts from Bransfield Strait, particularly if the effects of fractionation are taken into account. Thus, the LIL/HFS element ratios of Bransfield Strait lavas are much higher (compare Figs 5 and 6). $^{87}Sr/^{86}Sr$ ratios of Guaymas Basin samples range from 0.7024 to 0.7028 been normalized to give constant Zr levels (Fig. 7). Although only a crude approximation, this technique does circumvent the effects of low-pressure fractionation. Three observations can be drawn from Fig. 7:

(i) The back-arc basalts with the highest LIL/HFS element ratios are found in *narrow* basins associated with *mature* subduction zone systems.
(ii) Wider basins associated with youthful subduction zone systems (e.g. Parece Vela Basin) erupt basalts with low LIL/HFS element ratios.
(iii) The degree of LIL element enrichment in narrow ensialic basins such as the central Gulf of California is less pronounced, perhaps because of the absence of subducted slab.

The significance of these differences, in relation to the petrogenesis of back-arc basalts, is considered below.

LIL element enrichment: the role of the subducted slab

The close spatial and tectonic association between back-arc basins and island arcs is matched by a geochemical association of their magmatic products. The trace-element chemistry of back-arc basalts is transitional between ocean ridge and island-arc basalts (Gill 1976; Saunders & Tarney 1979). The main evidence for this conclusion is the LIL element enrichment of many back-arc basalts. However, LIL element enrichment is not unique to island-arc and back-arc magmas. For example, LIL-enriched tholeiites and alkali basalts are erupted on some segments of ocean ridge, and many intra-plate magmas are strongly LIL-enriched. Ocean ridge basalts which contain high abundances of K, Rb, Th, Ba and LREE are termed 'enriched' or E-type MORB (e.g. Sun *et al.* 1979; Tarney *et al.* 1980), and usually are erupted along elevated ridge segments (e.g. Iceland and the Azores Platform). However, unlike most island-arc and back-arc basalts, E-type MORB have enhanced abundances of Nb and Ta.

Basalts resembling E-type MORB do occur in back-arc basins, having been recovered from the Shikoku Basin during Leg 58 of the Deep Sea Drilling Project (Marsh *et al.* 1980). An E-type MORB analysis was reported from the Lau Basin by Gill (1976). In general, however, reports of such basalts from back-arc basins are rare, possibly because of inadequate sampling. The origin of the characteristic geochemistry

of island-arc and continental margin calc-alkaline magmas (i.e. LIL enrichment and HFS depletion relative to N-type MORB) is still disputed. It is apparently not possible to derive island-arc tholeiites or calc-alkaline basalts from mantle with trace-element and isotopic properties similar to the source of N-type MORB. Although high-level fractional crystallization and low-pressure partial melting may produce variations in the absolute abundances of trace elements, such processes cannot increase LIL/HFS element ratios to the extent observed, particularly at the high degrees of partial melting thought necessary to form tholeiitic melts ($\sim$ 10–15%: Green 1971). The problem is amplified by the elevated $^{87}Sr/^{86}Sr$ ratios seen in most subduction zone-related magmas, and the displacement to higher $^{87}Sr/^{86}Sr$ values from the $\varepsilon_{Sr}-\varepsilon_{Nd}$ mantle array (e.g. Hawkesworth *et al.* 1979).

Rather, it appears necessary to invoke enrichment or metasomatism of the mantle source of island-arc basalts with a LIL element enriched fluid phase, probably originating from the dehydrating subducted oceanic crust. Dehydration, rather than melting, of the subducted crust not only overcomes the thermal problems of melting cool, refractory eclogite, but also allows decoupling of the LIL elements from the HFS elements (Anderson *et al.* 1978; Saunders *et al.* 1980a). We have previously suggested that the LIL elements are carried from the descending slab within supercritical fluid phases, leaving the small, highly charged HFS elements (e.g. Zr, Ta, Nb, Ti) bonded within the residual crystal lattices. The net effect is to enrich the overlying mantle wedge in LIL elements, H_2O and other slab-derived volatiles, and ultimately to produce the characteristic high LIL/HFS ratios seen in arc basalts.

Furthermore, repeated production of arc magmas from within the mantle wedge will gradually extract *all* incompatible elements from the source region. Some trace elements, particularly the HFS elements may, under conditions of high P_{H_2O} and P_{O_2}, be stabilized in minor mineral phases (e.g. Zr in zircon, Nb and Ta in rutile). They may then be retained in the source during melting (Saunders *et al.* 1980a). However, at the high degrees of melting envisaged for island-arc tholeiite genesis, this model is less attractive than one whereby the low abundances of HFS elements are due to melting of a source which has undergone previous melting events. Thus, repeated melt extraction serves to deplete the source in both LIL and HFS incompatible elements, but, because of continuous LIL element transfer from the adjacent subducted slab, the net result is to decrease the abundances of the HFS elements more rapidly. In effect, the concentration of the LIL elements is buffered by the adjacent slab.

There may be a third factor to consider. Increasingly, evidence suggests that the upper mantle as a whole is subject to re-enrichment by metasomatic fluids, rich in Nb, Ta and LIL elements, derived perhaps from the deeper mantle (see Hanson 1977; Frey *et al.* 1974; Tarney *et al.* 1980; Bailey 1982). Where such enrichment is focused, E-type MORB or alkaline, ocean island (e.g. Azores) magmas are produced. It is interesting to speculate that the mantle wedge above an active subduction zone will not be subject to metasomatic re-enrichment from the deeper mantle, because the slab forms a physical barrier to ascending material. Consequently, the mantle wedge will not be replenished in Nb- or Ta-rich fluids, unless material is transferred laterally. Indeed, only at the open end of the wedge, furthest from the trench, would the effects of 'deep' mantle metasomatism be expected. This would account for the fact that although Nb- and Ta-enriched undersaturated magmas are associated with many continent-based arcs (e.g. Japan: Wood *et al.* 1980; Patagonia: Hawkesworth *et al.* 1979; Antarctic Peninsula: Saunders 1982), they occur well behind the calc-alkaline belt.

Most of the LIL element enrichment observed in island-arc basalts may, therefore, be linked to dehydration of altered oceanic crust in the subduction zone. Low-temperature alteration of basalts by seawater leads to increases in the abundances of most of the LIL elements seen enriched in island-arc basalts, with the possible exception of Th (Fig. 1). However, subducted sediments may also play an important role. Recent investigations of accretionary prisms off Japan, the Mariana Islands and Central America indicate that most sediment is not immediately scraped off the descending slab, but is carried down into the subduction zone (e.g. Watkins *et al.* 1981). Whether these subducted sediments are 'subcreted' beneath the overhanging plate margin (Karig & Kay 1981), taken into the deep mantle (White & Hofmann 1982), or involved in island-arc magma genesis (Kay 1980), is not clear. The problem is exacerbated by a paucity of chemical data for abyssal sediments.

The sediment input into a trench system comprises two main components: (i) clastic material derived from the adjacent arc and/or continental margin; and (ii) abyssal pelagic detritus carried on the oceanic crust. Both types of sediment tend to exhibit LIL element enrichment, but each has a distinctive Pb-isotopic characteristic (Sun 1980). Most continent-based arcs demonstrate the effects of crustal contamination in the pattern of their apparent crust–mantle Pb isotope mixing trends and in their high contents of LIL elements. However, it is inherently difficult to resolve the chemical effects of subduction of continent-derived material from that of crustal contamination of ascending magma. To eliminate the latter effect it is necessary to study ocean island arcs based on oceanic crust.

Unfortunately, a simple model of sediment subduction is not applicable to oceanic island arcs (Sun 1980). The Pb-isotopic compositions of volcanic rocks from Tonga (Sun 1980) and the Mariana Arc (Meijer 1976) are similar to N-type MORB, suggesting that the LIL element enrichment may not be due to sediment subduction. Nonetheless, recent Pb isotope studies of the South Sandwich Islands and the adjacent E Scotia Sea back-arc basin strongly support a model with small quantities of sediment being involved in magma genesis (Cohen & O'Nions 1982). In addition, preliminary modelling suggests that it is necessary to involve subducted pelagic sediments which have pronounced negative Ce anomalies, in order to produce the Ce anomalies observed in several island-arc suites (e.g. Dixon & Batiza 1979; Hole *et al.*, in press). The volumes of sediment envisaged ($<1\%$ of the island-arc basalt source) are sufficient to produce the observed enrichment in LIL elements.

The relevance of this discussion to back-arc magma genesis should now be apparent. The transitional nature of back-arc basalts, between N-type MORB and island-arc basalts, can be most readily explained by localized LIL element (and H_2O) enrichment of their source by fluids derived from the subducted slab. It is not necessary, because of the long time scales involved, that the fluids affecting the source are derived solely from contemporaneous subduction zones; it is more likely that the fluids come from earlier (?10-20 Ma earlier) episodes of subduction. Such a derivation would explain the increasing LIL-enrichment in basins associated with mature subduction zones. With time, the mantle wedge will become increasingly LIL-enriched, and basins developed at a later stage in the history of the subduction zone will tap markedly LIL-enriched material. Conversely, the crust of back-arc basins formed at an early stage of subduction history (e.g. the Parece Vela Basin) will be derived from mantle with only minor (if any) subduction-related LIL-enrichment. Continent-based arcs, because of their longevity, would be expected to overlie mantle wedges which have undergone the greatest degree of LIL-enrichment (which concurs with Dickinson's (1975) observation that continent-based arcs are more LIL-enriched than most oceanic arcs), and hence any associated ensialic back-arc basins (e.g. Bransfield Strait) would be expected to erupt enriched magmas, at least during the initial stages of formation.

As the basin opens, and the spreading centre separates from the island arc, the LIL element component of successive magmas will become less pronounced, through a combination of (i) removal of incompatible elements during magma extraction, and (ii) isolation of the source of the back-arc basalts from the zone of LIL element replenishment. Both of these mechanisms are important but most probably the second is more effective. This is simply because extraction of a tholeiitic melt would be expected to remove from the basalt source all incompatible elements, to approximately the same degree, regardless of whether they are LIL or HFS elements. Thus, the LIL/HFS elements should not vary significantly from basin to basin or from arc to arc; this is clearly not the case. Testing this hypothesis within a single basin is difficult, because of the lack of suitable data, although geochemical studies of the Mesozoic 'rocas verdes' marginal basin in S Chile do provide some supportive evidence. Mafic rocks (gabbros and dolerites) from the northern outcrop of the ophiolite at Sarmiento, where the basin was probably very narrow (~ 20 km), are LREE-enriched with abundances of Sr and Ba significantly higher than in N-type MORB (Saunders *et al.* 1979; Dalziel *et al.* 1974; Dalziel 1981). Further S, in the Tortuga ophiolite, the basalts are LREE-depleted, and are chemically indistinguishable from N-type MORB, although mafic sills from the margins of the complex chemically resemble dolerites from the Sarmiento ophiolite (Stern 1979). An interpretation which is consistent with these observations is that early melts were derived from LIL-enriched mantle, adjacent to the source of a Mesozoic calc-alkaline magmatic arc, whereas later melts were extracted from a chemically more depleted source (e.g. Tarney *et al.* 1981). A similar model has been proposed for the Gulf of California (Saunders *et al.* 1982a).

To summarize, available geochemical data for back-arc and island-arc basalts are consistent with a model where the mantle wedge beneath the arc is replenished with LIL elements and volatiles derived from the subducted,

dehydrating slab. Some of the LIL elements may be derived from subduction of sediments. The degree of LIL-enrichment within the mantle wedge, and in the subsequent basaltic and andesitic melts, will be proportional to the age of the subduction zone, older systems presumably having undergone more LIL enrichment. There is probably an upper limiting value to this enrichment, perhaps buffered by the rate and extent of magma extraction. Back-arc basin magmatism locally taps the LIL element-enriched source, producing a spectrum of basalt types from N-type MORB to calc-alkaline basalts. However, in the back-arc environment, the degree of LIL-enrichment is probably dependent not only on the maturity of the adjacent subduction zone, but also on the width of the basin. In addition, the model envisages melt extraction as being an important process in causing a net reduction in the abundances of HFS elements, particularly Nb and Ta, in the source of most arc, and some back-arc, magmas.

Finally, the implications of the geochemical results on models for back-arc extension will be considered. Essentially there are three basic models:

(i) Passive diapirism in response to regional stress within the lithosphere (Packham & Falvey 1971).
(ii) Active diapirism resulting from heat or fluids generated along the Benioff zone (Karig 1971, 1974).
(iii) Dynamic circulation, involving secondary convection cells driven by drag of the downgoing slab along the Benioff zone (Toksöz & Bird 1977).

The geochemical evidence, that the mantle source of arc magmas and of (initial) back-arc magmas has been metasomatized by hydrous LIL element-rich fluids, supports Karig's model of active diapirism. It is supposed that such fluids lower the mantle solidus, promote incipient melting, and ultimately destabilize the mantle above the Benioff zone sufficiently to allow diapiric uprise. Such movement would produce a circulation pattern in the back-arc region (Fig. 8) which is almost the reverse of that in the dynamic model of Toksöz & Bird (1977). Mechanical decoupling of the subducting slab from the overriding mantle may be enhanced if significant volumes of clay-rich sediment are incorporated within the subduction zone. In most back-arc basins the extensional phase begins with the volcanic arc being split by mantle diapirism. The arc is evidently the weak

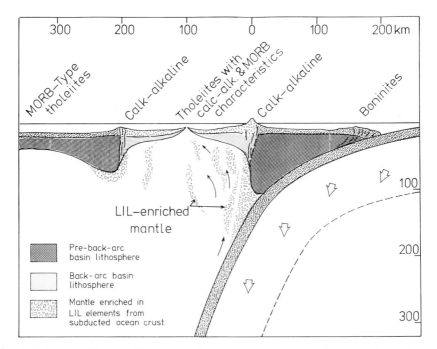

FIG. 8. Schematic diagram without vertical exaggeration, across an intra-oceanic arc–back-arc system, to illustrate the mantle destabilization model described in the text. Open arrows indicate motion vectors of descending plate. Rock types represent major basalt types erupted in the Mariana and Parece Vela Basins.

zone, capable of being exploited if plate motions permit extension.

The location of the heat and magma source within the back-arc region will be controlled by two main factors: (i) the rate of diapiric ascent; and (ii) the roll-back rate of the subducting oceanic crust. The diapir will rise initially beneath the arc, but the long time before it reaches the surface is sufficient for the arc to migrate laterally, provided there is a significant roll-back vector of the hinge zone of the subducting plate. For example, a diapir ascending at a rate of 1 cm yr^{-1} from a depth of 150 km would take almost 15 Ma to reach within 5 km of the surface, and in this time the arc could have migrated several hundred kilometres oceanwards. Thus the zone of initiation of such diapirs and the locus of their emplacement at the back-arc spreading axis become progressively separated, until the diapirs eventually revert to splitting the volcanic arc again and a new phase of back-arc spreading begins.

It is probable that the major process controlling the initiation of back-arc spreading is the relative motion rates of the subducting and overriding plates, as shown by Molnar & Atwater (1978). Diapirism alone is unlikely to promote back-arc extension, because to achieve diapiric uprise of sufficient magnitude requires considerable intrinsic density contrast between the diapir and the surrounding mantle (cf. Weaver & Tarney 1979). Rather, the injection of hydrous material from the subducting plate will lead to *potential* diapiric uprise, with the destabilized mantle taking advantage of the prevailing plate tectonic configuration.

Appendix

N-type MORB normalizing values used in various text figures are from Tarney *et al.* (1981), with minor modifications. All values are in parts per million:

Rb	1	P	570
Ba	12	Hf	2.5
Th	0.2	Zr	88
K	830	Eu	1.2
La	3	Ti	8400
Ce	10	Tb	0.71
Sr	136	Y	35
Nb	2.5	Yb	3.5
Ta	0.17	Ni	138
Nd	8	Cr	290

ACKNOWLEDGMENTS: We would like to thank Peter Kokelaar, whose comments improved the manuscript; Mrs Sheila Bishop, who typed the text, and Mr N. Sinclair-Jones, who draughted the text figures.

References

ANDERSON, R. N., DELONG, S. E. & SCHWARZ, W. H. 1978. Thermal model for subduction with dehydration in the downgoing slab. *J. Geol.* **86**, 731–9.

ATWATER, T. & MOLNAR, P. 1973. Relative motion of the Pacific and North American Plates deduced from sea-floor spreading in the Atlantic, Indian and South Pacific Oceans. *In*: KOVACH, R. L. & NUR, A. (eds) *Proc. Conf. Tectonic Problems in the San Andreas Fault System. Stanford Univ. Pub. Geol. Sci.* **13**, 136–48.

BAILEY, D. K. 1982. Mantle metasomatism—continuing chemical change within the Earth. *Nature, Lond.* **296**, 525–30.

BARKER, P. F. 1972. A spreading centre in the east Scotia Sea. *Earth planet. Sci. Lett.* **15**, 123–32.

—— & BURRELL, J. 1977. The opening of Drake Passage. *Mar. Geol.* **25**, 15–34.

—— & GRIFFITHS, D. H. 1972. The evolution of the Scotia Ridge and Scotia Sea. *Phil. Trans. R. Soc. London,* **A271**, 151–83.

—— & HILL, I. 1981. Back-arc extension in the Scotia Sea. *Phil. Trans. R. Soc. London,* **A300**, 249–62.

BATIZA, R., FUTA, K. & HEDGE, C. E. 1979. Trace element and strontium isotope characteristics of volcanic rocks from Isla Tortuga: a young seamount in the Gulf of California. *Earth planet. Sci. Lett.* **43**, 269–78.

BOUGAULT, H., TREUIL, M. & JORON, J.-L. 1978. Trace elements in basalts from 23°N and 36°N in the Atlantic Ocean: fractional crystallization, partial melting and heterogenity of the upper mantle. *In*: MELSON, W. G., RABINOWITZ, P. D. *et al.* (eds) *Init. Rep. Deep Sea drill. Proj.* **45**, 493–506. U.S. Government Printing Office, Washington, D.C.

COHEN, R. S. & O'NIONS, R. K. 1982. Identification of recycled continental material in the mantle from Sr, Nd and Pb isotope investigations. *Earth planet. Sci. Lett.* **61**, 73–84.

DALZIEL, I. W. D. 1981. Back-arc extension in the southern Andes: a review and critical reappraisal. *Phil. Trans. R. Soc. London,* **A300**, 319–55.

——, DEWIT, M. J. & PALMER, K. F. 1974. Fossil marginal basin in the southern Andes. *Nature, Lond.* **250**, 291–4.

DELANEY, J. R., MUENOW, D. W. & GRAHAM, D. G. 1978. Abundance and distribution of water, carbon and sulfur in the glassy rims of submarine pillow basalts. *Geochim. cosmochim. Acta* **42**, 581–94.

DICKINSON, W. R. 1975. Potash-depth (K-*h*) relations in continental margin and intra-oceanic magmatic arcs. *Geology*, **3**, 53–6.

—— & SNYDER, W. S. 1979. Geometry of subducted slabs related to San Andreas transform. *J. Geol.* **87**, 607–27.

DIXON, T. H. & BATIZA, R. 1979. Petrology or recent lavas in the northern Marianas: implications for the origin of island arc basalts. *Contrib. Mineral. Petrol.* **70**, 167–81.

FREY, F. A., BRYAN, W. B. & THOMPSON, G. 1974. Atlantic Ocean floor: geochemistry and petrology of basalts from Legs 2 and 3 of the Deep Sea Drilling Project. *J. geophys. Res.* **79**, 5507–27.

——, GREEN, D. H. & ROY, S. D. 1978. Integrated models of basalt petrogenesis: a study of quartz tholeiites and olivine melilites from south eastern Australia utilizing geochemical and experimental petrological data. *J. Petrol.* **19**, 463–513.

FRYER, P., SINTON, J. M. & PHILPOTTS, J. A. 1981. Basaltic glasses from the Mariana Trough. *In*: HUSSONG, D. M., UYEDA, S., *et al.* (eds) *Init. Rep. Deep Sea drill. Proj.* **60**, 601–9. U.S. Government Printing Office, Washington, D.C.

GARCIA, M. O., LIU, N. W. K. & MUENOW, D. W. 1979. Volatiles in submarine volcanic rocks from the Mariana Island arc and trough. *Geochim. cosmochim. Acta*, **43**, 305–12.

GAST, P. W. 1968. Trace element fractionation and the origin of tholeiitic and alkaline magma types. *Geochim. cosmochim. Acta*, **32**, 1057–86.

GILL, J. B. 1976. Composition and age of Lau Basin and Ridge volcanic rocks: implications for evolution of an interarc basin and remnant arc. *Bull. geol. Soc. Am.* **87**, 1384–95.

GREEN, D. H. 1971. Composition of basaltic magmas as indicators of conditions of origin: application to oceanic volcanism. *Phil. Trans. R. Soc. London*, **A268**, 707–25.

HANSON, G. N. 1977. Evolution of the suboceanic mantle. *J. geol. Soc. London* **134**, 235–54.

HART, S. R. 1971. K, Rb, Cs, Sr and Ba contents and Sr isotope ratios of ocean floor basalts. *Phil. Trans. R. Soc. London*, **A268**, 573–87.

——, ERLANK, A. J. & KABLE, E. J. D. 1974. Sea floor basalt alteration: some chemical and strontium isotopic effects. *Contrib. Mineral. Petrol.* **44**, 219–30.

——, GLASSLEY, W. E. & KARIG, D. E. 1972. Basalts and sea floor spreading behind the Mariana Island arc. *Earth planet. Sci. Lett.* **15**, 12–8.

HAWKESWORTH, C. J., NORRY, M. J., RODDICK, J. C., BAKER, P. E., FRANCIS, P. W. & THORPE, R. S. 1979. $^{143}Nd/^{144}Nd$, $^{87}Sr/^{86}Sr$, and incompatible element variations in calc-alkaline andesites and plateau lavas from South America. *Earth planet. Sci. Lett.* **42**, 45–57.

HAWKINS, J. W. 1976. Petrology and geochemistry of basaltic rocks of the Lau Basin. *Earth planet. Sci. Lett.* **28**, 283–97.

—— 1977. Petrologic and geochemical characteristics of marginal basin basalts. *In*: TALWANI, M. & PITMAN, W. C. (eds) *Island Arcs, Deep Sea Trenches, and Back-Arc Basins*, 355–65.

Maurice Ewing Series I, American Geophysical Union, Washington, D.C.

HERRON, E. M. & TUCHOLKE, B. E. 1976. Sea-floor magnetic patterns and basement structure in the southwestern Pacific. *In*: HOLLISTER, C. D., CRADDOCK, C. *et al.* (eds) *Init. Rep. Deep Sea drill. Proj.* 35, 263–78. U.S. Government Printing Office, Washington, D.C.

HOLE, M. J., SAUNDERS, A. D., MARRINER, G. F. & TARNEY, J. In press. Subduction of pelagic sediments: implications for the origin of Ce-anomalous basalts from the Mariana Islands. *J. geol. Soc. London*.

HUSSONG, D. M. & UYEDA, S. 1981. Tectonic processes and the history of the Mariana Arc: a synthesis of the results of Deep Sea Drilling Project Leg 60. *In*: HUSSONG, D. M., UYEDA, S., *et al.* (eds) *Init. Rep. Deep Sea drill. Proj.* 60, 909–29. U.S. Government Printing Office, Washington, D.C.

——, —— *et al.* (eds) 1981. *Init. Rep. Deep Sea drill. Proj.* 60. U.S. Government Printing Office, Washington, D.C. 929 pp.

IRVINE, T. N. & BARAGER, W. R. A. 1971. A guide to the chemical classification of the common igneous rocks. *Can. J. Earth Sci.* 8, 523–48.

JORON, J.-L., BOLLINGER, C., QUISEFIT, J. P., BOUGAULT, H. & TREUIL, M. 1979. Trace elements in Cretaceous basalts at 25°N in the Atlantic Ocean: alteration, mantle compositions, and magmatic processes. *In*: DONNELLY, T., FRANCHETEAU, J. *et al.* (eds) *Init. Rep. Deep Sea drill. Proj.* 51–53, 1087–99. U.S. Government Printing Office, Washington, D.C.

KARIG, D. E. 1970. Ridges and basins of the Tonga–Kermadec island arc system. *J. geophys. Res.* 75, 239–54.

—— 1971. Origin and development of marginal basins in the western Pacific. *J. geophys. Res.* 76, 2542–61.

—— 1974. Evolution of arc systems in the western Pacific. *In*: DONATH, F. A. (ed.) *Ann. Rev. Earth planet. Sci.* 2, 51–75.

—— & JENSKY, W. 1972. The proto-Gulf of California. *Earth planet. Sci. Lett.* 17, 169–74.

—— & KAY, R. W. 1981. Fate of sediments on the descending plate at convergent margins. *Phil. Trans. R. Soc. London*, A301, 233–51.

KAY, R. W. 1980. Volcanic arc magmas: implications of a melting-mixing model for element recycling in the crust–upper mantle system. *J. Geol.* 88, 497–522.

——, HUBBARD, N. J. & GAST, P. W. 1970. Chemical characteristics and origin of oceanic ridge volcanic rocks. *J. geophys. Res.* 75, 1585–1613.

KROENKE, L., SCOTT, R. *et al.* (eds) 1980. *Init. Rep. Deep Sea drill. Proj.* 59. U.S. Government Printing Office, Washington, D.C. 820 pp.

KUNO, H. 1966. Lateral variation of basalt magma across continental margins and island arcs. *Pap. geol. Surv. Can.* 66–15, 317–36.

LARSON, R. L. 1972. Bathymetry, magnetic anomalies, and plate tectonic history at the mouth of the Gulf of California. Bull. geol. Soc. Am. 83, 3345–60.

LAWVER, L. A. & HAWKINS, J. W. 1978. Diffuse magnetic anomalies in marginal basins: their possible tectonic and petrologic significance. *Tectonophysics*, 45, 323–39.

——, —— & SCLATER, J. G. 1976. Magnetic anomalies and crustal dilation in the Lau Basin. *Earth planet. Sci. Lett.* 33, 27–35.

LOPEZ, M.-M., PEREZ, R. J., URRUTIA, F. J., PAL, S. & TERRELL, D. J. 1978. Geochemistry and petrology of some volcanic rocks dredged from the Gulf of California. *Geochem. J.* 12, 127–32.

LUDDEN, J. N. & THOMPSON, G. 1979. An evaluation of the behaviour of the rare earth elements during the weathering of sea-floor basalt. *Earth planet. Sci. Lett.* 43, 85–92.

MARSH, N. G., SAUNDERS, A. D., TARNEY, J. & DICK, H. J. B. 1980. Geochemistry of basalts from the Shikoku and Daito Basins, Deep Sea Drilling Project Leg 58. *In*: DEVRIES KLEIN, G., KOBAYASHI, K. *et al.* (eds) *Init. Rep. Deep Sea drill. Proj.* 58, 805–42. U.S. Government Printing Office, Washington, D.C.

MATTEY, D. P., MARSH, N. G. & TARNEY, J. 1980. The geochemistry, mineralogy and petrology of basalts from the West Philippine and Parece Vela Basins and from the Palau-Kyushu and West Mariana Ridges, Deep Sea Drilling Project Leg 59. *In*: KROENKE, L., SCOTT, R. *et al.* (eds) *Init. Rep. Deep Sea drill. Proj.* 59, 753–800. U.S. Government Printing Office, Washington, D.C.

MEIJER, A. 1976. Pb and Sr isotopic data bearing on the origin of volcanic rocks from the Mariana island-arc system. *Bull. geol. Soc. Am.* 87, 1358–69.

MITCHELL, W. S. & AUMENTO, F. 1977. Uranium in oceanic rocks: Deep Sea Drilling Project Leg 37. *Can. J. Earth Sci.* 14, 794–808.

MOLNAR, P. & ATWATER, T. 1978. Interarc spreading and cordilleran tectonics as alternates related to the age of subducted oceanic spreading. *Earth planet. Sci. Lett.* 41, 330–40.

MOORE, D. G. & BUFFINGTON, E. C. 1968. Transform faulting and growth of the Gulf of California since the late Pliocene. *Science*, 161, 1238–41.

MROZOWSKI, C. L. & HAYES, D. E. 1979. The evolution of the Parace-Vela basin, eastern Philippine Sea. *Earth planet. Sci. Lett.* 46, 49–67.

MUENOW, D. W., LIU, N. W. K., GARCIA, M. O. & SAUNDERS, A. D. 1980. Volatiles in submarine volcanic rocks from the spreading axis of the East Scotia Sea back-arc basin. *Earth planet. Sci. Lett.* 47, 272–8.

O'NIONS, R. K., HAMILTON, P. J. & EVENSON, N. M. 1977. Variations in $^{143}Nd/^{144}Nd$ and $^{87}Sr/^{86}Sr$ ratios in oceanic basalts. *Earth planet. Sci. Lett.* 34, 13–22.

PACKHAM, G. H. & FALVEY, D. A. 1971. An hypothesis for the formation of marginal seas in the western Pacific. *Tectonophysics*, 11, 79–109.

PEARCE, J. A. 1975. Basalt geochemistry used to investigate past tectonic environments on Cyprus. *Tectonophysics*, 25, 41–67.

—— & CANN, J. R. 1973. Tectonic setting of basic

volcanic rocks using trace element analysis. *Earth planet. Sci. Lett.* **19**, 290–300.

——, ALABASTER, T., SHELTON, A. W. & SEARLE, M. P. 1981. The Oman Ophiolite as a Cretaceous arc-basin complex: evidence and implications. *Phil. Trans. R. Soc. London*, **A300**, 299–317.

ROACH, P. J. 1978. Geophysical investigation into the evolution of Bransfield Strait. Unpublished Ph.D. thesis, University of Birmingham.

SAUNDERS, A. D. 1982. Petrology and geochemistry of alkali basalts from Jason Peninsula, Oscar II Coast. *Bull. Br. Antarct. Surv.* **55**, 1–9.

—— 1983. The Gulf of California: geochemistry of basalts recovered during Leg 65 of the Deep Sea Drilling Project. *In*: ROBINSON, P., LEWIS, B. T. R. *et al.* (eds) *Init. Rep. Deep Sea drill. Proj.* **65**, 591–622. U.S. Government Printing Office, Washington, D.C.

——, FORNARI, D. J. & MORRISON, M. A. 1982a The composition and emplacement of basaltic magmas produced during the development of continental-margin basins: the Gulf of California, Mexico. *J. geol. Soc. London*, **139**, 335–46.

——, ——, JORON, J.-L., TARNEY, J. & TREUIL, M. 1982b. Geochemistry of basic igneous rocks, Gulf of California, Deep Sea Drilling Project Leg 64. *In*: CURRAY, J. R., MOORE, D. G. *et al.* (eds) *Init. Rep. Deep Sea drill. Proj.* **64**, 595–642. U.S. Government Printing Office, Washington, D.C.

—— & TARNEY, J. 1979. The geochemistry of basalts from a back-arc spreading centre in the East Scotia Sea. *Geochim. cosmochim. Acta*, **43**, 555–72.

——, —— & WEAVER, S. D. 1980a. Transverse geochemical variations across the Antarctic Peninsula: implications for the genesis of calc-alkaline magmas. *Earth planet. Sci. Lett.* **46**, 344–60.

——, —— MARSH, N. G. & WOOD, D. A. 1980b. Ophiolites as ocean crust or marginal basin crust: a geochemical approach. *In*: PANAYIOTOU, A. (ed.) *Proc. int. Ophiolite Conf.* 193–204. Nicosia, Cyprus.

——, ——, STERN, C. R. & DALZIEL, I. W. D. 1979. Geochemistry of Mesozoic marginal basin floor igneous rocks from southern Chile. *Bull. geol. Soc. Am.* **90**, 237–58.

——, WEAVER, S. D. & TARNEY, J. 1982c. The pattern of Antarctic Peninsula plutonism. *In*: CRADDOCK, C. (ed.) *Antarctic Geoscience*, 305–14. University of Wisconsin Press, Madison.

STERN, C. R. 1979. Open and closed system igneous fractionation within two Chilean ophiolites and the tectonic implication. *Contrib. Mineral. Petrol.* **68**, 243–58.

—— & ELTHON, D. 1979. Vertical variations in the effects of hydrothermal metamorphism in Chilean ophiolites: their implications for ocean floor metamorphism. *Tectonophysics*, **55**, 179–213.

SUN, S.-S. 1980. Lead isotopic study of young volcanic rocks from mid-ocean ridges, ocean islands and island arcs. *Phil. Trans. R. Soc. London*, **A297**, 409–45.

——, NESBITT, R. W. & SHARASKIN, A. YA. 1979. Geochemical characteristics of mid-ocean ridge basalts. *Earth planet. Sci. Lett.* **44**, 119–38.

TARNEY, J., SAUNDERS, A. D., MATTEY, D. P., WOOD, D. A. & MARSH, N. G. 1981. Geochemical aspects of back-arc spreading in the Scotia Sea and western Pacific. *Phil. Trans. R. Soc. London*, **A300**, 263–85.

——, WOOD, D. A., SAUNDERS, A. D., CANN, J. R. & VARET, J. 1980. Nature of mantle heterogeneity in the North Atlantic; evidence from Deep Sea Drilling. *Phil. Trans. R. Soc. London*, **A297**, 179–202.

TAYLOR, S. R. 1964. Abundance of chemical elements in the continental crust: a new table. *Geochim. cosmochim. Acta*, **28**, 1273–85.

TERRELL, D. J., PAL, S., LOPEZ, M. M. & PEREZ, R. J. 1979. Rare-earth elements in basalt samples, Gulf of California. *Chem. Geol.* **26**, 267–75.

TOKSÖZ, M. N. & BIRD, P. 1977. Formation and evolution of marginal basins and continental plateaus. *In*: TALWANI, M. & PITMAN, W. C. (eds) *Island Arcs, Deep Sea Trenches, and Back-Arc Basins*, 379–93. Maurice Ewing Series I, American Geophysical Union, Washington, D.C.

UYEDA, S. & MIYASHIRO, A. 1974. Plate tectonics and the Japanese Islands: a synthesis. *Bull. geol. Soc. Am.* **85**, 1159–70.

VERMA, S. 1982. Magnetic properties and incompatible element geochemistry of some igneous rocks from Deep Sea Drilling Project Leg 64. *In*: CURRAY, J. R., MOORE, D. G. *et al.* (eds) *Init. Rep. Deep Sea drill. Proj.* **64**, 667–73. U.S. Government Printing Office, Washington, D.C.

WATANABE, T., LANGSETH, M. G. & ANDERSON, R. N. 1977. Heat flow in back-arc basins of the western Pacific. *In*: TALWANI, M. & PITMAN, W. C. (eds) *Island Arcs, Deep Sea Trenches, and Back-Arc Basins*, 137–63. Maurice Ewing Series I, American Geophysical Union, Washington, D.C.

WATKINS, J. S., MCMILLEN, K. J., BACHMAN, S. B., SHIPLEY, T. H., MOORE, J. C. & ANGEVINE, C. 1981. Tectonic synthesis, Leg 66: transect and vicinity. *In*: WATKINS, J. S., MOORE, J. C. *et al.* (eds) *Init. Rep. Deep Sea drill. Proj.* **66**, 837–49. U.S. Government Printing Office, Washington, D.C.

WATTS, A. B. & WEISSEL, J. K. 1975. Tectonic history of the Shikoku marginal basin. *Earth planet. Sci. Lett.* **25**, 239–50.

WEAVER, B. L. & TARNEY, J. 1979. Thermal aspects of komatiite generation and greenstone belt models. *Nature, Lond.* **279**, 689–92.

WEAVER, S. D., SAUNDERS, A. D., PANKHURST, R. J. & TARNEY, J. 1979. A geochemical study of magmatism associated with the initial stages of back-arc spreading: the Quaternary volcanics of Bransfield Strait, from South Shetland Islands. *Contrib. Mineral. Petrol.* **68**, 151–69.

WEISSEL, J. K. 1977. Evolution of the Lau Basin by the growth of small plates. *In*: TALWANI, M. &

PITMAN, W. C. (eds) *Island Arcs, Deep Sea Trenches, and Back-Arc Basins*, 429–36. Maurice Ewing Series I, American Geophysical Union, Washington, D.C.
—— 1981. Magnetic lineations in marginal basins of the western Pacific. *Phil. Trans. R. Soc. London*, **A300**, 233–47.
WHITE, W. M. & HOFMANN, A. W. 1982. Sr and Nd isotope geochemistry of oceanic basalts and mantle evolution. *Nature, Lond.* **296**, 821–5.
WOOD, D. A., MARSH, N. G., TARNEY, J., JORON, J.-L., FRYER, P. & TREUIL, M. 1981. Geochemistry of igneous rocks recovered from a transect across the Mariana Trough, Arc, Fore-arc, and Trench, Sites 453 through 461, Deep Sea Drilling Project Leg 60. *In*: HUSSONG, D. M., UYEDA, S. *et al.* (eds) *Init. Rep. Deep Sea drill. Proj.* **60**, 611–45. U.S. Government Printing Office, Washington, D.C.
——, MATTEY, D. P., JORON, J.-L., MARSH, N. G., TARNEY, J. & TREUIL, M. 1980. A geochemical study of 17 selected samples from basement cores recovered at Sites 447, 448, 449, 450 and 451, Deep Sea Drilling Project Leg 59. *In*: KROENKE, L., SCOTT, R. *et al.* (eds) *Init. Rep. Deep Sea drill. Proj.* **59**, 743–52. U.S. Government Printing Office, Washington, D.C.

ANDREW D. SAUNDERS, Department of Geology, Bedford College, Regent's Park, London NW1 4NS, U.K. Present address: Department of Geology, Bennett Building, University of Leicester, Leicester LE1 7RH, U.K.
JOHN TARNEY, Department of Geology, Bennett Building, University of Leicester, Leicester LE1 7RH, U.K.

Characteristics and tectonic significance of supra-subduction zone ophiolites

J. A. Pearce, S. J. Lippard & S. Roberts

SUMMARY: Supra-subduction zone (SSZ) ophiolites have the geochemical characteristics of island arcs but the structure of oceanic crust and are thought to have formed by sea-floor spreading directly above subducted oceanic lithosphere. They differ from 'MORB' ophiolites not only in their geochemistry but also in the more depleted nature of their mantle sequences, the more common presence of podiform chromite deposits, and the crystallization of clinopyroxene before plagioclase which is reflected in the high abundance of wehrlite relative to troctolite in their cumulate sequences. Most of the best-preserved ophiolite complexes in orogenic belts are of this type.

Geological reconstructions suggest that most SSZ ophiolites formed during the initial stages of subduction prior to the development of any volcanic arc. Evidence from these ophiolites suggests that the first magma to form in response to intra-oceanic subduction is boninitic in composition, derived by partial melting of hydrated oceanic lithosphere in the 'mantle wedge'. As subduction proceeds, the magma composition changes to island-arc tholeiite, probably because the hydrated asthenosphere of the 'mantle wedge' eventually becomes the dominant mantle source. Other SSZ ophiolites formed in the early stages of back-arc spreading following splitting of a pre-existing arc. Nonetheless the more common mechanism for formation of SSZ ophiolites appears to have been *pre-arc* rather than back-arc spreading.

It is generally accepted that ophiolite complexes could have formed in a variety of tectonic settings: major ocean ridges, incipient ocean ridges, marginal basins, 'leaky' transform faults and (possibly) the roots of ocean islands, aseismic ridges and island arcs. In the few cases where a complex is autochthonous or parautochthonous, its original tectonic setting may be most easily deduced. Good examples are: Macquarie Island in the Tasman Sea (Griffiths & Varne 1972), which formed at an ocean ridge in a small ocean between Australia and an adjacent microcontinent, the Lord Howe Rise; the 'Rocas Verdes' ophiolites in the Andes (Dalziel *et al*. 1974), which formed in a back-arc basin behind an active continental margin; the Zambales Range of Luzon (Hawkins 1980), which may represent back-arc or fore-arc oceanic lithosphere thrust over its adjacent island arc; and the La Palma complex in the Canary Islands (Schmincke & Staudigel 1976), which could have formed during the initial stages of Atlantic sea-floor spreading. Usually, however, assigning an ophiolite complex to its likely setting of formation is more difficult, requiring information from a variety of sources, including its present geological setting (e.g. Dewey & Bird 1971), its internal structure (e.g. Moores & Jackson 1974), the nature and provenance of the overlying sediments (e.g. Robertson 1981), the geochemistry (e.g. Pearce & Cann 1973) and petrography (e.g. Cameron *et al*. 1979) of the igneous rocks; and the tectonics and timing of emplacement (e.g. Gealey 1977). The accumulated evidence from such studies has demonstrated that most of the Earth's well-preserved ophiolite complexes, and a significant proportion of its fragmentary complexes, formed in marginal basins (e.g. Upadhyay & Neale 1979). Clearly, therefore, these complexes can be a major source of information on the three-dimensional structure and composition of marginal basin lithosphere and on the magmatic and tectonic processes by which that lithosphere is formed. Before this information can be assessed, however, each complex has to be fitted more precisely into the context of marginal basin evolution. It is important to determine, for example, whether a given complex formed before or after the development of a volcanic arc and in the latter instance whether it formed in a fore-arc, inter-arc or back-arc setting. In this paper, we examine one important line of evidence for identifying the precise setting of marginal basin ophiolites: the presence or absence of a geochemical component from an underlying subduction zone. We term ophiolites containing this component 'supra-subduction zone' (SSZ) ophiolites and those not containing this component as 'mid-ocean ridge basalt' (MORB) ophiolites. Our aim is to present criteria by which SSZ ophiolites can be recognized and to assess the contribution that this ophiolite type can make towards our understanding of the development of marginal basins.

Trace-element characteristics of SSZ basalts

The nature of the subduction component that can be used to fingerprint SSZ ophiolites may be identified by comparing oceanic basalts of known non-SSZ and SSZ affinities. This comparison is made in Fig. 1 using MORB-normalized geochemical patterns for a selected range of trace elements.

Figure 1a shows the patterns that are obtained from basalts that are clearly unaffected by subduction. They range from flat patterns in N(normal)-type MORB to humped patterns in E(enriched)-type MORB and ocean islands. Since the shape of these patterns is little affected by differences in partial melting and fractional crystallization histories (Pearce 1983), the variations between the patterns are best explained in terms of heterogeneities in the

mantle source. The main incompatible element-depleted reservoir of convecting upper mantle is thus thought to yield the flat patterns of N-type MORB, whereas the humped patterns are thought to be derived from the incompatible element-enriched portions of this reservoir such as those associated with 'mantle plumes' (e.g. Schilling 1973; Tarney *et al.* 1980).

Figure 1a can be contrasted with Fig. 1b, which shows the patterns that are obtained from basalts that clearly are affected by subduction—i.e. basalts from island arcs. The patterns presented are typical examples from tholeiitic, high-K calc-alkaline and alkalic series and cover the spectrum of arc basalt compositions. The most distinctive feature common to all three types of pattern is the selective enrichment in certain elements (Sr, K, Rb, Ba, Th ± Ce ± Sm ± P) and a relative lack of enrichment in others (Ta, Nb, Hf, Zr, Ti, Y, Yb). This feature can be

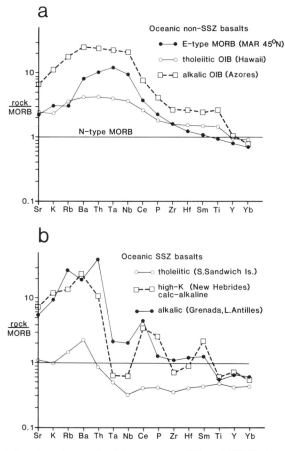

FIG. 1. Geochemical patterns (normalized to an average N-type MORB) for oceanic basalts from (a) non-SSZ, and (b) SSZ settings (from Pearce 1982).

attributed to the modification of the mantle source region for island-arc basalts by a 'subduction component', i.e. by aqueous and siliceous fluids derived from an underlying subduction zone (e.g. Best 1975; Hawkesworth *et al.* 1977; Saunders & Tarney 1979). Thus the shape of each pattern in Fig. 1b can be explained in terms of the two components in the mantle source: a 'mantle component' which represents the composition of the mantle prior to subduc-

tion and which would produce the flat-to-humped patterns of Fig. 1a; and the 'subduction component' which causes an additional selective enrichment in elements that are in high concentration in the fluids derived from subduction zone.

All three patterns in Fig. 1b can be broken down into their two components using the method described by Pearce (1983). The results are shown in Fig. 2. It is thus apparent

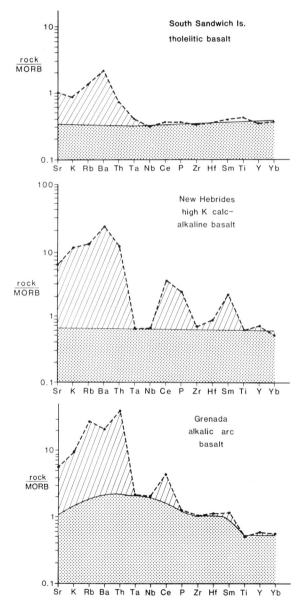

FIG. 2. Breakdown of the patterns for SSZ basalts into their mantle components (shaded) and subduction components (striped) (from Pearce 1983).

that the tholeiitic and calc-alkaline basalt patterns are made up of subduction components superimposed on a flat pattern indicating that their mantle components resembled an N-type MORB source; by contrast the alkalic basalt pattern is made up of a subduction component superimposed on a humped pattern indicating that its mantle component resembled an E-type MORB mantle source. The latter type of pattern is however very rare in present-day island arcs. It is also apparent from Fig. 2 that, compared with tholeiitic basalts, the subduction component in calc-alkaline basalts is both larger and more diverse chemically.

MORB-normalized patterns for marginal basin basalts have been discussed in detail by Saunders & Tarney (this volume). From their work, it is apparent that only some marginal basin basalts show this subduction component, the others usually having flat N-type MORB patterns. Clearly, only the former are potential analogues for SSZ ophiolites. Their precise setting will therefore depend on their temporal and spatial relationship with the underlying subduction zone, a point made by Saunders & Tarney and followed up later in this paper.

The selective enrichments characteristic of this subduction component also form the basis for a number of discrimination diagrams. It is apparent from the patterns in Fig. 1b that any ratio of an enriched to an unenriched element should provide an effective discriminant for the identification of SSZ character. However, most of the enriched elements are mobile during weathering and metamorphism and thus rarely applicable to the study of lavas from ophiolite complexes (Cann 1970). The only element in the pattern that is always enriched yet is usually stable during metamorphism up to at least greenschist facies is Th. A ratio such as Th/Ta is particularly useful, since these two elements behave in a coherent way in oceanic non-SSZ basalts and are only decoupled in the subduction environment. Published discrimination diagrams based on this ratio include: Th-Ta (Wood et al. 1979); Th-Hf/3-Ta (Wood et al. 1979; Wood 1980); Th/Yb-Ta/Yb (Pearce 1982); Hf/Th-Ta/Th (Noiret et al. 1981). Best known is the diagram of Wood (1980) which is reproduced as Fig. 3. It shows the field of volcanic-arc basalts (VAB) displaced towards the Th axis relative to the fields of mid-ocean ridge basalt (MORB) and within plate basalts (WPB). The compositional fields of basalts from some marginal basins have also been plotted on this diagram. Those plotting as 'MORB' are all derived from basins of back-arc type and

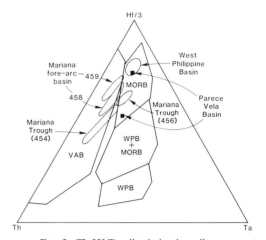

FIG. 3. Th-Hf-Ta discrimination diagram of Wood (1980) showing the separation of volcanic-arc basalts (VAB) from mid-ocean ridge basalts (MORB) and within plate basalts (WPB). Compositional fields for marginal basin basalts from DSDP cores (core numbers in parenthesis) are taken from Wood et al. (1981, 1982).

are assumed to have been derived from mantle diapirs containing (on this diagram) no detectable subduction component. Those plotting as 'VAB' include one basalt suite from the Mariana Trough (a back-arc basin) and both suites from the Mariana fore-arc basin.

The patterns in Fig. 1b are also characterized by low abundances of the elements not enriched by the subduction or (in the case of the alkalic basalt) mantle components. Thus the elements Ti, Y and Yb in all patterns, and Ta, Nb, Zr and Hf in all except the alkalic pattern, plot significantly below the values of the N-type MORB normalizing factor. This feature has been attributed to the hydrous conditions of melting which may cause more refractory mantle to melt (e.g. Green 1973), increase the degree of partial melting (e.g. Pearce & Norry 1979) or stabilize minor oxide phases in the melt residue (e.g. Dixon & Batiza 1979).

It also forms the basis for a set of discriminant diagrams that can be used to study the affinities of ophiolite lavas. Alteration is of minor importance since all the unenriched elements are usually immobile during weathering and metamorphism. Unlike the Th/Ta ratio, however, the abundances of these elements will vary with fractional crystallization, causing overlap between the magma types. The best discrimination is therefore achieved by plotting one of these elements against an independent index of fractional crystallization such as Cr or

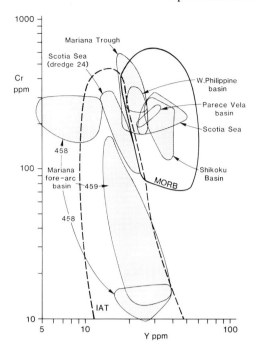

FIG. 4. Cr-Y discrimination diagram of Pearce (1982) showing the separation of mid-ocean ridge basalts (MORB) from island-arc tholeiites (IAT) (the latter field also includes calc-alkaline and alkalic basalts from oceanic arcs). The compositional fields for the marginal basin basalts are taken from Wood *et al.* (1981); Mattey *et al.* (1981); Marsh *et al.* (1980); Saunders & Tarney (1979); Wood *et al.* (1982); Sharaskin (1982); and Bougault *et al.* (1982).

Ni. Published diagrams of this type include Ti-Cr (Pearce 1975), Cr-Y (Pearce 1980), Ni-Y (Capedri *et al.* 1980), Cr/Ti-Ni (Beccaluva *et al.* 1979) and Ti-V (Shervais 1982). Figure 4 shows the Cr-Y diagram, in which island-arc tholeiites (IAT) are displaced towards lower Y values relative to MORB. Compositional fields for some marginal basins have also been plotted and, as in Fig. 3, fall into two groups. Those from most back-arc basins fall entirely within the MORB field, whereas those from one back-arc basin locality (dredge haul 24 from the Scotia Sea), and both of the fore-arc basin localities, plot within or to the left of the IAT field and therefore represent the best analogues for SSZ ophiolites. A further point of note is that the samples from site 458 of the Mariana fore-arc basin that plot to the left of the IAT field are those with boninitic, as opposed to tholeiitic, affinities.

Trace-element characteristics of SSZ ophiolites

Immobile trace-element data have been used by many authors to identify the tectonic environments of formation of ophiolite complexes and it is from their results that the concept of SSZ ophiolites has developed. These results are summarized below using the two discrimination diagrams of Figs 3 and 4 to compare the compositions of the various complexes and to demonstrate their subdivision into MORB and SSZ ophiolites.

Tethyan ophiolites

Data from Tethyan ophiolite complexes has been plotted in Figs 5 and 6a,b. Thus it is apparent that the Jurassic complexes of the W Mediterranean, from Calabria (JAP, unpublished data), Corsica and the N Apennines (Ferrara *et al.* 1976; Beccaluva *et al.* 1977; Venturelli *et al.* 1979, 1981), the French Alps (Lewis & Smewing 1980) and the Austrian Alps (Bickle & Pearce 1975), plot almost entirely within the MORB field of both diagrams, a result consistent with their inferred origin in small oceans of Red Sea type. The Jurassic complexes of Greece are more varied geochemically. The Othris ophiolite (Bickle & Nisbet 1972) plots within the MORB field. However, the Vourinos complex (Noiret *et al.* 1981; Beccaluva *et al.* 1984) plots within the volcanic-arc field in Fig. 5 and within and to the left of the IAT field in Fig. 6a. It is thus assumed to be of SSZ type and to have boninitic affinities. The Pindos complex (Capedri *et al.* 1980), plotted only in Fig. 6a, contains lavas of both boninitic island-arc and MORB affinities, indicating that more than one mantle source and possibly more than one eruptive setting may be represented in the samples analysed.

Further east, the Cretaceous ophiolite complexes of Cyprus (the Troodos Massif) and Oman (the Semail Nappe) both plot as SSZ ophiolites. The Troodos lavas (Desmons *et al.* 1980; JAP, unpublished data) plot clearly within the volcanic-arc fields in both Figs 5 and 6b. The lavas from the Semail thrust sheet (Alabaster *et al.* 1982) plot within the IAT field in Fig. 6b and generally within the volcanic-arc field in Fig. 5. However, it is also apparent in Fig. 5 that the lower (axis) lavas from this complex plot closer to, and partly overlap, the MORB field, whereas the upper lava units plot entirely within the volcanic-arc field. It can thus be inferred that the magnitude of the subduction zone component has increased with time during the formation of this complex. The Troodos Massif and the Semail Nappe mark the W and E extremities of a well-defined ophiolite belt termed the 'Croissant Ophiolitique Peri-Arabe' by Ricou (1971) that extends from Cyprus through Turkey and Iran to Oman. Other complexes of similar age within this belt include Baer Bassit in Syria (Parrot 1977), Hatay in Turkey (Delaloye & Wagner, in press) and the Zagros ophiolites of Iran (Desmons & Beccaluva 1983), all of which can also be chemically assigned to the SSZ group on the basis of the limited geochemical data so far available.

Other ophiolites in this eastern region include the Cretaceous complexes of Masirah Island (Abbotts 1981), the E Iranian ranges of Baluchistan (Desmons & Beccaluva 1983) and the Zangbo suture of Tibet (JAP, unpublished data), all of which have MORB affinities, as illustrated in Fig. 6b.

Caledonian ophiolites

Ophiolites are found throughout the Caledonides of the N Atlantic region in Norway, Scotland and Canada. In the Norwegian Caledonides, where detailed studies of ophiolites have so far only been carried out on the W coast around Bergen and in the Trondheim region, two groups have been recognized (Furnes *et al.* 1980): the first is of pre-Arenig age and includes complexes with well-developed stratigraphies such as Karmøy (Sturt *et al.* 1980) and Leka (Prestvik 1980); the second group is younger (?early to mid Ordovician), has incom-

plete stratigraphies and is associated with island-arc volcanic sequences and thick volcaniclastic sediments. On Fig. 6c, ophiolites of the first group all plot clearly within the IAT field and are therefore of SSZ type. The second group, however, plots entirely within the MORB field on this diagram and is therefore of MORB type.

Figure 6d includes data from the Ordovician Betts Cove and Bay of Islands ophiolites from Newfoundland. Lavas and dykes from the Betts Cove ophiolite (Coish & Church 1979) plot predominantly within and to the left of the IAT field indicating that the complex is of SSZ type and that it has boninitic affinities. However, the upper lava unit from this complex has MORB affinities indicating temporal changes in mantle source and possibly in eruptive setting during the development of this complex. The Betts Cove ophiolite belongs to a lineament which extends from Newfoundland to Quebec and includes the ophiolites of Mings Bight and Baie Verte in Newfoundland and Thetford Mines in Quebec, all of which have been shown on Ti-Cr and other diagrams to be of SSZ origin (Laurant 1975; Upadhyay & Neale 1979). By contrast the Bay of Islands complex of western Newfoundland (Suen *et al.* 1979), which is off-set from this lineament, plots predominantly in the MORB field of Fig. 6d.

Other ophiolites

Data from two W U.S.A. ophiolites are plotted in Fig. 6e. The Point Sal complex (Menzies *et al.* 1977) in the coast range of S California

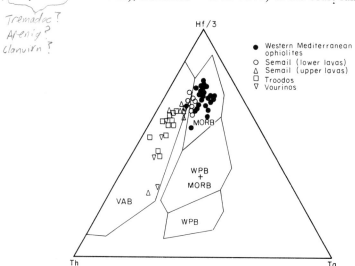

FIG. 5. Distribution of basalt analyses from some Tethyan ophiolites on the Th-Hf-Ta discrimination diagram (see text for sources of data).

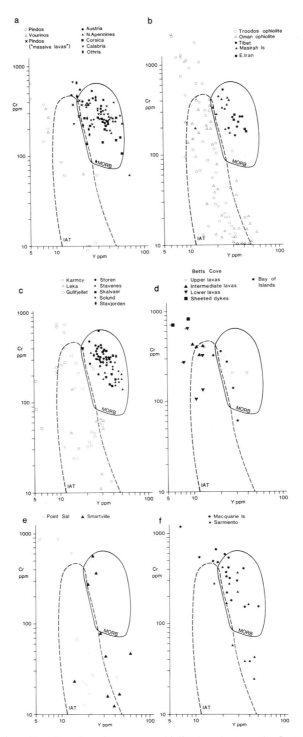

FIG. 6. Distribution of basalt analyses from some ophiolite complexes on the Cr-Y diagram (see text for sources of data).

plots within the IAT field thus classifying as a SSZ ophiolite. The Smartville complex (Menzies *et al.* 1980) in the W foothills of the Sierra Nevada plots in and below the MORB field, and thus does not classify unequivocally as either a SSZ or MORB ophiolite. Given the strong geological evidence for a back-arc setting of formation (Xenophontos & Bond 1978) it is possible that this complex is genuinely transitional between the two types, containing only a small subduction component.

Finally, Fig. 6f contains data from two ophiolites, Macquarie Island (Griffin & Varne 1980) and the Sarmiento complex of S Chile (Stern 1979), which are known to have formed in a small ocean and a back-arc basin respectively. Data from Macquarie Island plot predictably in the MORB field whereas data from the Sarmiento complex plot within and below the MORB field. Like the Smartville complex it is probable that the Sarmiento complex contains only a small subduction component. Other ophiolites of 'known affinities' include the Zambales and Palawan complexes of the Philippines (Hawkins 1980) and the various complexes of Papua-New Guinea (Davies & Smith 1971), all of which are thought to represent the crust of fore-arc basins that was detached as a result of continent–island-arc collisions (Gealey 1980). The limited geochemical data available for these complexes are of boninite and island-arc tholeiite compositions and hence indicate SSZ affinities.

The main conclusion that can be reached on the basis of these geochemical studies is that there is a bimodality of compositions, one group having clear MORB affinities, the other (the SSZ group) having clear island-arc affinities. The ophiolites that did not completely fit this pattern were Pindos and Betts Cove which contained lavas of both affinities and Smartville and Sarmiento which may be of transitional character.

Other characteristics of SSZ ophiolites

Having divided ophiolites into two groups on the basis of their chemical characteristics, it is now possible to compare the two groups of ophiolites on the basis of other criteria. The differences observed are essentially those recognized by Rocci *et al.* (1975) for Tethyan ophiolites, and are summarized in Table 1.

Cumulate sequences

As summarized in the second column of Table 1, SSZ ophiolites and MORB ophiolites generally exhibit distinct crystallization sequences. In SSZ ophiolites, clinopyroxene, and sometimes orthopyroxene, usually crystallizes before plagioclase, whereas the reverse is true of MORB ophiolites. Thus, in SSZ ophiolites, basal cumulate dunites are generally followed up-sequence by lherzolites, wehrlites, norites and gabbros. In MORB ophiolites, basal dunites are followed up-sequence by troctolites and gabbros. This contrast in the order of crystallization corresponds to that between boninites (ol → opx → cpx), island-arc tholeiites (ol → cpx → pl) and MORB (ol → pl → cpx) from present oceanic settings. A likely explanation is that primary magmas generated above subduction zones contain higher CaO/Al_2O_3 ratios than primary magmas generated outside the influence of subducted oceanic crust, a model which is strongly supported by the marked paucity of Ca-bearing clinopyroxene in the mantle residue of SSZ ophiolites.

A further characteristic of SSZ ophiolites, summarized in the fourth column of Table 1, is the common occurrence of cumulate chromite-dunite bodies (podiform chromites) within their mantle sequences. This discriminant is not perfect, since small podiform chromite bodies do exist within MORB ophiolites such as Othris, but it is nonetheless true that, on the basis of information so far available, all major podiform chromite deposits are found in SSZ ophiolites. The reason for this relationship between podiform chromite deposits and SSZ ophiolites is not fully understood, although it is thought that the presence of subduction-derived water in the melt could expand the olivine and spinel phase volumes and so lead to extensive crystallization of olivine and chromite.

Mantle sequences

The key difference between the ultramafic tectonites ('mantle sequences') of SSZ and MORB ophiolites is summarized in the third column of Table 1.

The predominant lithology present in SSZ ophiolites is harzburgite, which usually comprises between 80 and 90% of the total outcrop, the remainder including irregular lenticular masses of dunite and pockets of lherzolite and pyroxenite. Olivine (Fo_{90-92}) generally constitutes about 70% of the mode and orthopyroxene (En_{90-91}) reaches a maximum of 30%. Other minerals are clinopyroxene, which occurs as isolated grains of chrome diopside and averages less than 1% of the mode in most complexes, and chrome spinel, which generally has a modal abundance of about 2%. The dunite

TABLE 1. *Characteristics of some of the world's major ophiolite complexes (see text for references)*

Complex	1 Chemical signature Lower	Upper	2 Crystallization sequence	3 Mantle residue	4 Chromite pods?	5 Lava thickness (km)	6 Total crustal thickness (km)	7 Conformable cover	8 Age of formation	9 Age of emplacement	10 Inferred ophiolite type
Betts Cove	IAT	MORB	A C	—	—	1.0	>5	VS	OL	—	SSZ
Thetford Mines	IAT		(A)B	—	C	0.6	2–3	VS	OL	—	SSZ
Baer Bassit	IAT	IAT	A B	H	C	0.2	2–3	P	KM	KM	SSZ
Troodos	IAT	IAT	(A)BC	H	C	1.0	3–4	P	KM	KM	SSZ
Hatay	IAT	IAT	BC	H	C	0.5	3–4	P	KM	KM	SSZ
Vourinos		IAT	BC	H	—	0.5	3–4	P	JL	JL-M	SSZ
Papua-New Guinea	IAT	IAT	(A)BC(D)	H	C	—	>4.5	VS?	KU-TL	TL	SSZ
Semail	IAT^M	IAT	C	H	—	1.5	6–7	P	KM	KM	SSZ
Karmøy		IAT	C	—	—	1.1	5–6	S	OL?	—	SSZ
Point Sal	IAT	IAT	C	H	C	1.0	2.5	PS	KM	—	SSZ
Zambales		IAT		H	—	—	—	P	TL?	TU	SSZ
Sarmiento	MORB^I	CAB*	—	—	—	2.0	>4	VS	J	—	MORB/SSZ
Smartville	MORB^I	CAB*	D	—	—	1.5	>4	VS	J	J	MORB/SSZ
Bay of Islands	MORB	IAT	D	HL	(C)	0.3	3–4	S	OL	OM	MORB
Pindos	MORB	IAT	D	HL	(C)	0.4	2–3	P	JL	JL-M	MORB
Liguria	MORB		D	L	no	0.5	3	P	J	TL	MORB
Inzecca (Corsica)	MORB		D	L	no	1.0	2–3	P	J	TL	MORB
Xigaze (Tibet)	MORB		D	L	no	—	>3	P	KU	—	MORB
Othris	MORB		D	L	(C)	0.6	≥1	P	JL	—	MORB
Macquarie Is.	MORB		D	H?	—	1.5	5	P	TU	—	MORB

1. IAT, island-arc tholeiite; MORB, mid-ocean ridge basalt; CAB, calc-alkali basalt; * not part of ophiolite sequence; ^{M} slight MORB affinities; ^{I} slight IAT affinities.

2. A, ol → opx → cpx → plag; B, ol → cpx → opx → plag; C, ol → cpx → plag; D, ol → plag → cpx.

3. H, predominantly harzburgite; HL, harzburgite + lherzolite; L, predominantly lherzolite.

4. C, significant component; (C), minor component.

7. P, pelagic sediment; S, terrigenous sediment; VS, volcaniclastic sediment ± lava.

8. and 9. O, Ordovician; J, Jurassic; K, Cretaceous; T, Tertiary; L, lower; M, middle; U, upper.

10. SSZ, supra-subduction zone; MORB, mid-ocean ridge basalt.

—, no data available.

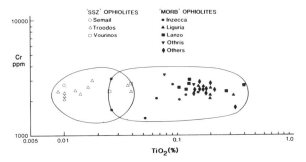

FIG. 7. Cr-TiO$_2$ plot for tectonized ultramafic rocks from some Tethyan ophiolites. The 'SSZ' ophiolite mantle residue is generally more residual (carries less TiO$_2$) than that of the 'MORB' ophiolites. Cr is used as a vertical axis for comparison with Fig. 8.

masses consist almost entirely of olivine together with 2–3% of small subhedral chromite grains.

The mantle sequences in MORB ophiolites contain both harzburgites and lherzolites. The lherzolites are typically four- or five-phase assemblages (ol + opx + cpx + sp ≪ plag) and these can grade into harzburgite when the modal proportion of clinopyroxene drops below 5%. In the mantle sequence of the Bay of Islands ophiolite, for example, the composition grades from a harzburgitic top to a lherzolitic base (Suen *et al.* 1979). The proportion of lherzolite relative to harzburgite varies from the Bay of Islands (harzburgite > lherzolite) to the W Mediterranean complexes (lherzolite > harzburgite). Dunite pods are found in MORB ophiolites, but gabbroic segregations are more common (Spray 1982).

Clearly this simple discriminant is not perfect since harzburgite can occur in both ophiolite types. Nonetheless the lherzolite does appear to be restricted to MORB ophiolites; moreover, the harzburgite of MORB ophiolites is generally less depleted than that of SSZ ophiolites, containing on average more Al, Ca and Ti (Fig. 7). The mantle sequence of SSZ ophiolites may therefore be considered as more 'residual', having been derived by higher degrees of melting of a similar source, or by similar degrees of melting of a less fertile source, than the mantle sequences of MORB ophiolites.

Ophiolite structure

Three aspects of ophiolite structure have been investigated as possible indicators of SSZ character (columns 5 and 6 of Table 1): the thickness of the crustal sequence; the thickness of the lava sequence; and the proportion and geological setting of rocks of intermediate and acid composition. None of these correlate well with ophiolite type, though all may permit important qualifications to be made regarding tectonic setting of formation.

Crustal thickness varies from 2 to 3 km to at least 7 km. Even though the former may be underestimates due to the difficulty of reconstructing fragmented sequences, it is apparent that complexes such as Pindos, Thetford, Inzecca, Troodos, Hatay and Vourinos are considerably thinner than average *in situ* ocean crust and also that this group includes both SSZ and MORB complexes. One contributory factor may be the very short time interval between the formation and emplacement of these complexes; another may be the fact that some of these complexes may have formed at incipient spreading centres or 'leaky' transform faults rather than at 'normal' ridge axes. Lava thickness shows a fair correlation with crustal thickness, bearing in mind that both sets of figures are approximate values.

A better discriminant may be the proportion of intermediate and acid rocks within the lava sequence and in high-level plutons. MORB ophiolites thought to have formed in major (as opposed to marginal) ocean basins contain only a tiny proportion of diorites and plagiogranites. By contrast, MORB ophiolites of apparent back-arc origin, such as Sarmiento and Smartville, are intruded by large bodies of diorite and granodiorite and are also directly overlain by lava sequences containing a significant proportion of andesites and rhyolites. Of the SSZ ophiolites, Troodos, Hatay and Vourinos do contain a greater proportion of intermediate and acid lavas than MORB ophiolites thought to have formed in major ocean basins (e.g. Miyashiro 1973; Beccaluva *et al.* 1984), although basic rocks still predominate. However, the Semail complex of Oman is intruded by diorite-plagiogranite bodies and also contains a significant proportion of andesites and rhyolites in its upper lava sequences.

Sedimentary sequences

It is commonly believed that all marginal basin ophiolites should be conformably overlain by sediment that contains a volcaniclastic component derived from an adjacent volcanic arc (Karig 1982).

As Table 1 shows, however, this need not be the case. Many of the Tethyan SSZ ophiolites—e.g. the Troodos Massif, Hatay, and the Semail Nappe—are overlain by ferro-manganoan sediments and chalks which contain no significant volcanogenic component. This result is not surprising as none of these complexes are related in space or time to subaerial arc volcanoes. The simplest explanation is that the spreading events that formed these complexes took place immediately after subduction and that the subduction event was too short-lived for any subaerial arc to develop and hence for any source of volcaniclastic sediments to be available. Ophiolites that were associated with longer-lived subduction events are, however, overlain by sediments or volcano-sedimentary sequences of arc derivation. For example, the Smartville ophiolite is overlain by an upper unit comprising basalt to andesite lava flows, breccias and tuffs, which is in turn overlain by volcaniclastic turbidites containing angular clasts of acid lava and quartz (Xenophontos & Bond 1978).

The sedimentary cover thus indicates the likely presence of an adjacent subaerial arc, but this is not a necessary criterion for the recognition of a SSZ ophiolite.

Evidence from SSZ ophiolites for the magmatic and tectonic evolution of marginal basins

Even though SSZ ophiolites form a distinct geochemical group, they exhibit a considerable diversity of lava stratigraphies, internal structures, geological settings and detailed geochemical and petrological characteristics. This diversity may reflect variations *between* marginal basins, due perhaps to differences in the various subduction vectors, or it may reflect variations *within* marginal basins, due perhaps to the preservation of different parts of marginal basin systems. Here we examine the second possibility by attempting to identify examples of (i) the initial stages of marginal basin spreading,

(ii) the transition from spreading to arc volcanism, and (iii) the transition from arc volcanism to back-arc spreading.

The initial stages of spreading

One ophiolite likely to represent the earliest stages of marginal basin evolution is the Troodos Massif of Cyprus. Despite the consensus that this complex formed in a marginal basin, its exact setting is still the subject of debate. At one extreme, Miyashiro (1973) favoured an origin in an island-arc setting; at the other, Smewing *et al.* (1975) favoured an origin by sea-floor spreading. Pearce (1975) presented an intermediate view in which the main complex formed by sea-floor spreading and the upper lava unit (the 'Upper Pillow Lavas') marked the initiation of island-arc volcanism. More recent work by Robinson *et al.* (1983) and Schmincke *et al.* (1983) has indicated that the volcanic stratigraphy may be more complex than the simple 2-fold (Lower–Upper Pillow Lava) subdivision hitherto envisaged. They also demonstrated the presence of a geochemical discontinuity near the base of their lava section which, they suggested, could mark the change from an initial arc to a later spreading event.

In making sense of these conflicting interpretations, it is important to distinguish between petrological-geochemical and geological evidence. The former indicates a supra-subduction zone origin, as detailed earlier, but does not actually distinguish between a spreading axis above a subduction zone and an arc edifice. To achieve this distinction, geological evidence is required, and here the sheeted dyke swarm, the thin crust and the pelagic nature of the volcanics and overlying sediments favour the spreading axis hypothesis of Gass *et al.* (1975) and others. Thus the combination of geological and geochemical evidence points to an origin by sea-floor spreading in a supra-subduction zone setting.

A further feature of complexes formed during the initial stages of spreading is the absence of a significant volcanic component in the overlying sediment, the absence of any major overlying or adjacent sequences of arc volcanics, and the absence of any major intrusions cross-cutting the ophiolite stratigraphy. It is reasonable, therefore, to propose that the spreading event that formed these complexes was neither preceded nor superceded by the construction of a volcanic arc. The absence of preceding arc volcanism can be explained if spreading, rather than arc volcanism, was the first subduction-related event in these areas. The absence of subsequent arc volcanism may be explained by the fact that subduction was clearly short-lived in these areas (as indicated by the short time interval between formation and emplacement of the complexes), so that we are seeing only the earliest stages of marginal basin evolution.

We therefore suggest that complexes of this type represent episodes of 'pre-arc' spreading. If the related subduction events had been longer-lived, these complexes would have become the crust of fore-arc or back-arc basins and been covered by volcanogenic sediments and lavas from the adjacent arcs.

Spreading-arc transition

The interpretation of the Troodos complex as the product of spreading rather than arc volcanism is further reinforced when one looks at complexes which do contain volcanic arc sequences. The Semail ophiolite of Oman may provide a good example of the transition from spreading to arc volcanism. The main ophiolite sequence is overlain by two lava sequences, both entirely submarine in charcter. The first, termed the Lasail Unit by Alabaster *et al.* (1980), occupies discrete volcanic centres along the ophiolite and is associated with high-level intrusions of gabbro-diorite-trondhjemite composition from which cone-sheet swarms emanate. The second, termed the Alley Unit by Alabaster *et al.* (1980), is best developed in graben structures between the volcanic centres. It, too, is associated with high-level intrusions and cone sheets. Both lava units are dominated by basalts, but the Lasail Unit contains significant proportions of andesites, dacites and rhyolites and the Alley Unit contains a small proportion of rhyolites. The time difference between the eruption of the ophiolite lava unit and that of the overlying lava units is, however, very small throughout the complex. Thus it is still not certain whether these upper lavas genuinely represent arc volcanism or whether they simply represent seamount volcanism of the type found in such regions as the Mariana fore-arc basin (Hussong & Uyeda 1982). What is very probable, however, is that the Lasail Unit was erupted from a chain of major submarine volcanic edifices that developed on marginal basin crust above a subduction zone.

A reconstruction of the marginal basin in which the Semail ophiolite is thought to have formed is presented by Pearce *et al.* (1981), and the details of the lava stratigraphy by Alabaster *et al.* (1982). In this reconstruction the main volcanic centres are spaced some 25 km apart and tend to be developed on 120°-trending

faults, which have been interpreted as transform-fault directions. These faults form planes of weakness that transect the marginal basin lithosphere and which can be exploited by magma that is generated after the oceanic crust is formed. The later Alley lavas, however, exploited planes of weakness that lay parallel to the direction (170°) of the original ridge axis.

Few, if any, other complexes are preserved in this stage of evolution. More usually, arc seamounts develop into major volcanic edifices and the underlying oceanic crust is not preserved or exposed. Perhaps the best example of a true 'arc ophiolite' is Canyon Mountain (Gerlach *et al.* 1981). However, it will be noted that the Canyon Mountain sequence is not an ophiolite *sensu stricto* in that it lacks a sheeted dyke complex and contains an acid-basic sill and lava complex and large volumes of high-level acid intrusions.

Arc–back-arc spreading transition

The clearest example of the transition between arc magmatism and back-arc spreading is provided by the Jurassic volcanics and 'Rocas Verdes' ophiolites of the southern Andes (e.g. Dalziel *et al.* 1974). The transition is marked by regional extensional faulting accompanied by eruption of mafic and acid lavas, the latter thought to be derived by crustal anatexis (Bruhn *et al.* 1978). This was followed by probable sea-floor spreading and formation of the Tortuga and Sarmiento complexes. Although exposure is not continuous, it has proved possible to recognize a temporal and spatial trend from calc-alkaline arc to MORB magmatism.

Ophiolite evidence for the initial stages of back-arc spreading in an oceanic environment is more ambiguous. Crawford *et al.* (1981) used evidence from the Pacific to suggest that the period between cessation of arc volcanism and initiation of back-arc spreading is characterized by the eruption of boninites. They thus considered that ophiolites representative of the initial stages of arc splitting should contain boninitic lavas and that, with time, lavas of MORB composition would be erupted. They cited the Betts Cove and SE Australian ophiolites as examples of crust formed in this setting.

Evidence from SSZ ophiolites for the petrogenetic evolution of marginal basins

By petrogenetic modelling of the geochemical variations observed in the Th-Hf-Ta and Cr-Y diagrams it is possible to place some constraints

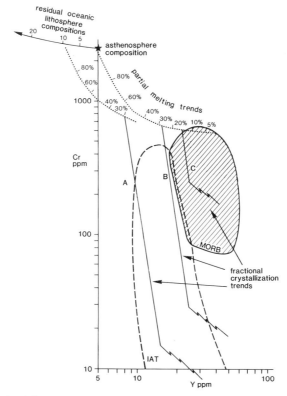

FIG. 8. Petrogenetic pathways on the Cr–Y diagram for 'MORB' ophiolites (pathway C), 'SSZ' ophiolites of island-arc tholeiitic affinities (pathway B) and 'SSZ' ophiolites of boninitic affinities (pathway A).

on the genesis of the SSZ ophiolites and, in particular, to explain the general trend from boninite to island-arc tholeiite to MORB compositions that appears to characterize marginal basin evolution.

Figure 8 illustrates how the differences in Y and Cr between MORB, IAT and boninites might be explained. The choice of Y and Cr is not arbitrary, but based on the fact that neither element appears to be significantly affected by the processes that cause heterogeneities in the convecting upper mantle. In consequence, the composition of the convecting upper mantle can be predicted from meteorite compositions, mantle xenoliths and estimates of primordial mantle, as approximately 2500 and 5 ppm for Cr and Y respectively. The compositions of primary magmas derived from this mantle composition can be modelled and presented as partial melting trends annotated according to the degree of melting. The exact trend will depend on the mantle mineralogy and rates of disappearance of the various phases, but should not differ greatly from the trend shown in Fig. 8 (modelled for a mantle of $ol_{0.6}$ $opx_{0.2}$ $cpx_{0.1}$ $pl_{0.1}$ in

which the phases melt in the ratio 3:1:4:4). Fractional crystallization of the primary magma can be represented by crystallization vectors. In Fig. 8, the steep vector represents crystallization of olivine + Cr spinel ± clinopyroxene and the shallower vector represents crystallization of olivine + Cr spinel + clinopyroxene + plagioclase. The combination of trends required to reach a given basalt composition can be termed the 'petrogenetic pathway' for that lava.

Three petrogenetic pathways are shown in Fig. 8. Pathway C represents the pathway for a typical basalt from a MORB ophiolite. Projection of the basalt composition back to the partial melting trend indicates that the primary magma was derived from about 15% partial melting of a convecting upper mantle source. Pathway B represents the pathway for the basalts from a typical SSZ ophiolite. Projection of its composition back to the partial melting trend indicates that the primary magma was derived from a greater degree of melting than that of the MORB ophiolite, perhaps 30%. Pathway C represents the pathway for a basalt from a typical boninite suite from an ophiolite complex. In this case, projection of the composition back to the partial melting trend would intersect the trend at some 80% melting, clearly an impossible value. It is thus necessary to propose that the source of these magmas was not the convecting upper mantle but the overlying sub-oceanic lithosphere. Having lost a basaltic component, the lithosphere will be depleted in Y relative to the convecting upper mantle, as modelled in Fig. 8. Melting of mantle of this composition could yield boninites (and perhaps also some IAT) after a geologically reasonable degree of melting. This model for the generation of boninites is essentially that proposed by Crawford *et al.* (1981) and provides a very plausible explanation of the Cr-Y covariations.

This model is also consistent with the variations observed in the Th-Hf-Ta diagram (Fig. 9). This diagram can be considered the petrogenetic opposite of the Cr-Y diagram in that it records heterogeneities in the convecting upper mantle and lithosphere yet is largely independent of variations due to fractional crystallization and partial melting. The predicted variation in the composition of the convecting upper mantle is denoted by the dashed line: mantle enrichment events in within plate settings cause increases in concentrations of Th and Ta relative to Hf (cf. Fig. 1a) and hence move the mantle composition towards the WPB field; similarly, mantle depletion events move the mantle composition towards the MORB field. Thus Pathway A, which represents the pathway for MORB ophiolites, is simply a line of depletion relative to an estimated bulk man-

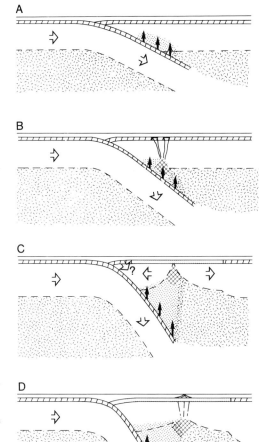

FIG. 10. Schematic model for the initial stages of evolution of at least some ensimatic marginal basins. Hydration of sub-oceanic lithosphere immediately follows subduction (A). Melting of this mantle source generates initial boninitic magmatism (B) which marks the start of the pre-arc spreading event that forms Troodos-type ophiolites (C). This is followed by submarine arc volcanism (D). In this model, the extent of pre-arc spreading will depend in part on 'roll back' of the subducting plate and 'subcretion' of the leading edge of the overriding plate.

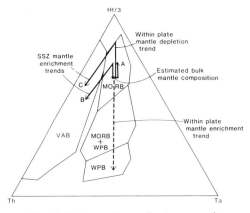

FIG. 9. Petrogenetic pathways on the Th-Hf-Ta diagram for 'MORB' ophiolites (pathway A), 'SSZ' ophiolites (pathway B) and 'SSZ' ophiolites of boninitic affinities .(pathway C).

tle composition. By contrast, mantle enrichment events in SSZ settings move the mantle composition along Th enrichment trends into the volcanic-arc field. Pathway B, which represents the pathway for island-arc tholeiites from SSZ ophiolites, therefore requires the addition of a subduction vector to Pathway A. To obtain Pathway C, which represents the pathway for boninite compositions from SSZ ophiolites, it is necessary first to invoke an additional depletion (relative to that of Pathway A) before the subduction component is added. This depleted mantle could be the lithosphere source required to explain the Cr-Y covariations.

This petrogenetic model is illustrated in its geological context in Fig. 10. It is suggested that subduction is followed by hydration of suboceanic lithosphere and that it is the melting of this source that gives rise to boninites and depleted IAT of ophiolites such as Vourinos and Troodos (e.g. Simonian & Gass 1978). Subsequently, hydrated asthenosphere from the mantle wedge is melted producing IAT compositions of ophiolites such as the Semail complex. The accumulation of magma of this composition to form an arc edifice rather than oceanic crust marks the transition from supra-subduction zone spreading to arc volcanism. Subsequently, this arc may split, perhaps generating further boninites, and back-arc spreading may take place. The lava may then eventually achieve MORB composition, being derived from mantle that contains no component from subducted oceanic lithosphere.

Conclusions

By plotting geochemical data from ophiolite lavas and dykes onto basalt discrimination diagrams, it has proved possible to divide ophiolite complexes into two groups: SSZ (supra-subduction zone) ophiolites which have the geochemical characteristics of island-arc tholeiites; and MORB ophiolites which have the geochemical characteristics of mid-ocean ridge basalts. Study of lavas from known environments suggests that present-day equivalents of SSZ ophiolites lie in fore-arc and parts of some back-arc basins. Equivalents of MORB ophiolites lie in incipient oceans, major oceans, 'leaky' transforms and most back-arc basins.

SSZ ophiolites are also characterized by strongly depleted, harzburgite mantle sequences containing podiform chromite deposits and by crystallization sequences in which olivine is followed by pyroxene rather than

plagioclase. They may also contain lavas of boninitic composition. They are sometimes overlain by arc volcanics and/or volcanogenic sediments but this is not a diagnostic feature. Ophiolites of this type are found in most, if not all, of the world's major Phanerozoic orogenic belts, where they commonly comprise the largest and best-preserved complexes.

It has been shown that many SSZ ophiolites were detached soon after their formation, often because they were related to the subduction of the crust of small oceans and soon became involved in continent–arc or continent–fore-arc collisions. They may therefore provide useful information on the earliest stages of marginal basin evolution. Most interesting are the Troodos-type complexes, which appear to have formed by some form of marginal basin spreading event, yet are neither preceded nor followed by significant arc volcanism. It is probable, therefore, that, in these cases, the first significant volcanic episode related to subduction is a spreading event, rather than the construction of an arc edifice as is commonly supposed. In more evolved arc–basin systems, the products of this event would be found as the basement to fore-arc basins. However, the process should be termed 'pre-arc' rather than 'fore-arc' spreading to signify that the arc was not in existence when spreading took place. Petrogenetic considerations indicate that the boninitic affinities of the lavas from these complexes may be due to the fact that hydrated lithosphere, rather than convecting upper mantle, is melted during the initial stages of subduction.

The passage from 'pre-arc' spreading to arc volcanism may be preserved in the Semail ophiolite. If so, the transition is marked by the accumulation of magma in discrete magma chambers which develop on fracture zones in the newly formed marginal basin crust. These chambers erupt magma of basic, intermediate and acid composition, as lava and cone-sheet swarms, to form a series of submarine volcanic centres; these may contain sheeted dyke swarms of limited extent at depth. If the analogy is correct, basaltic fissure eruptions can also take place in the depressions between the volcanic centres. Given a sufficiently long period of subduction, these centres could develop into subaerial volcanic islands of the type seen in many of the W Pacific intra-oceanic arcs.

Supra-subduction zone ophiolites may also form in the initial stages of arc splitting. If the interpretations of Crawford *et al.* (1981) are correct, magmas of boninitic affinities, this time derived from hydrated sub-arc lithosphere, are erupted initially, although magmas of MORB affinities may be erupted once a back-arc basin of significant size has evolved.

Finally, it must be emphasized that the model may not be generally applicable as plate geometries and plate motions, including roll-back of the subducted plate (Elsasser 1971; Molnar & Atwater 1978; Chase 1978), are highly variable. Nonetheless, the occurrence of SSZ ophiolite complexes of similar composition and internal structure in so many orogenic belts must indicate that sea-floor spreading in a supra-subduction zone setting has characterized many marginal basins in the past.

ACKNOWLEDGMENTS: We are grateful to Professor I. G. Gass, Drs M. Menzies, T. Alabaster, R. Hall, J. Malpas, P. Ryan, A. H. F. Robertson, A. G. Smith and many others for helpful discussions during the writing of this paper, to Professor J. R. Cann, Drs K. G. Cox and H.-U. Schmincke for reviewing the manuscript, to Drs P. Potts, O. Williams Thorpe and Mr J. Watson for analytical assistance, to Neil Mather and Helen Boxall for drawing the diagrams and to Marilyn Leggett for typing the manuscript. S. Roberts was supported in this work by an EEC studentship.

References

ABBOTTS, I. L. 1981. Masirah (Oman) ophiolite sheeted dykes and pillow lavas: geochemical evidence of the former ocean ridge environment. *Lithos*, **14**, 283–94.

ALABASTER, T., PEARCE, J. A., MALLICK, D. I. J. & ELBOUSHI, I. M. 1980. The volcanic stratigraphy and location of massive sulphide deposits in the Oman ophiolite. *In*: PANAYIOTOU, A. (ed.) *Ophiolites*, 751–7 Geol. Survey Dept. Cyprus.

——, —— & MALPAS, J. 1982. The volcanic stratigraphy and petrogenesis of the Oman ophiolite complex. *Contrib. Miner. Petrol.* **81**, 168–83.

BECCALUVA, L., OHNENSTETTER, D. & OHNENSTETTER, M. 1979. Geochemical discrimination between ocean floor and island-arc tholeiites—application to some ophiolites. *Can. J. Earth Sci.* **16**, 1874–82.

——, ——, —— & VENTURELLI, G. 1977. The trace element geochemistry of Corsican ophiolites. *Contrib. Miner. Petrol.* **64**, 11–31.

—— OHNENSTETTER, M., OHNENSTETTER, D. & PAUPY A. 1984. Two magmatic series with island-arc affinity within the Vourinos ophiolites. *Contrib. Miner. Petrol.* **85**, 253–71.

BEST, M. G. 1975. Amphibole-bearing cumulate inclusions, Grand Canyon, Arizona, and their bearing on silica-undersaturated hydrous magmas in the upper mantle. *J. Petrol.* **16**, 212–36.

BICKLE, M. J. & NISBET, E. G. 1972. The oceanic affinities of some alpine mafic rocks based on their Ti-Zr-Y contents. *J. geol. Soc. London*, **128**, 267–71.

—— & PEARCE, J. A. 1975. Oceanic mafic rocks in the Eastern Alps. *Contrib. Miner. Petrol.* **49**, 177–89.

BOUGAULT, H., MAURY, R. C., EL AZZOUZI, M., JORON, J.-L., COTTEN, J. & TREUIL, M. 1982. Tholeiites, basaltic andesites and andesites from Leg 60 sites: geochemistry, mineralogy and low partition coefficient elements. *In*: LEE, M. & POWELL, R. (eds) *Init. Rep. Deep Sea drill. Proj.* **60**, 657–74. U.S. Government Printing Office, Washington, D.C.

BRUHN, R. L., STERN, C. R. & DE WIT, M. J. 1978. Field and geochemical data bearing on the development of a Mesozoic volcano-tectonic rift zone and back-arc basin in southernmost South America. *Earth planet. Sci. Lett.* **41**, 32–46.

CAMERON, W. E., NISBET, E. G. & DIETRICH, V. J. 1979. Boninites, komatiites and ophiolitic basalts. *Nature, Lond.* **280**, 550–3.

CANN, J. R. 1970. Rb, Sr, Y, Zr and Nb in some ocean-floor basaltic rocks. *Earth planet. Sci. Lett.* **19**, 7–11.

CAPEDRI, S., VENTURELLI, G., BOCCHI, G., DOSTAL, J., GARUTI, G. & ROSSI, A. 1980. The geochemistry and petrogenesis of an ophiolitic sequence from Pindos, Greece. *Contrib. Miner. Petrol.* **74**, 189–200.

CHASE, C. G. 1978. Extension behind island arcs and motions relative to hot spots. *J. geophys. Res.* **83**, 5385–7.

COISH, R. A. & CHURCH, W. R. 1979. Igneous geochemistry of mafic rocks in the Betts Cove ophiolite, Newfoundland. *Contrib. Miner. Petrol.* **70**, 29–39.

CRAWFORD, A. J., BECCALUVA, L. & SERRI, G. 1981. Tectono-magmatic evolution of the West Philippine–Mariana region and the origin of boninites. *Earth planet. Sci. Lett.* **54**, 346–56.

DALZIEL, I. W. D., DE WIT, M. J. & PALMER, K. F. 1974. Fossil marginal basin in the southern Andes. *Nature, Lond.* **250**, 291–4.

DAVIES, H. L. & SMITH, E. I. 1971. Geology of Eastern Papua. *Bull. geol. Soc. Am.* **82**, 3299–312.

DELALOYE, M. & WAGNER, J.-J. 1984. Ophiolites and volcanic activity near the western edge of the Arabian plate. *In*: ROBERTSON, . H. F. & DIXON, J. (eds) *Tectonic Evolution of the Eastern Mediterranean. Spec. Publ. geol. Soc. London*, **17**, 225–33. Blackwell Scientific Publications, Oxford.

DESMONS, J. & BECCALUVA, L. 1983. Mid-ocean ridge and island arc affinities in ophiolites from Iran: Palaeogeographic implications. *Chem. Geol.* **39**, 39–63.

——, DELALOYE, M., DESMET, A., GAGNY, Cl., ROCCI, G. & VOLDET, P. 1980. Trace and rare earth element abundances in Troodos lavas and sheeted dikes, Cyprus. *Ofioliti*, **5**, 35–56.

DEWEY, J. F. & BIRD, J. 1971. Origin and emplacement of the ophiolite suite: Appalachian ophiolites in Newfoundland. *J. geophys. Res.* **76**, 3179–206.

DIXON, T. H. & BATIZA, R. 1979. Petrology and chemistry of recent lavas in the northern Marianas: implications for the origin of island arc basalts. *Contrib. Miner. Petrol.* **70**, 167–82.

ELSASSER, W. M. 1971. Sea-floor spreading as thermal convection. *J. geophys. Res.* **76**, 1101–12.

FERRARA, G., INNOCENTI, F., RICCI, C. A. & SERRI, G. 1976. Ocean-floor affinity of basalts from North Apennine ophiolites: geochemical evidence. *Chem. Geol.* **17**, 101–11.

FURNES, H., STURT, B. A. & GRIFFIN, W. L. 1980. Trace element geochemistry of metabasalts from the Karmøy ophiolite, southwest Norwegian Caledonides. *Earth planet. Sci. Lett.* **50**, 75–91.

GASS, I. G., NEARY, C. R., PLANT, J., ROBERTSON, A. H. F., SIMONIAN, K. O., SMEWING, J. D., SPOONER, E. T. C. & WILSON, R. A. M. 1975. Comments on 'The Troodos ophiolitic complex was probably formed in an island arc' by A. Miyashiro and subsequent correspondence by A. Hynes and A. Miyashiro. *Earth planet. Sci. Lett.* **25**, 236–8.

GEALEY, W. K. 1977. Ophiolite obduction and the geologic evolution of the Oman Mountains and adjacent areas. *Bull. geol. Soc. Am.* **88**, 1183–91.

—— 1980. Ophiolite obduction mechanism. *In*: PANAYIOTOU, A. (ed.) *Ophiolites*, 228–43. Geol. Survey Dept, Cyprus.

GERLACH, D. C., AVE'LALLEMENT, H. G. & LEEMAN, W. P. 1981. An island arc origin for the Canyon Mt ophiolite complex, E. Oregon, U.S.A. *Earth planet. Sci. Lett.* **53**, 255–65.

GREEN, D. H. 1973. Experimental melting studies on a model upper mantle composition at high pressure under water-saturated and water-undersaturated conditions. *Earth planet. Sci. Lett.* **19**, 27–53.

GRIFFIN, B. J. & VARNE, R. 1980. The Macquarie Island ophiolite complex: mid Tertiary oceanic lithosphere from a major ocean basin. *Chem. Geol.* **30**, 285–308.

GRIFFITHS, J. R. & VARNE, R. 1972. Evoltion of the Tasman Sea, Macquarie Ridge and Alpine Fault. *Nature (phys. Sci.)* **235**, 83–6.

HAWKESWORTH, C. J., O'NIONS, R. K., PANKHURST, R. J., HAMILTON, P. J. & EVENSEN, N. M. 1977. A geochemical study of island-arc and back-arc tholeiites from the Scotia Sea. *Earth planet. Sci. Lett.* **36**, 253–62.

HAWKINS, JR, W. 1980. Petrology of back-arc basins and island arcs: their possible role in the origin of ophiolites. *In*: PANAYIOTOU, A. (ed.) *Ophiolites*, 244–54. Geol. Survey Dept, Cyprus.

HUSSONG, D. M. & UYEDA, S. 1982. Tectonic processes and the history of the Mariana Arc: a synthesis of the results of Deep Sea Drilling Project Leg 60. *In*: LEE, M. & POWELL, R. (eds) *Init. Rep. Deep Sea drill. Proj.* **60**, 909–29. U.S. Government Printing Office, Washington, D.C.

KARIG, D. E. 1982. Initiation of subduction zones: implications for arc evolution and ophiolite development. *In*: LEGGETT, J. K. (ed.). *Trench–Forearc Geology. Spec. Publ. Geol. Soc. London*, **10**, 563–76. Blackwell Scientific Publications, Oxford.

LAURANT, R. 1975. Occurrences and origin of the ophiolites of southern Quebec, Northern Appalachians. *Can. J. Earth Sci.* **12**, 443–55.

LEWIS, A. D. & SMEWING, J. D. 1980. The Montgenevre Ophiolite (Hautes Alpes, France). Metamorphism and trace element geochemistry of the volcanic sequence. *Chem. Geol.* **28**, 291–306.

MARSH, N. G., SAUNDERS, A. D., TARNEY, J. & DICK, H. J. B. 1980. Geochemistry of basalts from the Shikoku and Daito Basins, Deep Sea Drilling Project Leg 58. *In*: STOUT, L. N. (ed.) *Init. Rep. Deep Sea drill. Proj.* **58**, 805–42. U.S. Government Printing Office, Washington, D.C.

MATTEY, D. P., MARSH, N. G. & TARNEY, J. 1981. The geochemistry, mineralogy and petrology of basalts from the West Philippine and Parece Vela Basins and from the Palau–Kyushu and West Mariana Ridges. *In*: ORLOFSKY, S. (ed.) *Init. Rep. Deep Sea drill. Proj.* **59**, 753–800. U.S. Government Printing Office, Washington, D.C.

MENZIES, M., BLANCHARD, D. & JACOBS, J. 1977. Rare earth and trace element geochemistry of metabasalts from the Point Sal ophiolite, California. *Earth planet. Sci. Lett.* **37**, 203–15.

——, —— & XENOPHONTOS, C. 1980. Genesis of the Smartville arc-ophiolite, Sierra Nevada foothills, California. *Am. J. Sci.* **280**, 329–44.

MIYASHIRO, A. 1973. The Troodos ophiolite complex was probably formed in an island arc. *Earth planet. Sci. Lett.* **19**, 218–24.

MOLNAR, P. & ATWATER, T. 1978. Inter-arc spreading and cordilleran tectonics as alternatives related to the age of subducted oceanic lithosphere. *Earth planet. Sci. Lett.* **41**, 330–40.

MOORES, E. M. & JACKSON, E. D. 1974. Ophiolites and oceanic crust. *Nature, Lond.* **250**, 136–9.

NOIRET, G., MONTIGNY, R. & ALLEGRE, C. J. 1981. Is the Vourinos Complex an island arc ophiolite? *Earth planet. Sci. Lett.* **56**, 375–86.

PARROT, J.F. 1977. Assemblage ophiolitique du Baer-Bassit et termes effusifs du volcano-sedimentaire. *Travaux et Documents de L'O.R.S.T.R.O.M.* **72**, 332 pp.

PEARCE, J. A. 1975. Basalt geochemistry used to investigate past tectonic environments on Cyprus. *Tectonophysics*, **25**, 41–67.

—— 1980. Geochemical evidence for the genesis and eruptive setting of lavas from Tethyan ophiolites. *In*: PANAYIOTOU, A. (ed.) *Ophiolites*, 261–72. Geol. Survey Dept, Cyprus.

—— 1982. Trace element characteristics of lavas from destructive plate boundaries. *In*: THORPE, R. S. (ed.) *Andesites*, 525–48. John Wiley and Sons.

—— 1983. Role of sub-continental lithosphere in magma genesis at active continental margins. *In*: HAWKESWORTH, C. J. & NURRY, M. J. (eds) *Continental Basalts and Mantle Xenoliths*, 230–49. Shiva Publishing, Nantwich.

——, ALABASTER, T., SHELTON, A. W. & SEARLE, M. P. 1981. The Oman ophiolite as a Cretaceous arc-basin complex: evidence and implications. *Philos. Trans. R. Soc. London*, **A300**, 299–317.

—— & CANN, J. R. 1973. Tectonic setting of basic

volcanic rocks determined using trace element analyses. *Earth planet. Sci. Lett.* **19**, 290–300.

—— & NORRY, M. J. 1979. Petrogenetic implications of Ti, Zr, Y and Nb variations in volcanic rocks. *Contrib. Miner. Petrol.* **69**, 33–47.

PRESTVIK, T. 1980. The Caledonian ophiolite complex of Leka, north central Norway. *In*: PANAYIOTOU, A. (ed.) *Ophiolites*, 555–66. Geol. Survey Dept, Cyprus.

RICOU, L. E. 1971. Le croissant ophiolitique periarabe, une ceinture de nappes mises en place au Crétacé supérieur. *Rev. Geogr. phys. Geol. dynam.* **18**, 327–49.

ROBERTSON, A. H. F. 1981. Metallogenesis on a Mesozoic passive continental margin, Anatalya Complex, southwest Turkey. *Earth planet. Sci. Lett.* **54**, 323–45.

ROBINSON, P. T., MELSON, W. G., O'HEARN, T. & SCHMINCKE, H.-U. 1983. Volcanic glass compositions of the Troodos Ophiolite, Cyprus. *Geology*, **11**, 400–4.

ROCCI, G., OHNENSTETTER, D. & OHNENSTETTER, M. 1975. La dualité des ophiolites tethysiennes. *Petrologie*, **1**, 172–4.

SAUNDERS, A. D. & TARNEY, J. 1979. The geochemistry of basalts from a back-arc spreading centre in the Scotia Sea. *Geochim. cosmochim. Acta*, **43**, 555–72.

SCHILLING, J.-G. 1973. Iceland mantle plume: Geochemical study of Reykjanes Ridge. *Nature, Lond.* **242**, 565–72.

SCHMINCKE, H.-U. & STAUDIGEL, H. 1976. Pillow lavas on central and eastern Atlantic Islands (La Palma, Gran Canaria, Porto Santo, Santa Maria) (Preliminary Report) *Bull. Soc. geol. Fr.* **7**, 871–83.

——, RAUTENSCHLEIN, M., ROBINSON, P. T. & MEHEGAN, J. M. 1983. Troodos extrusive series of Cyprus: a comparison with oceanic crust. *Geology*, **11**, 405–9.

SHARASKIN, A. YA. 1982. Petrography and geochemistry of basement rocks from five leg 60 sites. *In*: LEE, M. & POWELL, R. (eds) *Init. Rep. Deep Sea drill. Proj.* **60**, 647–56. U.S. Government Printing Office, Washington, D.C.

SHERVAIS, J. W. 1982. Ti-V plots and the petrogenesis of modern and ophiolitic lavas. *Earth planet. Sci. Lett.* **59**, 101–18.

SIMONIAN, K. O. & GASS, I. G., 1978. Arakapas fault belt, Cyprus: a fossil transform fault. *Bull. geol. Soc. Am.* **89**, 1220–30.

SMEWING, J. D., SIMONIAN, K. O. & GASS, I. G. 1975. Metabasalts from the Troodos Massif, Cyprus: genetic implication deduced from petrography and trace element geochemistry. *Contrib. Miner. Petrol.* **51**, 49–64.

SPRAY, J. G. 1982. Mafic segregations in ophiolite mantle sequences. *Nature, Lond.* **299**, 524–8.

STERN, C. 1979. Open and closed system igneous fractionation within two Chilean ophiolites and the tectonic implication. *Contrib. Miner. Petrol.* **68**, 243–58.

STURT, B. A., THON, A. & FURNES, H. 1980. The geology and preliminary geochemistry of the Karmøy ophiolite, S. W. Norway *In*: PANAYIOTOU, A. (ed.) *Ophiolites*, 538–54. Geol. Survey Dept, Cyprus.

SUEN, C. J., FREY, F. A., & MALPAS, J. 1979. Bay of Islands ophiolite suite, Newfoundland: petrologic and geochemical characteristics with emphasis on rare earth element geochemistry. *Earth planet. Sci. Lett.* **45**, 337–48.

TARNEY, J., WOOD, D. A., SAUNDERS, A. D., CANN, J. R. & VARET, J. 1980. Nature of mantle heterogeneity in the North Atlantic: evidence from deep sea drilling. *Philos. Trans. R. Soc. London*, **A297**, 179–202.

UPADHYAY, H. D. & NEALE, E. R. W. 1979. On the tectonic regime of ophiolite genesis. *Earth planet. Sci. Lett.* **43**, 93–102.

VENTURELLI, G., CAPEDRI, S., THORPE, R. S. & POTTS, P. J. 1979. Rare-earth and other element distribution in some ophiolitic metabasalts of Corsica, Western Mediterranean. *Chem. Geol.* **24**, 339–53.

——, THORPE, R. S. & POTTS, P. J. 1981. Rare earth and trace element characteristics of ophiolitic metabasalts from the Alpine-Apennine belt. *Earth planet. Sci. Lett.* **53**, 109–23.

WOOD, D. A. 1980. The application of a Th-Hf-Ta diagram to problems of tectonomagmatic classification and to establish the nature of crustal contamination of basaltic lavas of the British Tertiary Volcanic Province. *Earth planet. Sci. Lett.* **50**, 11–30.

——, JORON, J.-L. & TREUIL, M. 1979. A reappraisal of the use of trace elements to classify and discriminate between magma series erupted in different tectonic settings. *Earth planet. Sci. Lett.* **45**, 326–36.

——, MARSH, N. G., TARNEY, J., JORON, J.-L., FRYER, P. & TREUIL, M. 1982. Geochemistry of igneous rocks recovered from a transect across the Mariana trough, arc, fore-arc and trench, sites 453 through 461. *In*: LEE, M. & POWELL, R. (eds) *Init. Rep. Deep Sea drill. Proj.* **60**, 611–45. U.S. Government Printing Office, Washington, D.C.

——, MATTEY, D. P., JORON, J.-L., MARSH, N. G., TARNEY, J. & TREUIL, M. 1981. A geochemical study of 17 selected samples from basement cores recovered at sites 447, 448, 449, 450 and 451, Deep Sea Drilling Project Leg 59. *In*: ORLOFSKY, S. (ed.) *Init. Rep. Deep Sea drill. Proj.* **59**, 743–52. U.S. Government Printing Office, Washington, D.C.

XENOPHONTOS, C. & BOND, G. C. 1978. Petrology, sedimentation and palaeogeography of the Smartville terrane (Jurassic)—bearing on the genesis of the Smartville ophiolite. *In*: HOWELL, B. & McDOUGALL, I. (eds) *Mesozoic Symposium*. **2**, 292. Soc. Econ. Palaeontologists Mineralogists.

PEARCE, J. A.,* LIPPARD, S. J. & ROBERTS, S. Department of Earth Sciences, The Open University, Walton Hall, Milton Keynes MK7 6AA, U.K. *Present address: Department of Geology, University of Newcastle upon Tyne, Newcastle upon Tyne NE1 7RU, U.K.

WESTERN PACIFIC REGION

Marginal basins of the SW Pacific and the preservation and recognition of their ancient analogues: a review

E. C. Leitch

SUMMARY: Marginal basins of the SW Pacific floored by oceanic lithosphere comprise those formed by sea-floor spreading immediately behind active magmatic arcs (back-arc basins) and those created by the rifting of continental crust without obvious connection to an arc system (small ocean basins). The basins opened rapidly, range greatly in shape and size, and show diverse relations to sediment sources. Thick sediment piles mainly accumulated adjacent to emergent continental margins or active arcs, with thin sequences of pelagic sediments, ash, and fine grained turbidites on basin floors. Facies are commonly asymmetrically distributed. Basaltic magmatic activity, mild deformation and submarine erosion affect both types of basins. Ancient back-arc basins should be identifiable on the basis of their temporal relations to magmatic arcs and the volcanic influence in their sedimentary sequence, but distinguishing between small and major ocean basins is often difficult. Narrow marginal basins may collapse following accumulation of a thick sediment pile, but most basins close by subduction. Similarities are recognized between the present-day SW Pacific and the formative stage of the North American Cordillera. The Late Mesozoic palaeogeography of the Western Alps resembles the complex of small ocean basins and continental fragments west of the Norfolk Ridge, and the Ordovician palaeogeography of central Newfoundland bears close resemblance to the assemblage of back-arc basins, remnant arcs and active arc east from the Norfolk Basin.

This paper examines the marginal basins of the SW Pacific with the aim of recognizing diagnostic characters that will aid in the identification of their ancient analogues. Considerable information on these basins has been derived from the Deep Sea Drilling Project and from marine geophysical surveys. These data, allied with the range of marginal basin types, make the region a suitable one from which to derive criteria of wide applicability.

However, such derivation of criteria largely ignores any features that may result from the incorporation of marginal basins into fold belts. Thus the mechanisms whereby marginal basins are converted into stratotectonic units are also discussed. Several examples of marginal basins interpreted in ancient sequences are reviewed, to illustrate the scales on which analogies with the present-day SW Pacific can be drawn, and also the evidence used to support these interpretations.

Marginal basins of the SW Pacific

The SW Pacific region comprises a collage of continental blocks, island arcs, trenches, marginal basins, remnant arcs, seamount chains and fracture zones that lies between the major continental mass of Australia and the Pacific Ocean Basin (Fig. 1). It is a broad zone in which changes in the relative motion of the Pacific and Indian plates have been resolved along a complex series of rapidly evolving plate boundaries (Packham 1973; Packham & Andrews 1975; Crook & Belbin 1978). Despite recent investigations many of the sea-floor features remain poorly known and their characters disputed (Leitch 1981); only generalized tectonic reconstructions can be made.

The oldest oceanic sea-floor identified in the SW Pacific is of Late Cretaceous age and it was probably at about this time that the present cycle of fragmentation of the margin of E Gondwanaland commenced (Packham 1973). The shape and size of the marginal basins range widely as does their proximity to major terrigenous sediment sources. Bounding tectonic elements include continental masses, microcontinental blocks, active and remnant arcs and fracture zones. Following Karig (1971) and Packham & Falvey (1971), usage here of the term marginal basin is restricted to those basins floored by oceanic crust. These workers supported Carey (1958) in suggesting that the marginal basins of the W Pacific were formed by rifting at independent spreading centres removed from the main ocean basin, a conclusion clearly confirmed by magnetic anomaly patterns, palaeomagnetic results, and basement ages derived from deep-sea drilling. However, not all marginal basins are as closely linked to island arcs as these writers suggested and two major categories can be recognized; those that formed immediately behind active magmatic

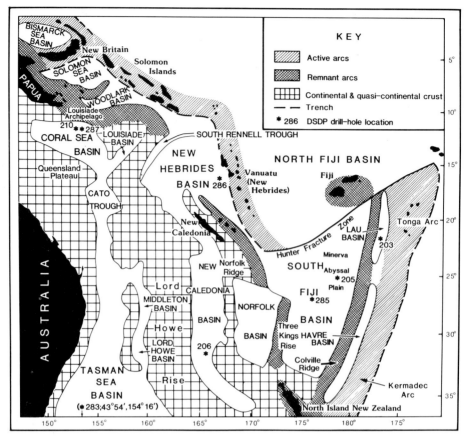

FIG. 1. Marginal basins of the SW Pacific and their bounding tectonic elements. (Five degrees of longitude is about 550 km.)

arcs (back-arc basins), and those that formed without any clear connection to an active arc (small ocean basins).

The apparent absence of clearly defined linear magnetic anomaly patterns in marginal basins led Karig (1971) to suggest that the oceanic lithosphere was generated from randomly distributed diapirs, rather than from a well-defined axial zone as occurs in major ocean basins. However, such absence may result from extensive hydrothermal alteration, relatively thick sediment cover, widespread off-axis magmatic activity and the presence of numerous transform faults (e.g. Lawver & Hawkins 1978). A well-developed pattern of axially symmetrical anomalies occurs in the Tasman Sea Basin (Hayes & Ringis 1973; Weissel & Hayes 1977; Shaw 1978), and linear anomaly patterns have recently been recognized in parts of other basins including the Coral Sea Basin (Weissel & Watts 1979), the Bismarck Sea Basin (Connelly 1976; Taylor

1979), the New Hebrides Basin (Weissel et al. 1982b), the Lau Basin (Weissel 1977), the Havre Basin (Malahoff et al. 1982), the North Fiji Basin (Malahoff et al. 1982) and the South Fiji Basin (Watts et al. 1977; Malahoff et al. 1982). However, in the basins closely associated with island arcs, no simple system of spreading normal to the trend of the arc has been identified. Taylor's (1979) model for the Bismarck Sea involves asymmetrical spreading across a plate boundary composed of two transforms, a leaky transform and a spreading segment. Opening of the Lau Basin (Weissel 1977) produced several small plates with the boundary between three being a ridge–transform–transform triple junction, and a three ridges triple junction has been recognized in the South Fiji Basin (Watts et al. 1977) and the North Fiji Basin (Falvey 1978). The distribution of early rift structures in the Tasman Sea Basin (Mutter & Jongsma 1978) suggests that initial opening patterns, of long transforms and irregular boundaries, become simpler with time.

Linear spreading rates within some marginal basins vary considerably, indicating proximity to the pole of opening. The angular opening rate of $4°$ Ma^{-1} calculated by Taylor (1979) for the Bismarck Sea yields spreading half-rates of about 6.5 cm yr^{-1}, and a rate of $1°$ Ma^{-1} in the Tasman Basin (Shaw 1978) indicates half-rates of about 1–3 cm yr^{-1}. Half-rates of 3–3.5 cm yr^{-1} have been estimated for the ridges of the South Fiji Basin (Watts *et al*. 1977; Malahoff *et al*. 1982), 3.5 cm yr^{-1} for the North Fiji Basin (Malahoff *et al*. 1982), 3.8 cm yr^{-1} for the Lau Basin (Weissel 1977), 3 cm yr^{-1} for the Woodlark Basin (Weissel *et al*. 1982a), and 2.7 cm yr^{-1} for the Havre Basin (Malahoff *et al*. 1982).

Information concerning basin type, age, and site of rifting are summarized in Table 1. Several basins ascribed to back-arc settings are queried because of the absence of a clearly related arc. Interpretation of the New Hebrides Basin as a back-arc structure is based on the reversal of polarity of the New Hebrides arc subsequent to basin formation (Karig & Mammerickx 1972). A similar situation holds with the Solomon Sea Basin, unless it developed behind a S-facing 'Woodlark–Louisiades arc' that was later split by the opening of the Woodlark Basin. However, the existence of this arc (Karig 1972), and hence also the tectonic position of the Woodlark Basin, remain to be demonstrated (Weissel *et al*. 1982a) (see Smith & Milsom, this volume). The narrow S Rennel Trough, considered to be a fracture zone by

Packham (1973) and Terrill (1975), has recently been interpreted as a spreading centre (Larue *et al*. 1977; Daniel *et al*. 1978).

The Tasman Sea and Coral Sea basins and, tentatively, the New Caledonia Basin, are classified as small ocean basins. Falvey & Mutter (1981) suggested that at the time of opening of the Tasman Sea Basin (80–55 Ma) a W-dipping convergent plate margin lay adjacent to the present-day eastern coastline of the North Island of New Zealand. A high-potassium calc-alkaline andesite-dacite-rhyolite suite of this age occurs in the Mt Somers area of the South Island (Oliver *et al*. 1979), but according to Falvey & Mutter (1981) these rocks occur beyond a transform boundary which terminated the convergent margin. Late Cretaceous rhyolite from Lord Howe Rise and a Cretaceous basalt-rhyolite assemblage from the North Island, are probably related to rifting rather than to plate convergence (Leitch 1975; Pirajno 1979) and Late Cretaceous sedimentary rocks fail to show any arc influence (Leitch 1975).

Taylor (1975) commented on the absence of an arc associated with the opening of the Coral Sea Basin. No arc rocks of appropriate age occur on the Queensland Plateau (Mutter 1977) and arc rocks of Late Cretaceous–Eocene age are absent from Eastern Papua and the Louisiades Archipelago (unless the low-potassium basalts of these regions (Davies & Smith 1971) comprise primitive arc basement rather than ocean floor) (see Smith & Milsom, this volume).

The age of the Norfolk Basin has not been

TABLE 1. *Marginal basins of the SW Pacific*

Basin	Type	Site of rifting	Age of rifting	Reference (age)
Tasman	Small ocean	Continental crust	76–55 Ma	Shaw (1978)
Cato	Small ocean	Continental crust	65–56 Ma	Falvey & Mutter (1981)
Coral Sea	Small ocean	Continental crust	62–56 Ma	Weissel & Watts (1979)
Louisiade	Small ocean	Continental crust	60–56 Ma	Weissel & Watts (1979)
Middleton	Small ocean	Continental crust	76–65 Ma	Shaw (1978)
Lord Howe	Small ocean	Continental crust	76–65 Ma	Shaw (1978)
New Caledonia	?Small ocean	?Continental crust	c. 76–55 Ma	Packham & Andrews (1975)
Norfolk	?Back arc	?Arc/continent boundary	?Early Miocene	(see text)
New Hebrides	?Back arc	?	55–42 Ma	Lapouille (1978)
Woodlark	?	?	3.5–0 Ma	Weissel *et al*. (1982a)
Solomon Sea	?Back arc	?	Eocene	Packham & Andrews (1975)
Bismarck	Back arc	Arc	3.5–0 Ma	Taylor (1979)
South Fiji	Back arc	Arc	36–25 Ma	Malahoff *et al*. (1982)
Havre	Back arc	Arc	2–0 Ma	Malahoff *et al*. (1982)
Lau	Back arc	Arc	3.5–0 Ma	Weissel (1977)
South Rennell	?	?	Oligocene	Larue *et al*. (1977)
North Fiji	Back arc	Arc	8–0 Ma	Malahoff *et al*. (1982)

established. Packham & Andrews (1975) favoured Eocene rifting but Davey (1982) and Malahoff *et al.* (1982) argued for a late Oligocene or more recent age and suggested that the basin is a back-arc structure which opened behind an arc now represented by the Three Kings Rise (see also Lapouille 1978). Interpretation of the rise as an extinct E-facing arc is supported by the westward younging of magnetic lineations in the adjacent part of the South Fiji Basin and the dredging of andesite from the rise (Davey 1982; Malahoff *et al.* 1982). Early Miocene activity is indicated by the Tertiary stratigraphy of northernmost New Zealand (Leitch 1970).

The age of the New Caledonia Basin has not been accurately established. The geology of New Caledonia suggests that the Norfolk Ridge is a continental fragment rather than a remnant arc, and the New Caledonia Basin possibly opened without associated arc activity.

Three points can be made from consideration of Table 1:

(i) Marginal basins have a short spreading history.
(ii) With the possible exception of the S Rennell Trough, rifting commenced within relatively thick crust, either in continental or island-arc settings, and not within areas of thin oceanic crust.
(iii) Back-arc basins that are presently spreading (Bismarck, Lau, Havre, North Fiji) are all associated with arcs that are volcanically active. They thus contrast with the basins of the NW Pacific that appear to have opened during periods of little or no arc volcanism (Scott *et al.* 1981).

Sedimentology

The sediments in marginal basins of the SW Pacific reflect the variable influences of spreading history, clastic sources, biogenic sediment production and water depth. Sedimentation adjacent to the emergent continental margins of the Tasman and Coral Sea basins, which resembles that at the margins of major ocean basins, has been described by Mutter (1977), Taylor & Falvey (1977), Mutter & Karner (1980) and Falvey & Mutter (1981). The sedimentation in marginal basins close to New Zealand has been described by Katz (1974) and Davey (1977). These areas are only briefly considered here and discussion concentrates on regions removed from major continental sources.

Insufficient data are available from any single

basin to determine the detailed distribution of sedimentary facies. However, the sedimentation within the basins can be inferred from DSDP sites (Fig. 2) and seismic reflection surveys.

The initial topography of marginal basins is irregular, with fault-bound ridges and troughs producing numerous small and narrow sediment ponds. Thick clastic deposits, largely from sediment gravity flows, dominate the early stage of basin opening. With widening of the basin these deposits are restricted to areas adjacent to emergent margins, especially close to magmatic arcs, and towards the basin centre pass laterally into distal silt turbidites which are in turn succeeded distally by nannofossil ooze. Sediments progressively bury the early irregularities of the basin floor and wide abyssal plains are characteristic of larger mature basins. Seismic reflection records indicate that major lithostratigraphic units in these basins are persistent even though they can be diachronous and show marked thickness variations. Decrease in the supply of clastic detritus, due to source degradation, waning volcanic activity, or the development of a new spreading system between basin and source, increases the area of pelagic deposition and oozes overlie fine clastic rocks. Where thermal contraction has resulted in basin floors sinking below the carbonate compensation depth, calcareous rocks are succeeded by abyssal clays.

Unconformities are present in the sequences of many of the marginal seas. Some breaks are restricted to a single basin or just part of a basin, but that of Early–Middle Oligocene age which was related to strong submarine currents, occurs over much of the SW Pacific (Kennett *et al.* 1972).

Source area evolution in the SW Pacific favours an asymmetrical distribution of clastic facies within the basins, similar to that in the marginal basins of the Mariana island-arc system (Karig & Moore 1975), rather than the symmetrical distribution indicated by Klein (1975). In the Tasman Sea Basin thicker sediment piles probably occur along the E Australian margin than along the Lord Howe Rise because, although non-axial breaching of the early rift valley complicated the geology of the rise (Mutter & Jongsma 1978), DSDP results (Burns *et al.* 1973) indicate that it was the site of marine deposition even before opening of the Tasman Sea Basin was completed. Nevertheless, at least some of the relatively thick (3000 m) sediment sequence in the New Caledonia Basin has been derived from submarine slumping off the Lord Howe Rise

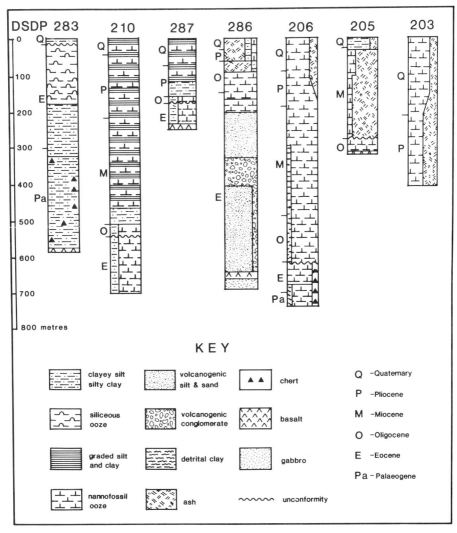

FIG. 2. DSDP sections from SW Pacific marginal basins. Key to sites: 283—Tasman Basin (Kennett *et al.* 1975); 210—Coral Sea Basin (Burns *et al.* 1973); 287—Coral Sea Basin (Andrews *et al.* 1975); 286—New Hebrides Basin (Andrews *et al.* 1975); 206—New Caledonia Basin (Burns *et al.* 1973); 205—South Fiji Basin (Burns *et al.* 1973); 203—Lau Basin (Burns *et al.* 1973).

(Bentz 1974). In back-arc basins, thicker clastic sediment piles are likely to occur along the active arc margin than along the remnant-arc margin facing the arc. Thick clastic accumulations on the margins of remnant arcs facing away from the rift formed on the margin of the active arc before rifting.

Source rejuvenation has been recognized in the sediments of the Coral Sea Basin and South Fiji Basin (Fig. 2). In the former, clays deposited below the carbonate compensation depth are overlain by a thick sequence of terrigenous silt and clay turbidites interbedded with nanno-fossil ooze. At DSDP Site 210 some 1900 turbidites occur in a sequence 470 m thick which was deposited in the last 10 Ma (Burns *et al.* 1973). In the South Fiji Basin intense middle Miocene volcanic activity is indicated by the thick sequence of tuffaceous rocks encountered at Sites 205 and 285. An extraneous source has also affected sedimentation in the southern part of the South Fiji Basin. Although drill-hole information is not available, the many seismic reflectors present indicate extensive clastic deposits derived from northern New Zealand (Packham & Terrill 1975). The southern part of the Havre

Basin also contains a thick terrigenous sediment sequence derived from this source (Karig 1970; Cronan *et al.*, this volume; Lewis & Pantin, this volume).

The volcanic seamounts and extensive fracture zones within the marginal basins provide sources for local accumulations of coarse clastic igneous material similar to those associated with these features in other oceanic basins.

Sedimentation rates in the SW Pacific basins range widely. Rates of 50–100 m Ma^{-1} are reported from the volcaniclastic sequence of the Lau Basin (Burns *et al.* 1973), and 200 m Ma^{-1} may have been reached in the sequences of coarse volcanic detritus in the New Hebrides Basin (Andrews *et al.* 1975). Very high rates also occurred in the basal section of DSDP Site 285 in the South Fiji Basin (Andrews *et al.* 1975). The turbidite sequence in the Coral Sea Basin accumulated at rates of 25–120 m Ma^{-1}, nannofossil oozes between 5 and 15 m Ma^{-1}, and abyssal clays at an average of about 6 m Ma^{-1} (Andrews *et al.* 1975).

Clastic detritus in the basins reflects the nature of the bounding elements; volcanogenic glass, crystal debris and rock fragments dominate in back-arc basins, whereas in small ocean basins quartz, feldspar and mica are the most prominent detrital phases. Hawkins (1974) suggested that on a carbonate-free basis the sediment in back-arc basins has a somewhat higher SiO_2 content (60–65%) than average abyssal sediments (c. 45%). Discrete layers of ash and tuff are confined to back-arc basins, but distal ash-fall material is scattered through the sedimentary sequence of the small ocean basins close to volcanic sources. Thus, although no volcanic component was recognized in sediments from DSDP Site 283 from the Tasman Sea Basin (Kennett *et al.* 1975), glass shards are a minor constituent of the Miocene and younger rocks of the New Caledonia Basin intersected at Site 206, and occur in trace amounts throughout the sequence here (Burns *et al.* 1973). Volcanic glass fragments also occur in the upper part of the Coral Sea sequence (Andrews *et al.* 1975).

Igneous activity

In the SW Pacific the opening of both back-arc and small ocean basins was preceded by igneous activity. Back-arc basins developed within the island-arc tholeiite and calc-alkaline rocks of the magmatic arcs (e.g. Gill 1981). Igneous activity preceding the opening of small ocean basins is represented by diverse units including rhyolitic rocks on the Lord Howe

Rise (van der Lingen 1973), small alkaline intrusions in SE New South Wales and Tasmania (McDougall & Wellman 1976), and possibly also the mainly allochthonous Cretaceous basaltic and silicic volcanic rocks in northern New Zealand (Leitch 1975).

Igneous processes in the generation of new crust in marginal basins are probably similar to those of accretion of oceanic lithosphere in major ocean basins. In the SW Pacific detailed data concerning the composition of marginal basin lithosphere are only available for the basalts of the Lau Basin (Hawkins 1974, 1976, 1977; Gill 1976). Basalts akin to mid-ocean ridge basalts (MORB) and those enriched in large ion lithophile elements both occur (see Saunders & Tarney, this volume). Much less complete information from the Woodlark Basin (Luyendyk *et al.* 1973) and the Coral Sea Basin (Stoeser 1975) suggests they are floored by rocks of mid-ocean ridge affinity.

Off-axis igneous activity within the marginal basins is indicated by seamounts in the Lau Basin (Hawkins 1974), the South Fiji Basin (Packham & Terrill 1975), and the Tasman Sea Basin (Vogt & Conolly 1971). Dolerite has intruded Middle Miocene sediments at DSDP Site 285 in the South Fiji Basin (Andrews *et al.* 1975).

Deformation

Acoustic basement in SW Pacific marginal basins, which normally corresponds to the basaltic oceanic crustal layer, usually has a rough topography resulting from normal faulting during spreading. Overlying sediments are mostly horizontal and continuous over large areas, indicating that there has been little subsequent deformation. Close to the basin margins the bedded sediments are commonly faulted and/or gently tilted, and small horst and graben structures locally affect basin floors. Packham & Terrill (1975) presented evidence of both local and regional faulting, and possibly folding, in the South Fiji Basin, and a 'rolling' topography in the New Caledonia Basin (Burns *et al.* 1973) may be the result of very open flexuring.

Close to active transform faults and fracture zones, the sea-floor topography is irregular and seismic reflectors are disturbed. Deformation of the Minerva Abyssal Plain in the north of the South Fiji Basin increases towards the Hunter Fracture Zone (Packham & Terrill 1975). In between the New Hebrides and Solomon Arcs and Papua several major fractures disrupt the floors of marginal basins producing widespread disturbance of sedimentary layers, giving rise to local sediment traps and complex depositional patterns (e.g. Daniel *et al*. 1977; Hamilton 1979).

Recognition of marginal basin deposits

The rocks of the marginal basins of the SW Pacific share many characteristics with those of the major ocean basins. The basements to both are subalkaline basalts, sedimentary sequences are generally thin and dominated by deep-water pelagic clays and oozes (except at basin margins), sequences commonly indicate progressive subsidence, many units are regionally extensive and most clastic material is fine grained. Local intercalations of extrusive basalts and their degradation products and breccias, mostly of basic and altered ultramafic material, that have accumulated close to active fracture zones are probably present

The earliest sediments deposited in back-arc basins are arc-derived, and hence include a substantial volcanic component. This commonly decreases up sequence, although influxes of pyroclastic debris periodically occur in pelagic-dominated sections and in most back-arc sequences at least a minor volcanic fraction is present throughout. Although volcanism commonly accompanies the rifting that leads to the development of both major and small ocean basins, this activity is short-lived and is not a continuing or important source of debris. Bimodal (rhyolite-basalt) volcanism characterizes initial ocean basin rifting whereas back-arc basins open adjacent to or within volcanic chains dominated by basic–intermediate rocks of island-arc tholeiite or calc-alkaline affinities. Subduction-related magmatism may accompany closure of small ocean basins but debris derived from this source will be present only in the youngest ocean floor sediment units.

Determinations of age relations between ancient magmatic arcs and adjacent sedimentary rocks can help to determine the depositional setting of the latter. Back-arc basin fill is younger than the onset of activity within the associated magmatic arc, commonly by tens of millions of years, whereas deposition within some major ocean basins commenced a hundred million years or more before the start of subduction and accompanying generation of a magmatic arc.

It appears almost impossible to distinguish between the sedimentary sequences of small ocean basins, like those in the Coral Sea and the Tasman Sea, and those of major ocean basins, on the basis of sediment type and composition alone. In the small ocean basins of the SW Pacific the tendency for one margin to subside soon after rifting is probably reflected in an asymmetry in both sediment thickness and facies, but such a pattern could also occur in major ocean basins. Where both flanks of a small ocean basin have largely subsided, as is the case for the New Caledonia Basin, there is an almost complete absence of terrigeneous material in the subsequent basin fill. Similar basinal sequences in old rocks might be interpreted as reflecting such subsidence.

Incorporation of marginal basin deposits into the geological record

Extant marginal basins are all relatively young components of the Earth's surface and older analogous features must have existed. Subduction is the only viable mechanism for closure of marginal basins of the size of those in the SW Pacific. Current views vary on the amount of sedimentary material removed from downgoing slabs (Scholl *et al*. 1980). Several plate margins once considered to be sites of major subduction accretion have been shown to have only a very small or no subduction complex (von Huene *et al*. 1980a,b; Hussong & Uyeda 1982), and it is possible that these complexes form only when thick sections of relatively buoyant material occur on the downgoing plate. Thus much of the sediment mantling oceanic lithosphere in marginal basins may be subducted rather than accreted and, irrespective of whether it forms parts of deep-seated metamorphic complexes, is recycled by melting, or is incorporated into the mantle, evidence of its original character is probably almost completely destroyed.

Destruction of marginal basin crust is presently taking place in the SW Pacific. The Solomon Sea Basin is being consumed at the New Britain and Solomons trenches. Further S the Woodlark Basin is being thrust beneath the Solomons Trench, and part of the New Hebrides Basin has disappeared below the New Hebrides Trench. No undoubted accretionary subduction complex has been determined along the convergent margin off the Solomon Islands or Vanuatu (Karig & Sharman 1975; Ibrahim *et al*. 1980), but seismic-reflection profiles (Hamilton 1979, Fig. 153) suggest that a small accretionary mass forms the inner wall of the New Britain Trench. Thus some marginal basin deposits are probably preserved by subduction accretion, presumably as thrust-repeated slices, commonly intensely deformed and perhaps, in places, constituting mélange.

If sufficient underthrusting proceeds for subduction-related magmas to be generated, then marginal basin deposits will be overlain by volcaniclastic strata derived from an active arc. The arc-derived strata may be (i) conformable with the basinal sequence if the supply of detritus was voluminous and any trench buried so that volcaniclastic material gained access to the floor of the marginal basin, (ii) separated from earlier deposits by a low-angle unconformity if the volcaniclastics were deposited within the trench, or (iii) unconformable if the clastics were trapped within slope basins in the outer part of the fore-arc complex. Alternatively, where closure occurs before the development of arc volcanism the marginal basin deposits may be overlain by diverse sequences of flysch with olistostromes.

Where subduction has not produced an accretionary prism, the sites of former marginal basins are probably marked only by narrow suture zones containing disrupted mafic and ultramafic rocks, pelagic deposits and possibly remnants of basin margin clastic accumulations (cf. Dewey 1977).

Packham & Falvey (1971) argued that large marginal basins such as the Tasman Sea Basin and the South Fiji Basin are a consequence of unusual plate interactions and suggested that small basins may have been more typical of earlier times. They also speculated that the base of thick sediment accumulations in narrow marginal basins might fuse, engendering the rise of granites, deformation of the sedimentary pile and low-pressure–high-temperature regional metamorphism, ultimately producing a fold belt. However, it has not been shown that large marginal basins are more prevalent today than they were in the past, and it has yet to be demonstrated that this mechanism for incorporation of marginal basin deposits into the geological record is viable.

Ancient analogues

A marginal basin origin has been proposed for many units now incorporated in fold belts. These range from ophiolite nappes to shredded suture-zone complexes, and from high-grade metamorphic terranes to comparatively well-preserved sedimentary sequences. The state of preservation depends in part on the tectonic history of the fold belt and in part on the level of exposure within it. Several examples of interpreted ancient analogues of marginal basins are discussed below, concentrating on less metamorphosed sedimentary sequences.

Comparisons between present-day marginal basins of the SW Pacific and the formative stages of major fold belts can be made on the scale of the whole assemblage of basins, or on the scale of the individual basins and their margins. The former comparison can be only general because of uncertainties in the palinspastic reconstruction of even the best-known belts.

If the small ocean basins of the SW Pacific (Table 1 and Fig. 1) were to close by subduction the result would be an assemblage of continental blocks, some topped by arc volcanics dating from the subduction event, separated by narrow tracts of deformed pelagic sediment, marginal clastic rocks, seamount remnants and possibly ophiolite slices. The width of the assemblage would depend on the importance of subduction accretion relative to subduction erosion, and the effects of strike-slip displacements. Abrupt along-strike discontinuities would result from the diverse shapes of the basins and both thrust zones and weakly deformed regions would characterize most convergent zones (cf. Dewey & Kidd 1974). Closure of the back-arc basins further E would produce an assemblage of narrow masses of calc-alkaline and tholeiitic arc material separating accreted back-arc basin floor. Churkin (1974) drew attention to the broad similarities between the interpreted evolution of the North American Cordillera and the evolution of the SW Pacific, and although subsequent investigations suggest that some Cordilleran terranes are truly exotic (e.g. Coney et al. 1980), and stress the importance of very large strike-slip movements, the analogy still holds.

The Late Mesozoic palaeogeography of the W Alps (Trumpy 1960) can be compared with the assemblage of small ocean basins between Australia and the New Caledonia–Norfolk Ridge–New Zealand region. The small oceanic basins of the Piemont and the Valais are analogous to the Tasman Sea and New Caledonia basins, and the Brianconnais Platform, perhaps to a diminutive Lord Howe Rise. The scarcity of arc debris in the Alpine Belt and indeed in much of the Tethyan realm, is noteworthy. In the Alps no volcanic arc debris is recorded prior to the onset of ocean basin closure, indicating that the basins opened without associated arcs. The subsequent small amounts of volcanic arc debris suggest that in most cases the opposed continental margins came together before underthrusting was sufficient to generate large volumes of arc magma.

Many individual basins and basin remnants have been likened to modern marginal basins, although in some cases the term has been used to include a broader range of structures than herein. In the Appalachian–Caledonian Orogen, units as diverse as the Highland Boundary Complex and the Dunnage Zone have been interpreted as marginal basin accumulations. Although there is a diversity of opinion (e.g. Henderson & Robertson 1982; Curry *et al.* 1982), much evidence suggests that the Highland Boundary Complex of Scotland formed between a continental mass beneath the Midland Valley (Graham & Upton 1978) and another beneath the S Highlands. The components of the complex, including mafic and ultramafic bodies, coarse clastic sedimentary rocks, chert, jasper, and argillite, could be fragments of the floor and margin of a narrow back-arc basin that opened behind a S-facing arc (Dewey 1974; Curry *et al.* 1982).

The central section (Dunnage and Gander Zones) of the Appalachian Orogen in northern Newfoundland contains a diverse assemblage of early Palaeozoic oceanic, island-arc and continental margin rocks (Strong 1977; Williams 1979), for which a variety of tectonic schemes has been devised (see Currie *et al.* 1980). Comparison of the inferred Ordovician palaeogeography of this region with the region W from the Kermadec Arc to the Norfolk Ridge reveals a basically similar assemblage of elements.

The western part of the Dunnage Zone includes parautochthonous ophiolites, which are associated with older and younger magmatic arc rocks (Mattinson 1975; Williams *et al.* 1976; Strong 1977), and are probably of marginal basin origin (Upadhyay *et al.* 1971). This suggests an arc–back-arc basin–remnant-arc assemblage comparable with the Kermadec Arc–Havre Basin–Colville Ridge assemblage. To the E the magmatic arc rocks give way to a thick sequence of volcaniclastic turbidites, olistostromes (including the Dunnage Mélange), intercalated mafic volcanic rocks and penecontemporaneous silicic igneous masses (e.g. McKerrow & Cocks 1981). The Ordovician evolution of this region as shown by Horne (1969, Fig. 5) portrays the situation anticipated at the rear of an arc soon after the onset of back-arc spreading. The Dunnage Mélange is considered to mark the initial rifting (E. C. Leitch & C. A. Cawood, unpublished data) and the igneous rocks probably represent peripheral later magmatism on the active arc (see Lorenz, this volume). The W slope of the Colville Ridge shows a very similar situation with thick sedimentary sequences, intrusive igneous bodies and common slumping on the upper slope (Packham 1978).

The western part of the Dunnage Zone is dominated by siltstone with intercalated sandstone turbidites, conglomerates, tuffs and manganiferous beds (Kay 1975; Pajari & Currie 1978) that in the E rest unconformably on ultramafic rocks, trondjhemite and pillow lavas of the Gander Ultramafic Belt (Currie *et al.* 1980). The sedimentary rocks are of oceanic facies and can be interpreted as the fill of a marginal basin in a tectonic position comparable to the South Fiji Basin. The analogy with the SW Pacific can be extended with the Gander Ultramafic Belt interpreted as a former remnant arc that could be compared with the Three Kings Ridge, and the Gander Group, which accumulated between the Ultramafic Belt and the old Avalon continent (Williams 1979), having accumulated in a basin similar to the Norfolk Basin which is flanked on the W by continental crust of the Norfolk Ridge. The gradation between the Gander Group and sedimentary rocks deposited W of the Ultramafic Belt results from the progressive subsidence of the remnant arc and its ultimate burial by marginal basin deposits. The pinching-out of the Dunnage Zone against the Cabot Promontory in southern Newfoundland (Williams 1979) may in part reflect the lateral termination of the inferred Ordovician arc-basin assemblage against older rocks, in a manner similar to the termination of the arcs and basins E of the Norfolk Ridge against New Zealand (Fig. 1).

Cas & Jones (1979) considered the middle Palaeozoic Hill End Trough in E Australia to be an ancient analogue of the southern termination of the Havre Basin. Both are flanked by arc volcanics, pass S into regions of intense silicic magmatism, and are atypical of the general SW Pacific pattern in containing a thick sequence of clastic rocks (see Lewis & Pantin, this volume). It is unclear if this part of the Havre Basin has an oceanic basement, and unlikely that the Hill End Trough is so floored. However, the volume of sediment that accumulates in marginal settings of this type suggests that they may be preferentially preserved in the stratigraphic record and hence be of greater significance than their present areal extent might suggest.

ACKNOWLEDGMENTS: The greater part of this paper was written during study leave from the University of Sydney. Thanks are extended to Professors E. R. Oxburgh and H. B. Whittington for making available to me facilities within the Department of Earth Sciences, University of Cambridge. Discussions with W. R. Dickinson, G. H. Packham, J. Rodgers, A. G. Smith and R. Trumpy are gratefully acknowledged.

References

ANDREWS, J. E., PACKHAM, G. H., EADE, J. V. *et al.* 1975. Part II: Site reports. *Init. Rep. Deep Sea drill. Proj.* **30**, 25–398. U.S. Government Printing Office, Washington, D.C.

BENTZ, F. P. 1974. Marine geology of the southern Lord Howe Rise, Southwest Pacific. *In*: BURK, C. A. & DRAKE, C. L. (eds) *The Geology of Continental Margins*, 537–47. Springer Verlag, Berlin.

BURNS, R. E., ANDREWS, J. E. VAN DER LINGEN, G. J. *et al.* 1973. Part I: Site Reports *Init. Rep. Deep Sea drill. Proj* **21**, 1–440. U.S. Government Printing Office, Washington, D.C.

CAREY, S. W. 1958. A tectonic approach to continental drift. *In*: CAREY, S. W. (ed.) *Continental Drift: A Symposium*, 177–355. University of Tasmania, Hobart.

CAS, R. A. F. & JONES, J. G. 1979. Paleozoic interarc basin in eastern Australia and a modern analogue. *N.Z. Jl Geol. Geophys.* **22**, 71–85.

CHURKIN, M. 1974. Deep sea drilling for landlubber geologists—the Southwest Pacific, an accordion plate tectonics analog for the Cordilleran Geosyncline. *Geology*, **2**, 339–42.

CONEY, P. J., JONES, D. L. & MONGER, J. W. H. 1980. Cordilleran suspect terranes. *Nature, Lond.* **288**, 329–33.

CONNELLY, J. B. 1976. Tectonic development of the Bismarck Sea based on gravity and magnetic modelling. *Geophys. J. R. astr. Soc.* **46**, 23–40.

CROOK, K. A. W. & BELBIN, L. 1978. The Southwest Pacific area during the last 90 million years. *J. geol. Soc. Aust.* **25**, 23–40.

CURRIE, K. L., PICKERILL, R. K. & PAJARI, G. E. 1980. An Early Paleozoic plate tectonic model of Newfoundland. *Earth planet. Sci. Lett.* **48**, 8–14.

CURRY, G. B., INGHAM, J. K., BLUCK, B. J. & WILLIAMS, A. 1982. The significance of a reliable Ordovician age for some Highland Border rocks in central Scotland. *J. geol. Soc. London*, **139**, 451–4.

DANIEL, J., JOUANNIC, C., LARUE, B. M. & RECY, J. 1977. Interpretation of D'Entrecasteaux Zone (north of New Caledonia). *International Symposium on Geodynamics in South West Pacific*, 117–24. Editions Technip, Paris.

——, ——, —— & —— 1978. Marine geology of eastern Coral Sea (eastern margin of Indo-Australian Plate, north of New Caledonia). *South Pacific Mar. geol. Notes*, **1**, 81–94.

DAVEY, F. J. 1977. Marine seismic measurements in the New Zealand region. *N.Z. Jl Geol. Geophys.* **20**, 719–77.

—— 1982. The structure of the South Fiji Basin. *Tectonophysics*, **87**, 65–107.

DAVIES, H. L. & SMITH, I. E. 1971. Geology of eastern Papua. *Bull. geol. Soc. Am.* **82**, 3299–312.

DEWEY, J. F. 1974. Continental margins and ophiolite obduction: Appalachian Caledonian system. *In*: BURK, C. A. & DRAKE, C. L. (eds) *The Geology of Continental Margins*, 933–50. Springer-Verlag, Berlin.

——, 1977. Suture zone complexities: a review. *Tectonophysics*, **40**, 54–67.

—— & KIDD, W. S. F. 1974. Continental collisions in the Appalachian–Caledonian orogenic belt: variations related to complete and incomplete suturing. *Geology*, **2**, 543–6.

FALVEY, D. A. 1978. Analysis of palaeomagnetic data from the New Hebrides. *Bull. Aust. Soc. explor. Geophys.* **9**, 117–23.

—— & MUTTER, J. C. 1981. Regional plate tectonics and the evolution of Australia's passive continental margins. *BMR J. Aust. Geol. Geophysics.* **6**, 1–29.

GILL, J. B. 1976. Composition and age of Lau Basin and Ridge volcanic rocks: implications for evolution of an interarc basin and remnant arc. *Bull geol. Soc. Am.* **87**, 1384–95.

—— 1981. *Orogenic Andesites and Plate Tectonics*, 390 pp. Springer-Verlag, Berlin.

GRAHAM, A. M. & UPTON, B. G. J. 1978. Gneisses in diatremes, Scottish Midland Valley: petrology and tectonic implications *J. geol. Soc. London*, **135**, 219–28.

HAMILTON, W. 1979. Tectonics of the Indonesian region. *Prof. Pap. U.S. geol. Surv.* **1078**, 1–345.

HAWKINS, J. W. 1974. Geology of the Lau Basin, a marginal sea behind the Tonga Arc. *In*: BURK, C. A. & DRAKE, C. L. (eds) *The Geology of Continental Margins*, 505–20. Springer-Verlag, Berlin.

—— 1976. Petrology and geochemistry of basaltic rocks of the Lau Basin. *Earth planet. Sci. Lett.* **28**, 283–97.

—— 1977. Petrologic and geochemical characteristics of marginal basin basalts. *In*: TALWANI, M. & PITMAN, W. C. (eds) *Island Arcs, Deep Sea Trenches, and Back-Arc Basins, Maurice Ewing Series*, **1**, 355–65. Am. Geophys. Union, Washington, D.C.

HAYES, D. E. & RINGIS, J. 1973. Seafloor spreading in the Tasman Sea. *Nature (phys. Sci.)* **243**, 454–8.

HENDERSON, W. G. & ROBERTSON, A. H. F. 1982. The Highland Border rocks and their relation to marginal basin development in the Scottish Caledonides. *J. geol. Soc. London*, **139**, 433–50.

HORNE, G. S. 1969. Early Ordovician chaotic deposits in the Central Volcanic Belt of northeastern Newfoundland. *Bull. geol. Soc. Am.* **80**, 2451–64.

VON HUENE, R., AUBOUIN, J., AZEMA, J. *et al.* 1980a. Leg 67: The Deep Sea Drilling Project Mid America Trench transect of Guatemala. *Bull. geol. Soc. Am.* **91**, 421–32.

——, LANGSETH, M., NASU, N. & OKADA, H. 1980b. Summary, Japan Trench transect. *Init. Rep. Deep Sea drill. Proj.* **56–7**, 473–88. U.S. Government Printing Office, Washington, D.C.

HUSSONG, D. & UYEDA, S. 1982. Tectonic processes and the history of the Mariana Arc: a synthesis of the results of Deep Sea Drilling Project Leg 60. *Init. Rep. Deep Sea drill. Proj.* **60**, 909–29. U.S. Government Printing Office, Washington, D.C.

IBRAHIM, A. K., PONTOISE, B., LATHAM, G., LARUE, M., CHEN, T., ISACKS, B., RECY, J. & LOUAT, R. 1980. Structure of the New Hebrides arc–trench system. *J. geophys. Res.* **85**, 253–66.

KARIG, D. E. 1970. Ridges and basins of the Tonga–Kermadec island arc system. *J. geophys. Res.* **75**, 239–54.

—— 1971. Origin and development of marginal basins in the western Pacific. *J. geophys. Res.* **76**, 2542–61.

—— 1972. Remnant arcs. *Bull. geol. Soc. Am.* **83**, 1057–67.

—— & MAMMERICKX, J. 1972. Tectonic framework of the New Hebrides Island Arc. *Mar. Geol.* **12**, 187–205.

—— & MOORE, G. F. 1975. Tectonically controlled sedimentation in marginal basins. *Earth planet. Sci. Lett.* **26**, 233–8.

—— & SHARMAN, G. F. 1975. Subduction and accretion in trenches. *Bull. geol. Soc. Am.* **86**, 377–89.

KATZ, H. R. 1974. Margins of the Southwest Pacific. *In*: BURK, C. A. & DRAKE, C. L. (eds). *The Geology of Continental Margins*, 349–65. Springer-Verlag, Berlin.

KAY, M. 1975. Campbellton sequence, manganiferous beds adjoining the Dunnage Melange, northeastern Newfoundland. *Bull. geol. Soc. Am.* **86**, 105–8.

KENNETT, J. P., BURNS, R. E., ANDREWS, J. E., CHURKIN, M., DAVIES, T. A., DUMITRICA, P., EDWARDS, A. R., GALEHOUSE, J. W., PACKHAM, G. H. & VAN DER LINGEN, G. J. 1972. Australian–Antarctic continental drift, paleocirculation changes and Oligocene deep sea erosion. *Nature (phys. Sci.)* **239**, 51–5.

KENNETT, J. P., HOUTZ, R. E., ANDREWS, P. B. *et al.* 1975. Site reports. *Init. Rep. Deep Sea drill. Proj.* **29**, 19–445. U.S. Government Printing Office, Washington, D.C

KLEIN, G. DE V. 1975. Sedimentary tectonics in southwest Pacific Marginal Basins based on Leg 30 Deep Sea Drilling Project cores from the South Fiji, Hebrides, and Coral Sea Basins. *Bull. geol. Soc. Am.* **86**, 1012–8.

LAPOUILLE, A. 1978. Southern New Hebrides Basin and western South Fiji Basin as a single marginal basin. *Bull. Aust. Soc. explor. Geophys.* **9**, 130–3.

LARUE, B. M., DANIEL, J., JOUANNIC, C. & RECY, J. 1977. The South Rennell Trough: evidence for a fossil spreading zone. *International Symposium on Geodynamics in South West Pacific*, 51–61. Editions Technip, Paris.

LAWVER, L. A. & HAWKINS, J. W. 1978. Diffuse magnetic anomalies in marginal basins: their possible tectonic and petrologic significance. *Tectonophysics*, **45**, 323–39.

LEITCH, E. C., 1970. Contributions to the geology of northernmost New Zealand. II. The stratigraphy of the North Cape district. *Trans. R. Soc. N.Z., Earth Sci.* **8**, 45–68.

—— 1975. Mesozoic–Middle Tertiary Development of northern New Zealand. *Bull. Aust. Soc. explor. Geophys.* **6**, 56–8.

—— 1981. Active margins in the Australasian region—problems for resolution. *In*: COOK, P. J., CROOK, K. A. W. & FRAKES, L. A. (eds) *The Future of Scientific Ocean Drilling in the Australasian Region*, 55–66. Consortium for Ocean Geosciences, Canberra.

VAN DER LINGEN, G. J. 1973. The Lord Howe Rise rhyolites. *Init. Rep. Deep Sea drill. Proj.* **21**, 523–39. U.S. Government Printing Office, Washington, D.C.

LUYENDYK, B. P., MACDONALD, K. C. & BRYAN, W. B. 1973. Rifting history of the Woodlark Basin in the southwest Pacific. *Bull. geol. Soc. Am.* **84**, 1125–34.

MALAHOFF, A., FEDEN, R. H. & FLEMING, H. S. 1982. Magnetic anomalies and tectonic fabric of marginal basins north of New Zealand. *J. geophys. Res.* **87**, 4109–25.

MATTINSON, J. M. 1975. Early Paleozoic ophiolite complexes of Newfoundland: isotopic ages of zircon. *Geology*, **3**, 181–3.

McDOUGALL, I. & WELLMAN, P. 1976. Potassium-argon ages for some Australian Mesozoic igneous rocks. *J. geol. Soc. Aust.* **23**, 1–9.

McKERROW, W. S. & COCKS, L. R. M. 1981. Stratigraphy of eastern Bay of Exploits. *Can. J. Earth Sci.* **18**, 751–64.

MUTTER, J. C. 1977. The Queensland Plateau. *Bur. Miner. Res. Bull.* **179**, 1–55.

—— & JONGSMA, D. 1978. The pattern of the pre-Tasman Sea rift system and the geometry of breakup. *Bull. Aust. Soc. explor. Geophys.* **9**, 70–5.

—— & KARNER, G. 1980. The continental margin off northeast Australia. *In*: HENDERSON, R. A. & STEPHENSON, P. J. (eds) *The Geology and Geophysics of Northeastern Austalia*, 47–69. Geol. Soc. Aust. (Queensland Div.), Brisbane.

OLIVER, P. J., MUMME, T. C., GRINDLEY, G. W. & VELLA, P. 1979. Paleomagnetism of the Upper Cretaceous Mount Somers Volcanics, Canterbury, New Zealand. *N.Z. Jl Geol. Geophys.* **22**, 199–212.

PACKHAM, G. H. 1973. A speculative Phanerozoic history of the South-west Pacific. *In*: COLEMAN, P. J. (ed.) *The Western Pacific: Island arcs, Marginal Seas, Geochemistry*, 369–88. Univ. of Western Australia, Perth.

—— 1978. Evolution of a simple island arc: the Lau–Tonga Ridge. *Bull. Aust. Soc. explor. Geophys.* **9**, 133–40.

—— & ANDREWS, J. E. 1975. Results of Leg 30 and the geologic history of the southwest Pacific arc and marginal sea complex. *Init. Rep. Deep Sea drill. Proj.* **30**, 691–705. U.S. Government Printing Office, Washington, D.C.

—— & FALVEY, D. A. 1971. An hypothesis for the formation of marginal seas in the Western Pacific. *Tectonophysics*, **11**, 79–109.

—— & TERRILL, A. 1975. Submarine geology of the South Fiji Basin. *Init. Rep. Deep Sea drill. Proj.* **30**, 617–33. U.S. Government Printing Office, Washington, D.C.

PAJARI, G. E. & CURRIE, K. L. 1978. The Davidsville and Gander Lake Groups of northeastern Newfoundland; a re-examination. *Can. J. Earth Sci.* **15**, 708–14.

PIRAJNO, F. 1979. Geology, geochemistry and mineralisation of a spilite–keratophyre association in Cretaceous flysch, East Cape area, New Zealand. *N.Z. Jl Geol. Geophys.* **22**, 307–28.

SCHOLL, D. W., VON HUENE, R., VALLIER, T. L. & HOWELL, D. G. 1980. Sedimentary masses and concepts about tectonic processes at underthrust ocean margins. *Geology*, **8**, 564–8.

SCOTT, R. B., KROENKE, L., ZAKARIADZE, G. & SHARASKIN, A. 1981. Evolution of the South Philippine Sea: Deep Sea Drilling Project Leg 59 Results. *Init. Rep. Deep Sea drill. Proj.* **59**, 803–15. U.S. Government Printing Office, Washington, D.C.

SHAW, R. D. 1978. Sea floor spreading in the Tasman Sea: A Lord Howe Rise–eastern Australian reconstruction. *Bull. Aust. Soc. explor. Geophys.* **9**, 75–81.

STOESER, D. B. 1975. Igneous rocks from Leg 30 of the Deep Sea Drilling Project. *Init. Rep. Deep Sea drill. Proj.* **30**, U.S. Government Printing Office, Washington, D.C.

STRONG, D. F. 1977. Volcanic regimes of the Newfoundland Appalachians. *Geol. Soc. Can. Spec. Paper*, **16**, 61–90.

TAYLOR, B. 1979. Bismarck Sea: evolution of a back-arc basin. *Geology*, **7**, 171–4.

TAYLOR, L. W. H. 1975. Depositional and tectonic patterns in the western Coral Sea. *Bull. Aust. Soc. explor. Geophys.* **6**, 33–5.

—— & FALVEY, D. A. 1977. Queensland Plateau and Coral Sea Basin: stratigraphy, structure and tectonics. *Aust. petrol. Explor. Assoc. J.* **17**, 13–29.

TERRILL, A. 1975. Depositional and tectonic patterns in the northern Lord Howe Rise–Mellish Rise area. *Bull. Aust. Soc. explor. Geophys.* **6**, 37–9.

TRUMPY, R. 1960. Paleotectonic evolution of the central and western Alps. *Bull. geol. Soc. Am.* **71**, 843–908.

UPADHYAY, H. D., DEWEY, J. F. & NAGLE, E. R. W. 1971. The Betts Cove ophiolite complex, Newfoundland: Appalachian oceanic crust and mantle. *Geol. Assoc. Can. Proc.* **24**, 27–34.

VOGT, P. R. & CONOLLY, J. R. 1971. Tasmantid Guyots, the age of the Tasman Basin and motions between the Australian plate and the mantle. *Bull. geol. Soc. Am.* **82**, 2577–84.

WATTS, A. B., WEISSEL, J. K. & DAVEY, F. J. 1977. Tectonic evolution of the South Fiji Marginal Basin. *In*: TALWANI, M. &. PITMAN, W. C. (eds) *Island Arcs, Deep Sea Trenches, and Back-Arc Basins, Maurice Ewing Series*, **1**, 419–27. Am. Geophys. Union, Washington, D.C.

WEISSEL, J. K. 1977. Evolution of the Lau Basin by the growth of small plates. *In*: TALWANI, M. & PITMAN, W. C. (eds) *Island Arcs, Deep Sea Trenches, and Back-Arc Basins, Maurice Ewing Series*, **1**, 429–36. Am. Geophys. Union, Washington, D.C.

—— & HAYES, D. E. 1977. Evolution of the Tasman Sea reappraised. *Earth planet. Sci. Lett.* **36**, 77–84.

—— & WATTS, A. B. 1979. Tectonic evolution of the Coral Sea Basin. *J. geophys. Res.* **84**, 4572–82.

——, TAYLOR, B. & KARNER, G. D. 1982a. The opening of the Woodlark Basin, subduction and the evolution of northern Melanesia since mid Pliocene time. *Tectonophysics*, **87**, 253–77.

——, WATTS, A. B. & LAPOUILLE, A. 1982b. Evidence for Late Paleocene to Late Eocene seafloor in the southern New Hebrides Basin. *Tectonophysics*, **87**, 243–51.

WILLIAMS, H. 1979. Appalachian Orogen in Canada. *Can. J. Earth Sci.* **16**, 792–807.

——, DALLMEYER, R. D. & WANLESS, R. K. 1976. Geochronology of the Twillingate Granite and Herring Neck Group, Notre Dame Bay, Newfoundland. *Can. J. Earth Sci.* **13**, 351–60.

E. C. LEITCH, Department of Geology and Geophysics, University of Sydney, NSW 2006, Australia.

Taupo–Rotorua Depression: an ensialic marginal basin of North Island, New Zealand

J. W. Cole

SUMMARY: New Zealand lies on the boundary between the Pacific and Indian plates. In the vicinity of North Island, oceanic crust of the Pacific plate is subducted obliquely beneath continental crust of the Indian plate to form the Taupo–Hikurangi arc–trench system.

The volcanic component of this system is the Taupo Volcanic Zone which comprises a young (<50,000 yr) andesite–dacite arc, to the west of which is the Taupo–Rotorua Depression, a basin filled with approximately 2 km of low-density volcanics. Most of these volcanics are rhyolitic (total volume > 10,000 km^3) and have been erupted in the last 0.6 Ma from the Rotorua, Okataina, Maroa and Taupo Volcanic Centres. Approximately 2 km^3 of high-alumina basalt has also been erupted from the Depression.

Normal faults are common in the Taupo–Rotorua Depression and some have been active in historic time when open fissures formed. No strike-slip movement has been recorded. This evidence, together with the presence of basalt dykes, suggests the area is an extensional ensialic marginal basin. A change in orientation of the faults, from 040° at Taupo to 090° at the N end of the Okataina Volcanic Centre, is considered to be a result of dextral strike-slip movement within the North Island Shear Belt superimposed on the extensional tectonics of the marginal basin. The resultant oblique extension would also account for the *en échelon* basalt dykes in the 1886 Tarawera rift.

Crustal thickness in the marginal basin is estimated to be 15 km compared to 30–35 km outside the basin. This may account for the high heat flow recorded (average may be up to 1575 mW m^{-2}) and for partial melting of the base of the crust to produce the voluminous rhyolitic volcanics. The high-alumina basalts are considered to be derived from the upper mantle. They form dykes in the crust which in places reach the surface to form a series of vents aligned parallel to the regional fault pattern. The basalts are considered to be intruded and extruded mainly as a response to the extensional tectonics.

The boundary between the Pacific and Indian plates is largely convergent (Fig. 1). Along the northern part of the boundary, the Indian plate is subducted beneath the Pacific plate to form the Solomon Islands and New Hebrides ensimatic arcs (Mitchell & Warden 1971; Karig & Mammerickx 1972), and in the central part the Pacific plate is subducted beneath the Indian plate to form the Tonga and Kermadec ensimatic arcs (Oliver & Isacks 1967; Karig 1970, 1971). The southern part of the boundary is marked by the Macquarie Ridge Complex (Hayes *et al*. 1972) which is partly convergent and partly transform (Fig. 1). Present-day shift pole for the two plates is given by Walcott (1978) as 60°S, 180° and by Chase (1978) as 62°S, 174°E. Rotation is estimated to be 1.27° Ma^{-1} (Sissons 1979).

In North Island, the Pacific plate is obliquely subducted beneath continental crust of the Indian plate (Fig. 2) to form the Taupo–Hikurangi arc–trench system (Cole & Lewis 1981). In South Island, continental crust of the Chatham Rise is obliquely obducted on to continental crust of the Indian plate, much of the movement being taken up by reverse-dextral movement on the Alpine Fault (Fig. 2).

Wellman (1964) considers rocks on either side of this fault to have a total offset of 450 km. In the southern part of South Island, oceanic crust of the Indian plate is obliquely subducted beneath continental crust of the Pacific plate (Christoffel 1971) with the boundary marked by the Puysegur Trench (Fig. 2).

Taupo–Hikurangi arc–trench system

The Taupo–Hikurangi arc–trench system extends from the Hikurangi Trough on the E side, to the Taupo Volcanic Zone on the W (Fig. 2). The Hikurangi Trough is not directly continuous with the Kermadec Trench but is slightly offset in terms of bathymetry, and the area between the two is seismically quiet (Eiby 1977; Ansell 1978). Katz (1974) and van der Lingen (1982) do not regard the trough as a subduction-related trench, but, as the Benioff zone dips at a shallow angle from this location (Adams & Ware 1977; Reyners 1980), Cole & Lewis (1981) regard it as the easternmost limit of the Indian plate.

For some 200 km W of the Hikurangi Trough the Benioff zone dips at about 5° until it

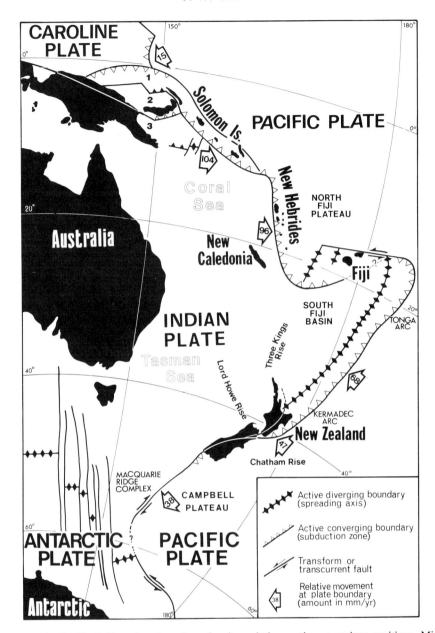

FIG. 1. The Pacific–Indian plate boundary showing relative motion at various positions. Minor plates in the N are: 1—North Bismarck plate; 2—South Bismarck plate; 3—Solomon Sea plate. Based on map by Wellman & McCracken (1979).

reaches a depth of approximately 25 km. The dip then changes to about 20° and the zone continues at this angle until it reaches a depth of 80 km under the Taupo Volcanic Zone. Here a second increase in dip occurs to about 55° and the zone continues to its maximum depth of 250 km (Reyners 1980).

Above the low-angled part of the Benioff zone is an accretionary prism up to 150 km wide and characterized by a series of imbricate thrust faults, along which movement becomes progressively more oblique (dextral) towards the W. On the outer edge of the prism is the accretionary slope, comprising a series of ridges and basins

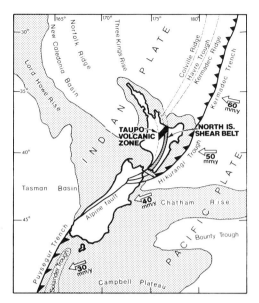

FIG. 2. Major components of the Pacific–Indian plate boundary in the New Zealand region. Stipple represents continental crust. Arrows show motion of Pacific plate relative to the Indian plate. Rates are from Walcott (1978).

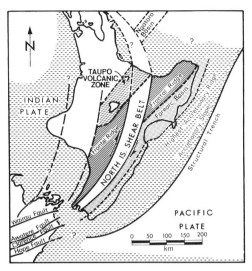

FIG. 3. Taupo–Hikurangi arc–trench system in the North Island. The offshore area of continental crust is shaded.

5–30 km wide and 10–60 km long. Sediments on the accretionary slope are progressively older away from the trench, with the oldest sediments forming the highest accretionary ridge. The inner part of the prism is structurally a fore-arc basin filled by thick Plio–Pleistocene sediments. These sediments are crossed by strike-slip faults which increase in number and in displacement westwards. The locations of the main features of the accretionary prism are shown on Fig. 3 (after Cole & Lewis 1981).

The prism is bound on the W by a frontal ridge composed of Upper Palaeozoic–Mesozoic greywackes and argillites. This ridge is undergoing rapid uplift. Present uplift rates are estimated to vary from 2 to 7 mm yr^{-1} along the axis of the ranges (Wellman 1967), and studies of summit height surfaces at the S end of North Island, give rates of 3.0–4.5 mm yr^{-1} (Ghani 1978).

Cutting obliquely across the frontal ridge is the North Island Shear Belt (Fig. 3), a zone of major dextral strike-slip faults. These faults are still active and have a combined dextral displacement of 14–18 mm yr^{-1} (Lensen 1975; Sissons 1979) and a reverse component of 4 mm yr^{-1} (Sissons 1979). This belt is regarded as the western limit of the movement caused by relative motions of the Pacific and Indian

plates. The oblique subduction shown in Figs 1 and 2 can therefore be resolved so that movement at the trench is almost entirely normal to the boundary, and movement along the shear belt is mainly strike-slip.

The Taupo Volcanic Zone (Healy 1962) extends NNE from Ohakune to White Island and comprises a main andesitic arc, to the W of which is the Taupo–Rotorua Depression (Fig. 4).

At the S end of the main andesite arc is the Tongariro Volcanic Centre which consists of four major andesite massifs (Cole 1978a). Much of the lava forming these massifs was erupted within the last 0.3 Ma (Cole 1981) from aligned vents within a series of NW-trending *en échelon* zones. The orientation of these zones is similar to that of aligned vents within an older andesite arc to the N (Fig. 4). Volcanoes in this arc are progressively younger southwards (Cole 1978b), so that the Tongariro Volcanic Centre may represent the youngest eruptive centre of this feature. Crossing the *en échelon* zones are a series of younger (<50,000 yr) vents from which predominantly olivine-bearing andesite, low-Si andesite and low-Al basalt have been erupted. These vents are aligned NNE, parallel to the major lineaments of the Taupo–Hikurangi system (Cole 1979), and it is these which are considered to form the southern part of the present-day main andesite arc.

Young andesite-dacite volcanoes also occur at the N end of the main andesite arc at Man-

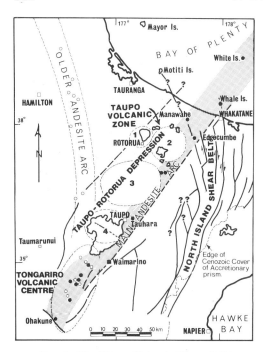

FIG. 4. The Taupo Volcanic Zone. Solid circles are locations of recent andesite or dacite volcanoes; open circles are andesite volcanoes of a NW-trending arc; solid square represents Waimarino olivine basalt. Rhyolitic volcanic centres are as follows: 1—Rotorua; 2—Okataina; 3—Maroa; 4—Taupo.

awahe, Whale Island and White Island (Duncan 1970; Cole 1979). Lavas are predominantly dacite but in March 1977 olivine-bearing andesite, similar to that erupted from the NNE-trending vents of the Tongariro Volcanic Centre, was erupted at White Island (Clark *et al.* 1979). In the middle section of the arc andesite-dacite lavas have only reached the surface at four localities (Fig. 4). At Tauhara, 8 km E of Taupo, magnesian olivine phenocrysts (Fo_{83}) occur in a dacite, suggesting that andesitic magma, perhaps similar to the young Tongariro lavas or the 1977 White Island lava, occurs at depth under this volcano.

The most primitive lava in the Taupo Volcanic Zone is the Waimarino olivine basalt which has been erupted in the last 20,000 yr (W. M. Hackett, pers. comm. 1982).

Taupo–Rotorua depression

Behind the main andesite arc is a series of linear faulted depressions which Grindley (1960) col-

lectively named the Taupo–Rotorua Depression. These extend from Lake Taupo in the S to Lake Rotorua in the N. The depresssion is filled with 2.0 ± 0.2 km of low-density, low-velocity, predominantly rhyolitic pyroclastics and lavas (T. A. Stern, pers. comm. 1982), which have been erupted from four rhyolitic volcanic centres: Rotorua, Okataina, Maroa and Taupo (Fig. 5). The volcanic centres are considered to be multiple calderas formed after voluminous eruptions spread thick ignimbrite sheets on either side of the Taupo Volcanic Zone. The maximum age of the ignimbrites clearly demonstrable as having been erupted from the Taupo–Rotorua Depression is 0.6 Ma (Cole 1981). Smaller volumes of rhyolite lava and tephra have subsequently been erupted within each centre. The total volume of rhyolitic volcanics is estimated to be $>10,000$ km^3 (Cole 1981).

Late in the history of Maroa, Taupo and Okataina volcanic centres there were small eruptions of high-alumina basalt (Cole 1972, 1979). This basalt has a total volume of about 2 km^3.

Lineaments in the Taupo–Rotorua Depression

Normal faulting

Normal faults, commonly with topographic expression, are abundant within the Taupo–Rotorua Depression (Fig. 6; for examples see also Hochstetter 1864; Grange 1937; Grindley 1960; Nairn 1971, 1976, 1981). Grindley (1960) considered that there were two episodes of faulting, in the Early and the Late Pleistocene, and that faulting in the latter episode formed, what he termed, the 'Taupo Fault Belt'. However, there is little evidence of a time break and it is more likely that faulting continued throughout the tectonic history of the depression. Many of the fault traces show displacements of 1 m or less, but others (e.g. Paeroa, Ngakuru, Whangamata and Whakaipo Faults; Fig. 5) have displacements of many tens of metres. No strike-slip movement has been identified.

Historic earth movements with open fissures have been recorded at three locations. Hochstetter (1864) observed a large and widely gaping fissure to the SE of Earthquake Flat (Fig. 6) and Thomas (1888) reported the re-opening of this and many other fissures in the area during the 1886 Tarawera eruption. These fissures were 5 cm to 2 m wide and sufficiently numerous 'to make it difficult to take a horse across country' (Smith 1886).

In August 1895, a series of NE-striking fis-

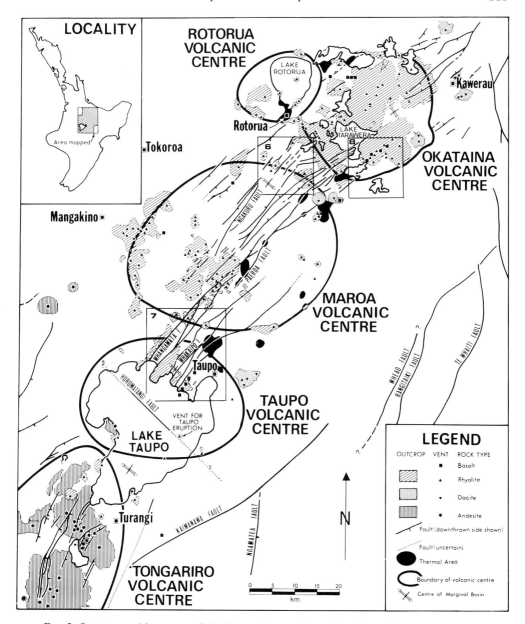

FIG. 5. Structure and lava types of the Taupo–Rotorua Depression. Boxes bound areas shown in Figs 6, 7 and 8.

sures opened up across the W flank of Maunganamu, which is a small rhyolite dome 5 km SE of Taupo. Many slips occurred and much damage was done to roads and tracks (Hill 1911; Henderson 1932). No vertical displacement was recorded.

From May to December 1922 more numerous earth movements occurred on the northern shores of Lake Taupo (Fig. 7). Fissures up to 30 cm wide, 1 m deep and tens of metres long were reported from a number of localities (Ward 1922). Vertical displacements ranged from 0.3 to 4 m (Fig. 7) and, by the end of the episode, total subsidence at the centre of Whakaipo Bay was 4 m relative to the level of the outlet of the lake (Grange 1932).

Fault movement is known to have occurred throughout the late Quaternary. Nairn (1971)

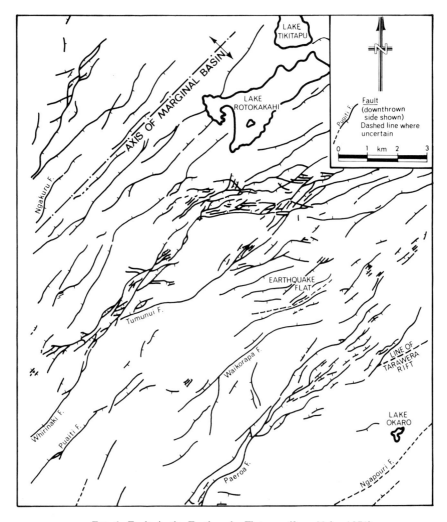

FIG. 6. Faults in the Earthquake Flat area (from Nairn 1971).

has demonstrated from the displacement of tephra layers that movement has occurred on several occasions during the last 42,000 yr. The relationship between near-surface and subsurface faulting has also been investigated in the Wairakei and Waiotapu geothermal fields (Grindley 1963, 1965). At these locations Grindley (1965) showed that there were few major faults at depth, but abundant faults at the surface. The fault planes were steeply dipping (80–85°) and their throw increased with depth indicating repeated movement throughout the accumulation of the volcanics.

Orientation of the normal faults changes progressively along the length of the Taupo–Rotorua Depression. In the northern part of the Taupo Volcanic Centre and in the Maroa Volcanic Centre the strike averages 040° (Fig. 5), but in the Okataina Volcanic Centre it is 057° and at the northern margin many faults trend 080–090° (Nairn 1981). Faults are not continuous throughout the depression and appear to occupy a series of *en échelon* fault belts. This is most clearly demonstrated in the Maroa Volcanic Centre (Fig. 5). Some NW-trending faults have also been recognized. Northey (1983) has described the NW-trending Horomatangi Fault across Lake Taupo (Fig. 5) and Modriniak & Studt (1959) have suggested, on the basis of gravity and magnetic data, that a NW-trending structure (Maroa graben) occurred in the southern part

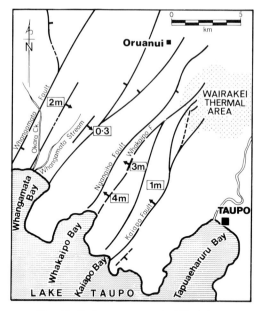

FIG. 7. Fault movement in Whakaipo—Whangamata Bay areas north of Lake Taupo from May to December 1922 (from Grange 1932).

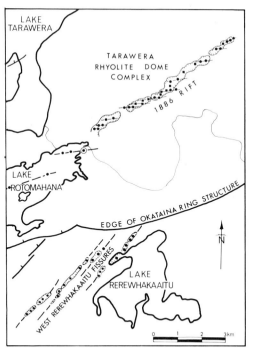

FIG. 8. *En échelon* arrangement of vents in the Tarawera–Lake Rerewhakaaitu area. Dots represent explosive vents, squares represent basalt dyke exposures (data from Nairn 1981 and Nairn & Cole 1981).

of the Maroa Volcanic Centre. However the latter feature has not been clearly identified from surface geology.

Alignment of basaltic vents

High-alumina basalt lavas have been erupted within the Maroa, Okataina and Taupo Volcanic Centres (Cole 1972, 1979). Vents are typically aligned parallel to the regional fault trend and probably represent the surface expressions of basaltic dykes.

At K Trig, 5 km W of Taupo, four small scoria cones are aligned parallel to NNE faults in the area (Fig. 5) and a magnetic anomaly within Lake Taupo, along strike to the S, probably marks the location of a fifth vent. Another basaltic vent alignment occurs at Rotokawau in the NW part of the Okataina Volcanic Centre (Fig. 5). High-alumina basalt was erupted from five craters approximately 4,000 yr ago. Three of the craters are linked and all five are aligned at 090°.

The most voluminous eruption (approximately 0.7 km³) of high-alumina basalt took place on 10 June 1886, from Mt Tarawera. Basaltic scoria was erupted from an *en échelon* series of dykes (trending between 073° and 080°) across the mountain (Fig. 8). The eruption lasted 4.5 h and in its final phase explosions formed the series of coalescing craters referred

to as the Tarawera rift (Cole 1970). This rift has a general trend of 057° (Fig. 8). Most dykes are less than 2 m wide and appear to have filled pure dilatational fractures without detectable horizontal or vertical shear displacement of the fissure walls (Nairn & Cole 1981). This evidence, together with the relatively small size of the eruption, suggests that the dykes were passively emplaced during or immediately after dilatation had occurred.

Contemporaneous with this activity, smaller amounts of basalt were erupted from Lake Rotomahana (Fig. 8) to the SW, and still further in this direction from craters at Waimangu (Nairn 1979). No dykes are exposed at either location but the trends of the vents are the same as those at Tarawera which strongly suggests that the 'feeder' dyke for the Tarawera eruption extends as far SW as Waimangu, along a total length of about 17 km.

To the S of Tarawera, a series of explosion vents, the West Rerewhakaaitu Fissures (Fig. 8), were formed about 10,000 yr ago. They have an *en échelon* arrangement, with vents aligned between 042° and 048°. Nairn (1981) considers that they result from phreatic explosions triggered by the intrusion of basalt dykes to a shallow depth. There are likely to be many other dykes in the Taupo–Rotorua Depression which have not reached the surface. This conclusion is supported by the occurrence of basaltic fragments in many rhyolitic tephras. Indeed there is evidence from some tephras that intrusion of a basaltic dyke or dykes may have triggered the rhyolitic eruption.

Other vent alignments

In the Okataina Volcanic Centre, Nairn (1981) recognizes an alignment of rhyolite domes along the Haroharo and Tarawera lineaments. Both are parallel to the 1886 rift (approximately 057°). Similar alignments occur in the southern part of Lake Taupo (Fig. 5). The lineaments probably reflect deep-seated faults which have acted as channelways for the magmas and there is some indication that the locations of the vents along these faults may be controlled by cross-fractures (Northey 1983).

Heat flow

The Taupo–Rotorua Depression is an area of particularly high heat flow, most of which reaches the surface through hydrothermal systems (Fig. 5). Some systems are large enough to form commercial geothermal fields, as at Wairakei (Fig. 7).

Total natural output for the Taupo Volcanic Zone, most of which is through the Taupo–Rotorua Depression, is estimated by Studt & Thompson (1969) to be 3.3×10^9 W and by Hochstein (1976) to be $4.0–6.3 \times 10^9$ W. Assuming an area of the Taupo Volcanic Zone of 4000 km^2, these values give an average range of heat flows of $825–1575$ mW m^{-2}. Normal continental crust is estimated by Lee & Uyeda (1965) to have an average heat flow value of 59.2 ± 21.8 mW m^{-2}, so Taupo heat flow may be up to 27 times higher than normal.

Hochstein (1976) has indicated that this high heat flow could be explained by a slab of molten material at 1200°C and a depth of 10 km, but this is not supported by studies of S-wave attenuation or P-wave travel time delays within the crust (Robinson *et al.* 1981), nor is it likely from geological considerations. Small, local reservoirs of magma could occur at higher levels within the crust, with little affect on P and S waves, but it is unlikely that these are basalt dykes. At Tarawera the dykes are only 1–2 m wide, so they are likely to solidify very quickly

with little heat contributed to the surrounding rocks. From equilibration studies, Ewart *et al.* (1971) suggested that the 'low-temperature' rhyolitic magmas of the Taupo Volcanic Zone accumulated at depths of 5–7 km, and, as the most recent rhyolitic eruption at Tarawera (Kaharoa eruption) took place only 600 yr ago (I. A. Nairn, pers. comm. 1982), some partially crystallized acidic magma may still be present. Such reservoirs, however, are likely to be small and are therefore unlikely to provide the amount of heat necessary to account for the total flow for the zone.

High heat flow in back-arc spreading areas is well known and is commonly related to rotational asthenosphere flow caused by the subduction (Andrews & Sleep 1974; Toksöz & Hsui 1978). It may well be that a similar explanation applies to New Zealand. Sissons (1979) considers that as a result of this flow the boundary between asthenosphere and lithosphere under the Taupo–Rotorua Depression may be unusually shallow, possibly at a depth of only 10–15 km. If so, hydrothermal circulation may penetrate the entire thickness of the lithosphere to provide a very efficient heat transfer mechanism and an explanation for the abnormally high heat flow values.

Discussion

The features of the Taupo–Hikurangi arc–trench system (Fig. 9) compare closely to the generalized framework of arc–trench systems described by Karig (1974), and the situation of the Taupo–Rotorua Depression in the arc–trench system is the same as for many oceanic marginal basins elsewhere in the W Pacific (e.g. Karig 1971; Packham & Falvey 1971; Weissel 1981). Karig (1970, 1971) considered the Taupo Volcanic Zone to be a direct continuation of the Lau–Havre Trough but there is a clear offset between the two structures (Fig. 2) and it is more likely that the Taupo Volcanic Zone is a separate structure (see Lewis & Pantin, this volume).

The Taupo–Rotorua Depression is regarded as an ensialic marginal basin (Fig. 4). Sissons (1979) has calculated that the spreading rate at the Bay of Plenty Coast is 7 mm yr^{-1} and considers that this amount is probably constant throughout the zone. No geodetic data on rates of extension are yet available for the main part of the Taupo–Rotorua Depression, but it seems likely from the open fissures formed during historic fault movement, and from evidence of basalt dyke intrusion, that the value quoted is a

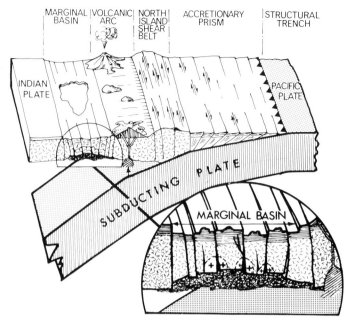

Fig. 9. Schematic model of the major features of the Taupo–Hikurangi arc-trench system.

minimum. However, even if a value of 7 mm yr^{-1} is adopted and the maximum age of volcanism is 0.6 Ma, total widening of the Depression would be 4.2 km. Some of this is obviously taken up by intrusion of basalt dykes, but some may well be due to crustal thinning, perhaps by listric fault development. Seismic refraction experiments indicate that the crust under the Taupo Volcanic Zone is about 15 km thick (T. A. Stern, pers. comm. 1982), compared to 30–35 km outside the zone (Thomson & Evison 1962). Precise levelling in the central part of the Taupo–Rotorua Depression has been undertaken since the 1950s but results are still being evaluated. Preliminary results, however, indicate that the Earthquake Flat area (Fig. 6) is sinking at about 8 mm yr^{-1} (P. M. Otway, pers. comm. 1982). This would be consistent with crustal thinning.

Extensional tectonics of the Taupo–Rotorua Depression are complicated in the N by the effects of the North Island Shear Belt. This belt deviates to the NW in the Bay of Plenty (Fig. 4) and almost certainly extends through the northern part of the Taupo Volcanic Zone. As this shear belt is active, it superimposes dextral strike-slip movement on to the extensional tectonics of the marginal basin. The combined effect of the two processes is to rotate the direction of maximum extensional stress from approximately 130° at Taupo to 180° at the

northern end of Okataina Volcanic Centre. The combined extension and rotation produces an *en échelon* arrangement of faults and fissures, such as that in the 1886 rift across Tarawera (Fig. 8).

The origin of the magmas of the marginal basin is uncertain although their bimodal composition suggests two sources. The high-alumina basalts are almost certainly derived from the mantle (Cole 1973, 1979, 1981; Ewart *et al.* 1977). Cole (1973) has suggested that the initial rise of basaltic magma could in part be due to the eruption of the voluminous rhyolitic volcanics. Large eruptions of ignimbrite would leave a potential void in the crust which would be compensated by caldera collapse at the surface and possibly also by diapiric rise of basaltic magma from the upper mantle. Once in the crust it is likely that the basalt will rise as dykes along the regional faults related to the extension from the back-arc spreading. Basaltic eruptions simply mark the locations at which basalt dykes reach the surface.

Most authors (e.g. Clark 1960a,b; Steiner 1963; Ewart 1963; Ewart & Stipp 1968; Cole 1979, 1981; Reid 1983) agree that the rhyolitic lavas are derived by partial melting of crustal rocks, probably Mesozoic greywacke-argillite basement. The main problem concerns the heat source necessary for partial melting. Cole (1981) considered that initial melting might have been due to hydration of the base of the crust, the 'water' having come from dehydration of the downgoing slab. But if the crust is 15 km thick and Sissons' (1979) interpretation of the lithosphere–asthenosphere boundary at 10–15 km is correct, then heat for the partial melting may be derived from the rotational asthenosphere flow. Possibly because of the extension, the partial melt is able to rise into reservoirs in the upper crust (5–7 km), where partial crystallization occurs. With continued extension, perhaps accompanied by basalt dyke intrusion, rhyolite eruption takes place. Vents for particularly large eruptions (e.g. Taupo Ignimbrite of 1820 yr BP: Vucetich & Pullar 1973; Walker 1980; Froggatt 1981) may be located at the intersection of NNE- and NW-trending faults such as on the Horomatangi Fault. These locations would provide the easiest access to the surface.

The Taupo–Rotorua Depression is a young (<1 Ma) ensialic marginal basin and the features described represent the early stages of development. Geodetic measurements indicate that both extension and subsidence are still taking place. This activity is likely to continue until there is a change in plate configuration or relative motion causing cessation of subduction in the Taupo–Hikurangi arc-trench system.

ACKNOWLEDGMENTS: I should like to thank my colleagues in the Geology Department and Institute of Geophysics, Victoria University of Wellington, in particular, Professor H. Wellman and Dr R. Korsch for the many discussions related to this paper. I should also like to thank many of our past and present research students, particularly B. Sissons, D. Northey and T. Stern for the discussions and valuable contributions they have made to my thinking on the subject. Dr R. Korsch reviewed an early draft of the paper. Drafting of the figures was undertaken by E. F. Hardy, to whom I am most grateful.

References

ADAMS, R. D. & WARE, D. E. 1977. Structural earthquakes beneath New Zealand; locations determined with a laterally inhomogeneous velocity model. *N.Z. Jl Geol. Geophys.* **20**, 59–83.

ANDREWS D. J. & SLEEP, N. H. 1974. Numerical modelling of tectonic flow behind island arcs. *Geophys. J. R. astr. Soc.* **28**, 235–51.

ANSELL, J. 1978. The location and mislocation of subcrustal earthquakes. *Phys. Earth planet. Int.* **17**, 1–6.

CHASE, C. G. 1978. Plate Kinematics: the Americas, East Africa and the rest of the world. *Earth planet. Sci. Lett.* **37**, 355–68.

CHRISTOFFEL, D. A. 1971. Motion of the New Zealand and Alpine Fault deduced from the pattern of seafloor spreading. *In*: COLLINS, B. W. & FRASER, R. (eds) *Recent Crustal Movements. Bull. R. Soc. N.Z.* **9**, 25–30.

CLARK, R. H. 1960a. Petrology of the Volcanic Rocks of Tongariro subdivision. Appendix 2. *In*: GREGG, D. R. (ed.) *The Geology of Tongariro Subdivision. Bull. N.Z. geol. Surv.* n.s. **40**, 107–23.

—— 1960b. Andesitic lavas of the North Island, New Zealand. *Proc. 21st int. Geol. Cong. Norden*, **23**, 121–31.

——, COLE, J. W., NAIRN, I. A. & WOOD, C. P. 1979. Magmatic eruption of White Island Volcano, New Zealand, December 1976–April 1977. *N.Z. Jl Geol. Geophys.* **22**, 175–90.

COLE, J. W. 1970. Structure and eruptive history of the Tarawera Volcanic Complex *N.Z. Jl Geol. Geophys.* **13**, 879–903.

—— 1972. Distribution of high-alumina basalts in Taupo Volcanic Zone. *Publ. Geol. Dept Victoria Univ. Wellington*, **1**, 1–15.

—— 1973. High-alumina basalts of Taupo Volcanic Zone, New Zealand. *Lithos*, **6**, 53–64.

—— 1978a. Andesites of the Tongariro Volcanic Centre, North Island, New Zealand. *J. Volcan. geoth. Res. Amsterdam*, **3**, 121–53.

—— 1978b. Distribution, petrography and chemistry of Kiwitahi and Maungatautari Volcanics, North Island, New Zealand. *N.Z. Jl Geol. Geophys.* **21**, 143–53.

—— 1979. Structure, petrology and genesis of Cenozoic Volcanism, Taupo Volcanic Zone, New Zealand—a review. *N.Z. Jl Geol. Geophys.* **22**, 631–57.

—— 1981. Genesis of lavas of the Taupo Volcanic Zone, North Island, New Zealand. *J. Volcan. geoth. Res. Amsterdam*, **10**, 317–37.

—— & LEWIS, K. B. 1981. Evolution of the Taupo-Hikurangi Subduction System. *Tectonophysics*, **72**, 1–21.

DUNCAN, A. R. 1970. Eastern Bay of Plenty volcanoes. Unpublished Ph.D. thesis, Victoria University of Wellington.

EIBY, G. A. 1977. The function of the main New Zealand and Kermadec seismic regions. *Symp. Int. Geodynamique du Sud—Ouest Pacifique Noumea-Nouvelle Caledonie 27 Aout–2 Sept. 1976*, 167–78. Editions Technip, Paris.

EWART, A. 1963. Petrology and petrogenesis of the Quaternary pumice ash in the Taupo area, New Zealand. *J. Petrol.* **4**, 392–431.

—— & STIPP, J. J. 1968. Petrogenesis of the volcanic rocks of the Central North Island, New Zealand, as indicated by a study of $^{87}Sr/^{86}Sr$ ratios, and Sr, Rb, K, U and Th abundances. *Geochim. cosmochim. Acta*, **32**, 699–735.

——, BROTHERS, R. N. & MATEEN, A. 1977. An outline of the geology and geochemistry and the possible petrogenetic evolution of the volcanic rocks of the Tonga-Kermadec-New Zealand Island arc. *J. Volcan. geoth. Res. Amsterdam*, **2**, 205–50.

——, GREEN, D. C., CARMICHAEL, I. S. E. & BROWN, F. H. 1971. Voluminous low temperature rhyolitic magmas in New Zealand. *Contr. Miner. Petrol.* **33**, 128–44.

FROGGATT, P. C. 1981. Stratigraphy and nature of Taupo Pumice Formation. *N.Z. Jl Geol. Geophys.* **24**, 231–48.

GHANI, M. A. 1978. Late Cenozoic vertical crustal movements in the southern North Island, New Zealand. *N.Z. Jl Geol. Geophys.* **21**, 117–25.

GRANGE, L. I. 1932. Rents and faults formed during earthquake of 1922 in Taupo district *N.Z. Jl Sci. Technol.* **14**, 139–41.

—— 1937. The geology of the Rotorua–Taupo Subdivision. *Bull. N.Z. geol. Surv.* n.s. **37**, 138 pp.

GRINDLEY, G. W. 1960. Sheet 8—Taupo. *Geological Map of New Zealand. 1:250,000.* New Zealand Department of Scientific and Industrial Research, Wellington.

—— 1963. Geology and structure of Waiotapu Geothermal Field. *In: Waiotapu Geothermal Field. Bull. N.Z. Dept Sci. ind. Res.* **155**, 10–26.

—— 1965. The geology, structure and exploitation of the Wairakei geothermal field, Taupo, New Zealand. *Bull. N.Z. geol. Surv.* n.s. **75**, 131 pp.

HAYES, D. E., TALWANI, M. & CHRISTOFFEL, D. A. 1972. The Macquarie Ridge Complex. *In: ADIE, R. J. (ed.) Antarctic Geology and Geophysics*, 767–72. R. J. Int. Un. Geol. Sci. Ser. B. No. 1 Universitets folaget, Oslo.

HEALY, J. 1962. Structure and volcanism in the Taupo Volcanic Zone, New Zealand. *In: Crust of the Pacific Basin. Geophys. Am. geophys. Un. Monogr.* **6**, 151–7.

HENDERSON, J. 1932. Earthquakes in New Zealand. *N.Z. Jl Sci. Technol.* **14**, 129–39.

HILL, H. 1911. Napier to Runanga and the Taupo Plateau. *Trans. N.Z. Inst.* **43**, 288–96.

HOCHSTEIN, M. P. 1976. Estimation of geothermal resource. *In: NATHAN, S. (ed.) Volcanic and Geothermal Geology of the Central North Island New Zealand*, 44–8. Excursion Guide No. 55A and 56C, 25th int. Geol. Congress.

VON HOCHSTETTER, F. 1864. *Geologie von Neu-Seeland—Beiträge zur Geologie der Provinzen Auckland und Nelson. Novara Expedition Geol. Theil 1* (Transl. by C. A. FLEMING, 1959, Government Printer).

KARIG, D. E. 1970. Ridges and basins of the Tonga-Kermadec island arc system. *J. geophys. Res.* **75**, 239–54.

—— 1971. Origin and development of marginal basins in the Western Pacific. *J. geophys. Res.* **76**, 2542–61.

—— 1974. Evolution of arc systems in the Western Pacific. *Ann. Rev. Earth planet. Sci.* **2**, 51–75.

—— & MAMMERICKX, J. 1972. Tectonic framework of the New Hebrides Island arc. *Mar. Geol.* **12**, 187–205.

KATZ, H. R. 1974. Margins of the southwest Pacific.

In: BURK, C. A. & DRAKE, C. L. (eds) *The Geology of Continental Margins*, 549–65. Springer-Verlag, New York.

LEE, W. H. K. & UYEDA, S. 1965. Review of heat flow data. *In:* LEE, W. H. K. (ed.) *Terrestrial Heat Flow. Geophys. Monogr. Am. geophys. Un.* **8**, 87–190.

LENSEN, G. J. 1975. Earth deformation studies in New Zealand. *Tectonophysics*, **29**, 541–51.

VAN DER LINGEN, G. J. 1982. Development of the North Island subduction system, New Zealand. *In:* LEGGETT, J. K. (ed.) *Trench-Forearc Geology.* Spec. Pub. geol. Soc. London, **10**, 259–72. Blackwell Scientific Publications, Oxford.

MITCHELL, A. H. G. & WARDEN, A. J. 1971. Geological evolution of the New Hebrides Island arc. *J. geol. Soc. London*, **127**, 501–29.

MODRINIAK, N. & STUDT, F. E. 1959. Geological structure and volcanism of the Taupo-Tarawera district. *N.Z. Jl Geol. Geophys.* **2**, 654–84.

NAIRN, I. A. 1971. Studies of the earthquake Flat Breccia Formation and other unwelded pyroclastic flow deposits of the Central Volcanic Region, New Zealand. Unpublished MSc. thesis, Victoria University of Wellington.

—— 1976. Late Quaternary faulting in the Taupo Volcanic Zone. *In:* NATHAN, S. (ed.) *Geology of the Central North Island, New Zealand*, 26–30. Excursion Guide No. 55A and 56A, 25th int. Geol. Congress.

—— 1979. Rotomahana-Waimangu eruption, 1886: base surge and basalt magma. *N.Z. Jl Geol. Geophys.* **22**, 363–78.

—— 1981. Some studies of the geology, volcanic history and geothermal resources of the Okataina Volcanic Centre, Taupo Volcanic Zone, New Zealand. Unpublished Ph.D. thesis, Victoria University of Wellington.

—— & COLE, J. W. 1981. Basalt dikes in the late 1886 Tarawera Rift. *N.Z. Jl Geol. Geophys.* **24**, 585–92.

NORTHEY, D. J. Seismic studies of the structure beneath Lake Taupo. Unpublished Ph.D. thesis, Victoria University of Wellington.

OLIVER, J. & ISACKS, B. 1967. Deep earthquake zones, anomalous structures in the upper mantle and lithosphere. *J. geophys. Res.* **72**, 4259–75.

PACKHAM, G. H. & FALVEY, D. A. 1971. An hypothesis for the formation of marginal seas in the Western Pacific. *Tectonophysics*, **11**, 79–110.

REID, F. W. Origin of rhyolitic rocks of the Taupo Volcanic Zone, New Zealand. *J. Volcan. geoth. Res. Amsterdam*, **15**, 315–38.

REYNERS, M. 1980. A microearthquake study of the plate boundary, North Island, New Zealand. *Geophys. J. R. astr. Soc.* **63**, 1–22.

ROBINSON, R., SMITH, E. G. C. & LATTER, J. H. 1981. Seismic studies of the crust under the hydrothermal area of Taupo Volcanic Zone, New Zealand. *J. Volcan. geoth. Res. Amsterdam*, **9**, 253–67.

SISSONS, B. A. 1979. The horizontal kinematics of the North Island of New Zealand. Unpublished Ph.D. thesis, Victoria University of Wellington.

SMITH, S. P. 1886. *The Eruption of Tarawera: a report*

to the Surveyor-General. Government Printer, Wellington. 84 pp.

STEINER, A. 1963. Crystallisation behaviour and origin of the acidic ignimbrite and rhyolite magma in the North Island of New Zealand. *Bull. Volcanol.* **25**, 217–41.

STUDT, F. E. & THOMPSON, G. E. K. 1969. Geothermal heat flow in the North Island of New Zealand. *N.Z. Jl Geol. Geophys.* **12**, 673–83.

THOMAS, A. P. W. 1888. *Report on the Eruption of Tarawera and Rotomahana, New Zealand*. Government Printer, Wellington. 77 pp.

THOMSON, A. A. & EVISON, F. F. 1962. Thickness of the Earth's crust in New Zealand. *N.Z. Jl Geol. Geophys.* **1**, 29–45.

TOKSÖZ, M. N. & HSUI, A. T. 1978. Numerical studies of back-arc convection and the formation of marginal basins. *Tectonophysics*, **50**, 177–96.

VUCETICH, C. G. & PULLAR, W. A. 1973. Holocene Tephra formations erupted in the Taupo area, and interbedded tephras from other volcanic sources. *N.Z. Jl Geol. Geophys.* **16**, 745–80.

WALCOTT, R. I. 1978. Present tectonics and Late Cenozoic evolution of New Zealand. *Geophys. J. R. astr. Soc.* **52**, 137–64.

WALKER, G. P. L. 1980. The Taupo Plinian pumice: product of the most powerful known (ultraplinian) eruption? *J. Volcan. geoth. Res. Amsterdam*, **8**, 69–94.

WARD, R. H. 1922. A note on the significance of the recent subsidence of the shore of Lake Taupo. *N.Z. Jl Sci. Technol.* **5**, 280–1.

WEISSEL, J. K. 1981. Magnetic lineations in marginal basins of the western Pacific. *Philos. Trans. R. Soc. London*, **A300**, 223–47.

WELLMAN, H. W. 1964. Age of the Alpine Fault New Zealand. *In*: SUNDARAM, R. K. (ed.) *Rock Deformation and Tectonics*, 148–62. Proc. Section 4 int. Geol. Congress, India.

—— 1967. Report on studies relating to Quaternary diastrophism in New Zealand. Minutes of the working group meeting for the Neotectonic study of the Pacific regions. Appendix 5. *Quat. Res.* (Japan) **6**, 34–6.

WELLMAN, P. & McCRACKEN, H. M. 1979. *BMR Earth Science Atlas: Plate Tectonics of the Australian Region*. Scale 1: 20,000,000. Bureau of Mineral Research, Canberra.

J. W. COLE, Department of Geology, Victoria University, Private Bag, Wellington, New Zealand.

Intersection of a marginal basin with a continent: structure and sediments of the Bay of Plenty, New Zealand

K. B. Lewis & H. M. Pantin

SUMMARY: West of the Kermadec Trench and Arc, the Havre Trough is an actively spreading, oceanic marginal basin. Bathymetric, magnetic and seismic data indicate that it terminates against the lower slope off the Bay of Plenty, North Island, New Zealand. It is replaced on the adjacent upper slope, and on land, by the ensialic marginal basin known as the Taupo Volcanic Zone, which is part of the Hikurangi Plate Boundary System. The Taupo Volcanic Zone is offset about 50 km eastwards from the Havre Trough, by transform faults on the middle and lower slope. It is intersected on the upper slope, and its bounding faults are offset, by the northern end of the North Island Shear Belt.

The central part of the Bay of Plenty, which is crossed by the marginal basins, is characterized by a network of active faults, and by arc and back-arc volcanism. Sedimentation here is highly variable and the result of local balances between a continuous, rapid, terrigenous supply from rising, frontal ridge mountains to the E, and an intermittent supply of mainly rhyolitic, pyroclastic debris from the Taupo Volcanic Zone. This contrasts with the smooth, non-volcanic, eastern side of the bay where there is a sigmoidal, prograding shelf and slope of terrigenous silt on the northward plunging frontal ridge. It also contrasts with the western side of the bay where an extinct, Upper Miocene to Lower Pleistocene, arc–back-arc system has been eroded and is covered by a veneer of relict gravel and volcanogenic sand on the shelf, and buried by sandy mud on the slope.

The Bay of Plenty, off North Island, New Zealand, is in a critical structural position. Here, volcanic and structural elements of a late Neogene to present-day oceanic subduction system, including the tholeiitic volcanism of the Kermadec Ridge (Brothers & Hawke 1981) and the back-arc spreading of the Havre Trough (Malahoff *et al.* 1982), abut against the NE margin of the New Zealand continental block (Fig. 1). The arc–back-arc relationships continue on land in the ensialic Taupo Volcanic Zone (Cole, this volume), which is associated with subduction and accretion occurring 250 km to the E, along the Hikurangi Trough (Lewis 1980). Dextral strike-slip faulting in the North Island Shear Belt caused the rotation, during late Neogene times, of fore-arc elements of the Taupo–Hikurangi System into their present alignment with the Havre–Kermadec System. The two systems have very different structures and histories (Cole & Lewis 1981; van der Lingen 1982; Ballance *et al.* 1982). The North Island Shear Belt, the Taupo Volcanic Zone and the Havre Trough now meet and interact in the Bay of Plenty.

In this paper we collate bathymetric, seismic, magnetic and sediment sample data to determine the structural, volcanic and sedimentological relationships in this critical region. We interpret the results in terms of the relationships between an ensialic arc–back-arc system and a contiguous intra-oceanic arc–back-arc system.

Bathymetry

The bathymetry of the Bay of Plenty (Fig. 2; Pantin *et al.* 1973) provides a wealth of information on structural character and trends, not available from the sparse geophysical data. The continental shelf is smooth in the E but its surface is interrupted in the centre and W by numerous small islands, stacks, and submerged pinnacles. The continental slope is smooth in both the E and W, but diversified in the centre by rounded knolls and by NNE-trending valleys, scarps and ridges.

The boundary between the smooth eastern zone and the irregular central zone is a NNE-trending lineament, clearly marked on the shelf by the 26 m high, W-facing Motuhora Scarp, on the upper slope by the zig-zag White Island Canyon, and on the lower slope by the White Island Trough. This lineament lies on an extrapolation of the eastern boundary of the downthrown Taupo Volcanic Zone, and is interpreted as an extension of the fault or faults defining the eastern margin of the zone. It apparently ends at the Raukumara Plain, at the base of the slope.

Most of the ridges, knolls, islands and stacks to the W of this lineament are known or presumed to be volcanic. Immediately to the W of the lineament there are several andesite-dacite volcanoes. The crater remnant of Motuhora (Whale Island) protrudes from the inner shelf, while the active volcano of White Island, and the

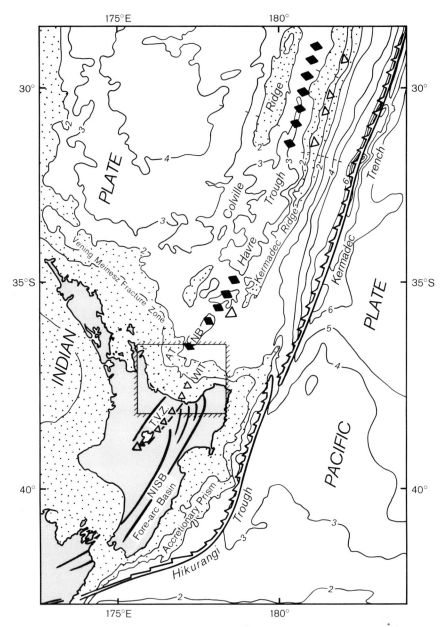

FIG. 1. Major elements of the Indian–Pacific Plate Boundary in the vicinity of northern New Zealand, showing the structural position of the Bay of Plenty (boxed) at the intersection of the Havre Trough Marginal Basin with the ensialic Taupo Volcanic Zone (TVZ) and the North Island Shear Belt (NISB). Open triangles indicate arc volcanism. Diamonds indicate back-arc spreading. Serrated ornament indicates underthrusting of the Pacific Plate beneath the feather-edge of the Indian Plate. Thick lines indicate the area of dextral strike-slip faulting. Contours are depths below sealevel in 1000 m intervals, with light stippling to 2000 m. WIT = White Island Trough; NB = Ngatoro Basin; AT = Alderman Trough.

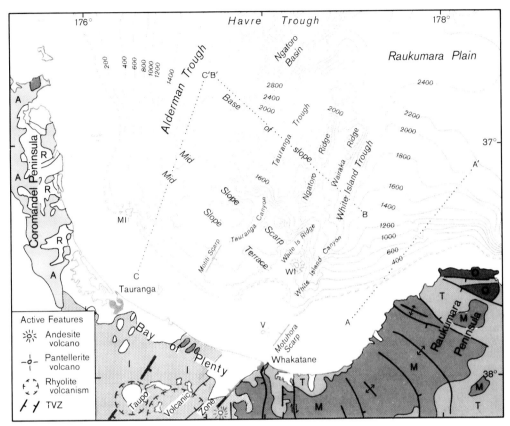

FIG. 2. Bathymetry and simplified onshore geology of the Bay of Plenty. Offshore, bathymetric contours are at 200 m intervals and the broken hatched line indicates the shelf break. Onshore, O = Matakaoa ophiolites; M = Mesozoic greywackes; T = Tertiary marine sediments; A = Miocene andesites; R = Miocene–Lower Pleistocene rhyolites; I = Quaternary ignimbrites; Unstippled = Late Quaternary marine and fluviatile deposits; WI = White Island; V = Motuhora (volcanic remnant); MI = Mayor Island (pantellerite volcano). Also shown are the positions of seismic profiles (A-A′, B-B′ and C-C′) illustrated in Fig. 4.

neighbouring Volkner Rocks, rise from the White Island Ridge (on the upper slope). They form a continuation of the line of andesite-dacite volcanoes in the Taupo Volcanic Zone (see Cole, this volume, Fig. 4). On the lower slope the NNE-trending Ngatoro Ridge is offset westward from the White Island Ridge and, like the White Island Trough, it fades into the flat floor of the Raukumara Plain.

Most of the central zone, between the White Island–Ngatoro Ridge and the Alderman Trough, is studded by numerous circular, elliptical, crescentic and irregular knolls. The basal diameter of the knolls ranges from 1 to 5 km, and the relief from 150 to 700 m. One— Mahina Knoll at 37°21′S, 177°06′E—even has a central crater 1.5 km across and 150 m deep. All are considered to be volcanic.

The continental slope in the central zone has

two other NNE-trending lineaments. One continues the trend of the W margin of the Taupo Volcanic Zone. On the upper slope, it is marked by the 60 m high, E-facing Motiti Scarp, on the mid-slope by the zig-zag Tauranga Canyon and on the lower slope by the flat-floored Tauranga Trough. The other NNE-trending lineament, the Alderman Trough, incises the lower slope and marks the western boundary of the central zone.

Across the slope, transverse to the other geomorphic trends, there is a conspicuous mid-slope scarp, above which is a mid-slope terrace. Most of the NNE-trending ridges, canyons and troughs stop, are offset, or change character at this scarp. A similar, linear discontinuity of bathymetric features occurs at the base of the slope.

Beyond the base of the slope, at a depth of

about 2000 m, there is the flat floor of the Raukumara Plain in the E, and the undulating southern end of the Havre Trough in the W. The most conspicuous feature of the Havre Trough is the 3300 m deep Ngatoro Basin.

Geomagnetic data

Geomagnetism, like bathymetry, can indicate important structural trends. Magnetic data for the Bay of Plenty were collected along tracks 5 km apart, reduced and contoured (Roberts 1967), and synthesized with data from adjacent areas (Davey & Robinson 1978). The data, although sparse, are adequate to distinguish the high magnetic relief of the intra-oceanic marginal basin beyond the base of the slope, the moderate magnetic relief with several major anomaly belts on the volcano-studded continental margin of the central and western Bay of Plenty, and the low magnetic relief with a single large bipolar anomaly field on the fore-arc of the eastern Bay of Plenty (Fig. 3).

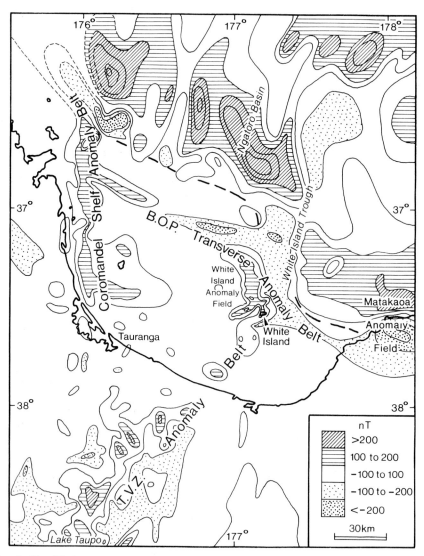

FIG. 3. Simplified magnetic anomalies in the Bay of Plenty. The boundary between low magnetic relief of the continental margin and high magnetic relief of the oceanic system is indicated by a heavy dashed line. Contours at 100 nT (nanotesla) intervals. (Adapted from Davey & Robinson 1978.)

The onshore geology of the Bay of Plenty (Fig. 2) consists mainly of indurated Mesozoic flysch, soft Neogene marine sediments, abundant Quaternary acid pyroclastic debris and Neogene rhyolitic lavas, andesite-dacite volcanoes and rare basalts (Healy *et al.* 1964a; Kingma 1965; Schofield 1967). In general, basic and intermediate igneous rocks are more highly magnetic than acid igneous rocks and sediments (Hatherton 1954; Hunt & Smith 1982). Thus, at least some of the complex anomalies in the Bay of Plenty are interpreted as due to bodies of intermediate (and rare basic) volcanic rock.

Volcanic-arc anomalies

In the centre of the Bay of Plenty, the White Island Anomaly Field is a group of mainly positive magnetic anomalies that coincides with the White Island Ridge and adjacent knolls. Anomalies range from +700 nT to −200 nT (Roberts 1967). The field lies at the northern end of the Taupo Volcanic Zone Anomaly Belt, a NNE-trending line of anomalies associated with the line of andesite-dacite volcanoes that extends for 250 km SSW of White Island (Cole, this volume).

In general, large positive anomalies lie over the northern flank of conical bathymetric features and, in the case of Volkner Rocks, White Island and Motuhora, there are also small negative anomalies to the S (Roberts 1967). This anomaly pattern indicates simple magnetic bodies, magnetized when the geomagnetic field was the same as at present.

Isolated back-arc anomalies

In the central and western parts of the Bay of Plenty, there are numerous small, isolated anomalies, crossed by only one or two survey tracks (Roberts 1967). Some are bipolar but mostly only a positive or negative anomaly was detected. The mid-slope terrace is almost devoid of anomalies. On the shelf, some very localized, high-amplitude anomalies may indicate steep-sided volcanic plugs left by wave erosion (during lower sealevel) of an extrusive pile. Broader anomalies on the lower slope may indicate complete volcanic bodies. On both shelf and slope, some anomalies correlated with known bathymetric features. However, commonly there is no bathymetric expression of a magnetic anomaly, perhaps indicating truncation by wave erosion on the shelf, or burial by sediment on the slope. Conversely, there are several islets on the shelf and several knolls on

the magnetic slope for which no signature was detected. The lack of anomalies may reflect the wide spacing of survey tracks relative to the size of the structures, or indicate that the bathymetric feature is constructed of sedimentary rocks or acid igneous rocks with low magnetic susceptibilities.

A frontal ridge ophiolite anomaly

In the fore-arc area of the eastern Bay of Plenty, the low geomagnetic relief is interrupted by one broad bipolar anomaly, the Matakaoa Anomaly Field. It consists of a major positive anomaly (+400 nT: Roberts 1967) extending E–W on the inner shelf off the northern end of the Raukumara Peninsula, together with a parallel negative anomaly (−300 nT: Davey & Robinson 1978) onshore to the S. This large, simple, bipolar magnetic anomaly, typical of a large igneous body with steep contacts, coincides with the fault-bound outcrop of the Lower Cretaceous to Middle Eocene Matakaoa Volcanics (Kingma 1965; Strong 1976, 1980). These rocks are an ophiolite sequence, interpreted as allochthonous, dismembered oceanic seamounts, obducted on to greywacke foreland in the Mid-Tertiary (Ballance & Spörli 1979; Pirajno 1980; Brothers & Delaloye 1982). Since the magnetic anomaly is truncated just W of the coast, it is possible that the Matakaoa Volcanics are truncated just offshore by a N–S fault, similar to faults seen onshore.

A transverse anomaly belt

The Bay of Plenty Transverse Anomaly Belt is an elongate, low-amplitude magnetic feature extending between the peninsulas on either side of the bay and crossing both the fore-arc and the back-arc. It is positive (up to +150 nT: Roberts 1967) on the slope off the Coromandel Peninsula, and negative (up to −200 nT: Roberts 1967) across the rest of the bay. On the eastern (fore-arc) side of the bay, it crosses the continental margin diagonally and appears to continue the WNW structural trend of the onshore Cretaceous strata of the Raukumara Peninsula (Kingma 1965). The anomaly belt then trends northwards for some 60 km, roughly along the line of the White Island Trough, before continuing WNW along the outer edge of the mid-slope terrace in the central (back-arc) part of the bay. Thus, it appears to be dextrally offset at the lineament between the fore-arc and the back-arc basin. Seismic evidence (see below) suggests that the Bay of Plenty Transverse Anomaly Belt is due to a basement high, possibly of

Mesozoic rocks that include magnetically susceptible igneous units.

Intra-oceanic marginal basin anomalies

About 20 km N of the western part of the Transverse Anomaly Belt, there is a change, from the low magnetic relief of the continental slope, to a high magnetic relief associated with the 0–1.8 Ma old oceanic crust in the Havre Trough (Malahoff *et al*. 1982).

Extinct arc anomalies

The Coromandel Shelf Anomaly Belt, on the western side of the bay, is an irregular, N–S trending, zone of localized, high-amplitude positive and negative anomalies. Anomaly peaks range from +1300 nT to −700 nT (Roberts 1967). At some places anomalies correspond with known igneous outcrops, at others they do not. The many, very sharply peaked anomalies may indicate volcanic cones that have been wave-eroded on the shelf, leaving only steep-sided plugs. Several sharp positive anomalies with little or no associated negative anomaly (e.g. 37°07′S, 175°58′E; 37°26′S, 176°14′E: see Roberts 1967) may indicate plugs magnetized when the geomagnetic field was much the same as at present. Conversely, negative anomalies (e.g. 36°39′S, 176°03′E; 37°07′S, 176°06′E: see Roberts 1967) perhaps indicate remanent magnetism formed when the geomagnetic field was the reverse of that at present. The Coromandel Shelf Anomaly Belt lies E of, and parallel with, a line, on the eastern Coromandel coast, of Upper Miocene to Lower Pleistocene rhyolitic domes and flows, which become younger towards the S (Kear 1959; Healy *et al*. 1964a; Rutherford 1978). The belt may indicate andesitic arc volcanism E of the rhyolites. It extends to the coast near Tauranga and is tentatively traced southwards to intersect the Taupo Volcanic Zone Anomaly Belt near Lake Taupo. There are no active volcanoes along this line so that the anomalies are considered to represent an extinct arc whose activity immediately preceded that of the Taupo Volcanic Zone (Ballance 1976; Cole & Lewis 1981).

Seismic data

Seismic reflection data are available along about 2000 km of ship's track in the Bay of Plenty. Early single-channel data, collected by Lamont Geological Observatory, are mainly from the lower slope (Davey 1977). Later multichannel data, collected by oil companies, from the shelf and upper slope, are available on open file at N.Z. Oceanographic Institute, Wellington, and N.Z. Geological Survey, Lower Hutt.

The offshore frontal ridge

Profiles to the E of the White Island Canyon and Trough show a smooth prograding sequence of clinoform surfaces (Fig. 4 A-A′). Some layers are sigmoidal lenses, others are terminated updip by toplap or erosional truncation. The whole seismic configuration could be classified as complex sigmoid-oblique (Vail *et al.* 1977).

Within the seismic configuration, four depositional sequences are identified in the slope deposits (more than 3 km thick). The oldest sequence (1 on Fig. 4 A-A′) is of divergent reflectors indicating slope-front strata that thicken away from a buried shelf break. This is followed by a sequence of onlap and then regression of the shelf break (2 in Fig. 4 A-A′), with strata terminated updip by toplap and passing downdip to bottomset strata, indicating deposition from high-energy into low-energy environments. This prograding sequence is buried by onlapping fill (3 in Fig. 4 A-A′), which is at least 1 km thick on the lower slope at the edge of the Raukumara Plain. The top sequence (4 in Fig. 4 A-A′) is of sigmoidal lenses of sediment that are thickest on the upper slope and onlap the shelf. Surface strata that dip at only 1–2° are deformed by penecontemporaneous sheet-slumping similar to that occurring elsewhere on the Hikurangi Margin (Lewis 1971). On the upper slope there is a filled channel, perhaps indicating erosion during a phase of lower sea-level and filling during a subsequent phase of rising and high sealevel.

The repeated onlaps on this margin represent several major relative rises of sealevel, which are considered to indicate continuing tectonic downwarping on which eustatic and sediment supply changes are superimposed.

Beneath the shelf, the older depositional sequences are folded, together with a core of reflection-free 'basement' strata, into an outer anticline and a nearshore shallow synclinal basin. This configuration can be correlated with the Cretaceous 'greywacke basement' and Neogene basin-fill which crop out on the adjacent peninsula. A nearshore anticline correlates with the WNW-trending Transverse (magnetic) Anomaly Belt, and appears to be a lateral continuation of a WNW-trending anticline in Cretaceous rocks onshore.

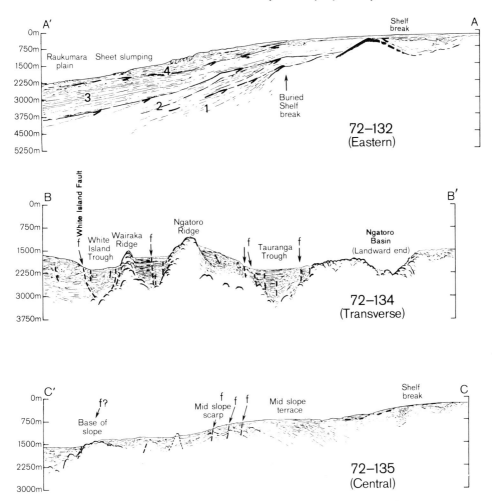

FIG. 4. Tracings of continuous seismic reflection profiles from the Bay of Plenty continental margin. Profile locations are shown in Fig. 2. Seawater depths indicated at 750 m (1 s two-way travel time) intervals. A-A′ is obliquely downslope on the eastern margin off the frontal ridge of the Raukumara Peninsula. Thickened lines indicate unconformities. B-B′ is across the bay on the lower slope. Broken lines and 'f' at faults. C-C′ is downslope in the western part of the central volcanic zone. Broken lines and 'f' at faults, and heavy broken line at unconformity. (Profiles courtesy of Mobil Exploration and Producing Services Inc, Dallas, Texas, U.S.A.)

The boundary between the frontal ridge and the volcanic zone

Profiles across the Bay of Plenty show the smooth, regular, depositional sequences of the eastern margin pinching out towards the up-thrown side of a major NNE-trending fault (Fig. 4 B-B′) (Davey 1977). This fault, here named the White Island Fault, is the lineament defined by the Motuhora Scarp, the White Island Canyon and the eastern side of the White Island Trough. It clearly separates the non-volcanic frontal ridge of the eastern Bay of

Plenty from the seaward extension of the Taupo Volcanic Zone.

The amount of throw on the White Island Fault is difficult to determine, because of the different nature of the reflectors on the two sides. However, tentative correlations suggest a throw much less than the 2 km estimated on land (Healy *et al.* 1964a). Strike-slip movement is indicated only by speculative evidence of dextral offset of the Transverse Anomaly Belt (Fig. 3). The bathymetrically determined zig-zag course of the White Island Canyon and landward end of the White Island Trough reflect the complex interaction between the NNE-trending White Island Fault and the curving, N–NW-trending faults of the northern end of the North Island Shear Belt.

The volcanic central zone

Profiles across the central part of the Bay of Plenty (Fig. 4 B-B') (Davey 1977) show that most of the ridges produce strong, parabolic, seismically opaque reflections, typical of volcanic basement. They show that most of the troughs, including the Alderman Trough in the far W, have at least 1 km of fill, with parallel and, or, contorted (slump-modified) seismic reflectors, overlying reflectors considered to be volcanic. The exception is the Ngatoro Basin, which has a bare volcanic floor at the base of the slope (Fig. 4 B-B'), and only thin or uncertain indications of sediment fill further N (Davey 1977, Fig. 32). Faulted sedimentary layers occur in the White Island Trough, through the middle of the perched trough between the Wairaka Ridge and the Ngatoro Ridge, and on each side of the Tauranga Trough, where faulting may have triggered instability in sediments on the flanks of the adjacent ridges (Fig. 4 B-B'). A major slump on the flanks of the Ngatoro Ridge, and a buried slump deposit along the opposite side of the Tauranga Trough, are each more than 500 m thick and are therefore distinct in form from the thin sheet-slumping that occurs on the eastern margin.

The NNE trend of the volcanic ridges is probably also controlled by faulting. The ridges could be horsts of pre-sedimentary volcanic basement, but indications of strong reflectors downlapping on to sedimentary strata (e.g. Ngatoro Ridge in Fig. 4 B-B') suggest that the ridges are actively growing, elongate accumulations of volcanic material erupted from fissures. The seismic data cannot show such fissures, but their positions can be inferred to lie along the ridge axes.

The Mid-Slope Terrace, at right-angles to the other structures, is the surface of a broad anticline, which has an unconformity and eroded fill on the landward flank, and faulting along the axis and seaward flank (Fig. 4 C-C'). The faulting, with downthrow to the NE, defines the position of the Mid-Slope Scarp. The faulted fold axis coincides with the position of the Transverse Anomaly Belt. On the upper slope, smaller anticlines have been erosionally truncated and the synclines between them have been infilled by a surface layer that is everywhere less than 150 m thick. Parabolic reflec-

tors from steep-sided knolls project through the sediment cover, both above and below the Mid-Slope Terrace. Just above the base of the slope, conformable strata thin towards, and onlap, reflection-free outcrops, the most seaward of which is steep and possibly faulted on its seaward side.

The extinct volcanic margin

Published seismic data (Davey 1977), and open-file oil company reports, indicate that the smooth bathymetry of the NW margin of the Bay of Plenty, off the Coromandel Peninsula, overlies a buried, irregular erosional unconformity. This contact separates a draped upper sequence, 0–300 m thick, of 'divergent-fill' strata that thin over underlying highs and thicken in depressions, from a folded and faulted older sequence that includes volcanic basement-type reflectors. At the base of the slope there is a basement ridge and on its seaward side a faulted contact with the NW margin of the Alderman Trough.

Sediment cover

A synthesis of 200 analysed sediment samples and over 5000 Admiralty Chart notations, adapted from Doyle *et al.* (1979), is presented in Fig. 5. Additional information on offshore tephra stratigraphy, rates of sedimentation and sedimentary history for the last 30,000 yr has been obtained from piston cores from the eastern Bay of Plenty (Kohn & Glasby 1978).

On the continental shelf, the lithology of the sediments shows the same bipartite division, between a non-volcanic eastern zone and a volcanic central and western zone, that is evident in the bathymetric and geophysical data. However, the division is blurred by the movements of sediment.

The eastern shelf is blanketed by a prism of Holocene sediments, predominantly derived from the adjacent greywacke ranges. The prism, labelled the Eastern Silt Prism (Fig. 5), is thickest and most silty on the mid-shelf. Here, Holocene ($<$ 10,000 yr) sediments are estimated, on the basis of tephrachronology, to be about 20 m thick (Kohn & Glasby 1978); greywacke-derived silt and mud (Glasby 1975) constitute more than 90%. These beds thin to less than 4 m on the outer shelf where modified beach deposits, of mainly greywacked-derived pebbly sand, radiocarbon dated at 12,000–13,000 yr, are found in short cores and exposed at the seabed (Kohn & Glasby 1978).

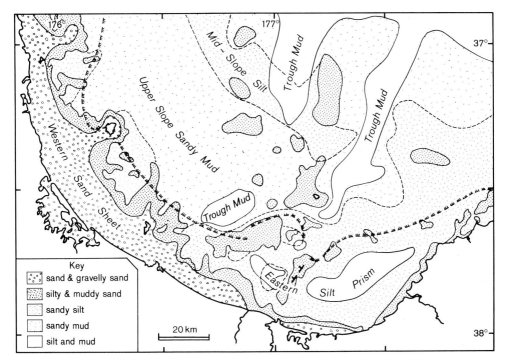

FIG. 5. Surface sediments in the Bay of Plenty (adapted from Doyle *et al.* 1979). Sediment classification after Folk 1974. Solid contours at 90, 50 and 10% sand in the< 2 mm fraction. Broken line at 67% silt in the mud fraction. Broken hatched line at shelf break.

The Holocene prism also thins towards the shore, within 3 km of which it becomes predominantly quartz-rich detrital sand with some greywacke pebbles. Except for well-developed, white rhyolitic ash layers, the Holocene prism in the eastern Bay of Plenty is essentially similar in form, thickness and composition to the Holocene prism in fore-arc areas elsewhere on the Hikurangi Margin (Lewis 1973).

The Eastern Silt Prism extends W of the White Island Fault, into the offshore extension of the Taupo Volcanic Zone. In the vicinity of the Motuhora Scarp there is a NNE-trending belt of impure sand and sandy silt, which may be a consequence of the constriction of coast-parallel currents at the fault scarp.

In the western half of the bay, the shelf sediments are predominantly feldspathic sand with abundant vesiculated glass and bubble-wall shards, interspersed with large patches of shelly and pebbly sand (Godfriaux 1973). They are labelled the Western Sand Sheet (Fig. 5). The arenaceous fraction is considered to come mainly from the Taupo Volcanic Zone, although sand in pocket beaches on the indented Coromandel coast is locally derived (Schofield 1970). The pebbles, generally

rounded, are of rhyolite and andesite, apparently eroded from what are now small islands, stacks and submerged pinnacles when the breaker zone was well seaward of its present position. Holocene deposition must be slow or absent. Patches of sandy silt and sandy mud occur on the outer shelf between the larger islands. These represent the feather-edge of an Upper Slope Sandy Mud facies (Fig. 5).

On the continental slope there is no clear division between a volcanic and a non-volcanic regime. Much of the slope is blanketed by sandy mud, the largest area, in the western part of the bay, being labelled Upper Slope Sandy Mud (Fig. 5). White, rhyolitic ash layers, correlated with dated eruptions in the onshore Taupo Volcanic Zone, indicate rates of deposition that are generally between 0.1 and 0.2 m $(1000 \text{ yr})^{-1}$. Many of the knolls in the area protrude from the mud blanket, and have a covering of muddy sand, or gravelly sand. The sand fraction is feldspathic with abundant rhyolitic shards, derived from the onshore Taupo Volcanic Zone. The gravel fraction is autochthonous andesite and shell debris with allochthonous pumice clasts (Duncan 1970). However, on knolls within about 12 km of White Island, the surface sediment contains dark-grey andesitic ash, which is also found in cores up to 60 km NE of White Island (Kohn & Glasby 1978).

Along the Mid-Slope Scarp and on the mid-slope E of the White Island Fault, there is a broad belt of sandy silt, labelled the Mid-Slope Silt (Fig. 5), and again one of the principal constituents is rhyolitic tephra. However, tephra is less abundant in the Trough Mud (Fig. 5) that fills each of the major NNE-trending depressions. Beyond the base of the slope there is calcilutite in the southern end of the Havre Trough but terrigenous mud, with a significant ash component, in the troughs to the E and on the Raukumara Plain (McDougall 1975; Cronan *et al.*, this volume).

Discussion on the structural and sedimentological relationships

The structural and sedimentological complexity of the Bay of Plenty reflects its position at the rapidly evolving junction between an intra-oceanic marginal basin and a virtually contiguous ensialic marginal basin. For each parameter considered there are differences between (i)

the eastern, non-volcanic, frontal ridge, (ii) the central, active, volcanic marginal basin, and (iii) the western extinct arc. In each of these three zones there are also downslope changes from a continental to an oceanic system.

The non-volcanic, downwarping frontal ridge

On the eastern margin of the Bay of Plenty, there has been considerable Neogene tectonic deformation and rapid Holocene sedimentation, but there are no local volcanic vents and only a subordinate input of volcanogenic sediment.

A correlation, between seismic data, magnetic data and onshore geology, suggests that folds trend northwestward across the shelf, continuing the change in trend of folds and faults, from a regional NNE trend to a NW trend, that occurs at the northern end of the North Island Shear Belt (Fig. 6). The same data suggest that the allochthonous ophiolitic Matakaoa Volcanics are limited to the extreme tip of the Raukumara Peninsula and do not continue far offshore in any direction.

If it is assumed that the erosional unconformities in probable Neogene strata underlying

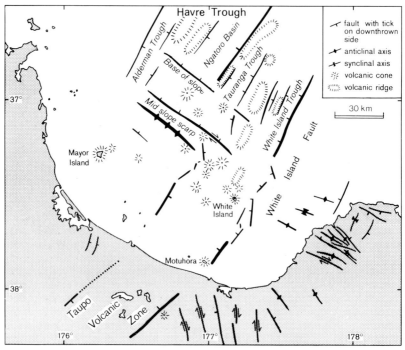

FIG. 6. Major structural features of the Bay of Plenty showing inferred trends of faults, fold axes, volcanic cones (including knolls) and volcanic ridges. Onshore information from Healy *et al.* (1964a) and Kingma (1965).

the smooth slope N of the Raukumara Peninsula were caused by sea-surface wave action, then there must have been downwarping of the continental margin towards the N. Such downwarping was less marked towards the White Island Fault. It may be inferred that similar deformation is still active. If it is further assumed that the shelf break formed at a uniform level (c. 120 m below present sealevel) during the last major glacio-eustatic lowering of sealevel about 20,000 yr ago, then the present variation in depth, from 100 m immediately adjacent to the White Island Fault to 160 m off the Raukumara Peninsula (Pantin et al. 1973), can be interpreted to be the result of subsequent tectonic deformation. The downwarping offshore contrasts with rapid uplift onshore, illustrated by a Last Interglacial Age marine terrace that is 274 m above sealevel at the northern end of the Raukumara Peninsula (Gibb 1981) and dips almost to sealevel at the boundary with the Taupo Volcanic Zone (Chappell 1975; Iso et al. 1982). It is inferred that the northern end of the frontal ridge of the Hikurangi Margin is a growing anticlinal structure plunging towards the N. Rapid seaward tilting may be a partial cause of the sheet-slumping on the mid-slope.

The rapid deposition of detrital silts and sands on the eastern margin reflects a source in the high, Mesozoic greywacke ranges of the Raukumara Peninsula. The rivers that drain the ranges have catchments that are a factor of 10 larger, and suspended sediments yields that approach a factor of 100 larger, than the rivers that drain the Taupo Volcanic Zone and ignimbrite plains to the W (Adams 1979). This results in burial, by the Holocene silt prism, of the transgressive sands and gravels that are exposed further W on the mid-shelf.

Offshore continuation of the Taupo Volcanic Zone

On the shelf and upper slope, inward-facing fault scarps define a graben at the extrapolated position of the downthrown Taupo Volcanic Zone (Fig. 6). Further offshore the margins of the zone are defined by subparallel zig-zag canyons. Canyon erosion or, at least, reduced deposition, probably occurred along structural depressions as a result of the passage of turbidity currents. It is envisaged that, during periods of lower sealevel, longshore-moving sediments were trapped at fault scarps near the shelf break, where the increased sediment loading precipitated periodic failure. This resulted in autosuspension currents (Pantin 1979, 1982) which travelled along downsloping fault-angle

depressions. The angular zig-zag changes of course in the canyons are considered to indicate interaction of the northern end of the North Island Shear Belt with the normal faults that define the graben; Cole (this volume) describes the same interaction of faults onshore. There are no indications of strike-slip movement in the onshore Taupo Volcanic Zone (Cole, this volume) so that the magnetic evidence, presented here, of a 60 km dextral offset of a basement high, and a similar apparent offset of the Vening Meinesz Fracture Zone (Karig 1970), require further investigation.

The major subsidence in the Taupo–Rotorua region (Cole, this volume) decreases towards the N, and inliers of Mesozoic greywacke basement and Mid-Pleistocene marine strata crop out near the coast (Fig. 2; Healy et al. 1964b). Offshore, the evidence from shelf break depths is equivocal. If it is again assumed that the shelf break was everywhere formed at about 120 m below present sealevel during the last glacial maximum, the bathymetry (Pantin et al. 1973) indicates subsidence of about 50 m in the E, on the promontory between Motuhora and White Island, and no detectable deformation to the W. However, subsidence relative to the sea floor on either side is indicated by the inward-facing scarps and by tentative correlations across bounding faults from seismic data.

The andesite-dacite volcanism of the eastern side of the Taupo Volcanic Zone obviously continues as far as the activity on White Island and Volkner Rocks, which project from the upper slope. It is clearly defined by a magnetic anomaly belt. Some of the scattered knolls on the upper slope to the W of White Island also have magnetic signatures, and samples collected from them have all been andesitic (Duncan 1970). Some knolls have no associated magnetic anomalies, but no rock samples have been collected from these. All of these knolls occur in relatively the same back-arc position as the rhyolitic volcanoes on the adjacent land, and the supposed basaltic sea floor in the intra-oceanic back-arc basin to the N. The back-arc andesite volcanism may reflect conditions and processes peculiar to the setting at the continental edge.

On the lower slope, the arc–back-arc pattern is difficult to recognize. The Ngatoro Ridge may possibly represent an offset continuation of the andesite arc, although no samples are available. It has only a small negative magnetic anomaly. The Tauranga Trough is in a back-arc position. The White Island Fault continues to define the eastern boundary of the seaward extension of the Taupo Volcanic Zone, but a continuation of the western boundary is more difficult to demonstrate in the Tauranga Trough.

The faults, ridges and troughs of the offshore continuation of the Taupo Volcanic Zone appear, on the basis of the present sparse data, to fade into the flat floor of the Raukumara Plain.

The relationship of the ensialic and intra-oceanic marginal basins

The trenches, frontal ridges, Benioff zones and gravity anomalies of the Havre–Kermadec and Taupo–Hikurangi systems are not contiguous (Hatherton & Syms 1975; Eiby 1977; van der Lingen 1982), and the nature and location of the boundary between the systems is difficult to determine.

The presently available data suggest that the southernmost part of the Havre Trough lies parallel with, but to the W of, the northernmost part of the Taupo Volcanic Zone (Fig. 6). The major bathymetric expression of the southern Havre Trough, the Ngatoro Basin, ends abruptly at the base of the slope, as do associated ridges and large, oceanic-type, magnetic anomalies. The only continuation of extension into the lower slope is in the far W, at the Alderman Trough. However, there is evidence, from the pantellerite volcanism at Mayor Island, that incipient rifting extends landward as far as the shelf break (Cole 1978).

Thus, there is overlap between the two systems. The main change, from dominance of extension in the Taupo marginal basin to dominance in the Havre marginal basin, takes place on the mid- and lower slope. In this position there are two faults or fault zones perpendicular to the axial trend of the basins, one at the Mid-Slope Scarp and one, less clearly defined, at the base-of-slope discontinuity (Fig. 6). It is inferred that these have acted as transform faults between the axis of spreading in the Havre Trough (Malahoff *et al*. 1982) and that in the Taupo Volcanic Zone (Sissons 1979). The axes are offset by about 50 km. The reason for an overlap of structure across the transform boundary may be related to the present rapid evolution of the Taupo Volcanic Zone (Cole, this volume).

Although they have approximately the same trend, the inferred transform faults in the Bay of Plenty are unrelated to the major discontinuity off northern New Zealand known as the Vening Meinesz Fracture Zone (Fig. 1). This is truncated NE of the Coromandel Peninsula and appears to predate the opening of the Havre Trough (Malahoff *et al*. 1982). The similar trend may be a function of the trend of the plate margin of the northern North Island. The anticline on the mid-slope may also be related to margin processes that predate the 0.6 Ma (Cole, this volume) Taupo Volcanic Zone, since this fold appears to be offset by the White Island Fault.

Sedimentation in the active volcanic zone

In the Bay of Plenty, sedimentation in the off-shore ensialic marginal basin is highly variable. At any place it is primarily the result of a local balance between the supply of terrigenous, detrital, quartz-rich, sand and mud from the frontal ridge, and volcanogenic shards and feldspar from the rhyolitic pyroclastic eruptions of the onshore Taupo Volcanic Zone.

The supply of detrital sediment is almost continous, whereas the supply of volcanogenic sediment is episodic. Vesiculated glass and bubble-wall shards indicate fallout from plinean clouds that blew offshore (Healy *et al*. 1964b; Pullar *et al*. 1977; Kohn & Glasby 1978). Widely dispersed finer grained, less vesiculated shards may have been derived from phreatoplinean eruptions, probably from Lake Taupo (Self & Sparks 1978), there being no evidence for such eruptions offshore. Some volcanogenic material may also be derived from pyroclastic (ignimbrite) flows that reached the coast (Nairn 1972), although there is as yet no clear indication of the fate of the flows when they entered the sea. A suggestion that some flows exploded in major, secondary, 'ventless' eruptions on contact with seawater (Walker 1979) is probably stratigraphically and physically untenable (Nairn 1981); certainly pyroclastic flows do enter the sea without exploding (Sparks *et al*. 1980). Except in the immediate vicinity of White Island, the conspicuous andesite-dacite volcanoes are insignificant contributors of pyroclastic debris, but erosion of volcanic cones has produced a locally significant supply of epiclastic gravel. Fallout directly from rhyolitic eruptions, and rapid fluviatile flushing of loose material soon after eruptions, probably account for the major input of volcanogenic material. The present river input is minor, much of the Taupo Volcanic Zone being drained by a river system that flows to the sea on the W coast of North Island. The shelf sediments in the central Bay of Plenty include large areas, particularly on the inner shelf and near the shelf break, that are predominantly volcanogenic sand with autochthonous shell fragments and allochthonous gravel. The ash and shell debris could be relict or modern, but rounded igneous pebbles almost certainly remain from the time when the surf zone migrated across the shelf during the post-glacial rise of sealevel. Thus, the shelf sand is considered to be a transgressive sand sheet which has been subsequently modified by the addition of shell debris and tephra, and by reworking and redistribution during storms.

In the eastern part of the shelf volcanic zone, the episodic supply of ash is subordinate and incorporated as discrete layers into sequences that reflect the otherwise continuous, rapid influx of detrital silt from the large rivers that drain the main ranges. The transgressive Western Sand Sheet is buried by the Eastern Silt Prism, which overlaps the Motuhora Scarp. Eddying of coast-parallel currents at the scarp is considered to limit mud deposition and produce the band of sandier sediment on either side of the scarp. A line of geothermal activity W of the scarp (Duncan & Pantin 1969) has no detectable effect on surface sediments (Glasby 1975). In the western part of the volcanic zone detrital silt has settled only on the outer shelf.

The sediments on the slope of the central Bay of Plenty reflect two different modes of transport as well as two different sources. The voluminous rhyolitic tephra eruptions in the Taupo Volcanic Zone (Cole, this volume), and minor andesite-dacite eruptions from White Island, periodically produce a rain of lithic, crystal and vitric ash which settles through the water column. Much of this material, although subject to dispersal by tidal and geostrophic currents, settles in a more or less continuous mantle, covering the entire slope, including the knolls and troughs. Conversely, the detrital mud reaches the sea at point sources along the coast where much of it flocculates and settles to the lower part of the water column. Here, much remains in dilute suspension, or is resuspended during storms, to move as density currents, down the slope as underflows, or perhaps away from the slope as interflows on top of denser, colder water layers. Underflows tend to carry the mud into the troughs that incise the slope, and around the ridges and knolls that protrude from it. Thus, the ash on the ridges and knolls is buried by mud more slowly, and has more time for mixing by bioturbation, than the ash in the troughs. Because of this, the sand/mud ratio at a particular site is a rough inverse measure of the overall rate of deposition.

The origin of the mid-slope silt belt is less obvious but it may be related to some intermediate geostrophic flow at depths of 1000–1500 m, or to internal waves between water masses that inhibit the settling of clay-sized material at that depth.

Although the mud in the troughs may result largely from deposition from dilute suspension, the incised canyons on the upper slope, and horizontal fill in the lower slope troughs and parts of the Raukumara Plain, suggest that high-density turbidity currents have played a significant roll in slope sedimentation during periods of lower sealevel.

The thin veneer of calcilutite in the southern Havre Trough indicates a site without major terrigenous input, whereas the shard-rich mud on the Raukumara Plain to the E is downcurrent from detrital sources in the eastern Bay of Plenty and downwind from the tephra eruptions of the Taupo Volcanic Zone.

The western eroded and buried extinct arc

Erosion of the continental shelf on the western side of the Bay of Plenty has planed an extinct volcanic arc. The outer part of the arc has been buried on the upper slope. This arc, which is probably of uppermost Miocene to Lower Pleistocene age becoming younger southwards (Kear 1959; Ballance 1976; Rutherford 1978), intersects the Taupo Volcanic Zone in the vicinity of Lake Taupo.

The wide dispersal of volcanogenic sand, derived from the Taupo Volcanic Zone, northwards along the shelf of the western Bay of Plenty is difficult to explain under present conditions. Like the associated but locally derived gravels, it is probably relict from the last period of glacially lowered sealevel, when a phase of intense rhyolitic volcanic activity coincided with a phase of severe erosion in a cold, seasonally wet climate (Schofield 1965; Hume *et al.* 1975). At that time a sharp atmospheric thermal gradient probably existed to the S, giving rise to frequent intense, cyclonic storms with concomitant southerly gales. The littoral drift so engendered would have carried volcanogenic material northwards, away from the centre of the bay (Harray & Healy 1978).

The sandy mud that buries an eroded and probably downwarped surface on the continental slope onlaps the outer shelf in large patches. By analogy with examples around the British Isles (Belderson & Stride 1966), the patches are probably growing laterally as they blanket the rough, relict or palimpsest gravelly sands away from islands, where water flows are least constricted or disturbed.

ACKNOWLEDGMENTS: We wish to express our gratitude to Dr J. W. Cole of Victoria University of Wellington; Dr F. J. Davey of Geophysics Division, D.S.I.R., Wellington; Mr J. V. Eade of N.Z.O.I., Wellington; and Dr D. Hamilton of Bristol

University, U.K., all of whom have willingly discussed the subject of this paper and offered valuable criticism of the manuscript.

We are particularly indebted to Dr B. P. Kokelaar for his thorough editing and many helpful suggestions.

References

ADAMS, J. A. 1979. Sediment loads of North Island rivers, New Zealand—a reconnaissance. *J. Hydro.* **18**, 36–48.

BALLANCE, P. F. 1976. Evolution of the Upper Cenozoic magmatic arc and plate boundary in northern New Zealand. *Earth planet. Sci. Lett.* **28**, 356–70.

—— & SPÖRLI, K. B. 1979. Northland Allochthon. *J. R. Soc. N.Z.* **9**, 259–75.

——, PETTINGA, J. R. & WEBB, C. 1982. A model of the Cenozoic evolution of Northern New Zealand and adjacent areas of the Southwest Pacific. *Tectonophysics*, **87**, 37–48.

BELDERSON, R. H. & STRIDE, A. H. 1966. Tidal current fashioning of a basal bed. *Mar. Geol.* **4**, 237–57.

BROTHERS, R. N. & DELALOYE, M. 1982. Obducted ophiolites of North Island, New Zealand: origin, age, emplacement and tectonic implications for Tertiary and Quaternary volcanicity. *N.Z. Jl Geol. Geophys.* **25**, 257–74.

—— & HAWKE, M. M. 1981. The tholeiitic Kermadec volcanic suite: additional field and petrological data including iron-enriched plagioclase feldspars. *N.Z. Jl Geol. Geophys.* **24**, 167–75.

CHAPPELL, J. 1975. Upper Quaternary warping and uplift rates in the Bay of Plenty and West Coast, North Island, New Zealand. *N.Z. Jl Geol. Geophys.* **18**, 129–55.

COLE, J. W. 1978. Tectonic setting of Mayor Island volcano (Note). *N.Z. Jl Geol. Geophys.* **21**, 645–7.

—— & LEWIS, K. B. 1981. Evolution of the Taupo–Hikurangi Subduction System. *Tectonophysics*, **72**, 1–21.

DAVEY, F. J. 1977. Marine seismic measurements in the New Zealand Region. *N.Z. Jl Geol. Geophys.* **20**, 719–77.

—— & ROBINSON, A. C. 1978. Cook (1st Edn) *Magnetic Total Force Anomaly Map, Oceanic Series 1:1,000,000*. Dept Sci. Industr. Res., Wellington.

DOYLE, A. C., CARTER, L., GLASBY, G. P. & LEWIS, K. B. 1979. Bay of Plenty Sediments. *New Zealand Oceanographic Institute Chart, Coastal Series, 1:200,000*. Dept Sci. Industr. Res., Wellington.

DUNCAN A. R. 1970. Petrology of rock samples from seamounts near White Island, Bay of Plenty. *N.Z. Jl Geol. Geophys.* **13**, 690–6.

—— & PANTIN, H. M. 1969. Evidence for submarine geothermal activity in the Bay of Plenty. *N.Z. Jl mar. Freshwater Res.* **3**, 602–6.

EIBY, G. A. 1977. The junction of the main New Zealand and Kermadec seismic regions. *International Symposium on Geodynamics in Southwest Pacific*, 167–77, Editions Technip, Paris.

FOLK, R. L. 1974. *Petrology of Sedimentary Rocks*. Hemphill Pub. Co., Austin, Texas, 182 pp.

GIBB, J. G. 1981. Coastal hazard mapping as a planning technique for Waiapu County, East Coast, North Island, New Zealand. He ripoata whakature mo nga whenua papa-a-tai o te rohe o te kaunihera o Waiapu-Tairawhiti. *Water Soil techn. Publ.* **21**, 63 pp.

GLASBY, G. P. 1975. Geochemical dispersion patterns associated with submarine geothermal activity in the Bay of Plenty, New Zealand. *Geochem. J.* **9**, 125–38.

GODFRIAUX, B. L. 1973. Sediment variability in offshore sampling areas. *N.Z. Jl mar. Freshwater Res.* **7**, 329–9.

HARRAY, K. G. & HEALY, T. R. 1978. Beach erosion at Waihi Beach, Bay of Plenty, New Zealand. *N.Z. Jl mar. Freshwater Res.* **12**, 99–107.

HATHERTON, T. 1954. The permanent magnetisation of horizontal volcanic sheets. *J. geophys. Res.* **59**, 223–32.

—— & SYMS, M. 1975. Junction of Kermadec and Hikurangi negative gravity anomalies (Note). *N.Z. Jl Geol. Geophys.* **5**, 753–6.

HEALY, J., SCHOFIELD, J. C. & THOMPSON, B. N. 1964a. Sheet 5—Rotorua (1st Edn). *Geological Map of New Zealand 1:250,000* Dept Sci. Industr. Res., Wellington.

——, VUCETICH, C. G. & PULLAR, W. A. 1964b. Stratigraphy and chronology of Late Quaternary volcanic ash in Taupo, Rotorua, and Gisborne Districts. *N.Z. geol. Surv. Bull.* **73**, 87 pp.

HUME, T. M., SHERWOOD, A. M. & NELSON, C. S. 1975. Alluvial sedimentology of the Upper Pleistocene Hinuera Formation, Hamilton Basin, New Zealand. *J. R. Soc. N.Z.* **5**, 421–62.

HUNT, T. M. & SMITH, E. G. C. 1982. Magnetisation of New Zealand igneous rocks. *J. R. Soc. N.Z.* **12**, 159–80.

ISO, N., OKADA, A., OTA, Y. & YOSHIKAWA, T. 1982. Fission-track ages of Late Pleistocene tephra on the Bay of Plenty coast, North Island, New Zealand. *N.Z. Jl Geol. Geophys.* **25**, 295–303.

KARIG, D. E. 1970. Kermadec Arc–New Zealand tectonic confluence. *N.Z. Jl Geol. Geophys.* **13**, 21–9.

KEAR, D. 1959. Stratigraphy of New Zealand's Cenozoic volcanism north-west of the volcanic belt. *N.Z. Jl Geol. Geophys.* **2**, 578–89.

KINGMA, J. T. 1965. Sheet 6—East Cape (1st Edn) *Geological Map of New Zealand 1:250,000*. Dept Sci. Industr. Res., Wellington.

KOHN, B. P. & GLASBY, G. P. 1978. Tephra distribution and sedimentation rates in the Bay of Plenty, New Zealand. *N.Z. Jl Geol. Geophys.* **21**, 49–70.

LEWIS, K. B. 1971. Slumping on a continental slope inclined at 1–4° *Sedimentology*, 16, 97–110.

—— 1973. Sediments on the continental shelf and slope between Napier and Castlepoint, New Zealand. *N.Z. Jl mar. Freshwater Res.* **7**, 183–208.

—— 1980. Quaternary sedimentation on the Hikurangi oblique-subduction and transform margin, New Zealand. *In*: BALLANCE, P. F. & READING, H. G. (eds) *Sedimentation in Oblique-slip Mobile Zones. Spec. Publ. Int. Assoc. Sediment.* **4**, 171–89.

VAN DER LINGEN, G. J. 1982. Development of the North Island subduction system, New Zealand. *In*: LEGGETT, J. K. (ed.) *Trench–Forearc Geology Geol. Soc. London, Spec. Publ.* **10**, 259–72. Blackwell Scientific Publications, Oxford.

MALAHOFF, A., FEDEN, R. H. & FLEMING, H. S. 1982. Magnetic anomalies and tectonic fabric of marginal basins north of New Zealand. *J. geophys. Res.* **87**, 4109–25.

MCDOUGALL, J. C. 1975. Cook Sediments. *New Zealand Oceanographic Institute Chart, Oceanic Series 1:1,000,000.* Dept. Sci. Industr. Res., Wellington.

NAIRN, I. A. 1972. Rotoehu Ash and the Rotoiti Breccia Formation, Taupo Volcanic Zone, New Zealand. *N.Z. Jl Geol. Geophys.* **15**, 251–61.

—— 1981. Some studies of the geology, volcanic history and geothermal resources of the Okataina Volcanic Center, Taupo Volcanic Zone, New Zealand. Unpublished Ph.D. thesis, Victoria Univ., Wellington.

PANTIN, H. M. 1979. Interaction between velocity and effective density in turbidity flow: phase-plane analysis, with criteria for autosuspension. *Mar. Geol.* **31**, 59–99.

—— 1982. Comments on: Experimental test of autosuspension. *Earth Surf. Processes Landforms*, **7**, 503–5.

——, HERZER, R. H. & GLASBY, G. P. 1973. Bay of Plenty bathymetry. *New Zealand Oceanographic Institute Chart, Coastal Series 1:200,000.* Dept Sci. Industr. Res., Wellington.

PIRAJNO, F. 1980. Subseafloor mineralisation in rocks of the Matakaoa Volcanics around Lottin Point, East Cape, New Zealand. *N.Z. Jl Geol. Geophys.* **23**, 313–34.

PULLAR, W. A., KOHN, B. P. & COX, J. E. 1977. Airfall Kaharoa Ash and Taupo Pumice, and sea-rafted Loisels Pumice, Taupo Pumice and Leigh Pumice in northern and eastern parts of the North Island, New Zealand. *N.Z. Jl Geol. Geophys.* **20**, 697–717.

ROBERTS, N. L. 1967. Magnetic survey of Bay of Plenty. *N.Z. Dept Sci. Industr. Res. Geophys. Div. Rep.* **40**, 4 pp.

RUTHERFORD, N. F. 1978. Fission-track age and trace element geochemistry of some Minden Rhyolite obsidians. *N.Z. Jl Geol. Geophys.* **21**, 443–8.

SCHOFIELD, J. C. 1965. The Hinuera Formation and associated Quaternary events. *N.Z. Jl Geol. Geophys.* **8**, 772–91.

—— 1967. Sheet 3 Auckland (1st Edn). *Geological Map of New Zealand 1:250,000.* Dept Sci. Industr. Res., Wellington.

—— 1970. Coastal sands of Northland and Auckland. *N.Z. Jl Geol. Geophys.* **13**, 767–824.

SELF, S. & SPARKS, R. S. J. 1978. Characteristics of widespread pyroclastic deposits formed by the interaction of silicic magma and water. *Bull. Volcanol.* **41**, 196–212.

SISSONS, B. A. 1979. The horizontal kinematics of the North Island of New Zealand. Unpublished Ph.D. thesis, Victoria Univ., Wellington.

SPARKS, R. S. J., SIGURDSSON, H. & CAREY, S. N. 1980. The entrance of pyroclastic flows into the sea. I. Oceanographic and geologic evidence from Dominica, Lesser Antilles. *J. Volcan. geoth. Res. Amsterdam*, **7**, 87–96.

STRONG, C. P. 1976. Cretaceous foraminifera from the Matakaoa Volcanic Group. *N.Z. Jl Geol. Geophys.* **19**, 140–3.

—— 1980. Early Paleogene foraminifera from Matakaoa Volcanic Group (Note). *N.Z. Jl Geol. Geophys.* **23**, 267–72.

VAIL, P. R., MITCHUM JR, R. M., TODD, R. G., WIDMIER, J. M., THOMPSON III, S., SANGREE, J. B., BUBB, J. N. & HATLELID, W. G. 1977. Seismic stratigraphy and global changes of sea level. *In*: PAYTON, C. E. (ed.) *Seismic Stratigraphy—Applications to Hydrocarbon Exploration. Mem. Am. Ass. Petrol. Geol.* **26**, 49–212.

WALKER, G. P. L. 1979. A volcanic ash generated by explosions where ignimbrite entered the sea. *Nature, Lond.* **281**, 642–6.

K. B. LEWIS, New Zealand Oceanographic Institute, P.O. Box 12–346, Wellington North, New Zealand.

H. M. PANTIN, British Geological Survey, Nicker Hill, Keyworth, Nottingham NG12 5GG, U. K.

Hydrothermal and volcaniclastic sedimentation on the Tonga–Kermadec Ridge and in its adjacent marginal basins

D. S. Cronan, S. A. Moorby, G. P. Glasby, K. Knedler, J. Thomson & R. Hodkinson

SUMMARY: Analyses of over 180 sediment samples from the Tonga–Kermadec Ridge and adjacent marginal basins demonstrate volcaniclastic, biogenic, lithogenous and hydrothermal influences on them. The volcaniclastic component decreases westwards from the Tonga–Kermadec Ridge where volcanism is presently active. Biogenic components are predominant in sediment throughout most of the region, but a significant lithogenous component becomes apparent in the S, in the vicinity of New Zealand. Hydrothermal influences have been detected in two areas: (i) on the Tonga–Kermadec Ridge where Mn oxide crusts represent the end member of a hydrothermal fractionation sequence, earlier members of which may occur beneath the sea-floor; and (ii) in the Lau Basin where Fe and Mn are widely dispersed in the sediments and where sulphide deposits may occur in the immediate vicinity of hydrothermal vents.

In contrast to mid-ocean ridges, island arcs and marginal basins have been little studied with regard to hydrothermal sedimentation. Early reports of localized hydrothermal deposits developed in such settings include those of the E Indian and SW Pacific Oceans (Zelenov 1964; Ferguson & Lambert 1972; Bertine & Keene 1975) and the Mediterranean (Honnorez 1969; Bonatti *et al.* 1972; Smith & Cronan 1975). In these settings, volcaniclastic sediments have a much wider distribution than hydrothermal deposits.

In order to outline areas of possible submarine hydrothermal activity in the SW Pacific, Cronan & Thompson (1978) carried out a regional geochemical survey based on analysis of all the sediment samples available from the region. These data, together with an appraisal of the likely loci of submarine volcanic activity in the region based on tectonic information, led to the delineation of several areas of greatest potential for hydrothermal sedimentation within the SW Pacific (Cronan 1983). One of these areas was the Tonga–Kermadec Ridge, and the adjacent marginal basins, the Lau Basin- and Havre Trough (Fig. 1).

As part of a joint research programme by Imperial College and the New Zealand Oceanographic Institute, two cruises to the Tonga–Kermadec Ridge and its adjacent marginal basins were carried out by the N.Z.O.I. research vessel, Tangaroa, in 1981 and in 1982. During this work, sediments were recovered from over 180 stations (Fig. 2), by coring and dredging. A wide variety of

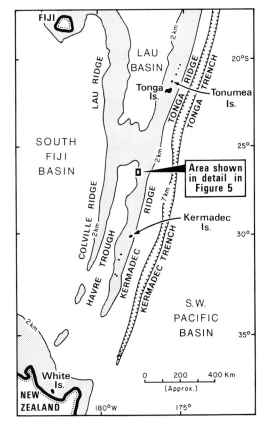

FIG. 1. Simplified bathymetric map of the SW Pacific Ocean showing the main topographic features. Bathymetric contours are in km.

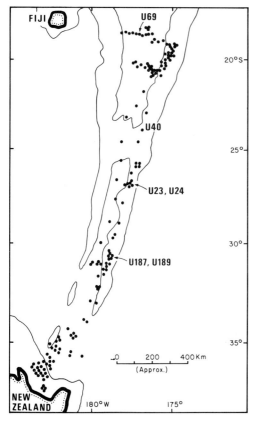

FIG. 2. Map showing location of sample stations (dots) used in this study.

Volcanic Zone of the North Island, New Zealand (Cole, this volume). On the W side of the Lau Basin and Havre Trough lies the remnant arc represented by the Lau and Colville Ridges (Karig 1970). The average depth of the back-arc basins, 2500 m, is in marked contrast to average sea-floor depths, in excess of 4000 m, in the South Fiji and SW Pacific Basins to the W and E (Fig. 1). Another difference is the more rugged topography and thinner sediment cover of the back-arc basins.

Both the Lau Basin and the Havre Trough consist of many ridges and small basins with an overall relief of 500–1500 m, probably representing faulted blocks (Karig 1970; Hawkins 1974). The acoustic basement in the Lau Basin consists of basalts of MOR and island-arc types (Saunders & Tarney, this volume). The Lau Basin is subject to frequent shallow-focus seismic activity and is an area of comparatively high heat flow (Sclater *et al.* 1972; Hawkins 1974). The age of the Lau Basin has been estimated as about 5 Ma by Gill (1976) and Lawver *et al.* (1976), who suggested that the magnetic anomaly patterns did not show any unambiguous evidence of symmetrical spreading of normal mid-ocean ridge type. However, magnetic surveys of the Lau Basin by Weissel (1977, 1982), Cherkis (1980) and Malahoff *et al.* (1982) revealed the presence of symmetrical magnetic anomalies, although there is not complete agreement between these authors on the exact location of the spreading centre or centres. The magnetic anomaly patterns obtained in the Havre Trough by Malahoff *et al.* (1982) led these authors to suggest that spreading also occurred in this basin, but that it commenced more recently (2 Ma ago) than in the Lau Basin and with a lower half-spreading rate of 27 mm yr^{-1} compared with the 38 mm yr^{-1} calculated by Weissel (1977) for the Lau Basin.

sediment types was found, including hydrothermal Mn crusts at a number of localities. A preliminary report on the latter has been published elsewhere (Cronan *et al.* 1982). In the present paper, we attempt to describe the sediment characteristics of the whole region studied, and to discuss volcanic influences, both clastic and hydrothermal, on them.

Geology and structure

The Lau Basin–Havre Trough–Tonga–Kermadec Ridge area forms part of the SW Pacific island-arc–back-arc system. The volcanic Tonga and Kermadec Islands delineate the active volcanic arc (Fig. 1), generated by the descent of the Pacific Plate in the E beneath the overriding Austral-Indian Plate in the W, the zone of subduction being marked by the Tonga and Kermadec Trenches. To the W of the active volcanic arc, the back-arc basin system comprises the Lau Basin, Havre Trough and Taupo

Sediment distribution

Sediment cover in much of the Lau Basin and Havre Trough is very thin. Seismic reflection profiles (Hawkins 1974) show generally less than 100 m of sediment cover (< 0.1 s acoustic penetration), with a few narrow ponded areas displaying 0.1–0.2 s penetration. Because of their generally rugged nature and their comparatively young age, sediment cover is probably not continuous in the marginal basins, and may be absent altogether along parts of the Colville Ridge (Karig 1970). Along the Tonga–Kermadec Ridge and its western flanks, however, sediment cover is up to 700 m thick.

In this area, many ridges, seamounts and scarps, some of which are probably fault-controlled (Hawkins 1974), tend to pond the sediment and probably restrict sediment input to the Lau Basin and Havre Trough.

Methods of sediment analysis

Shipboard smear-slide analysis was carried out on all surface sediment samples recovered. Where these were from gravity cores the sample represented the uppermost 1 cm of sediment. However, pipe-dredged sediment samples represent a homogenized sample of the uppermost 25–30 cm. Smear-slide estimates of abundances were based on areal estimates with the help of a comparison chart for visual percentage estimation (Terry & Chilingar 1955). The accuracy of this method may approach ± 2% for distinctive minor constituents (e.g. volcanic glass and Fe oxides), but for major components ± 10–20% is the best which can be expected (Burns *et al*. 1973). Estimation of nannofossil abundance is particularly difficult and can only be regarded as semi-quantitative. Chemical determination of Ca is a much better guide to carbonate content in carbonate-rich sediments. Wherever possible, sediments were classified according to the standard DSDP classification scheme (see Burns *et al*. 1973).

The uppermost sediment recovered at each site was analysed chemically by inductively coupled plasma spectrometry, after the samples had been air-dried, crushed, digested in a mixture of hydrofluoric, perchloric and nitric acids and taken up in 1 M HCl. Levels of analytical precision are given in Table 1. Accuracy was checked by analysis of standard reference materials; no systematic errors were observed.

Surface sediment lithology

A smoothed pattern of regional variation in the volcanogenic component of the sediments, based on smear-slide analysis, is shown in Fig. 3. No attempt has been made to try to indicate areas where sediment cover is absent. The volcanogenic component of the sediments decreases very rapidly westwards from the Tonga–Kermadec Ridge. Sediments on the flanks and upper slopes of the ridge generally contain 75–100% volcanic debris, of fine sand to silt size, comprising acidic glass shards, calcic plagioclase and clinopyroxene with minor amounts of opaques. The low amounts or absence of included biogenic debris indicate

rapid accumulation of the volcanic material. The westward decrease in the volcanogenic component of the sediments is less marked in the northern Lau Basin than further S. This may be because in the northern part of the Lau Basin the SE trade winds are more persistent than in the S and a greater westward dispersion of volcanic material might therefore be expected (Karig 1970). Alternatively, or in addition, this pattern may reflect an increase in the rate of generation of volcaniclastic debris northwards along the volcanic arc, or the greater westward penetration of such material by gravity flows due to more favourable topographic controls. Presently there is insufficient data to show which of these possibilities is most important.

Amounts of volcanic glass in the sediments fall to lower than 5% towards the western edge of the Lau Basin and most of the Havre Trough. In the southern part of the Havre Trough amounts increase slightly away from the volcanic arc, probably as a result of volcanic activity both on the North Island of New Zealand and at White Island (see Lewis & Pantin, this volume). The distribution pattern and the silicic nature of the volcanic glass, however, confirm that the main source of volcanic debris throughout the region is the volcanic activity along the Tonga–Kermadec Ridge (Ewart *et al*. 1977).

Much of the sediment in the area W of 177°W is carbonate ooze, consisting predominantly of nannofossils, with a total carbonate content of up to 75%. Siliceous microfossils are mostly either absent or present only in trace amounts, but in the northern Lau Basin radiolaria form as much as 5–8% of the biogenic component of the sediment. No part of this area lies below the carbonate compensation depth, which, generally in this region, lies at about 4300 m (Berger *et al*. 1976). Small, isolated basins in the Havre Trough exceed this depth but none was sampled. The rapid increase in the volcanogenic component, from W to E across the back-arc basins, is such that at the eastern edges of the basins and lower slopes of the volcanic arc, the surface sediments recovered had a carbonate content of only 20–40%. Although the flanks and upper slopes of the Tonga–Kermadec Ridge are dominated by volcanic silts and sands, in the N, where the ridge is wider and there are many coral islands, calcareous sands occur in the shallowest waters (Karig 1970). In the extreme S of the area, near New Zealand, terrigenous material becomes a major sediment component.

The regional variation in the reddish-brown

D. S. Cronan et al.

TABLE 1. *Mean chemical composition (with standard deviation) of sediments from different regions of the Tonga–Kermadec Ridge and associated marginal basins*

	Ca (%)	$CaCO_3$ (%)	Al (%)	Fe (%)	Mn (%)	Ni (ppm)	Cu (ppm)	Zn (ppm)
White Island area (9 stations)	3.6 ± 1.6	7.4 ± 2.1	7.0 ± 0.90	3.9 ± 0.8	0.057 ± 0.02	57 ± 18	41 ± 13	58 ± 5
Havre Trough (61 stations)	14.8 ± 7.7	35.7 ± 18	6.7 ± 0.97	2.9 ± 1.3	0.19 ± 0.12	35 ± 25	69 ± 41	66 ± 10
Lau Basin (44 stations)	14.9 ± 9.0	36.0 ± 21	7.4 ± 1.1	8.8 ± 1.2	0.94 ± 0.56	68 ± 36	210 ± 47	165 ± 34
Kermadec Ridge (16 stations)	11.8 ± 5.6	28.2 ± 12	7.1 ± 1.3	4.3 ± 1.5	0.19 ± 0.07	23 ± 19	71 ± 43	107 ± 24
Tonga Ridge (47 stations)	10.2 ± 6.4	24.2 ± 14	8.7 ± 2.4	6.8 ± 2.4	0.20 ± 0.13	24 ± 16	125 ± 41	116 ± 27
Analytical precision	±2%	—	±3%	±2%	±2%	±4%	±2%	±5%

Data are expressed on a $CaCO_3$-free basis (except for Ca).

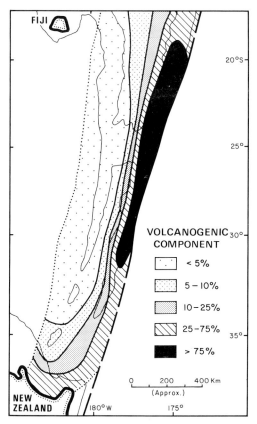

FIG. 3. Regional variation in the volcanogenic component of surface sediments in the SW Pacific, based on smear-slide estimates.

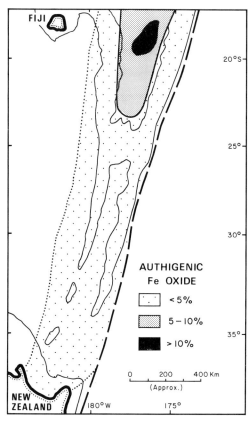

FIG. 4. Regional variation in the authigenic Fe-oxide component of surface sediments in the SW Pacific, based on smear-slide estimates.

semi-opaque Fe-oxide component of the sediments is shown in Fig. 4. In view of the relatively small amounts present, no attempt has been made to divide the material into the various sub-classes recognized in metalliferous sediments by Quilty *et al.* (1976). Throughout much of the region, Fe oxides are absent or present only in trace amounts. This is particularly so in sediments where volcaniclastic or terrigenous material is a major component. Only in the Lau Basin does the Fe-oxide component of the surface sediment regularly exceed 5%. In the central Lau Basin, the Fe-oxide component exceeds 10%. However, Fe oxides were not a dominant sediment component at any site, maximum amounts being of the order of only 15%.

Manganese oxide crusts were recovered by pipe dredging at three sites along the Tonga–Kermadec Ridge. At stations U40 and U187 (Fig. 2) small amounts of platy crusts were recovered, but at stations U23, U24 and

U189 large hauls were recovered of both platy material and blocky crusts up to 5 cm or more in thickness. Details of the morphology and geochemistry of some of the samples are given in Cronan *et al.* (1982) and Moorby *et al.* (1984). On the basis of their chemical composition and accumulation rate, Cronan *et al.* (1982) have shown that the deposits are of hydrothermal origin.

A detailed bathymetric survey and sampling operation was carried out at 27°S on the Tonga–Kermadec Ridge (Fig. 5). In this area, Mn crusts were only recovered at stations U23 and U24, suggesting that the deposits are very localized. No Mn enrichment was found in the sediments recovered from the other sites shown in Fig. 5. Stations U23 and U24 lie on the lower flanks of a feature which is 400–500 m high and forms part of a NNE–SSW-trending ridge (Fig. 5). This ridge lies WNW of the main part of the volcanic arc, which shoals in places to within a

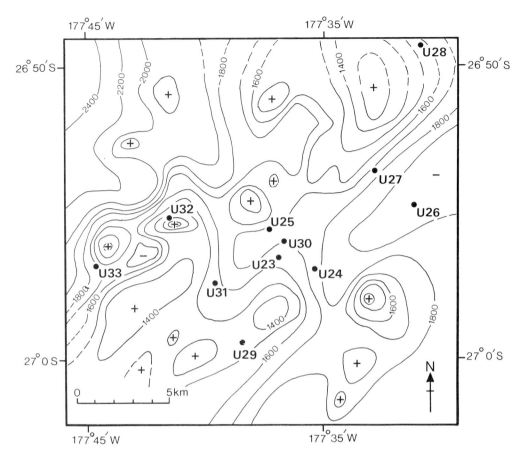

FIG. 5. Detailed bathymetry of the Tonga–Kermadec Ridge in the vicinity of stations U23 and U24. (Contours are in metres below sea level.)

few hundred metres of sealevel. The topography of the survey area probably results from the effects of local submarine activity superimposed upon the normal-fault scarp which defines the western margin of the volcanic arc (Hawkins 1974).

No detailed surveys were carried out at the other two localities at which Mn crusts were recovered. However, both sites lie in positions analogous to that of stations U23 and U24, on the western flanks of the Tonga–Kermadec Ridge, in water depths of 1500–1900 m on the lower flanks of seamounts 400–1000 m high.

Only a few of the Mn crust samples had a substrate attached. These substrates comprised indurated sediment or small, rounded pumice fragments, indicating growth on sediment or talus rather than on a hardrock substrate. By contrast, hydrothermal crusts recovered from oceanic spreading centres tend to be associated with basalt substrates, because of the largely sediment-free nature of these environments. This is an important difference since it implies that whilst many hydrothermal Mn deposits at oceanic spreading centres precipitate out from the water column on to exposed basalt close to hydrothermal vents, some of those described here may result from precipitation at the sediment–water interface from hydrothermal solutions slowly upwelling through a sediment or talus blanket. This has important implications with regard to the possible occurrence and location of other hydrothermal deposits in the area and is discussed below.

Distribution of elements in surface sediments

Cronan & Thompson (1978) have shown that, in SW Pacific sediments, certain element distribution patterns closely reflect specific sediment sources. Calcium, for example, reflects the distribution of biogenic $CaCO_3$, Al reflects the distribution of lithogenous and volcaniclastic materials, and Mn reflects the distribution of authigenic and/or hydrothermal deposits. The average concentrations of several elements in different parts of the study area have therefore been compiled (Table 1) in an attempt to refine conclusions concerning their provenance based on lithological data.

The distribution of Ca and Al in the sediments closely reflects the variations in the carbonate and volcanogenic component of the sediments (Fig. 3). The Fe/Al ratio of the sediments (Fig. 6) clearly shows the influence of terrigenous input in the S and the presence of abundant Fe oxides in central Lau Basin sediments. However, the Fe/Al ratio does not seem to be markedly affected by the variations in the amount of volcanogenic debris present (Fig. 3), although there is a tendency, particularly in the northern parts of the region, for Fe/Al to be

low in sediments on the Tonga–Kermadec Ridge and its flanks.

Manganese levels within the surface sediments of the region are very low, with the exception of the central Lau Basin area (Table 1). Manganese averages less than 0.06% adjacent to White Island (Table 1) compared with around 0.2% for the Tonga–Kermadec Ridge and Havre Trough sediments, and a mean of 0.94% (2.8% max) in central Lau Basin sediments. Possible causes of this enrichment of Mn in the sediments of the Lau Basin, of up to 10-fold compared with the remainder of the region, include diagenetic remobilization, hydrothermal activity, or an increase in the hydrogenous component of the sediment. These are evaluated below.

Sediment accumulation rates

Griffen *et al.* (1972) estimated a sediment accumulation rate of 2–3 cm 10^{-3} yr for one core from the Lau Basin, but found considerable differences in accumulation rates in a suite of 1–2 m gravity cores from the basin. They noted, however, that their data were consistent with rapid sediment accumulation, of the order of several centimetres per thousand years. In an attempt to discern possible hydrothermal influences on the sediments, two cores have been selected for radiometric dating by $^{230}Th_{excess}$ and $^{231}Pa_{excess}$ methods (Thomson 1982), and for detailed chemical analysis. Core U69 was taken about 50 km W of the supposed spreading centre in the Lau Basin, and the other, U49, was collected from the axial rift, near where the spreading centre was thought to be on the basis of coincidence of the magnetic anomaly data of Weissel (1982) and Malahoff *et al.* (1982).

The radiochemical data for the two cores are presented in Table 2 and $^{230}Th_{excess}$ and $^{231}Pa_{excess}$ decay plots for U69 in Fig. 7. The Th contents in both cores are low and this leads to high $^{230}Th/^{232}Th$ activity ratios in U69. Similar observations were made by Griffin *et al.* (1972) in the cores they studied. The Th/U ratio is also low by comparison with the average value of 5.6 usually observed in deep-sea sediments (Heye 1969). This fact, together with the high $^{234}U/^{238}U$ activity ratio, suggests U enrichment in the sediments.

Whilst Fig. 7 illustrates that a satisfactory estimate of sediment accumulation rate may be made for U69, another aspect of this data is noteworthy. The expected value of the initial $^{230}Th_{excess}/^{231}Pa_{excess}$ activity ratio, from supply by the U content of seawater, is 10.6. The val-

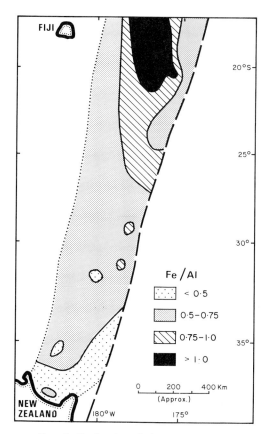

FIG. 6. Regional variation in the Fe/Al ratio of surface sediments in the SW Pacific Ocean.

TABLE 2. Uranium, thorium and protactinium data for cores U49 and U69, Lau Basin

Depth (cm)	U (ppm)	Th (ppm)	Th/U	$^{234}U/^{238}U$ activity ratio	$^{230}Th/^{232}Th$ activity ratio	^{234}U (dpm g^{-1})	^{230}Th (dpm g^{-1})	$^{230}Th_{excess}$ (dpm g^{-1})	$^{231}Pa_{excess}$ (dpm g^{-1})
Core U69									
2	0.71 ± 0.03	0.60 ± 0.08	0.85 ± 0.12	*	106 ± 13	—	15.6 ± 0.4	15.0 ± 0.4	2.17 ± 0.14
12	0.65 ± 0.02	0.40 ± 0.05	0.62 ± 0.08	1.19 ± 0.05	136 ± 18	0.58 ± 0.02	13.0 ± 0.4	12.4 ± 0.4	1.57 ± 0.09
25	0.54 ± 0.03	0.59 ± 0.05	1.09 ± 0.11	1.19 ± 0.08	68 ± 5	0.48 ± 0.02	9.7 ± 0.2	9.2 ± 0.2	1.01 ± 0.05
40	0.54 ± 0.03	0.54 ± 0.04	1.00 ± 0.09	*	69 ± 5	—	9.1 ± 0.2	8.6 ± 0.2	1.00 ± 0.06
55	0.57 ± 0.03	0.52 ± 0.04	0.91 ± 0.09	1.06 ± 0.06	71 ± 5	0.46 ± 0.02	8.9 ± 0.2	8.4 ± 0.2	0.86 ± 0.05
75	0.66 ± 0.03	0.49 ± 0.07	0.74 ± 0.11	*	82 ± 12	—	9.8 ± 0.3	9.3 ± 0.3	0.76 ± 0.05
Core U49									
13	0.27 ± 0.01	0.23 ± 0.02	0.85 ± 0.08	1.09 ± 0.07	15 ± 2	0.22 ± 0.01	0.80 ± 0.03	0.58 ± 0.03	—
33	0.22 ± 0.01	0.27 ± 0.02	1.2 ± 0.1	0.98 ± 0.06	9.6 ± 0.7	0.16 ± 0.01	0.64 ± 0.02	0.48 ± 0.02	—
57	0.37 ± 0.01	0.26 ± 0.03	0.70 ± 0.08	1.13 ± 0.06	23 ± 2	0.31 ± 0.01	1.44 ± 0.04	1.13 ± 0.04	—

* Spectra contaminated with ^{231}Pa.

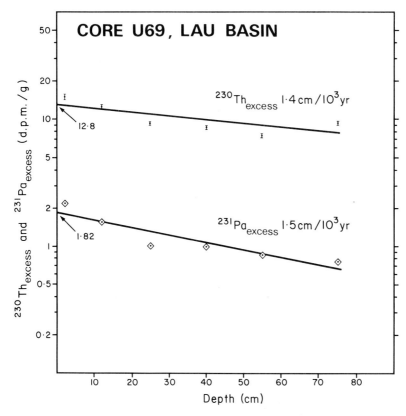

FIG. 7. Variation with depth of $^{230}\text{Th}_{\text{excess}}$ and $^{231}\text{Pa}_{\text{excess}}$ in core U69 from the Lau Basin.

ues in core U69 are considerably lower than this, showing an enrichment in $^{231}\text{Pa}_{\text{excess}}$. This situation is unusual, but has been noted in Mn nodules (Turekian & Chan 1971) and mid-ocean ridge sediments (Cochran 1982). In fact, the estimated $^{231}\text{Pa}_{\text{excess}}$ and $^{230}\text{Th}_{\text{excess}}$ in the sediment column of U69, relative to the calculated supply from a water column of 2444 m depth, are 2.9 and 1.7 times too high respectively. In other words, both are high, but $^{231}\text{Pa}_{\text{excess}}$ more so than $^{230}\text{Th}_{\text{excess}}$. Since the water column is the most probable source of the $^{230}\text{Th}_{\text{excess}}$ and $^{231}\text{Pa}_{\text{excess}}$, it is tempting to invoke the scavenging properties of the sediments as the cause. Cochran (1982) has speculated that enhanced removal of these two radionuclides from the water column in the vicinity of ridge crests may be related to effective scavenging by precipitation of hydrothermally supplied Mn.

It was not possible to obtain a meaningful sediment accumulation rate for core U49 because the $^{230}\text{Th}_{\text{excess}}$ values are very low. Additionally, this core appears to have accumu-

lated at an uneven rate, reflected by the presence of volcaniclastic-rich layers (probably rapidly deposited) and $CaCO_3$-rich layers (probably deposited more slowly). As a rough guide, the accumulation rate for U49 may be as much as 10 times higher than that for U69, on the basis of their $^{230}\text{Th}_{\text{excess}}$ contents. This estimate, however, has to be viewed with extreme caution in view of the limitations of the data.

Metal accumulation rates

The chemical composition of core U69 was found to be reasonably constant with depth (Table 3), permitting metal accumulation rates to be calculated using the following equation:

Element accumulation rate $= [M] \cdot \rho \cdot S,$

where $[M]$ = mean element concentration (average of six samples), ρ = dry *in situ* sediment density (measured as 0.59 g cm^{-3} for core U69 sediments), and S = sediment accumulation rate.

TABLE 3. *Bulk chemical composition of cores U49 and U69 from the Lau Basin*

Core	Sample interval (cm)	CaCO₃ (%)	Al (%)	Mn (%)	Fe (%)	V	Co	Ni	Cu	Zn
								ppm		
U69	0–2	55	2.2	0.65	2.8	98	18	41	121	59
	11–13	50	2.9	0.64	3.4	126	20	31	126	61
	24–26	45	2.3	0.63	3.2	126	20	30	107	54
	39–41	41	3.9	0.62	4.1	169	25	47	121	57
	54–56	45	3.4	0.62	3.6	141	21	44	113	54
	74–76	55	2.7	0.72	3.2	120	20	37	112	60
	Average	48	2.9	0.65	3.4	130	21	38	117	58
U49	Average (6 samples)	15	5.8	0.38	6.6	300	30	21	140	83

Taking a value of S from Fig. 7 of 1.5 cm 10^{-3} yr, the metal accumulation rates obtained are given in Table 4, together with other published values for comparison.

It can be seen from Table 4 that the accumulation rates of Mn and Fe are higher than typical rates for Pacific pelagic clays and approach some of the values reported for sediments at active ridge crests. Co, Ni and Zn show values similar to pelagic clay, but V and Cu show somewhat higher values. There is therefore some evidence to suggest an additional supply of Mn and Fe to the sediment in core U69, over and above a normal authigenic supply, but evidence for the other elements is inconclusive.

Discussion

The lithological information, on the nature and distribution of volcanic activity in the region investigated, enables us to consider the influence of volcanism on its sediments.

The gross influence of subaerial volcanism on the sediments of the region is clearly seen in the distribution of volcaniclastic material in them. As might be expected, this influence is strongest in the vicinity of the Tonga Ridge, where some sediments are almost entirely of volcanic origin. The influence is somewhat less on the Kermadec Ridge where fewer subaerial volcanoes occur, and it decreases in a westerly direction, away from both ridges. Volcaniclastic particles can be transported through the atmosphere by winds, within the water column suspended in currents, and along the sea floor in the form of slides, slumps and turbidity currents. Sediment gravity flows are of greatest importance in the immediate vicinity of the volcanic arc, as exemplified by the exceptionally flat, ponded rather than blanketing, nature of the sediments in the Tofua Trough between the two arms of the Tonga island chain.

Regional hydrothermal influences on the sediments are not as easy to discern as volcaniclastic influences. However, hydrothermal Mn crusts have been formed at several locations along the Tonga–Kermadec Ridge and crusts of similar composition also occur on the island of Tonumea in the Tonga Islands (Table 5). These

TABLE 4. *Comparison of metal accumulation rates in core U69 with values for other localities*

	Metal accumulation rate (μg cm^{-2} 10^{-3} yr)				
	U69 (this study)	Pacific pelagic clay*	EPR crest (Leinen & Stakes 1979)	EPR crest (Dymond & Veeh 1975)	TAG area (Scott et al. 1978)
Mn	5500	700	5400	5800–28000	3500
Fe	18000	7700	29000	11000–82000	49500
V	80	32	—	—	—
Co	12	15	—	—	120
Ni	30	30	80	56–160	235
Cu	90	48	160	50	214
Zn	27	24	170	22	—

*Data for pelagic clay obtained from Cronan (1969) (element abundances) and Krishnaswami (1976) (values for ρ and S).

TABLE 5. *Chemical composition of Mn oxide deposits from Tonumea Island and from station U23 on the Tonga–Kermadec Ridge (see Figs 1 and 2 for locations)*

	Mn (%)	Fe (%)	Ca (%)	Al (%)	Co (ppm)	Ni (ppm)	Cu (ppm)	Zn (ppm)
Tonumea Island Mn deposit	59	0.36	1.3	0.75	102	51	100	130
U23/1 (average of 5 samples of hydrothermal layers)	48	0.05	1.5	0.10	7	80	27	90

findings indicate that the volcanic arc has been, and still is, the site of sporadic hydrothermal activity.

It has become apparent in recent years that submarine hydrothermal Mn deposits are the final precipitates from submarine hydrothermal systems, earlier, higher temperature precipitates from which include massive sulphides, Fe silicates and Fe oxides, in that order of deposition (Cronan 1980, and references therein). In this respect the Tonga–Kermadec Ridge hydrothermal deposits show similarities with those seen in the TAG area in the North Atlantic, where Mn crusts are the only surface hydrothermal deposit, and the surrounding sediments show little Mn enrichment (Cronan *et al.* 1979). These fields differ from those discovered on the East Pacific Rise and Galapagos Rift. On the Galapagos Rift, for example (Malahoff *et al.* 1983), higher temperature hydrothermal deposits, dominated by sulphides, occur at the surface and are surrounded by Mn-enriched sediments. If sulphides and silicates have been precipitated by hydrothermal circulation along the Tonga–Kermadec Ridge, they are therefore most likely to occur at depth.

Hydrothermal influences elsewhere in the region are not as easy to discern as those on the Tonga–Kermadec Ridge. The main areas where such influences would be expected are associated with crustal extension in the back-arc basins. There is good evidence for crustal extension in the Lau Basin, but no definite geological or geophysical evidence exists to suggest that such a process occurs at the present time in the Havre Trough. This is in agreement with our geochemical data which are compatible with the likelihood of submarine hydrothermal activity in the Lau Basin, but provide no evidence for such activity in the Havre Trough. This conclusion is based on the observation that elements normally widely dispersed around spreading ridge segments on the World Mid-Ocean Ridge System, such as Mn and Fe, are much more abundant in the Lau Basin than in the Havre Trough. Indeed, the values of these

elements in the latter, particularly Fe, are amongst the lowest recorded in the whole SW Pacific region.

Manganese and Fe are higher, on average, in the Lau Basin than in any of the other areas considered (Table 1; Fig. 4). It is well known, however, that Fe and Mn concentrations in sediments are not, in themselves, unequivocal evidence of hydrothermal supply of these metals. Elevated Mn values, for example, may be caused in surface sediments by diagenetic remobilization of Mn from shallow depth, and high Fe values may reflect an increase in the importance of volcanogenic material or its low-temperature weathering products. The Lau Basin does not differ significantly in its surface sediment lithology from the Havre Trough. The type and amount of biogenic carbonate received appear to be similar on both basin floors. We can see no reason, therefore, why diagenetic remobilization of Mn might be more marked in the Lau Basin than in the Havre Trough. Furthermore, Mn levels at depth in Lau Basin sediments (Table 3 and unpublished data) provide little evidence for its upward remobilization. At comparable distances from the volcanic arc, volcanogenic material is more predominant in Lau Basin sediments than in those from the Havre Trough. If this material were the source of the Fe enrichment in Lau Basin sediments then, by analogy, an area of Fe enrichment would be expected in the Havre Trough also, albeit somewhat smaller and closer to the volcanic arc. No such area exists.

The metal accumulation rates obtained for the Lau Basin core, U69, demonstrate Fe and Mn accumulation above pelagic Pacific rates and near to East Pacific Rise crest rates. However, U69 was taken about 50 km from where the spreading centre was thought to be. In core U49, from the axial rift, sedimentation rates are higher and metal accumulation rates, based on the average composition of the core (Table 3), are also higher.

Thus, available evidence indicates that there is a hydrothermal contribution of Mn and Fe to the Lau Basin sediments. Hydrothermal enrichments of Fe and Mn occur in sediments in the vicinity of spreading ridges such as the East Pacific Rise and Galapagos Rift, which also exhibit sulphide deposition. The axial rift of the Lau Basin, therefore, might also be a favourable area to seek hydrothermal sulphide deposits.

Conclusions

1 Surface sediment lithology indicates a large volcanogenic contribution to the sediments of the study area, which decreases in abundance in a westerly direction. Other influences on the sediments are predominantly biogenic, and lithogenous in the S of the area. Semi-opaque Fe oxides occur in trace amounts in sediments throughout much of the region, but increase to as much as 15% of the sediment in the Lau Basin.

2 Hydrothermal Mn-oxide crusts were recovered at a number of stations on the Tonga–Kermadec Ridge, and also on land in the Tonga Islands. No Mn enrichment was detected in their associated sediments, suggesting that the crusts represent the end member of a submarine hydrothermal fractionation sequence, earlier members of which may be present at depth.

3 Element distribution trends in the sediments of the study area closely reflect their lithology. Ca and Al reflect biogenic and detrital components respectively, and the Fe/Al ratio highlights the contribution of Fe oxides. Manganese tends to be low throughout much of the area, but is enriched up to 10-fold in the sediments of the Lau Basin.

4 Radiometric dating of one core, taken 50 km W of the supposed spreading centre in the Lau Basin, gave an accumulation rate of 1.4–1.5 cm 10^{-3} yr, from which accumulation rates of 18000 μg cm^{-2} 10^{-3} yr Fe and 5500 μg cm^{-2} 10^{-3} yr Mn have been calculated. These approach some of the accumulation rate values for these elements reported on hydrothermally active mid-ocean ridge crests, suggesting a hydrothermal input of these metals to the Lau Basin. If this is so, sulphide deposits might be expected to occur in the vicinity of the hydrothermal vents in the Lau Basin.

ACKNOWLEDGMENTS: We thank that master and Crew of R.V. Tangaroa and scientific staff of the N.Z.O.I. for the collection of the samples described in this work. Financial support was received from N.E.R.C.

References

BERGER, W. H., ADELSECK, C. G. & MAYER, L. A. 1976. Distribution of carbonate in surface sediments of the Pacific Ocean. *J. geophys. Res.* **81**, 2617–27.

BERTINE, K. K. & KEEN, J. B. 1975. Submarine barite-opal rocks of hydrothermal origin. *Science*, **188**, 150–2.

BONATTI, E., HONNOREZ, J., JOENSUU, O. & RYDELL, H. S. 1972. Submarine iron deposits from the Mediterranean Sea. *In*: STANLEY, D. J. (ed.) *Symposium on Sedimentation in the Mediterranean Sea* 701–10. Hutchinson & Ross, Stroudsburg, Pennsylvania, U.S.A.

BURNS, R. E., ANDREWS, J. E., CHURKIN, M. *et al.* 1973. *Init. Rep. Deep Sea drill. Proj.* **21**, 5–16. U.S. Government Printing Office, Washington, D.C.

CHERKIS, N. Z. 1980. Aeromagnetic investigations and sea floor spreading history in the Lau Basin and Northern Fiji Plateau. *In: Symposium on Petroleum Potential in Island Arcs, Small Ocean Basins, Submerged Margins and Related Areas,* 37–46. United Nations ESCAP, CCOP/SOPAC Tech. Bull. No. 3. D.S.I.R., Wellington, New Zealand.

COCHRAN, J. K. 1982. The ocean chemstry of the U-

and Th- series nuclides. *In*: IVANOVICH, M. & HARMON, R. S. (eds) *Uranium Series Disequilibrium: Application to Environmental Problems*, 384–430. Oxford University Press, Oxford.

CRONAN, D. S. 1969. Average abundances of Mn, Fe, Ni, Co, Cu, Pb, Mo, V, Cr, Ti and P in Pacific pelagic clays. *Geochim. cosmochim. Acta*, **33**, 1562–5.

—— 1980. *Underwater Minerals*, 362 pp. Academic Press, London.

—— 1983. *Metalliferous Sediments in the CCOP/ SOPAC Region of the S.W. Pacific, with Particular Reference to Geochemical Exploration for the Deposits*. United Nations ESCAP, CCOP/ SOPAC Tech. Bull. No 4. Suva, Fiji.

—— & THOMPSON, B. 1978. Regional geochemical reconnaissance survey for submarine metalliferous sediments in the southwestern Pacific Ocean—a preliminary note. *Trans. Instn Min. Metall.* **87**, B87–9.

——, RONA, P. A. & SHEARME, S. 1979. Metal enrichments in sediments from the TAG Hydrothermal Field. *Mar. Mining*, **2**, 79–89.

——, GLASBY, G. P., MOORBY, S. A., THOMPSON, J., KNEDLER, K. E. & McDOUGALL, J. C. 1982. A submarine hydrothermal manganese deposit from the south west Pacific island-arc. *Nature, Lond.* **298**, 456–8.

DYMOND, J. & VEEH, H. H. 1975. Metal accumulation rates in the south east Pacific and the origin of metalliferous sediments. *Earth planet. Sci. Lett.* **28**, 13–22.

EWART, A., BROTHERS, R. N. & MATEEN, A. 1977. An outline of the geology and geochemistry, and the possible petrogenetic evolution of the volcanic rocks of the Tonga–Kermadec–New Zealand island arc. *J. Volcan. geoth. Res. Amsterdam*, **2**, 205–51.

FERGUSON, J. & LAMBERT, I. B. 1972. Volcanic exhalations and metal enrichments at Matupi Harbour, New Britain, T.P.N.G. *Econ. Geol.* **67**, 25–37.

GILL, J. B. 1976. Composition and age of Lau Basin and Ridge volcanic rocks: implications for evolution of an interarc basin and remnant arc. *Bull. geol. Soc. Am.* **87**, 1384–95.

GRIFFIN, J. J., KOIDE, M., HOHNDORF, A., HAWKINS, J. W. & GOLDBERG, E. D. 1972. Sediments of the Lau Basin—rapidly accumulating volcanic deposits. *Deep Sea Res. oceanogr. Abstr.* **19**, 139–48.

HAWKINS, J. W. 1974. Geology of the Lau Basin, a marginal sea behind the Tonga Arc. *In*: BURK, C. A. & DRAKE, C. L. (eds) *The Geology of Continental Margins*, 505–20. Springer Verlag, New York.

HEYE, D. 1969. Uranium, thorium and radium in ocean water and deep-sea sediments. *Earth planet. Sci. Lett.* **6**, 112–6.

HONNOREZ, J. 1969. La formation actuelle d'un gisement sous-marin de sulfures fumarolliens à Volcano (mer tyrrhenienne). I. Les mineraux sulfures des tufs immerges a faible profondeur. *Mineralium Deposita*, **4**, 114–31.

KARIG, D. E. 1970. Ridges and basins of the Tonga–Kermadec island arc systems. *J. geophys. Res.* **75**, 239–54.

KRISHNASWAMI, S. 1976. Authigenic transition elements in Pacific pelagic clays. *Geochim. cosmochim. Acta*, **40**, 425–34.

LAWVER, L. A., HAWKINS, J. W. & SCLATER, J. G. 1976. Magnetic anomalies and crustal dilation in the Lau Basin. *Earth planet. Sci. Lett.* **33**, 27–35.

LEINEN, M. & STAKES, D. 1979. Metal accumulation rates in the central equatorial Pacific during Cenozoic time. *Bull. geol. Soc. Am.* **90**, 357–75.

MALAHOFF, A., FEDEN, R. H. & FLEMING, H. S. 1982. Magnetic anomalies and tectonic fabric of marginal basins north of New Zealand. *J. geophys. Res.* **87**, 4109–25.

——, EMBLEY, R. W., CRONAN, D. S. & SKIRROW, R. 1983. The geological setting and chemistry of hydrothermal sulphides and associated deposits from the Galapagos Rift at 86°W. *Mar. Mining*, **4**, 123–49.

MOORBY, S. A., CRONAN, D. S. & GLASBY, G. P. 1984. Geochemistry of hydrothermal Mn oxide deposits from the S.W. Pacific island arc. *Geochim. cosmochim. Acta*, **48** (in press).

QUILTY, P. G., SACHS, H., BENSON, W. E., VALLIER, T. L. & BLECHSCHMIDT, G. 1976. Sedimentologic history, Leg 34 Deep Sea Drilling Project. *In*: YEATS, R. S., HART, S. R. *et al.* (eds) *Init. Rep Deep Sea drill. Proj.* **34**, 779–94. U.S. Government Printing Office, Washington, D.C.

SCLATER, J. G., HAWKINS, J. W., MAMMMERICKX, J. & CHASE, C. G. 1972. Crustal extension between Tonga and Lau Ridges: petrological and geophysical evidence. *Bull. geol. Soc. Am.* **83**, 505–18.

SCOTT, M. R., SCOTT, R. B., MORSE, J. W., BETZER, P. R., BUTLER, L. W. & RONA, P. A. 1978. Metal-enriched sediments from the TAG Hydrothermal Field. *Nature, Lond.* **276**, 811–3.

SMITH, P. A. & CRONAN, D. S. 1975. The dispersion of metals associated with an active-submarine exhalative deposit. *Proc. 3rd Oceanol. Int. Conf. Brighton, U.K.* 111–3.

TERRY, R. D. & CHILINGAR, G. V. 1955. Summary of 'Concerning some additional aids in studying sedimentary formations' by M. S. Shvetsov. *J. sediment Petrol.* **25**, 229–34.

THOMSON, J. 1982. A total dissolution method for determination of the alpha-emitting isotopes of uranium and thorium in deep sea sediments. *Anal. Chim. Acta*, **142**, 259–68.

TUREKIAN, K. K. & CHAN, L. H. 1971. The marine geochemistry of the uranium isotopes, ^{230}Th and ^{231}Pa. *In*: BRUNFELT, A. C. & STEINNES, E. (eds) *Activation Analysis in Geochemistry and Cosmochemistry*, 311–20. Universitetsforlaget, Oslo.

WEISSEL, J. K. 1977. Evolution of the Lau Basin by the growth of small plates. *In*: TALWANI, M. & PITMAN, W. C. (eds) *Island Arcs, Deep Sea Trenches, and Back-Arc Basins, Maurice Ewing Series*, **1**, 429–36. Am. Geophys. Union, Washington, D.C.

—— 1982. Magnetic lineations in marginal basins of the western Pacific. *Phil. Trans. R. Soc. London*, **A300**, 223–47.

ZELENOV, K. K. 1964. Iron and manganese in exhalations of the Submarine Banu Wuhu volcano (Indonesia). *Dokl. Akad. Nauk SSSR*, **155**, 1317–20.

D. S. CRONAN, R. HODKINSON & S. A. MOORBY, Applied Geochemistry Research Group, Geology Department, Imperial College, London SW7 2BP, U.K.

G. P. GLASBY, New Zealand Oceanographic Institute, P.O. Box 12-346, Wellington, New Zealand.

K. KNEDLER, Department of Chemistry, Victoria University of Wellington, Wellington, New Zealand.

J. THOMSON, Institute of Oceanographic Sciences, Wormley, Godalming, Surrey, U.K.

Volcano-tectonic evolution of Fiji and adjoining marginal basins

H. Colley & W. H. Hindle

SUMMARY: From the Eocene to the Middle Miocene, Fiji was part of a N-facing Outer Melanesia arc system, stretching from Papua New Guinea to Tonga, and was dominated by tholeiitic arc volcanism. Oligocene back-arc spreading to the S of Fiji led to the formation of the Minerva Plain (South Fiji Basin). Reorganization of the plate boundaries in Outer Melanesia during the Middle Miocene fractured the simple arc system and caused polarity reversal in arc segments W of Fiji. Fiji, the major yield point in the break-up, experienced a compressive event followed by progressive isolation from a subduction regime as arc segments were rotated away from the region. This led to asthenospheric melting with a decreasing subduction component, and a consequent change in Fiji volcanism from arc andesites and tholeiites to alkalic ocean island basalts. During the Upper Miocene to Lower Pliocene, rotation of the Vanuatu arc segment caused opening of the Fiji Plateau marginal basin. This was accompanied by widespread, chemically diverse volcanism in Fiji, in which contamination of rising magma by pre-existing crust may have been an important process. The most recent phase of arc rotation resulted in opening of the Lau Basin between Fiji and Tonga, and effected the final divorce of Fiji from a subduction influence with commencement of ocean island basalt volcanism in the Middle Pliocene.

Outer Melanesia (Fig. 1), including the island chains of Vanuatu, Fiji, Lau and Tonga, is an area of considerable tectonic complexity in which the trench–arc–marginal basin association does not show a regular pattern. Within this region, the Fiji Platform is an area of shallow water, generally less than 1 km deep, upon which the major Fiji islands are situated. The shallow water extends southwards along the Lau island chain and southwestward along the Hunter–Kandavu Ridge. The Fiji Platform is flanked by a number of marginal basins; the Lau Basin to the E, the Fiji Plateau (North Fiji Basin) to the N and W, and the Minerva Plain (South Fiji Basin) to the SW.

The main aim of this paper is to relate volcanic events in Fiji to developments taking place in the marginal basins.

Characteristics of the marginal basins

Although the general characteristics of the Lau Basin, Fiji Plateau, and Minerva Plain are reasonably well known, the manner in which the basins opened continues to be widely debated. Most of the major differences between the proposed models stem from the various interpretations of the zones of shallow seismicity, which, taken in conjuction with diffuse magnetic patterns, have been regarded to represent either spreading centres or transform faults.

Lau Basin

The Lau Basin is the youngest of the marginal basins, for which, on petrological grounds, Gill (1976b) suggested an age of 5 Ma, and Weissel (1977), on the basis of recognition of magnetic anomaly 2', gave an age of 3.5 Ma. DSDP site 203 was drilled to 409 m in the southern part of the basin (Fig. 1) and did not reach acoustic basement; the oldest sediments were of Mid-Pliocene age at 3.0–3.5 Ma (Burns & Andrews 1973). Using seismic refraction data from the southern part of the basin and sedimentation rates derived from DSDP 203 studies, Katz (1978) gave an age of 8.0–8.5 Ma.

High heat flow values (average 2.0 HFU), thin sediment cover (about 100 m increasing to 700 m in marginal areas), and rugged topography are further evidence for a young basin. Seismic refraction profiles (Shor et al. 1971) suggest an oceanic crustal thickness between 6 and 10 km.

Since the Lau Basin was described as a marginal basin by Karig (1970), a number of hypotheses have been proposed for the manner of its extension. Most show various combinations of spreading ridges and orthogonal transforms forming triple junctions (Sclater et al. 1972; Weissel 1977; Weissel 1981). Estimates of half-rates of opening vary from 1–2 (Sclater et al. 1972) to 3.8 cm yr^{-1} (Weissel 1977).

A radically different interpretation for the Lau Basin has been propounded by Katz

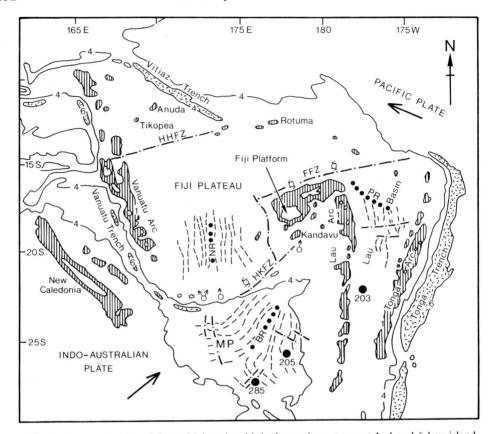

FIG. 1. The morphology of Outer Melanesia with bathymetric contours at 1, 4 and 6 km; island chains (vertical stripes); trenches (stipple); fracture zones (dot-dashed); rises (dotted); magnetic lineations (dashed); focal mechanisms (circles with arrows showing direction of movement); DSDP sites 203, 205 and 285; and present-day plate spreading directions (large arrows). BR = Bounty Ridge; FFZ = Fiji Fracture Zone; HHFZ = Hazel Holme Fracture Zone; HKFZ = Hunter–Kandavu Ridge (Fracture Zone); MP = Minerva Plain (South Fiji Basin); NR = Nova Rise; PR = Peggy Ridge.

(1977) who envisages the basin as a collapse structure in quasi-continental crust, flanked by rising island arcs (Lau and Tonga). Young basin basalts represent 'piercement' of this older crust which is being progressively 'oceanized'.

Fiji Plateau

A number of studies (e.g. Chase 1971; Mac-Donald *et al.* 1973; Luyendyk *et al.* 1974) have demonstrated major differences in the Fiji Plateau N and S of the Hazel Holme Fracture Zone (Fig. 1). To the N, low heat flow values, thicker sediment cover and subdued topography suggest a basin of considerable age, possibly dating back to the Mesozoic. To the S, high heat flow values, in the range 1.5–5.6 HFU (MacDonald *et al.* 1973; Watanabe *et al.* 1977),

more rugged topography, and magnetic lineations possibly dating back to anomaly 4, indicate a much younger basin. Chase (1971) has proposed a maximum age of 10 Ma, and Falvey (1975) and Malahoff *et al.* (1982a) give an age of 7.8 Ma to coincide with anomaly 4. In accordance with the young age, sediment cover is thin, ranging from 100 to 1200 m close to the Vanuatu arc (Chase 1971; Shor *et al.* 1971). Seismic refraction studies indicate an oceanic crustal thickness of about 8 km (Solomon & Biehler 1969; Shor *et al.* 1971).

There is no consensus of opinion on the location of spreading centres within the Fiji Plateau, because of the irregular pattern of heat flow values, the diffuse nature of magnetic anomalies, and the varying interpretation of shallow seismicity zones. Chase (1971), Luyendyk *et al.* (1974) and Malahoff *et al.* (1982a) favour a N–S spreading ridge at 173 °E (Nova Rise, see Fig. 1). A more complex situation, with a ridge–fault–fault triple junction in the SE portion of the Fiji Plateau, is presented by Packham (1982). Spreading half-rate estimates are 3.0–3.9 (Chase 1971; Luyendyk *et al.* 1974) and 4.75 cm yr^{-1} (Falvey 1975).

South Fiji Basin

The South Fiji Basin can be divided into two morphological units about the Cook Fracture Zone: the Kupe Abyssal Plain to the S and the Minerva Plain to the N. Only the latter is discussed in this paper. The Minerva Plain is the oldest of the marginal basins in Outer Melanesia. It has magnetic anomaly ages of 35 (anomaly 12) to 28 Ma (anomaly 7A) and appears to have been inactive since the Lower Miocene (Watts *et al.* 1977). Sediments from the lower part of DSDP site 205 (Fig. 1) have been dated as Upper Oligocene, with an age around 29–30 Ma (Burns & Andrews 1973). Sediment cover is variable, ranging from 0.5 to 1.0 km in thickness, and heat flow values are generally below 1.0 HFU (MacDonald *et al.* 1973). Seismic refraction profiles show a total crustal thickness of 8–9 km (Shor *et al.* 1971).

Shipborne and airborne magnetic surveys (Watts *et al.* 1977; Weissel 1981; Malahoff *et al.* 1982a) suggest opening of the basin, between 35 and 28 Ma, along a ridge–ridge–ridge triple junction at half-rates of 2.6–3.4 cm yr^{-1}. The Bounty Ridge (Fig. 1) spreading centre has the best-defined magnetic anomalies, and termination of these lineations against the Hunter–Kandavu Ridge suggests subduction of the Minerva Plain beneath this ridge. From the present-day direction of movement of the Indo-Australian plate (Fig. 1), it is apparent that the Hunter–Kandavu Ridge is close to a critical angle where a slight change in the relative direction of plate movement could result in a switch from subduction to transform faulting. It seems likely that changes in tectonic style occurred along this ridge in the late Cenozoic. It is significant that Johnson & Molnar (1972) locate focal mechanisms with both strike-slip movement and thrusting, along or close to the ridge (Fig. 1).

Geological history of Fiji

The geological history of Fiji is apparently restricted to the Cenozoic Era; the oldest rocks known are of Upper Eocene age and the youngest are subaerial ash falls on Taveuni, < 2000 yr old (Rodda 1974). There are three distinct stages, which reflect major changes in the geological evolution of the islands.

Stage 1. Upper Eocene–Middle Miocene

Early Tertiary rocks in Fiji, restricted to the southern part of Viti Levu (Fig. 2), are assigned principally to the Wainimala Group and the Singatoka Group.

The Wainimala Group consists of various volcaniclastic sediments interbedded with submarine lava flows and breccias of basalt and dacite, and their metamorphosed equivalents, spilite and keratophyre. Reef limestones within the sequence have been dated as Tertiary *b* (Cole 1960) and Tertiary *e–f* (Hirst 1965). Wainimala Group rocks exhibit zeolite to greenschist facies metamorphism and Gill (1970) has shown them to have the chemical characteristics of arc tholeiites.

The Singatoka Group, which crops out in SW Viti Levu, was formerly interpreted as part of an island-arc succession (Houtz 1960; Rodda & Band 1967). However, Colley (1984) interprets the group as the upper part of an ophiolite suite in which the widespread pelagic sediments, such as foraminiferal oozes, cherts, red clays and Fe- and Mn-rich sediments, with interbeds of fine grained turbidite and polymict lapillistone, are taken to represent Layer 1 of oceanic lithosphere. Beneath these sediments, and regarded as Layer 2, are thick sequences of pillow lava cut by diabase dykes and including pockets of gabbro. The lavas have REE patterns (Fig. 4) typical of ocean-floor tholeiites although, overall, their chemistry shows affinities with both arc and ocean-floor tholeiites.

The Singatoka Group was emplaced against the Wainimala Group arc rocks along low-angle, arcuate thrust faults (Fig. 2). A Tertiary *e–f* age for the Singatoka Group (Skiba 1964) suggests that it is an obducted portion of the South Fiji Basin (Colley 1984).

Stage 2. Middle Miocene–Middle Pliocene

This, most complex, period of Fiji's history began with a phase of deformation—the Tholo Orogeny—which was unusually intense for Outer Melanesia. The strata of the Wainimala Group, which occur in a belt curving from ENE to NNW (Fig. 2), were deformed into a series of folds with axes parallel to the curve of the belt. Within some fold cores there are synorogenic tonalite-gabbro bodies (Tholo Plutonic Suite), ranging in age from 11 to 7 Ma.

During the waning stages of the orogeny, and

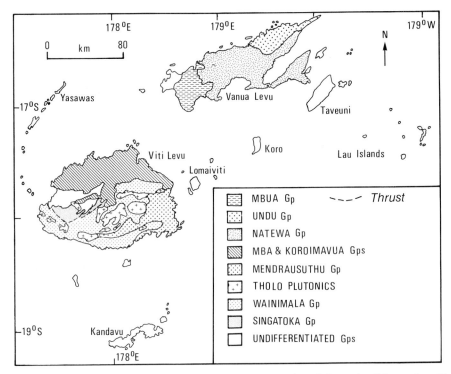

FIG. 2. The geology of the Fiji Platform showing the distribution of the major lithostratigraphic units. Age ranges are shown in Table 1.

the following period, a vigorous resurgence of volcanic activity and associated sedimentation occurred over a wide area of the Fiji Platform. In northern Viti Levu the Mba and Koroimavua Groups are largely composed of basic submarine lava flows and volcaniclastic sediments of shoshonitic and tholeiitic affinity (Gill 1970; Seeley & Searle 1970). Similar rocks extend eastwards into the Lomaiviti island group (Fig. 2). In SE Viti Levu the Mendrausuthu Group consists of the Namosi Andesites, of calc-alkaline composition (Gill 1970), and derived sediments.

During this stage, the large island of Vanua Levu was formed by eruption of arc tholeiite basic andesites of the Natewa Group (Hindle 1976), low-K dacites and rhyodacites of the Undu Group (Colley & Rice 1975), and very minor calc-alkaline andesites of the Nararo Group (Hindle 1976).

Low-K andesites and tholeiitic basic andesites were also erupted during this stage in the Lau island group, the former ranging in age from 9 to 6 Ma and the latter from 3.9 to 3.5 Ma (Gill 1976b).

Stage 3. Middle Pliocene–Recent

During the latest stage of Fiji's history there was a remarkable change in the style of volcanism. Ocean island basalts were erupted along major NNW and NE-trending fissures in SW Vanua Levu (Mbua Group), on Koro and Taveuni, and on a number of the Lau islands (Hindle & Colley 1981). Although this period is dominated by basaltic volcanism, high-K calc-alkaline andesites were erupted in the Upper Pliocene–Recent period on Kandavu (Woodrow 1980). These andesites may represent a phase of subduction along the Hunter–Kandavu Ridge.

Correlation of events in Fiji with marginal basin formation

The three stages in the geological history of Fiji reflect events in the formation of the marginal basins of Outer Melanesia.

Table 1 shows an age correlation between formation of the Wainimala and Singatoka

TABLE 1. *Correlation of major lithostrati-graphic units in Fiji with marginal basin opening in Outer Melanesia*

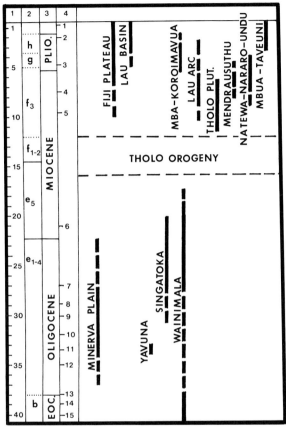

Column 1: age (in Ma); 2: Indonesian letter stage; 3: period; 4: magnetic anomalies. Solid lines indicate measured age range and dashed lines are possible extensions of that range.

Groups and the formation of the Minerva Plain, the coincidence of the opening of the Fiji Plateau with the major Upper Miocene volcanism on the Fiji Platform, and the initiation of spreading in the Lau Basin with the onset of ocean island basalt volcanism in Fiji.

Such correlations strongly suggest that tectonic events in marginal basins can be linked to volcano-tectonic events in the neighbouring arcs, and that arc development represents a complex interplay of extensional tectonics and subduction.

Stage 1

It is assumed that throughout the early Tertiary a relatively simple trench–arc–marginal basin system existed in Outer Melanesia (Fig. 3a). The trench was located on the Pacific side of the arc, with subduction of the Pacific plate beneath the Indian plate in a SW direction.

During this stage there appears to have been a switch of volcanic activity, from the arc to the marginal basins and back to the arc. The earliest recorded activity in Outer Melanesia is of Upper Eocene–Lower Oligocene age and occurs in Fiji (Cole 1960; Rodda & Band 1967), on Eua in Tonga (Ladd 1970), and in Vanuatu (Coleman 1969). In contrast, in the Middle Oligocene no significant volcanism occurred in Fiji (Rodda 1974) and Vanuatu (Carney & MacFarlane 1977) and this was the time of principal extension of the Minerva Plain (Table 1). Finally, in the Upper Oligocene, with

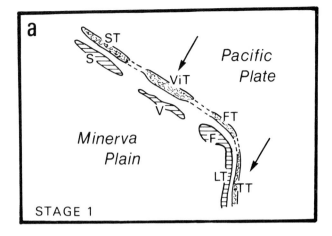

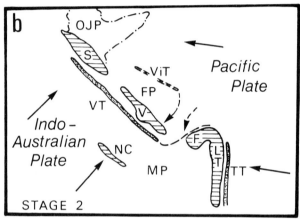

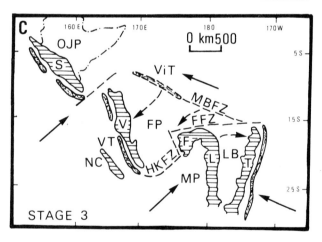

FIG. 3. Schematic plate tectonic reconstruction for Outer Melanesia during the Tertiary. *Stage 1* (Eocene–Middle Miocene): subduction of the Pacific plate southwestwards beneath the proto-Melanesian arcs; opening of the Minerva Plain behind the arcs. *Stage 2* (Middle Miocene–Middle Pliocene): arc polarity reversal with subduction of the Indo-Australian plate northeastwards; orogeny in Fiji; opening of the Fiji Plateau with development of a complex transform system between the Vanuatu and Tonga Arcs. *Stage 3* (Middle Pliocene–Recent): continued opening of the Fiji Plateau; opening of the Lau Basin; growth of the Vanuatu–Tonga transform system (after Colley & Greenbaum 1980). F = Fiji Platform; FFZ = Fiji Fracture Zone; FP = Fiji Plateau; FT = Fiji Trench (postulated); HKFZ = Hunter–Kandavu Fracture Zone; L = Lau Arc; LB = Lau Basin; LT = proto-Lau–Tonga Arc; MBFZ = Melanesian Border Fracture Zone; MP = Minerva Plain (South Fiji Basin); NC = New Caledonia; OJP = Ontong Java Plateau; S = Solomon Arc; ST = proto-Solomon Trench; T = Tonga Arc; TT = Tonga Trench; V = Vanuatu Arc; VT = Vanuatu Trench; ViT = Vitiaz Trench.

the cessation of Minerva Plain extension, renewed volcanic activity occurred in the arc. Thick sequences of volcaniclastic material of Tertiary *e–f* age occur in the upper part of the Wainimala Group in Fiji (Hirst 1965) and on Santo and Malekula in Vanuatu (Mitchell & Warden 1971).

There are significant differences in chemistry between rocks erupted in the marginal basin and those in the arc. Diabase from DSDP site 285A on the Minerva Plain is of basaltic andesite composition (Table 2, Column 1), with high TiO_2, typical of mid-oceanic ridge basalts, but Ni and Cr values intermediate between mid-oceanic ridge basalts and arc tholeiites (Stoeser 1975). Basalts of the Singatoka Group (Table 2, Column 2), which is regarded an an obducted portion of the Minerva Plain (Colley 1984) have a major-element chemistry, and REE pattern with slight LREE depletion (Fig. 4a), typical of mid-oceanic ridge basalt (MORB).

Rocks from the arc sequence in Fiji (Wainimala Group) form a bimodal arc tholeiite suite (Gill 1970) which is dominated by basaltic andesite (Table 2, Column 3), with subordinate rhyodacite (Table 2, Column 4). The major elements, apart from TiO_2, are similar for the Wainimala basaltic andesite and the Singatoka (Minerva Plain) basaltic andesite. Trace elements, however, show significant differences with the Wainimala rock showing lower Ni and Cr, and a more fractionated REE pattern (Fig. 4b), although the last may reflect seawater alteration and/or burial metamorphism, rather than primary magmatic differences.

TABLE 2. *Representative analyses of Stage 1 igneous rocks*

	1	2	3	4
SiO_2	52.1	45.65	53.30	72.34
TiO_2	2.5	1.34	1.09	0.42
Al_2O_3	15.8	15.17	17.18	13.88
Fe_2O_3	4.3	2.15	3.82	1.38
FeO	4.5	7.10	4.95	1.59
MnO	0.54	0.15	0.25	0.08
MgO	6.8	12.58	4.05	0.69
CaO	8.3	9.79	7.33	1.49
Na_2O	3.5	2.60	3.77	6.27
K_2O	0.62	0.45	0.60	0.87
P_2O_5	0.29	0.12	0.24	0.10
L.O.I.	0.77	2.95	2.85	1.02
Total	100.02	100.05	99.63	100.13
Rb	—	17	5	9
Sr	150	285	186	155
Ba	20	—	98	228
Zr	100	—	76	150
Ni	30	—	6	<5
Cr	70	—	7	2
Y	50	—	29	52

Column 1: Diabase, DSDP site 285A; 2: olivine basalt, Nandi River, Viti Levu; 3: Wainimala basaltic andesite (average of six); 4: Wainimala rhyodacite (average of four).

Data from Colley (unpublished); Gill (1970); Gill & Stork (1979); Stoeser (1975).

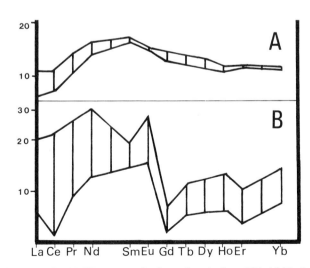

FIG. 4. Chondrite-normalized REE patterns for Stage 1 rocks from Fiji. (a) Tholeiitic basalts from the Singatoka Group (range of three samples); (b) arc tholeiite basic andesites from the Wainimala Group (range of four samples).

Stage 2

In the Middle–Upper Miocene, plate boundary reorganization in Outer Melanesia led to fracturing of the existing arc system and polarity reversal in arcs to the W of Fiji (Fig. 3b). As a consequence, a complex transform system formed between Vanuatu and Tonga. Subsequent movement along this system facilitated the opening of the Fiji Plateau and the Lau Basin and progressively isolated Fiji from a subduction-dominated tectonic regime.

In Fiji, the early part of the stage, probably between 14 and 12 Ma, was marked by emplacement of the Singatoka Group ophiolite suite and downbuckling and folding of the Wainimala Group volcanic pile during the Tholo Orogeny. Temperatures at the base of the downbuckled pile were sufficient to cause melting and intrusion of synorogenic tonalite-gabbro bodies.

The opening of the Fiji Plateau, around 8 Ma, coincided with the period of most intense volcanism in Fiji. This commenced around 8–7 Ma (Table 1) with the eruption of Natewa Group basaltic andesites in Vanua Levu (Table 3, Column 1) and Mendrausuthu Group calc-alkaline andesites in Viti Levu (Table 3, Column 2). This was closely followed by emplacement of the Mba and Koroimavua shoshonitic rocks (Table 3, Column 3) and arc tholeiite volcanism in northern Viti Levu, and Undu Group dacite-rhyodacite volcanism in NE Vanua Levu (Table 3, Column 4).

This volcanism in Fiji has been taken to represent arc activity related to subduction along the Tonga Trench (Gill & Gorton 1973; Gill 1976a). However, its varied character with no coherent K–h relationship, the large arc–trench gap (minimum of 250 km), and a lack of arc geometry in a geographical sense, cast doubt on the validity of a simple subduction model. In the late Miocene, the only part of Fiji that resembles an island arc is Lau, where andesitic volcanism occurred between 9 and 6 Ma (Gill 1976b). Compared to the volcanic activity to the W in most of the Fiji islands, Lau volcanism is very minor.

In considering this important volcanic phase in Fiji the following observations are relevant:

(i) With fracturing of the Outer Melanesia arc in the Middle–Upper Miocene, the influence of subduction would wane in Fiji, although melting of subducted lithosphere beneath Fiji would continue after the break-up.

(ii) Magmas produced by the melting of subducted lithosphere beneath Fiji would have to pass through hot crust, thickened by downbuckling and deformation. Fluids in this crust, derived from the original submarine volcanic pile, would probably contain elevated contents of elements such as Si, K, Ba, Sr and Cu, as a result of rock–water interactions (Humphris & Thompson 1978; Mottl & Holland 1978), and rising magmas would be subject to contamination.

(iii) The evolving transform system between the Vanuatu and Tonga arcs, produced by the Upper Miocene extensional tectonic regime, would bisect the Fiji region.

Of these observations the last is regarded as the most important in leading to the intense and widespread Upper Miocene volcanism in Fiji. Without the fractures the thickened crust would probably have restricted the passage of magmas

TABLE 3. *Representative analyses of Stage 2 volcanic rocks*

	1	2	3	4	5
SiO_2	52.50	58.4	49.18	72.71	49.5
TiO_2	0.83	0.7	0.65	0.50	1.2
Al_2O_3	17.78	17.4	15.88	14.09	15.5
Fe_2O_3	3.51	7.0*	5.16	2.43	3.9
FeO	5.29	—	4.05	1.11	6.2
MnO	0.18	0.2	0.18	0.06	0.1
MgO	5.14	3.4	6.76	0.61	6.7
CaO	9.78	7.5	10.34	2.75	11.3
Na_2O	2.88	3.9	2.51	4.82	2.7
K_2O	0.61	1.2	3.24	0.74	0.3
P_2O_5	0.16	0.2	0.43	0.11	0.1
L.O.I.	1.41	—	1.89	0.86	1.5
Total	100.07	99.9	100.27	100.79	99.9
Rb	15	21	67	9	—
Sr	273	545	1193	100	150
Ba	323	388	644	117	—
Zr	29	99	52	146	—
Ni	41	8	32	2	90
Cr	135	19	88	3	300
Y	25	22	16	43	—

*Total Fe as Fe_2O_3.

Column 1: Basaltic andesite, Vanua Levu (average of eleven); 2: calc-alkaline andesite, Namosi, Mendrausuthu Range, Viti Levu; 3: basic shoshonite (absarokite), Viti Levu (average of twelve); 4: rhyodacite, Undu Peninsula, Vanua Levu (average of five); 5: basalt, Fiji Plateau.

Data from Mineral Resources Division, Fiji (unpublished); Colley & Rice (1975); Gill (1976a); Gill & McDougall (1973); Hawkins (1977).

and hence limited the volcanism. However, the fracture system facilitated the movement of magma, with the result that volcanism was intense. Support for this proposal is provided by the marked NE–SW linearity of Vanua Levu (Fig. 2), and the fact that many of the Upper Miocene–Lower Pliocene volcanic centres in Fiji are located along NE–SW or ENE–WSW trends. Such trends are parallel to major transform features that are presently active (e.g. Hunter–Kandavu Ridge; Fiji Fracture Zone; Hazel Holme Fracture Zone; see Fig. 1), though on Viti Levu the probability of anticlockwise rotation of around 30° (James & Falvey 1978; Malahoff *et al.* 1982b) in the last 4–5 Ma has to be considered.

Where the crust was thin in Fiji (e.g Vanua Levu) basaltic andesite of tholeiitic composition (Table 3, Column 1) was rapidly erupted along the fractures. This magma is regarded as the closest approximation to parental magma produced by melting of the subducted slab beneath Fiji. Where such magma passed through thickened crust, as on Viti Levu, slow ascent, increased differentiation, and contamination under fluid-rich conditions led to eruption of shoshonitic (Table 3, Column 3) and calc-alkaline andesite (Table 3, Column 2) magmas.

Although magmas erupted during Stage 2 in Fiji had an arc-like chemistry (Table 3), the tectonic conditions which allowed their eruption also led to the formation, by extension, of the Fiji Plateau marginal basin. The little data available on basalts from the floor of the Fiji Plateau (Table 3, Column 5) suggest that they are chemically transitional between arc tholeiites and mid-ocean ridge tholeiites (Hawkins 1977).

Whether extension of the Fiji Plateau is continuing at the present time is debatable. Geographically the Fiji Plateau can be regarded as a back-arc basin to the Vanuatu Arc and in the models of back-arc spreading of Chase (1978) and Uyeda & Kanamori (1979) the Fiji Plateau is inactive. However, this is not consistent with the formation of very young extensional troughs at the southern end of Vanuatu (Dugas *et al.* 1977; Coudert *et al.* 1981), or with the development of the central intra-arc basin in Vanuatu with its active basalt volcanism (Colley & Warden 1974; Gorton 1977). It seems that incipient arc splitting and spreading is currently taking place in Vanuatu.

Stage 3

In the Middle Pliocene, extension commenced at the eastern end of the Van-uatu–Tonga transform system with the opening of the Lau Basin (Fig. 3c).

Lawver & Hawkins (1978) used the Lau Basin as a model for marginal basin formation in which 'disorganized' opening results from point-source magmatism along a number of short-lived ridges. The age of Seatura volcano in SW Vanua Levu—2.9–3.3 Ma (Hindle & Colley 1981)—is similar to that for initial opening of the Lau Basin—3.0–3.5 Ma (Burns & Andrews 1973; Weissel 1977)—which suggests that magmatism was not restricted to the basin but also occurred on the Fiji Platform. In addition, the Seatura volcano is centred on a NNW-trending fissure system which is parallel to the axis of the Lau Basin. Thus the volcano may represent an arrested stage of opening, with the main locus of volcanism eventually being established between Lau and Tonga where the crust was thinner. Comparison of Seatura basalt (Table 4, Column 1) and Lau Basin basalt (Table 4, Column 3) shows the former to have higher Ti, K, Rb, Sr, Ba and Zr, and lower Ni and Cr. The Seatura volcanism resembles ocean

TABLE 4. *Representative analyses of Stage 3 volcanic rocks*

	1	2	3	4
SiO_2	48.02	49.81	48.8	60.17
TiO_2	2.34	2.78	1.0	0.54
Al_2O_3	15.73	16.97	16.2	16.84
Fe_2O_3	3.71	4.20	1.6	1.70
FeO	6.99	7.30	7.2	3.56
MnO	0.17	0.26	0.2	0.11
MgO	6.71	4.23	9.3	3.18
CaO	9.43	7.63	12.8	5.96
Na_2O	3.63	4.58	2.2	4.22
K_2O	1.03	1.04	0.12	2.63
P_2O_5	0.44	0.70	0.07	0.28
L.O.I.	1.78	0.23	—	0.99
Total	99.98	99.73	99.49	100.18
Rb	18	20	2	35
Sr	495	460	97	1470
Ba	250	260	<2	—
Zr	193	180	90	—
Ni	60	—	199	—
Cr	105	—	459	—
Y	29	32	16	—

Column 1: Basalt, Seatura Volcano, Vanua Levu (average of seven); 2: basalt, Taveuni; 3: basalt, Lau Basin; 4: calc-alkaline andesite, Kandavu (average of two).

Data from Gill (1976a); Hawkins (1977); Woodrow (1980); Hindle & Colley (1981); Hindle (unpublished).

island activity and the Lau Basin volcanism is more closely allied to abyssal tholeiite activity. The low Ti values in Lau Basin basalts probably indicate high-pH_2O conditions close to the Tonga Trench subduction zone which would stabilize Ti phases in the mantle (Hellman & Green 1979).

The most recent volcanism in Fiji occurred along NE–SW fractures. In the Lau Basin, fractures with this orientation have been interpreted as both transforms and spreading centres. Basalts from Taveuni (Table 4, Column 2) and Koro are similar to Seatura basalts but with increased Ti, suggesting progressive devolatilization of the mantle beneath Fiji. However, this simple pattern is complicated by the Upper Pliocene–Recent eruption on Kandavu of high-K calc-alkaline andesites (Table 4, Column 4). This volcanism possibly reflects a phase of northward subduction of the Minerva Plain beneath the Hunter–Kandavu Ridge. In support of this, Johnson & Molnar (1972) determined a focal mechanism solution with northward thrusting, just S of Kandavu (Fig. 1).

The eruption of ocean island basalts in Fiji from about 3.5 Ma onwards indicates melting of an asthenospheric source containing no subduction component. It is probable that active subduction beneath Fiji stopped at around 14 Ma with the break-up of the Outer Melanesia arc, so it seems to have taken approximately 10 Ma before volcanism without a subduction-type chemistry was established.

Conclusions

Although more data are required, particularly on the basaltic basement rocks of the marginal basins, some general conclusions concerning the volcano-tectonic evolution of Fiji can be made:

(i) Extensional tectonic events leading to marginal basin development affect the adjacent arcs and may produce phases of intense volcanism. In a simplistic subduction model such increases in activity may be attributed erroneously to an increase in the rate of subduction, rather than to establishment of an extensional tectonic regime.

(ii) Rocks with a subduction-type chemistry may be produced for a considerable time after cessation of active subduction; in Fiji this period was about 10 Ma.

(iii) Magmatism in the marginal basins, especially the Fiji Plateau and Lau Basin, seems to be concentrated along short-lived fissures in the manner described by Lawver & Hawkins (1978). These fissures may develop in the adjacent arcs as well as in the basin. Thus, Stage 3 ocean island basalt volcanism in Fiji could represent arrested development of Lau Basin opening, and the extensional troughs in the Vanuatu arc may signal a new phase of spreading of the Fiji Plateau.

ACKNOWLEDGMENTS: For part of the work H. Colley was recipient of a NERC research grant. Dr P. Harvey (Nottingham University) and Dr N. Walsh (King's College, London) are thanked for providing analyses under NERC assisted schemes. We are indebted to the Director of the Mineral Resources Division, Mr H. Plummer, for assistance in Fiji including access to unpublished data. We also thank Dr P. Kokelaar (Ulster Polytechnic) for comment and criticism of the original draft.

References

BURNS, R. E. & ANDREWS, J. E. 1973. Regional aspects of deep sea drilling in the Southwest Pacific. *In*: BURNS, R. E. *et al*. (eds). *Init. Rep. Deep Sea drill. Proj.* **21**, 987–1006. U.S. Government Printing Office, Washington, D.C.

CARNEY, J. N. & MACFARLANE, A. 1977. Volcano-tectonic events and pre-Pliocene crustal extension in the New Hebrides. *In*: *Geodynamics in South-West Pacific*, 91–104. Editions Technip, Paris.

CHASE, C. G. 1971. Tectonic history of the Fiji Plateau. *Bull. geol. Soc. Am.* **82**, 3087–110.

—— 1978. Extension behind island arcs and motions relative to hot spots. *J. geophys. Res.* **83**, 5385–7.

COLE, W. S. 1960. Upper Eocene and Oligocene larger foraminifera from Viti Levu, Fiji. *Prof. Pap. U.S. geol. Surv.* **374-A**, 7 pp.

COLEMAN, P. J. 1969. Derived Eocene larger foraminifera on Maewo, eastern New Hebrides and their Southwest Pacific implications. *Ann. Rept. New Hebrides geol. Surv. 1967*, 36–7.

COLLEY, H. 1984. An ophiolite suite in Fiji. *In*: GASS, I. G., LIPPARD, S. J. & SHELTON, A. W. (eds) *Ophiolites and Oceanic Lithosphere. Spec. Publ. geol. Soc. London*, **13**, 333–40. Blackwell Scientific Publications, Oxford.

—— & GREENBAUM, D. 1980. The mineral deposits and metallogenesis of the Fiji Platform. *Econ. Geol.* **75**, 807–29.

—— & RICE, C. M. 1975. A Kuroko-type deposit in Fiji. *Econ. Geol.* **70**, 1373–86.

—— & WARDEN, A. J. 1974. Petrology of the New Hebrides. *Bull. geol. Soc. Am.* **85**, 1635–46.

COUDERT, E., ISACKS, B. L., BARAZANGI, M., LOUAT,

R., CARDWELL, R., CHEN, A., DUBOIS, J., LATHAM, G. & PONTOISE, B. 1981. Spatial distribution and mechanisms of earthquakes in the southern New Hebrides arc from a temporary land and ocean bottom seismic network and from worldwide observations. *J. geophys. Res.* **86**, 5905–25.

DUGAS, F., DUBOIS, J., LAPOUILLE, A., LOUAT, R. & RAVENNE, C. 1977. Structural characteristics and tectonics of an active island arc: the New Hebrides. *In: Geodynamics in South-West Pacific*, 79–89. Editions Technip, Paris.

FALVEY, D. 1975. Arc reversals and a tectonic model for the North Fiji Basin. *Bull. Aust. Soc. explor. Geophys.* **6**, 47–9.

GILL, J. B. 1970. Geochemistry of Viti Levu, Fiji and its evolution as an island arc. *Contrib. Miner. Petrol.* **27**, 179–203.

—— 1976a. From island arc to oceanic islands; Fiji, southwestern Pacific. *Geology*, **4**, 123–6.

—— 1976b. Composition and age of Lau basin and ridge volcanic rocks: implications for evolution of an interarc basin and remnant arc. *Bull. geol. Soc. Am.* **87**, 1384–95.

—— & GORTON, M. P. 1973. A proposed geological and geochemical history of eastern Melanesia. *In: COLEMAN, P. J. (ed.) The Western Pacific: Island Arcs, Marginal Seas, and Geochemistry*, 459–67. University of Western Australia Press, Perth.

—— & MCDOUGALL, I. 1973. Biostratigraphic and geological significance of Miocene–Pliocene volcanism in Fiji. *Nature, Lond.* **241**, 176–80.

—— & STORK, A. L. 1979. Miocene low-K dacites and trondhjemites of Fiji. *In: BAKER, F. (ed.) Trondhjemites, Dacites, and Related Rocks*, 629–49. Elsevier, Amsterdam.

GORTON, M. P. 1977. The geochemistry and origin of Quaternary volcanism in the New Hebrides. *Geochim. cosmochim. Acta*, **41**, 1257–70.

HAWKINS, J. W. 1977. Petrologic and geochemical characteristics of marginal basin basalts. *In: TALWANI, M. & PITMAN, W. C. (eds) Island Arcs, Deep Sea Trenches, and Back-Arc Basins*, 355–65. Am. Geophys. Union, Washington, D.C.

HELLMAN, P. L. & GREEN, T. H. 1979. The high pressure crystallization of staurolite in hydrous mafic compositions. *Contrib. Miner. Petrol.* **68**, 369–72.

HINDLE, W. H. 1976. The geology of west-central Vanua Levu. *Bull. Miner. Resour. Div. Fiji*, **1**, 76 pp.

—— & COLLEY, H. 1981. An oceanic volcano in an island arc setting—Seatura Volcano, Fiji. *Geol. Mag.* **118**, 1–12.

HIRST, J. A. 1965. Geology of east and north-east Viti Levu. *Bull. geol. Surv. Fiji*, **12**, 51 pp.

HOUTZ, R. E. 1960. Geology of Singatoka area, Viti Levu. *Bull. geol. Surv. Fiji*, **6**, 19 pp.

HUMPHRIS, S. E. & THOMPSON, G. 1978. Trace element mobility during hydrothermal alteration of oceanic basalts. *Geochim. cosmochim. Acta*, **42**, 127–36.

JAMES, A. & FALVEY, D. A. 1978. Analysis of palaeomagnetic data from Viti Levu, Fiji. *Bull. Aust. Soc. explor. Geophys.* **9**, 115–7.

JOHNSON, T. & MOLNAR, P. 1972. Focal mechanisms and plate tectonics of the southwest Pacific. *J. geophys. Res.* **77**, 5000–32.

KARIG, D. E. 1970. Ridges and basins of the Tonga–Kermadec island arc system. *J. geophys. Res.* **75**, 239–54.

KATZ, H. R. 1977. The Lau Basin: a collapse structure between rising island arcs. *In: Geodynamics in South-West Pacific*, 165–6. Editions Technip, Paris.

—— 1978. Composition and age of Lau basin and ridge volcanic rocks: implications for evolution of interarc basic and remnant arc: Discussion. *Bull. geol. Soc. Am.* **89**, 1118–9.

LADD, H. S. 1970. Eocene molluscks from Eua, Tonga. *Prof. Pap. U.S. geol. Surv.* **640-C**, 11 pp.

LAWVER, L. A. & HAWKINS, J. W. 1978. Diffuse magnetic anomalies in marginal basins: their possible tectonic and petrolgic significance. *Tectonophysics*, **45**, 323–39.

LUYENDYK, B. P., BRYAN, W. B. & JEZEK, P. A. 1974. Shallow structure of the New Hebrides island arc. *Bull. geol. Soc. Am.* **85**, 1287–300.

MACDONALD, K. C., LUYENDYK, B. P. & VON HERZEN, R. 1973. Heat flow and plate boundaries in Melanesia. *J. geophys. Res.* **78**, 2537–46.

MALAHOFF, A., FEDEN, R. H. & FLEMING, H. S. 1982a. Magnetic anomalies and tectonic fabric of marginal basins north of New Zealand. *J. geophys. Res.* **87**, 4109–25.

——, HAMMOND, S. R., NAUGHTON, J. J., KEELING, D. L. & RICHMOND, R. N. 1982b. Geophysical evidence for post-Miocene rotation of the island of Viti Levu, Fiji and its relationship to the tectonic development of the North Fiji Basin. *Earth planet. Sci. Lett.* **57**, 398–414.

MITCHELL, A. H. G. & WARDEN, A. J. 1971. Geological evolution of the New Hebrides island arc. *J. geol. Soc. London*, **127**, 501–29.

MOTTL, M. J. & HOLLAND, H. D. 1978. Chemical exchange during hydrothermal alteration of basalt by seawater. I. Experimental results for major and minor components of seawater. *Geochim. cosmochim. Acta*, **42**, 1103–15.

PACKHAM, G. H. 1982. Foreward to papers on the tectonics of the southwest Pacific region. *Tectonophysics*, **87**, 1–10.

RODDA, P. 1974. Fiji. *In: SPENCER, A. N. (ed.) Mesozoic–Cainozoic Orogenic Belts. Spec. Publ. geol. Soc. London*, **4**, 425–32. Blackwell Scientific Publications, Oxford.

—— & BAND, R. B. 1967. Geology of Viti Levu. *Rep. geol. Surv. Fiji, 1966*, 8–16.

SCLATER, J. G., HAWKINS, J. W., MAMMERICKX, J. & CHASE, C. G. 1972. Crustal extension between the Tonga and Lau ridges; petrological and geophysical evidence. *Bull. geol. Soc Am.* **83**, 505–18.

SEELEY, J. B. & SEARLE, E. J. 1970. Geology of the Rakiraki district, Viti Levu, Fiji. *N.Z. Jl Geol. Geophys.* **13**, 52–71.

SHOR, C. G., KIRK, H. K. & MENARD, H. W. 1971. Crustal structure of the Melanesian area. *J. geophys. Res.* **76**, 2562–8.

SKIBA, W. J. 1964. Geological studies in southwest Viti Levu. *Mem. geol. Surv. Fiji*, **1**, 56 pp.

SOLOMON, S., & BIEHLER, S. 1969. Crustal structure from gravity anomalies in the Southwest Pacific. *J. geophys. Res.* **74**, 6696–701.

STOESER, D. B. 1975. Igneous rocks from leg 30 of the Deep Sea Drilling Project. *In*: PACKHAM, G. H. & ANDREWS, J. E. (eds) *Init. Rep. Deep Sea drill. Proj.* **30**, 401–14. U.S. Government Printing Office, Washington, D.C.

UYEDA, S. & KANAMORI, H. 1979. Back-arc opening and the mode of subduction. *J. geophys. Res.* **84**, 1049–61.

WATANABE, T., LANSETH, M. G. & ANDERSON, R. N. 1977. Heat flow in back-arc basins of the Western Pacific. *In*: TALWANI, M. & PITMAN, W. C. (eds) *Island Arcs, Deep Sea Trenches, and Back-Arc Basins*, 137–61. Am. Geophys. Union, Washington, D.C.

WATTS, A. B., WEISSEL, J. K. & LARSON, R. L. 1977. Sea-floor spreading in marginal basins of the western pacific. *Tectonophysics*, **37**, 167–81.

WEISSEL, J. K. 1977. Evolution of the Lau Basin by growth of small plates. *In*: TALWANI, M. & PITMAN, W. C. (eds) *Island Arcs, Deep Sea Trenches, and Back-Arc Basins*, 429–36. Am. Geophys. Union, Washington, D.C.

—— 1981. Magnetic lineations in marginal basins of the western Pacific. *Phil. Trans. R. Soc. London*, **A300**, 223–47.

WOODROW, P. J. 1980. Geology of Kandavu. *Bull. Miner. Resour. Dept Fiji*, **7**, 31 pp.

H. COLLEY, Department of Geology and Physical Sciences, Oxford Polytechnic, Headington, Oxford OX3 0BP, U.K.

W. H. HINDLE, Brimfield Hall, Brimfield, Ludlow, Shropshire, U.K.

Late Cenozoic volcanism and extension in Eastern Papua

I. E. Smith & J. S. Milsom

SUMMARY: The sea floor around eastern New Guinea is divided into a number of deep basins, separated by submarine ridges which are capped in places by islands of metamorphic, volcanic or coralline rock. The area has been subject to phases of extension since at least the Palaeocene, when the Coral Sea basin was formed. There have also been major compressive events, including the thrust emplacement of an ophiolite, the Papuan Ultramafic Belt, on to the Papuan Peninsula in the Oligocene. This peninsula, on the E of New Guinea, has been volcanically active from the Middle Miocene to the present day, and there have been eruptions on many of the surrounding islands and in the marine basins. Extension-related volcanic rocks include low-K tholeiites dredged from the floor of the Woodlark Basin and a peralkaline rhyolite association on islands near the E end of the peninsula. Volcanic rock types usually regarded as indicators of subduction, including andesites and high-K trachybasalts (shoshonites), are common.

During the Neogene, two phases of extension with associated igneous activity can be interpreted; the first in the Middle and Late Miocene, and the second from the Middle Pliocene to the present day. Only in the first phase was the formation of marginal basins related to subduction. The later phase is seen as part of the response of the complexly fragmented Melanesian area to changes in relative motions of the surrounding major plates. Sea-floor spreading is currently occurring in the Woodlark Basin, and the post-Miocene calc-alkaline and shoshonitic rocks of the Papuan Peninsula and offshore islands reflect reactivation of subduction-modified mantle under this tensional regime, and not renewed subduction.

The geology of the eastern part of New Guinea, and of the surrounding marine basins and archipelagos, reflects past and continuing interaction along plate boundaries which have varied in both function and position according to the relative movements of the Pacific and Indo-Australian plates. The main features of the area are shown in Fig. 1. This paper is concerned chiefly with the Woodlark Basin and with the Papuan Peninsula and its largely submerged eastern continuations, the Woodlark and Pocklington Rises (Fig. 2). This entire area is here termed Eastern Papua.

In one of the earliest papers to consider subduction-related extensional processes, Karig (1972) interpreted the arcuate ridges of Eastern Papua as remnant arcs. Active subduction-type volcanism is widespread in the area and sea-floor is currently being formed in the Woodlark Basin (Weissel *et al.* 1982), but there is no associated Benioff zone. It seems that there have been two main extensional phases and that only the first, in the Middle and Late Miocene, was accompanied by subduction.

Pre-Miocene evolution of Eastern Papua

By the beginning of the Miocene, the general outlines of the present-day Papuan Peninsula had begun to appear. The Coral Sea had opened, separating a part of the N Australian margin from the continent (Weissel & Watts 1979), and Late Mesozoic mafic and ultramafic (ophiolitic) rocks, which now form the Papuan Ultramafic Belt (PUB), had been thrust on to the detached segment from the N and E, probably in the Oligocene (Davies 1980). The PUB is considered to be the frontal portion of an Early Tertiary island arc (Davies & Smith 1971; Pieters 1978; Davies 1980). It is cut and overlain by intrusions and extrusions of 'island-arc' rocks of Eocene age which are chemically distinct from the ophiolite (Jaques & Chappell 1980). Metamorphism of the sialic rocks along the axis of the peninsula, which grade into unmetamorphosed Cretaceous to Eocene sediments near Port Moresby, can probably be attributed to the Oligocene collision (Pieters 1978). Similar metamorphic rocks make up most of the islands of the Louisiade Archipelago on the Pocklington Rise (Smith 1973). However, between these two areas of metamorphic rock, much of the S and E of the peninsula is composed of Cretaceous–Eocene submarine tholeiitic basalts and associated intrusions, which have been referred to as the Milne Ophiolite (Hamilton 1979). Since there is no direct evidence for large masses of ultramafic rock, and gravity gradients are generally low (Milsom 1973), the term Milne Basic

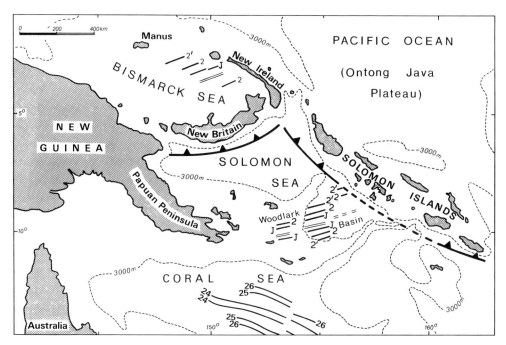

FIG. 1. Eastern New Guinea and the Solomon Islands, showing the currently active sites of subduction (heavy lines with triangles indicating overthrust side) and axes of sea-floor spreading (parallel thin lines). Magnetic lineations in the Woodlark Basin after Weissel *et al.* (1982), in the Bismarck Sea after Taylor (1979) and in the Coral Sea after Weissel & Watts (1979).

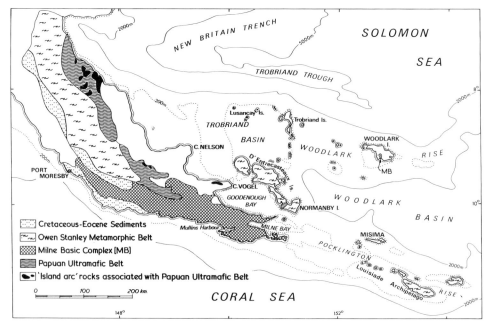

FIG. 2. Pre-Miocene geology of Eastern Papua, after Dow (1977).

Complex seems more appropriate. The only pre-Miocene rocks known on the Woodlark Rise are Eocene tholeiitic basalts which have been tentatively correlated with those of the Papuan Peninsula.

There is no indication as to whether the Woodlark and Pocklington Rises existed as discrete units in the Early Miocene, or that there was a basin between them. The Solomon Sea almost certainly existed at that time, either as a back-arc basin to the arc represented by the rocks of the PUB, or as a fragment of oceanic crust contemporaneous with the PUB basalts. Major uplift in Eastern Papua began late in the Oligocene (Davies & Smith 1971).

First volcano-tectonic phase

Middle and Late Miocene basic to intermediate volcanics are widely distributed throughout New Guinea, and are thought to be relics of a short period of southward subduction beneath the Australasian continental margin (Dow 1977). Hamilton (1979) identified a shallow trench, parallel to the N coast of W New Guinea, as the surface trace of the subduction zone, and suggested a similar origin for the Trobriand Trough, N of the Woodlark Rise

(Fig. 2). Many of the rock units of Eastern Papua can be related to such subduction (Fig. 3). These include basaltic agglomerates, tuffs and lavas in the W part of the Papuan Peninsula (Talama Volcanics: Brown 1977), andesites near Cape Ward Hunt (Iauga Formation: Paterson & Kicinski 1956) and smaller outcrops of basaltic and andesitic pyroclastic deposits near Port Moresby (Kore Volcanics: Pieters 1978). In all these areas, Miocene ages have been determined from fossils in associated sediments. Granodiorites in the NW of the peninsula have radiometric ages of 12–15 Ma (Page & McDougall 1972) and pyroxene andesite from the W end of the Louisiade Archipelago dated at about 11 Ma (Smith 1973).

Rather different Middle Miocene volcanics of the eastern part of the Papuan Peninsula (Smith & Davies 1976) are predominantly high-K basaltic subaerial lavas and agglomerates with minor submarine flows. Those known as the Fife Bay Volcanics, which crop out S of Mullins Harbour, have been dated at 12.6 Ma, and parts of Cloudy Bay Volcanics, which occur further W along the S coast of the peninsula, are sufficiently similar for a similar age to be assumed (Smith & Davies 1976). Comparable high-K basalts, dated at 11.2 Ma, occur on

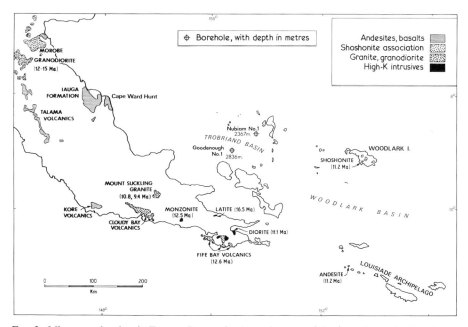

FIG. 3. Miocene volcanism in Eastern Papua. Geology after Dow (1977); radiometric dates for Mt Suckling and the Fife Bay Volcanics from Smith & Davies (1976), for the Morobe Granodiorite from Page & McDougall (1972), for Woodlark Island from Ashley & Flood (1981) and for the Louisiade Archipelago from Smith (1973).

Woodlark Island (Ashley & Flood 1981). Plutons and dyke swarms which intrude the eastern part of the Milne Basic Complex are considered on their chemistry to be the intrusive equivalents of the Fife Bay Volcanics and have been dated at 12–16 Ma (Smith 1972). The plutons are associated with strong magnetic and gravity anomalies and were regarded by Milsom & Smith (1975) as evidence for crustal thickening by intrusion in a tensional environment.

This eastern group of volcanics and intrusives constitutes an association of under- and oversaturated mildly alkaline rocks which has been termed shoshonitic (Joplin 1968). Their relatively low TiO_2 distinguishes them from the alkaline rocks which are commonly found in continental rift systems. Morrison (1980) notes that shoshonites are normally found closely associated with calc-alkaline rocks, and that they are characteristic of areas where there has been oblique convergence followed by 'flipping' of the subduction zone or complete cessation of subduction.

The Middle Miocene igneous activity seems to have been accompanied by emergence of the Papuan Peninsula above sealevel. Smith & Davies (1976) note that the sedimentary record in the E of the peninsula indicates that uplift began in the Middle or Late Oligocene and that a landmass was being rapidly eroded by the Late Miocene. Further W, emergence occurred in the Early Miocene (Pieters 1978).

During the Miocene the E–W fault-bound Trobriand Basin, defined by seismic surveys, developed N of the Papuan Peninsula (Fig. 3). The maximum thickness of sediments estimated from seismic records is > 5000 m, but two boreholes, drilled on structural highs, bottomed in Middle Miocene volcaniclastics at 2367 and 2863 m below sealevel (Tjhin 1976). The volcaniclastics lie close to the seismically defined basement and may be regarded as indicating basin initiation in the Middle Miocene or a little earlier. The volcaniclastics contain a distinctive pale-green clinopyroxene which is characteristic of the Fife Bay Volcanics. In both holes, about 700 m of marine shales and marls overlie the volcaniclastics, but shallow-water conditions prevailed from the Late Miocene onwards and fluviatile and deltaic sediments were deposited.

In the W part of the Woodlark Basin, Luyendyk *et al.* (1973) determined 200 m of sediments which were rifted during the latest phase of extension. They suggested an age of 20 Ma for the earlier (proto-Woodlark) basin, within which the sediments were deposited, on the basis of postulated changes in the Pacific spreading regime at that time.

We consider that the patterns of volcanic activity, uplift, subsidence and extension in Eastern Papua, coupled with the evidence for southward subduction beneath central New Guinea in the Miocene, suggest that there was also southward subduction beneath the Papuan Peninsula. The alternative of N-directed subduction (Karig 1972) seems less likely, as this would involve subduction in opposite directions in Eastern Papua and central New Guinea, and also because the 56–62 Ma magnetic anomalies in the Coral Sea (Weissel & Watts 1979) are symmetrically disposed about the central axis of the basin so that little, if any, of the basin crust can have been destroyed. The Trobriand Basin is interpreted as an intra-arc marginal basin which formed as a consequence of subduction, although there was insufficient extension for the development of new oceanic lithosphere.

Volcanic and tectonic quiescence

Smith (1970) recorded a raised Plio-Pleistocene erosional and depositional surface in the eastern part of the Papuan Peninsula. It marks a period of stability during which the land surface was eroded to a relief of generally less than 100 m over wide areas at the margins of the emergent mountain chain. The surface implies that the very considerable Neogene uplift took place in at least two stages.

Luyendyk *et al.* (1973) suggested, on the basis of the rifted sediment pile noted above, that extension in the Woodlark Basin was also in two stages, with a hiatus in the Late Miocene and Pliocene. This is reflected in the Trobriand Basin, where the transition from deep marine to deltaic sedimentation is marked by lignites in the Late Miocene; slow subsidence continued but deep-water conditions were never re-established (Tjhin 1976).

The available radiometric dates (Figs 3 and 4) suggest that there was a period of volcanic quiescence between 9 and 6 Ma. It is possible that this gap will be bridged by future determinations, but Brown (1977) and Pieters (1978) both note the existence of two distinct phases separated by a layer of non-volcanogenic sediments deposited in the Late Miocene or Early Pliocene. The volcanic and tectonic quiescence probably marks the interval between the cessation of subduction and the establishment of a purely tensional regime. It is not necessary that the breaks were simultaneous in all processes, as the times taken for the deep-seated causes to give rise to the various near-surface effects would presumably have been very different.

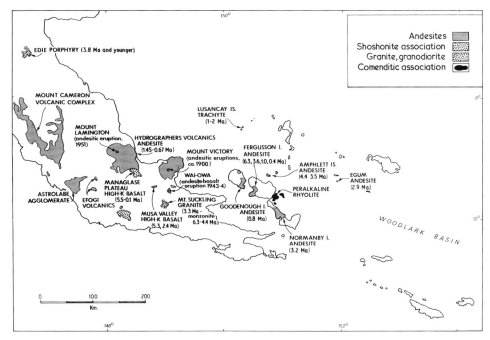

FIG. 4. Post-Miocene volcanism in Eastern Papua. Geology after Dow (1977); radiometric dates for Mt Suckling, Managlase Plateau, Musa Valley and Hydrographers Volcanics from Smith & Davies (1976), for Fergusson Island, Amphlett Islands and Egum from Smith & Compston (1982), for Lusancay Island from Smith *et al.* (1979) and for Edie Porphyry from Page & McDougall (1972).

Second volcano-tectonic phase

Extension is probably the most important aspect of the second tectonic phase and is most clearly demonstrated by the sea-floor spreading in the Woodlark Basin. Weissel *et al.* (1982) deduced that this spreading began at about 3.5 Ma in the extreme E and progressively later further W. There are also three deep, fault-bound depressions towards the E end of the Papuan Peninsula. The oldest of the three is probably Mullins Harbour (Fig. 2), which is almost entirely filled with sediment and is defined clearly only by gravity measurements (Milsom 1973). Milne Bay is described by Jongsma (1972) as a graben, of unknown age, which has subsided almost 800 m in the last 18,000 years with the accumulation of about 450 m of sediment. Probably the youngest of the three is Goodenough Bay. It is seismically active (Fig. 5) and, although it is surrounded by high and rapidly eroding land masses, the 200 m isobath is very close to the shore and the maximum depth is more than 2 km.

All three basins can be regarded as due to propagation of the Woodlark Basin rifting into the Papuan Peninsula. Their subsidence has been accompanied by uplift of the adjacent land masses. The Plio-Pleistocene surface, described generally by Smith (1970) and in the Milne Bay area in more detail by Smith & Simpson (1972), is now raised as much as 600 m above sealevel.

Another effect of rift propagation has been the eruption of mildly peralkaline rhyolites (comendites) in the D'Entrecasteaux Islands near the E end of Goodenough Bay (Fig. 4). The latest eruptions probably occurred less than 1000 yr ago and the volcanoes are regarded as still active (Smith 1982). The lavas, similar to those associated with continental rifts, are anomalous in having been produced in close spatial and temporal juxtaposition to subduction-type andesites (Smith 1976).

Volcanic rocks dated at 1–2 Ma in the Lusancay Islands (Fig. 4), on the N side of the Trobriand Basin, are also unusual. They are exceptionally potassic trachytes (K_2O 5–7%, Na_2O 2–4% and SiO_2 63–66%) with high contents of large ion lithophile elements (Smith *et al.* 1979).

The remainder of the volcanic rocks of this second volcano-tectonic phase in Eastern Papua are similar to the calc-alkaline and shoshonitic rocks of the first phase. The oldest

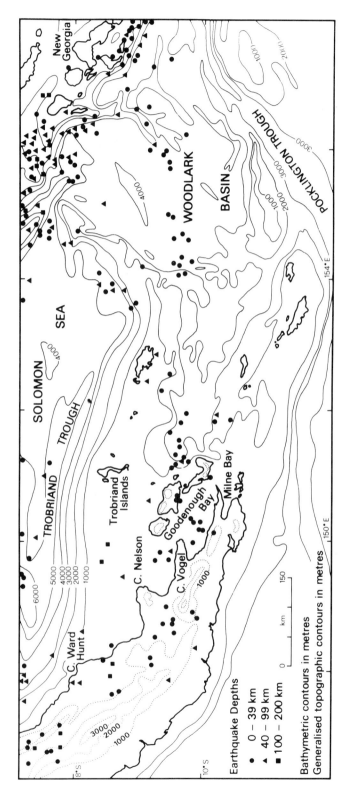

Fıɢ. 5. Seismicity of Eastern Papua, after Ripper (1982).

second-phase andesites recorded, at 6 Ma, occur on islands immediately N of the D'Entrecasteaux Group, and the oldest shoshonites have been dated at 5.5 Ma. However, most of the ages determined are considerably younger and calc-alkaline rocks have been erupted in the last 40 yr at Wai-owa and at Mt Lamington (Taylor 1958). There is no evidence of any evolution with time from one type to the other, but there has been a tendency for andesites to be erupted in a northern belt and shoshonites in a southern belt (Smith 1982). The granite intrusive complex of Mt Suckling (Fig. 4) occupies a rather ambiguous place in the dating scheme, with some ages around 10 Ma and others at 6.3 Ma or less. Smith & Davies (1976) believed that the earlier dates might record the initial emplacement of the complex and the later ones the commencement of uplift. The Edie Porphyry, which intrudes the Middle Miocene Morobe granodiorite in the NW of the Papuan Peninsula, is 3.8 Ma or younger (Page & McDougall 1972).

Seismicity

As noted above, the seismic activity associated with the spreading centre in the Woodlark Basin appears to extend into Goodenough Bay, which may be currently rifting (Fig. 5). The Woodlark Basin seismic pattern is complicated by the existence of a 'quiet zone' from 152°E to 154°E, and by a zone of shallow shocks along the rise to the N. Weissel *et al.* (1982) suggest that the latter zone may indicate decoupling of the crust of the Solomon Sea from that of the much younger Woodlark Basin as they react in different ways to collision with the Solomons Arc.

Current extensional movements in the Woodlark Basin seem to be seismically well defined, but there are few indications of subduction in the seismicity of Eastern Papua. All the focal depths determined in the Woodlark Basin, and most of those of the peninsula, are less than 70 km. However, rare focal depths of 180 km, S and W of Cape Ward Hunt in the extreme NW of the peninsula, have been interpreted as defining a very weakly active Benioff zone, dipping to the SW (Ripper 1982). This interpretation poses some problems. Firstly, it is hard to envisage current southward subduction in the region where the surface trace of such a zone should lie—i.e. in the extreme W of the Solomon Sea and the N coast ranges of New Guinea. Secondly, it does not explain the volcanic activity in Eastern Papua, since the epicentres of the deeper events, which supposedly define the zone, all lie beyond the W limits of outcrops of volcanics of the second phase. Hamilton's (1979) association, of the second-phase volcanoes with slow southward subduction beneath the central and eastern parts of the Papuan Peninsula, encounters the objection that the only deeper-focus earthquakes near there, which might indicate a subducting slab, have been two isolated events at *c.* 140 km beneath the Lusancay Islands. These were located very close to the supposed trench and well to the N of the andesitic activity.

Discussion

Here a model is presented to explain the volcano-tectonic evolution of Eastern Papua. Fundamental to this is the supposition that subduction-related magmas can be generated long after subduction has ceased. Johnson *et al.* (1978) suggested such an origin for the Late Cenozoic volcanism in both central New Guinea and Eastern Papua, but the subduction event was considered to be Cretaceous rather than Miocene in both areas.

It is proposed that after the Oligocene collision of the Papuan Ultramafic Belt with the Cretaceous–Eocene sediments of the Papuan Peninsula, the convergence of the Indo-Australian Plate and the lithosphere to the N was accommodated by southward-directed subduction during much of the Miocene. Intra-arc basins developed locally, one remnant, in a position rather far forward on the arc, being the Trobriand Basin. The Middle and Late Miocene igneous rocks are related to subduction. The shoshonitic magmatism seems to have been concentrated where there is evidence of Miocene extension, along the margins of the Trobriand Basin and its poorly defined continuation into the proto-Woodlark Basin. This suggests that extension may have been one of the factors which influenced the magma composition. The importance of the spatial correlation, between the Middle Miocene shoshonites and outcrops of the Milne Basic Complex, is impossible to determine without information on the early history of this complex. We believe that the Benioff zone interpreted (Ripper 1982) in the W part of the Papuan Peninsula marks lithosphere subducted during the Miocene, and that its absence further E is probably due to Miocene rifting which disrupted and attenuated the subducted slab.

Subduction beneath New Guinea, including Eastern Papua, ceased near the end of the

Miocene, and continuing convergence in Eastern Papua since that time has been taken up by subduction northwards beneath the New Britain Arc. However, at about 3.5 Ma or shortly before, the Woodlark Basin began to open in response to stress caused by changes in motion of the Pacific Plate (Weissel *et al*. 1982). Fresh basaltic glass dredged from near the central rift is compositionally similar to basalts from the mid-ocean ridges (Luyendyk *et al*. 1973). The opening of this marginal basin is not subduction-related and it is not an intra-arc basin in the usual sense.

One effect of the W-propagating rift was the eruption of the peralkaline comenditic volcanic rocks in the eastern D'Entrecasteaux Islands. In this tensional regime a second ascent of magma from the subduction-modified mantle beneath Eastern Papua occurred. Evidently fragmental relics of Middle Miocene subduction retain sufficient rigidity to give rise to earthquakes at depths of more than 100 km. The two shocks recorded beneath the Lusancay Islands are attributed to such fragments, the presence of which determined the composition of the trachytic rocks emplaced there. Elsewhere, the post-Miocene volcanic rocks are very similar to those which were erupted during the original subduction event.

The problems of interpreting the volcano-tectonic history of Eastern Papua since the Miocene are apparent from the diverse conclusions (see e.g. Karig 1972; Johnson *et al*. 1978). Such differences concerning a relatively brief geological history emphasize the problems of interpretation of older marginal basins. If the Woodlark Basin extension and Eastern Papuan volcanism were to cease in the near future, and convergence were to continue between Australasia and the Pacific, then the evidence that the basin had not been a subduction-related back- or intra-arc feature would be obliterated within a few million years.

ACKNOWLEDGMENTS: We thank Ian Ripper and Jeff Weissel for providing copies of papers on Eastern Papua and the Woodlark Basin prior to publication.

References

ASHLEY, P. M. &. FLOOD, R. H. 1981. Low-K tholeiites and high-K igneous rocks from Woodlark Island, Papua New Guinea. *J. geol. Soc. Aust.* **28**, 227–40.

BROWN, C. M. 1977. Explanatory notes, Sheet SC/55-2, Yule (with 1:250,000 geological map). *Bur. Min. Resour. Aust. Bull.* Canberra.

DAVIES, H. L. 1980. Crustal structure and emplacement of ophiolite in southeastern Papua New Guinea. *Colloques Internat. CNRS. No. 272—Association mafiques ultramafiques dans les orogenes.* Grenoble 1977, 17–33.

—— & SMITH, I. E. 1971. Geology of eastern Papua. *Bull. geol. Soc. Am.* **82**, 3299–312

DOW, D. B. 1977. Geological synthesis of Papua New Guinea. *Bur. Min. Resour. Aust. Bull.* **201**.

HAMILTON, W. 1979. Tectonics of the Indonesian Region. *Prof. Pap. U.S. geol. Surv.* **1078**, 345 pp.

JAQUES, A. L. & CHAPPELL, B. W. 1980. Petrology and trace-element geochemistry of the Papuan Ultramafic Belt. *Contrib. Miner. Petrol.* **75**, 55–70.

JOHNSON, R. W., MACKENZIE, D. E. & SMITH, I. E. 1978. Delayed partial melting of subduction modified mantle in Papua New Guinea. *Tectonophysics*, **46**, 197–216.

JONGSMA, D. 1972. Marine geology of Milne Bay, eastern Papua. *In: Geological Papers, 1969. Bur. Min. Resour. Aust. Bull.* **125**, 35–54.

JOPLIN, G. A. 1968. The shoshonite association: a review. *J. geol. Soc. Aust.* **15**, 257–94.

KARIG, D. 1972. Remnant Arcs. *Bull. geol. Soc. Am.* **83**, 1057–68.

LUYENDYK, B. P., MACDONALD, K. C. & BRYAN, W. B. 1973. Rifting history of the Woodlark Basin in the southwest Pacific. *Bull. geol. Soc. Am.* **84**, 1125–34.

MILSOM, J. S. 1973. Gravity field of the Papuan Peninsula. *Geol. Mijnbouw*, **52**, 13–20.

—— & SMITH, I. E. 1975. Southeastern Papua: generation of thick crust in a tensional environment *Geology*, **3**, 117–20.

MORRISON, G. W. 1980. Characteristics and tectonic setting of the shoshonite rock association. *Lithos*, **13**, 97–108.

PAGE, R. W. & MCDOUGALL, I. 1972. Ages of mineralization of gold and porphyry copper deposits in the New Guinea highlands. *Econ. Geol.* **67**, 1065–74.

PATERSON, S. J. & KICINSKI, F. M. 1956. Account of the geology and petroleum prospects of the Cape Vogel Basin, Papua. *In: Papers on Tertiary micropalaeontology. Bur. Min. Resour. Aust. Rept.* **25**, 47–70.

PIETERS, P. E. 1978. Explanatory Notes, Sheets SC/55-6, -7 and -11 Port Moresby, Kalo, Aroa, (with 1:250,000 geological map). *Bur. Min. Resour. Aust. Bull.* Canberra.

RIPPER, I. D. 1982. Seismicity of the Indo-Australian/Solomon Sea plate boundary in the southeast Papua region. *Tectonophysics*, **87**, 355–70.

SMITH, I. E. 1970. Late Cenozoic uplift and geomorphology in south-eastern Papua. *Search Sydney*, **1**, 222–5.

—— 1972. High-potassium intrusives from southeastern Papua. *Contrib. Miner. Petrol.* **34**, 167–76.

—— 1973. Geology of the Calvados chain, southeastern Papua. *Bur. Min. Resour. Aust. Bull.* **139**, 59–66.

—— 1976. Peralkaline rhyolites from the D'Entrecasteaux Islands, Papua New Guinea. *In*: JOHNSON, R. W. (ed.) *Volcanism in Australasia*, 275–85. Elsevier, Amsterdam.

——· 1982. Volcanic evolution in eastern Papua. *Tectonophysics*, **87**, 315–34.

—— & COMPSTON, W. 1982. Strontium isotopes in Cenozoic volcanic rocks from southeastern Papua New Guinea. *Lithos,* **15**, 199–206.

—— & DAVIES, H. L. 1976. Geology of the southeast Papuan mainland. *Bur. Min. Resour. Aust. Bull.* **165**, 86 pp.

—— & SIMPSON, C. J. 1972. Late Cenozoic uplift in the Milne Bay area, eastern Papua New Guinea. *Bull. Min. Resour. Aust. Bull.* **125**, 29–35.

——, TAYLOR, S. R. & JOHNSON, R. W. 1979. REE-fractionated trachytes and dacites from Papua New Guinea and their relationship to andesite petrogenesis. *Contrib. Miner. Petrol.* **69**, 227–33.

TAYLOR, B. 1979. Bismarck Sea: evolution of a back-arc basin. *Geology*, **7**, 171–4.

TAYLOR, G. A. 1958. The 1951 eruption of Mt Lamington, Papua. *Bur. Min. Resour. Aust. Bull.* **38**, 117 pp.

TJHIN, K. J. 1976. Trobriand Basin exploration. *Aust. Pet. Explor. Assoc. J.* **16**, 81–90.

WEISSEL, J. K. & WATTS, A. B. 1979. Tectonic evolution of the Coral Sea basin. *J. geophys. Res.* **84**, 4572–82.

—— , TAYLOR, B. & KARNER, G. D. 1982. Opening of the Woodlark Basin, subduction of the Woodlark spreading system and the evolution of northern Melanesia since mid-Pliocene time. *Tectonophysics*, **87**, 253–77.

I. E. M. SMITH, Department of Geology, University of Auckland, Private Bag, Auckland, New Zealand.

J. S. MILSOM, Department of Geology, University College London, Gower Street, London WC1E 6BT, U.K.

Heat flow and magmatism in the NW Pacific back-arc basins

P. M. Sychev & A. Y. Sharaskin

SUMMARY: High heat flow is a characteristic feature of a number of back-arc basins of the NW Pacific Ocean. Temperature calculations related to the geological and geophysical data reveal that the sources of the high heat flow are located at shallow depths in the upper mantle and probably are zones of partial melting. From the location of these zones, the character and composition of magmas, and the evolution of the basins, it is proposed that injection of magmas, by a magmafracting (magmarupture) mechanism, is the main deep process within the back-arc basins. It is assumed that 'primitive' ultrabasic magma rises from great depths in the mantle along vertical or sub-vertical channels and is then distributed nearly horizontally close to the lower boundary of the crust, under the back-arc basins. The initial cause of the magmatic process is apparently a stepped gravity differentiation, beginning perhaps on the core–mantle boundary, and not the supposed subduction of oceanic lithosphere.

There are many back-arc basins in the NW Pacific Ocean. Although they have been closely studied their origin and evolution remain debatable. Heat-flow data reflect tectonic and magmatic processes and are important in the discussion of deep processes in back-arc basins. An understanding of the causes of high heat flow allows discussion of the deep processes in the Earth's crust and upper mantle. This paper concentrates on the relationships between heat flow and magmatism.

Anomalously high heat flow in the back-arc basins and sources of excess heat

With a few exceptions, anomalously high heat flow, approximately 1.5–2 times higher than the 'mean' value for tectonically stable regions ($c.$ 60 mW m^{-2}*), is observed in the NW Pacific back-arc basins (Fig. 1). For example, in the back-arc basins of the Japan and Okhotsk Seas a fairly uniform high heat flow of mean value 90–95 mW m^{-2} is recorded (Sychev et $al.$ 1982), and very high heat flow, up to 180 mW m^{-2}, occurs in the Ryukyu Basin (Watanabe et $al.$ 1977).

High heat flow is also observed in the comparatively shallow-water troughs of the marginal seas. In the Tatar Strait, the northern extension of the Japan Sea, the mean value is 110 mW m^{-2} and the maximum value is 173 mW m^{-2}, and high heat flow is observed in the Deryugin and Tinro Troughs in the Okhotsk Sea (Sychev et $al.$ 1983).

Heat-flow distribution is less uniform in the

* HFU = 1 μcal cm^{-2} s $\simeq$ 42 mW m^{-2}.

back-arc basins of the Bering and Philippine Seas. The Komandorsky (Kamchatka) Basin is characterized by high average heat flow of 135 mW m^{-2}, whereas in the Aleutian Basin values are near to normal at 50–70 mW m^{-2} (Marshall 1978; Smirnov & Sugrobov 1979). A more complex pattern is observed in the Philippine Sea. In its western part the average heat flow is near to the 'normal' value, although the range of individual values is wide. In the eastern part of the basin, the heat flow is higher, especially in the Shikoku Basin. In the Mariana Trough the heat flow reaches 377 mW m^{-2} in a general background of low values (Uyeda 1980).

Anomalously high heat flow is commonly attributed to crustal thinning in the deep-water areas of marginal seas. However, there are shallow-water troughs where high heat flow is observed and deep-water basins where the heat flow is near to normal (e.g. the western part of the Philippine Sea). Also, local high anomalies occur within shallow seas and on submarine rises.

In many cases the transition from high to mean or low heat flow values is sharp, in less than several tens of kilometres, suggesting that the heat sources are at shallow depths. In order to estimate deep temperatures, numerous calculations have been made using Poisson's equation for a stationary field (Smirnov & Sugrobov 1980; Sychev et $al.$ 1983). According to these calculations, the 1100–1200°C isotherms, at which partial melting of the upper mantle can occur, are at depths of about 10–40 km in the Komandorsky and Kuril Basins.

Calculations to estimate the depth of the lower boundary of excess temperatures, and hence the thickness of the anomalously hot zone, are of interest. Results from the Kuril Basin

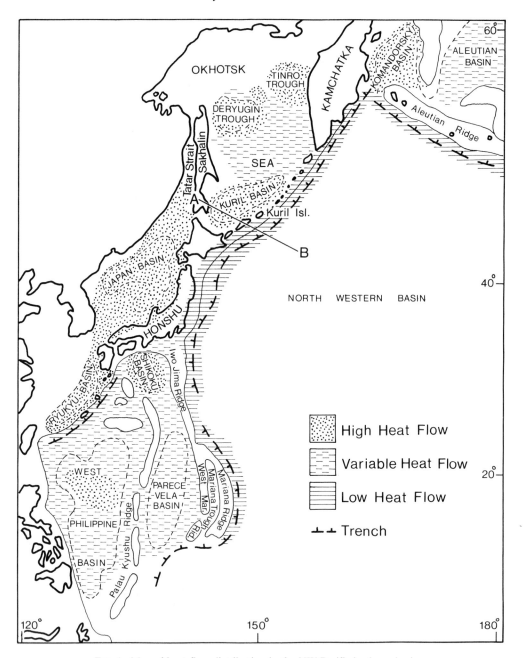

FIG. 1. Map of heat-flow distribution in the NW Pacific back-arc basins.

(Soinov & Soloviev 1978) are shown in Fig. 2. It appears that the temperature disturbance is restricted to depths shallower than 150–200 km beneath the basin. Since this solution is limiting, the thickness of the excess temperatures zone may be considerably smaller. From calculation of a two-dimensional stationary equation, the zone of high heat release lies approximately within an interval of 20–70 km (Tuyezov *et al.* 1982).

The calculated shallow position of excess heat sources is commonly in agreement with other geological and geophysical data. In particular, the existence of a layer of high elec-

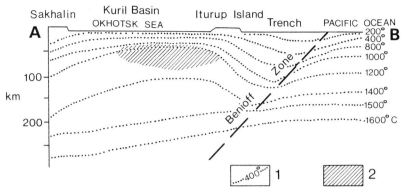

FIG. 2. Stationary temperature distribution in the upper mantle of the Kuril Arc and Basin (adapted from Soinov & Soloviev 1978). Location of the cross-section is shown on Fig. 1 (A–B). 1—isotherms; 2—excess temperatures zone.

troconductivity coincident with a zone of partial melting is proposed in the Japan and Shikoku Basins (Honkura 1974, 1975). Velocity inhomogeneities are observed at both shallow and great depths, including a possible velocity decrease at more than 1000 km (Julian & Sengupta 1973; Barazangi *et al.* 1975; Evans *et al.* 1978; Kirahara 1980; Boldyrev & Kats 1982). Of particular interest is that between 11 and 30 km in the Kuril Basin, immediately beneath the Moho discontinuity, there are layers where P-wave velocities decrease to 6.8–7.0 km s^{-1}, and the entire interval is characterized by anomalously high attenuation and absorption which indicate the existence of a zone of partial melting (Starshinova 1980).

With regard to the interpretation of gravity data, it has been suggested that positive gravity anomalies represent a considerable (Karig 1971) or small (Gainanov *et al.* 1974) density deficit in the mantle under back-arc basins. The deficit may be due to thermal expansion. However, the situation is complicated as positive background in the back-arc basins is required by the anomalous geoidal heights (e.g. Watts & Leeds 1977) that result from deep density inhomogeneities. Taking this factor into account, relatively dense shallow upper-mantle layers probably occur under some back-arc basins at least. Relatively dense mantle layers perhaps account for the great depths of the Philippine Sea.

Magmatism and some aspects of the geological history of marginal basins

Generally a close relationship is observed between the heat flow value and the age of tectonomagmatic activity (Polyak & Smirnov 1968). High heat flow indicates recent tectonomagmatic activity and low heat flow older activity. Therefore data on the age of magmatism provide evidence of the origin and duration of the sources of excess heat.

Magmatism occurred over a long time interval in the region. Within a number of island arcs the latest period of intense magmatic activity occurred in the Miocene. Magmatism at that time was extensive, occurring in western regions of Japan (Matsuda *et al.* 1967), in Sakhalin (Melnikov 1970), along the Aleutian Ridge (Hein *et al.* 1978), and the Izu–Bonin–Mariana (Scott *et al.* 1980) and Ryukyu island arcs (Bowin & Reynolds 1975). In places it has continued to the present.

Information on the age of magmatism in the back-arc basins is more limited because of the thick sediments that fill the basins. Cores from the Komandorsky Basin proved Middle Oligocene basalt overlain by Miocene and younger sediments (Scholl & Creager 1973). Evidence of magmatism has not been recorded from the deep part of the Okhotsk Sea. In the Japan Sea, rocks dredged from the slopes of volcanic seamounts and at rises of the acoustic basement give ages of 30–2.5 Ma (Sakhno *et al.* 1976). In the Japan Basin interlayers of volcanic rocks in the sediments indicate that magmatic activity waned from Middle or Upper Miocene to Holocene times (Vasilkovsky 1978). In the Philippine Sea magmatic activity has been determined with deep-sea drilling (Ingle 1975; Scott *et al.* 1980). Here the age gradually decreases from *c.* 50 Ma in the West Philippine Basin, to 25–17 Ma in the Parece Vela Basin and to 6–7 Ma in the Mariana Trough.

The above data indicate that the high heat flow is related to Miocene magmatism. In considering the magmatism in the Philippine Sea, certain patterns emerge. Firstly, it was episodic, with distinct maxima and minima (Scott & Kroenke 1980). Secondly, space–time relations can be defined between inter- and back-arc basins and island arcs; the minima at island arcs coincide with the maxima (spreading) in the adjacent back-arc basins. For example, the end of the Kyushu–Palau island-arc volcanism coincided with the early period of magmatism in the Parece Vela Basin (Scott *et al*. 1980), although it is possible that this correlation is not sufficiently distinct (Hussong & Uyeda 1981). However, in the northern part of the Philippine Sea, Honza (1981) confirms alternation of island-arc and inter-arc magmatism in the Ogasawara Ridge and the Shikoku Basin, with the island-arc activity preceding the inter-arc. Such episodic and alternating behaviour is difficult to reconcile with the continuity of the subduction process. The depth of origin and the nature of the sources of inter-arc and island-arc magmas are also problematic.

Layer 2 in the back-arc basins is composed of low-K tholeiites similar to those of 'normal' oceanic crust, so their sources in the mantle should be similar (e.g. Sharaskin *et al*. 1981; Bougault *et al*. 1982). Recent petrological data suggest that the mantle melting and separation of magmas that form abyssal tholeiites take place at a depth of not more than 60–70 km (Yoder 1976; Bender *et al*. 1978; Dmitriev *et al*. 1979). The basalts of the West Philippine and Parece Vela Basins were formed from lherzolitic mantle at temperatures near to 1270° and pressures of not more than 10 kbar (Zakariadze *et al*. 1980). The composition of chilled glass in basalts of the Shikoku Basin (Hole 442) suggests that the initial melt originated at depths of 10–15 km (Nesterenko & Suschevskaya 1981). This supports the proposal that excess heat sources lie at shallow depth. It is also noted that these basalts formed intermittently over a long period.

The alkaline olivine basalts of island arcs originate at a great depth and, having different isotopic characteristics, they ought to be related to a different mantle source. It is commonly assumed that the subducted oceanic lithosphere is the source, but if this were so there are a number of consequences which strongly contradict the observed data. Firstly, tholeiitic, calc-alkaline and alkaline magmas of island arcs are formed within a rather wide depth interval, and if contamination is derived from the upper part then the shallowest, tholeiitic melts should be most contaminated and the deeper calc-alkaline magmas should reflect more closely the typical mantle. However, the opposite pattern is observed. For instance, in the Japanese island arc the initial ratio $^{87}Sr/^{86}Sr$ does not exceed 0.704 in alkaline basalts of the back zone and nowhere exceeds 0.7038 in tholeiites of the frontal zone. It could be assumed that, having reached a proper depth and having passed through the magmatic cycle, all the sialic components return to the Earth's crust. However, even if this happened, isotopic exchange reactions should take place between the sialic components and the mantle rocks which should eventually result in the isotopic ratio being levelled to values corresponding to the intermixture ratio. However, the available evidence is to the contrary, and suggests that the isotopic character of the mantle that generates abyssal tholeiites and identical basalts of inter-arc basins originated and evolved independently, starting from 1.5 to 1.7 Ga (Tatsumoto 1978; Sun 1980).

On the other hand, vertical geochemical inhomogeneity of the mantle is shown by basalts of oceanic islands. Their $^{87}Sr/^{86}Sr$ ratios vary within 0.7024–0.7060, the mean value for the Pacific islands being 0.7037 (Hedge 1978). The same variation limits and mean value of this ratio are characteristic of the rocks of most island arcs.

These considerations make us reject the idea that island-arc magmas originated under the effect of subducting oceanic lithosphere or due to its direct melting. An alternative explanation is that the source of these melts is purely mantle with a primitive composition.

Finally, features of the evolution of the NW Pacific back-arc basins are discussed. It seems expedient to remark that the problem of the origin of back-arc basins is far from resolved, not even for the Philippine Sea (see e.g. Mrozovski *et al*. 1982). In addition to the idea of spreading in back-arc basins, other proposals have been made, such as the separation of parts of the ocean by newly formed island arcs (Vasilkovsky 1978; Sychev 1979), and the destruction of land areas with the development of marine basins in their places (Puscharovsky *et al*. 1977; Artyushkov *et al*. 1979). These debatable problems are not considered but certain general features of the structure and evolution of these basins are discussed in this paper.

A common feature of back-arc basins is the infill of sediments. Especially thick sequences are found in the Aleutian and Kuril Basins and in the Tatar Strait where they are at least 5 km thick. Acoustic characteristics of the sediments and their relation to the surrounding areas suggest that their lower part is at least of Cretaceous age (Sychev 1979). Seismic studies in the Japan Sea indicate that in Miocene times this basin had dimensions and depths similar to the modern examples (Ludwig *et al.* 1975). The even and persistent bedding of the sediments suggests that they were deposited in stable conditions over a long time and were accommodated by gradual downwarping. In many cases the basement to the sediments is composed of basalts similar to the oceanic Layer 2. Locally in the Okhotsk and Japan Seas the basement is uplifted to crop out on the sea floor. The interbedding of Neogene sediments and volcanics indicates that at least in places this basement is newly formed by intrusions of basalts in the lower part of the sedimentary sequence (Sychev & Snegovskoy 1976; Sychev 1979).

A possible mechanism of deep processes in the upper mantle of back-arc basins

Understanding of the mechanisms of deep processes in back-arc basins requires resolution of the problem of the excess heat sources. Different thermo-mechanical models have been proposed to account for the existing thermal anomalies. The fundamental process in the models is considered to be frictional heating of rocks due to the subduction of the oceanic lithosphere, which is assumed to cause either rise of diapirs (e.g. Karig 1971) or secondary convective cells (e.g. Sleep & Toksöz 1973). However, there is great doubt as to whether such processes are possible. In particular, it is unlikely that a large amount of heat can be generated by friction. The amount of energy necessary to cause the thermal anomalies in back-arc basins is an order or more higher than that which can be generated by means of the subduction (Artyushkov 1981). Hence another much more powerful source of heat, capable of maintaining high temperatures in the upper mantle for a long time, is necessary. Further, a frictional heat source cannot explain the pronounced localization of excess heat zones in the mantle, their high temperature and mosaic distribution, and the character and composition of the magmas. Also it is difficult to reconcile the rise of diapirs, or the existence of ascending convective currents, with the data on stable downwarping of the back-arc basins.

The number of possible means of heat transfer in the Earth's interior is limited. Rapid heat supply is necessary to form distinctly localized partial melting zones such as those assumed to exist under the back-arc basins. The most probable mechanism to satisfy these conditions is magmatic injection. The effects of extrusive and intrusive magmatism are widespread and the ability of magma to intrude horizontally for long distances is impressive. Sills, 50–300 m thick, extending for 750–950 km have been recorded (Feoktistov 1978). Theoretical calculations suggest that slightly overheated basaltic melt ($1300°C$) can intrude narrow (up to 10 m) fractures for hundreds of kilometres due to the magmafracting (magmarupture) mechanism (Popov 1972). To discuss satisfactorily a model of deep processes the adopted assumptions should be considered.

For periods of < 1000 yr crustal and upper-mantle rocks can be treated as an elastic solid medium whose tensile strength is estimated to lie within $1–10 \times 10^6 \, N \, m^{-2}$, which corresponds to a stress difference of 10–100 bar (Magnitsky 1965). As a result, under relatively short-time mechanical actions, the formation of brittle ruptures is possible within a wide depth range, probably to 700 km.

Viscosity of magmas decreases with increasing temperature and pressure and can be < 10 poise (Kushiro *et al.* 1976; Lebedev & Khitarov 1979). At depths of tens or hundreds of kilometres magmas would be extremely fluid.

Boundaries in the upper mantle are commonly interpreted as being caused by phase transition. But physical experiments show that chemical composition should also change at these boundaries. More gradual changes in the upper mantle possibly occur with depth (Levykin & Vavakin 1978; Liu 1980; Hales 1981) which would be related to changes of density.

High electrical conductivity is associated with partial melting zones in which the liquid fraction does not exceed 7–10%. But the high electrical conductivity is possible only if the liquid fraction is distributed in a network of interconnected channels (Shankland & Waff 1977).

Using these observations, it is possible to construct a model of deep processes (Fig. 3). This model implies that high-temperature melts of presumed ultrabasic primitive magma rise from local low-velocity zones at depths of approximately 200–300 km. This rise is caused by buoyancy wherein the excess pressure (ΔP) $= \Delta \rho g H$ where $\Delta \rho$ is the density difference, g the gravitational constant, and H is the height of the magma column (or thickness of the magma layer). It should be noted that it is not necessary to assume the existence of a continuous magma layer. If the magma is distributed in an interconnected system of fractures, then the

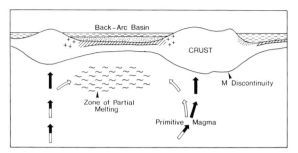

FIG. 3. Model of deep processes in the
upper mantle of the back-arc and island-arc
system (not to scale, explanation in text).

value of excess pressure at the upper edge of
the layer will be the same.

If, for instance, it is assumed that $\Delta\rho =$
0.1 g cm^{-3} and $H = 50$ km, then the hydrostatic excess pressure and the tension stresses are
approximately equal to about 500 bar at the
upper boundary of this layer or column and, as
this exceeds the strength of the host medium,
tension fractures can form. So, solid upper-
mantle rocks appear not to be a serious obstacle
for magmas where density is less than that of
the surrounding medium.

Stresses over an area of excess pressure can
form not only vertical but also inclined zones of
decreased strength along which the magma can
intrude into overlying layers (Roberts 1970),
and magma may rise along the surface of the
Benioff zone. This is partly supported by analysis of the mechanism of intermediate-depth
earthquakes, some of which can be directly related to the ruptures formed by intruding magma
(Robson *et al*. 1968). Detailed analysis of both
relatively deep (Balakina & Golubeva 1979;
Balakina 1981) and shallow earthquakes
(Takagi 1972; Hayakawa & Iizuka 1976)
commonly indicate that some are not due to
shear but are dilational.

The rise of magma is limited by its density
difference with the surrounding medium. Basaltic magma being of a much lower density than
the upper mantle will intrude into the crust and
can flow out on to the surface. In contrast,
ultrabasic magma can intrude the crust only in
special cases, as its density is greater than that of
the crust. The model (Fig. 3) assumes that the
initial melts are of predominantly ultrabasic
composition. As this magma rises the pressure
and temperature decrease and lighter basaltic
fractions separate from it and intrude the crust.
At the base of the crust, the ultrabasic magma
will, if its density is between that of the upper
mantle and crust, move horizontally by the
magmafracting (magmarupture) mechanism

(Popov 1972). But since the Earth's crust
thickens due to a continuous supply of basic
and relatively light components in the places
where magma rises, the ultrabasic magma will
be deflected to the shallowest position of the
upper-mantle surface, which occurs where the
crust is thinnest—i.e. the back-arc basins. The
vertical and lateral distribution of upper-mantle
densities sensitively controls the paths of
ultrabasic magma. Therefore the paths and
depths where ultrabasic magma moves under
the Earth's crust of the back-arc basins can be
variable.

It can hardly be expected that one can
observe this horizontal movement of magma,
but such a possibility should not be excluded.
The loci of shallow seismicity in the Kuril
back-arc region have been observed to shift
gradually from the Kuril Island Arc, where they
are deeper, to beneath the Kuril Basin. The
duration of this shift is estimated to be 1 yr
(Baranov & Lobkovsky 1980), which agrees
with data on the high velocity of magma movement.

The intrusion of ultrabasic magma below the
back-arc basins transfers the heat which results
in high heat flow observed at the sea floors. The
tholeiitic basalts of back-arc basins are considered to originate by interaction of the
ultrabasic magma and the enclosing mantle,
with partial melting of the latter. As basalts
have significantly lower density than the
upper-mantle rocks, it is unlikely that they can
accumulate in large quantities in the partial
melting zone. It is more likely that as they
accumulate, even in relatively small portions,
tension fractures would be formed along which
basalt will intrude the crust. Similar to the
intrusions of ultrabasic magmas, the formation
of tholeiitic basalts is considered to be a continuous process which lags behind the volcanism of the nearby island arcs.

The ultrabasic magma that is intruded under

the back-arc basins and has density equal to that of the enclosing rocks, constitutes excess mass, but the equilibrium is rapidly re-established in such cases by downwarping of the underlying layers. At the same time, partial melting zones which had been formed, will cool and disappear. This should contribute to the downwarping of both the zone itself and the overlying layers. It is proposed that this process explains the anomalous great depths in the West Philippine Sea and part of the downwarping of back-arc basins in general. The existence of low-velocity zones in the upper mantle does not contradict this proposal because the velocity decrease can be related to the influence of a liquid phase of intruded ultrabasic magmas without a significant density decrease.

Considering the velocity and other physical property inhomogeneities beneath island-arc regions, up to considerable depths (probably more than 1000 km), it is logical to assume that the ultrabasic magmas also have a deep source. This source may be due to gravity differentiation at the core-mantle boundary (Artyushkov 1968, 1979). Heat energy release due to gravity differentiation, and its stepped transfer as relatively light melts along weakened zones, are apparently the fundamental causes of the magmatic process (Sychev 1973, 1979). An interrupted, pulse-like character of magmatism at the surface would result from relatively light differentiates (in the upper mantle at least) rising in batches as they accumulate. The excess

pressure formed in this movement increases in the uppermost layers to the level at which tension fractures would form. The fractures appear only during magma movement along them. Later they may be represented by zones of decreased strength which subsequent magma intrusion would favour. The inclined zones of weakness may originate from stresses in the base of upper mantle or deeper (Kodama & Suzuki 1977; Suzuki et al. 1978).

Conclusion

The principal deep process in the back-arc upper mantle may be called mantle trap magmatism by analogy with trap magmatism in the Earth's crust. This model implies that ultrabasic magmatism is a powerful phenomenon in the upper mantle.

The process of magmatic injection concurs with the geological and geophysical data, including interrelation of island-arc and back-arc magmatism. Some problems of the magma injection process still demand explanation. It is hoped that this paper will stimulate further investigations in this trend to elucidate the deep processes.

ACKNOWLEDGMENTS: The authors are grateful to M. S. Fedorishin for translating the paper from Russian into English, to T. F. Eliseeva for her most useful assistance, and to Dr P. Kokelaar for editorial revision.

References

ARTYUSHKOV, E. V. 1968. Gravitational convection in the Earth's interior. *Izv. Akad. Nauk SSSR*, **9**, 3–17 (in Russian).
—— 1979. *Geodynamics*. Nauka, Moscow, 328 pp. (in Russian).
—— 1981. Mechanism of formation of active margin. *J. Oceanol. Acta, 26th Geol. Congr., Paris*, 245–50.
——, SHLEZINGER, A. E. & YANSHIN, A. L. 1979. Basic types and mechanisms of structures formation on lithospheric plates. I. Continental platforms. *Bulleten moscovskogo obschestva ispytaniya prirody. Otdel Geologii*, **54**, 8–30 (in Russian).
BALAKINA, L. M. 1981. Mechanism of intermediate seismic foci in the Kuril–Kamchatka focal zone. *Izv. Akad. Nauk SSSR*, **8**, 3–24 (in Russian).
—— & GOLUBEVA, N. V. 1979. Peculiarities of a mechanism of deep seismic foci in the Japan and Okhotsk Seas. *Izv. Akad. Nauk SSSR*, **9**, 3–21 (in Russian).
BARANOV, B. V. & LOBKOVSKY L. I. 1980. Shallow focus seismicity in the rear of the Kuril Island

Arc and its connection with Zavaritsky–Benioff zone. *Dokl. Akad. Nauk SSSR*, **225**, 67–71 (in Russian).
BARAZANGI, M., PENNINGTON, W. & ISACKS, B. 1975. Global study of seismic wave attenuation in the upper mantle behind island arcs using P-waves. *J. geophys. Res.* **80**, 1079–92.
BENDER, J. F., HEDGES, F. N. & BENCE, A. E. 1978. Petrogenesis of basalts from the project FAMOUS area: experimental study from 0 to 15 kbar. *Earth planet. Sci. Lett.* **41**, 277–302.
BOLDYREV, S. A. & KATS, S. A. 1982. Three-dimensional velocity model of the upper mantle of the Asia-to-Pacific transition zone. *Vulkanologiya i Seismologiya*, **2**, 80–95 (in Russian).
BOUGAULT, H., MAURY, R. C., ELAZZOURI, M., JORON, J.-L., COTTEN, J. & TREUIL, M. 1982. Tholeiites, basaltic andesites and andesites from Leg 80 Sites: geochemistry, mineralogy and low partition coefficient elements. *In*: HUSSONG, D. M., UYEDA, S. *et al.* (eds) *Init. Rep. Deep Sea drill. Proj.* **60**, 657–78. U.S. Government Printing Office, Washington, D.C.

BOWIN, C. & REYNOLDS, P. H. 1975. Radiometric ages from Ryukyu Arc Region and an $^{40}Ar/^{39}Ar$ age from biotite dacite in Okinawa. *Earth planet. Sci. Lett.* **27**, 363–70.

DMITRIEV, L. V., SOBOLEV, A. V. & SUSCHEVSKAYA, N. M. 1979. The primary melt of the oceanic tholeiite and upper mantle composition. *In*: TALWANI, M., HARRISON, C. G. & HAYES, D. E. (eds) *Deep Drilling Results in the Atlantic Ocean: Ocean Crust, M. Ewing Series* **2**, 302–13. AGU, Washington.

EVANS, J. R., SUYEHIRO, R. & SACKS, I. S. 1978. Mantle structure beneath the Japan Sea, a reexamination. *Geophys. Res. Lett. Washington*, **5**, 487–90.

FEOKTISTOV, G. D. 1978. *Petrology and Conditions of Trap Sills Formation.* Novosibirsk, Nauka, 168 pp. (in Russian).

GAINANOV, A. G., PAVLOV, YU. A., STROEV, P. A., SYCHEV, P. M. & TUYEZOV, I. K. 1974. *Anomalous Gravity Fields of the Far East Marginal Basins and the Adjacent Pacific.* Novosibirsk, Nauka, 108 pp. (in Russian).

HALES, A. L. 1981. The upper mantle velocity distribution. *Phys. Earth planet. Inter.* **25**, 1–11.

HAYAKAWA, M. & IIZUKA, S. 1976. A mechanism to explain the changes in Vp/Vs. *J. Seism. Soc. Japan*, **29**, 339–53.

HEDGE, C. 1978. Strontium isotopes in basalts from the Pacific Ocean basin. *Earth planet. Sci. Lett.* **38**, 88–94.

HEIN, J. R., SCHOLL, D. W. & MILLER, J. 1978. Episodes of Aleutian Ridge explosive volcanism. *Science*, **199**, 137–41.

HONKURA, J. 1974. Electrical conductivity anomalies beneath the Japan Arc. *J. Geomag. Geoelect. Kyoto*, **26**, 147–71.

—— 1975. Partial melting and electrical conductivity anomalies beneath the Japan and Philippine Seas. *Phys. Earth planet. Inter.* **10**, 128–34.

HONZA, E. 1981. The geological settings of the Ogasawara and the Northern Mariana Arcs—concluding remarks. *In*: HONZA, E., INOUE, E. & ISHIHARA, T. (eds) *Geological Investigation of the Ogasawara (Bonin) and Northern Mariana Arcs, Cruise Report* **14**, 159–64. Geol. Surv. Japan.

HUSSONG, D. M. & UYEDA, S. 1981. Tectonic process and the history of the Mariana Arc: a synthesis of the results of Deep Sea Drilling Project Leg 60. *In*: HUSSONG, D. M., UYEDA, S. et al. (eds) *Init. Rep. Deep Sea drill. Proj.* **60**, 909–29. U.S. Government Printing Office, Washington, D.C.

INGLE, JR, J. C. 1975. Summary of Late Paleogene–Neogene insular stratigraphy, paleobathymetry, and correlations, Philippine Sea and Japan Sea region. *In*: INGLE, J. C., KARIG, D. E. et al. (eds) *Init. Rep. Deep Sea drill. Proj.* **31**, 837–55. U.S. Government Printing Office, Washington, D.C.

JULIAN, B. R. & SENGUPTA, M. K. 1973. Seismic travel time evidence for lateral inhomogeneity in the deep mantle. *Nature, Lond.* **242**, 443–7.

KARIG, D. E. 1971. Origin and development of mar-

ginal basins in the Western Pacific. *J. geophys. Res.* **76**, 2542–61.

KIRAHARA, K. 1980. Three-dimensional shear velocity structure beneath the Japan Islands and its tectonic implications. *J. Phys. Earth.* **28**, 221–41.

KODAMA, K. & SUZUKI, Y. 1977. Formation of the deep earthquake zone due to mantle diapirism. *Bull. geol. Surv. Japan*, **28**, 795–810 (in Japanese, abstr. in English).

KUSHIRO, I., YODER, H. S. & MYSEN, B. O. 1976. Viscosities of basalt and andesite melts at high pressures. *J. geophys. Res.* **81**, 6351–6.

LEBEDEV, E. B. & KHITAROV, N. I. 1979. *Physical Properties of Magmatic Melts.* Nauka, Moscow, 200 pp. (in Russian).

LEVYKIN, A. U. & VAVAKIN, C. C. 1978. Investigations of the elastic wave velocities and rocks and minerals density at pressures up to 20 kbar and temperatures up to 500°C. *Izv. Akad. Nauk SSSR*, **5**, 42–51 (in Russian).

LIU, L. 1980. On the interpretation of mantle discontinuities. *Phys. Earth planet. Inter.* **23**, 332–6.

LUDWIG, W. J., MURAUCHI, S. & HOUTZ, R. E. 1975. Sediments and structure of the Japan Sea. *Bull. geol. Soc. Am.* **86**, 651–64.

MAGNITSKY, V. A. 1965. *Internal Structure and Earth's Physics.* Nedra, Moscow, 380 pp. (in Russian).

MARSHALL, B. V. 1978. Recent heat flow measurements in the Aleutian Basin, Bering Sea. *Eos*, **59**, 384 pp.

MATSUDA, T., NAKAMURA, K. & SUGIMURA, A. 1967. Late Cenozoic orogeny in Japan. *Tectonophysics*, **4**, 349–66.

MELNIKOV, O. A. 1970. *History of South Sakhalin Structure Formation in the Paleogene and Neogene.* Nauka, Moscow, 170 pp. (in Russian).

MROZOWSKI, C. L., LEWIS, S. D. & HAYES, D. E. 1982. Complexities in the Cenozoic evolution of the West Philippine Basin. *Tectonophysics*, **82**, 1–24.

NESTERENKO, G. V. & SUSCHEVSKAYA, N. M. 1981. Basaltic glasses of the hole 442B (the Philippine Sea). *Geokhimiya*, **9**, 1380–5 (in Russian).

POLYAK, B. G. & SMIRNOV, YA. B. 1968. Relationship between terrestrial heat flow and tectonics of continents. *Geotektonika*, **4**, 205–13 (in Russian).

POPOV, V. S. 1972. Estimations of the velocity of basic dykes and sills intrusions. *Geokhimiya*, **6**, 713–8 (in Russian).

PUSCHAROVSKY, YU. M., MELANKHOLINA, E. N., RAZNITSIN, YU. N. & SCHMIDT, O. A. 1977. Comparative tectonics of the Bering, Japan and Okhotsk Seas. *Geotektonika*, **5**, 83–94 (in Russian).

ROBERTS, J. G. 1970. Magma intrusions in brittle rocks. *In*: NEWALL, N. G. & RAST, N. (eds) *Mechanism of Igneous Intrusion*, 230–83. Gallery Press, Liverpool.

ROBSON, G. R., BARR, K. G. & LUNA, L. C. 1968. Extension failure: an earthquake mechanism. *Nature, Lond.* **218**, 28–32.

SAKHNO, V. G., VRZHOSEK, A. A. & MOISEENKO, V.

G. 1976. Peculiarities of the composition of marginal sea floor lavas. *In*: BEVZENKO, P. E. (ed.) *Igneous Rocks of Asia*, 19–30. Vladivostok, (in Russian).

SCHOLL, D. W. & CREAGER, J. S. 1973. Geologic synthesis of Leg 19 (DSDP) results; Far North Pacific and Aleutian Ridge and Bering Sea. *In*: CREAGER, J. S. & SCHOLL, D. W. *et al.* (eds) *Init. Rep. Deep Sea drill. Proj.* **19**, 897–913. U.S. Government Printing Office, Washington, D.C.

SCOTT, R. & KROENKE, L. 1980. Evolution of back-arc spreading and arc volcanism in the Philippine Sea: Interpretation of Leg 59 DSDP Results. *In*: HAYES, D. E. (ed.) *The Tectonic and Geologic Evolution of Southeast Asian Seas and Islands*, 283–91. AGU, Washington, D.C.

——, ——, ZAKARADZE, G. & SHARASKIN, A. 1980. Evolution of the South Philippine Sea, Deep Sea Drilling Project Leg 59 results. *In*: KROENKE, L., SCOTT, R. *et al.* (eds) *Init. Rep. Deep Sea drill. Proj.* **59**, 803–15. U.S. Government Printing Office, Washington, D.C.

SHANKLAND, T. J. & WAFF, H. S. 1977. Partial melting and electric conductivity anomalies in upper mantle. *J. geophys. Res.* **81**, 5260–6.

SHARASKIN, A. Y., BOGDANOV, N. A. & ZAKARIADZE, G. S. 1981. Geochemistry and timing of the marginal basins and arc magmatism in the Philippine Sea. *Phil. Trans. R. Soc. London*, **A300**, 287–97.

SLEEP, N. & TOKSÖZ, M. N. 1973. Evolution of marginal basins. *Nature, Lond.* **233**, 548–50.

SMIRNOV, YA. B. & SUGROBOV, V. M. 1979. Terrestrial heat flow in the Kuril–Kamchatka and Aleutian provinces. I. Heat flow and tectonics. *Vulkanologiya i Seismologiya*, **1**, 59–73 (in Russian).

—— & —— 1980. Terrestrial heat flow in Kuril–Kamchatka and Aleutian provinces. III. The estimations of deep temperatures and lithospheric density. *Vulkanologiya i Seismologiya*, **2**, 3–18 (in Russian).

SOINOV, V. V. & SOLOVIEV, V. N. 1978. Stationary model of the upper mantle temperatures in the Okhotsk Sea region. *In*: KRASNY, M. L. (ed.) *Geophysical Field of the Asia—Pacific Transition Zone*, 53–6. Yuzhno-Sakhalinsk (in Russian).

STARSHINOVA, E. A. 1980. Crust and upper mantle structure inhomogeneities in the Okhotsk Sea. *Dokl. Akad. Nauk SSSR*, **255**, 1339–43 (in Russian).

SUN, S.-S. 1980. Lead isotopic study of young volcanic rocks from mid-ocean ridges, ocean islands and island arcs. *Phil. Trans. R. Soc. London*, **A297**, 409–46.

SUZUKI, Y., KODAMA, K. & MITSUTANI, T. 1978. The

formation of intermediate and deep earthquake zone in relation to the geologic development of East Asia since the Mesozoic. *J. Phys. Earth.* **26**, S579–84.

SYCHEV, P. M. 1973. Upper mantle structure and nature of deep processes in island arc–trench system. *Tectonophysics*, **19**, 343–59.

—— 1979. *Deep and Surface Tectonic Processes of the Northwestern Pacific Mobile Belt*. Nauka, Moscow, 208 pp. (in Russian).

—— & SNEGOVSKOY, S. S. 1976. Abyssal depressions of the Okhotsk, Japan and Bering Seas. *Pacific Geol.* **11**, 57–80.

——, SOINOV, V. V., VESELOV, O. V. & VOLKOVA, N. A. 1983. Heat flow and geodynamics of the northwestern Pacific. *In*: HILDE, T. W. C. & UYEDA, S. (eds) *Geodynamics of the Western Pacific–Indonesian Region*, 237–47. AGU, Washington, D.C.

TAKAGI, S. 1972. Do earthquakes occur due to stress? *Pap. Meteorol. Geophys.* **23**, 1–19.

TATSUMOTO, M. 1978. Isotopic composition of lead in oceanic basalts and its implications to mantle evolution. *Earth planet. Sci. Lett.* **38**, 63–87.

TUYEZOV, I. K., VESELOV, O. V., EPANESHNIKOV, V. D. & LIPINA, E. N. 1982. Geothermics of Western Pacific. *Tikhookeanskaya Geologiya*, **3**, 90–100.

UYEDA, S. 1980. Review of heat flow studies in the Eastern Asia and Western Pacific region. *UN ESCAP, CCOP/SOPAC Tech. Bull.* **3**, 153–69.

VASILKOVSKY, N. P. 1978. Problems of the origin and geological history of the Sea of Japan. *In*: TUYEZOV, I. K. (ed.) *Peculiar Features of Geological Structure of the Sea of Japan Floor*. Nauka, Moscow, 215–45 (in Russian).

WATANABE, T., LANGSETH, N. G. & ANDERSON, R. N. 1977. Heat flow in back-arc basins of the Western Pacific. *In*: TALWANI, M. & PITMAN, W. C. (eds) *Island Arcs, Deep-Sea Trenches, and Back-Arc Basins*, **3**, 137–61. AGU, Washington D.C.

WATTS, A. B. & LEEDS, A. R. 1977. Gravimetric geoid in the northwestern Pacific Ocean. *Geophys. J. R. astr. Soc.* **50**, 249–77.

YODER, JR, H. S. 1976. *Generation of Basaltic Magma*, 231 pp. National Academy of Science, Washington, D.C.

ZAKARIADZE, G. S., DMITRIEV, L. V., SOBOLEV, A. V. & SUSCHEVSKAYA, N. M. 1980. Petrology of basalts of holes 447A, 449 and 450, South Philippine Sea transect, Deep Sea Drilling Project Leg 59. *In*: KROENKE, L., SCOTT, R. *et al.* (eds) *Init. Rep. Deep Sea drill. Proj.* **59**, 669–80. U.S. Government Printing Office, Washington, D.C.

P. M. SYCHEV, Sakhalin Complex Scientific Research Institute, Far East Science Centre, U.S.S.R. Academy of Sciences, Novoalexandrovsk 694050, Sakhalin, U.S.S.R.

A. Y. SHARASKIN, Vernadsky Institute of Geochemistry, Academy of Sciences of the U.S.S.R., Vorobyevskoe Shausse 47-a, 117334, Moscow B-334, U.S.S.R.

SOUTH AMERICA & ANTARCTICA

Spreading-subsidence and generation of ensialic marginal basins: an example from the early Cretaceous of central Chile

G. Åberg, L. Aguirre, B. Levi & J. O. Nyström

SUMMARY: During the early Cretaceous, intracontinental rifting, spreading and subsidence led to the formation of an ensialic trough in central Chile. This trough is interpreted as an aborted marginal basin since no oceanic crust was generated. However, mantle-derived material, represented by flood basalts with low initial $^{87}Sr/^{86}Sr$ ratio and evolved geochemical characteristics, was deposited in the basin. Rb-Sr dating shows that the lavas suffered burial metamorphism shortly after their extrusion. The flood basalts were extruded during an episode of slow oceanic spreading in the Pacific, and subsequent folding, granitoid intrusion and uplift are correlated with an episode of fast oceanic spreading and plate subduction. The intracontinental events related to these two episodes constitute a cycle, repetition of which might account for the geological evolution of central Chile during the Mesozoic and the Palaeogene.

The western margin of South America, during the Cretaceous, was characterized by basin formation with eruption of mantle-derived basalts. Some basins developed into marginal basins proper, with generation of oceanic crust, others were aborted. The type of basin was determined by the rate and volume of upwelling mantle material, and continental crust thickness. Rapid upwelling gave rise to back-arc spreading and marginal basins with primitive basalts and ocean-floor metamorphism or burial metamorphism reflecting steep thermal gradients. Slow upwelling produced ensialic basins with evolved basalts affected by burial metamorphism with less steep thermal gradients, as in central Chile during the early Cretaceous.

The main geological structure in central Chile (approximately 27–35°S) is a N-trending synclinorium (Fig. 1), 200 km wide and 800 km long, comprising predominantly volcanic rocks of early Jurassic to Palaeogene age. Both flanks of the synclinorium rest on a Palaeozoic pre-Andean basement. This ensialic synclinorium is composed of several stratigraphical-structural units separated by unconformities. The cumulative thickness of the units in the W flank exceeds 20 km, and in the E flank 10 km. The units are intruded by granitoids distributed in N-trending belts which are successively younger towards the E.

A coupled action of plate subduction and ensialic spreading-subsidence, acting alternately or simultaneously although at different rates, was proposed by Levi & Aguirre (1981) to explain the geological record of central Chile. The spreading-subsidence mechanism envisaged is similar to that of the oceanic crustal extension and subsidence in the Icelandic rift zone (Bodvarson & Walker 1964; Pálmason 1973), although in central Chile it would have operated within the margin of a continental plate. The formation of the synclinorium is consistent with a cyclic recurrence of spreading-subsidence which produced ensialic troughs where large volumes of volcanic rocks were deposited. Such a

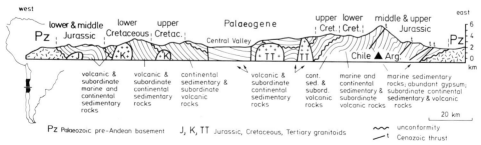

FIG. 1. Schematic cross-section through the ensialic synclinorium in central Chile at c. 33°S (slightly modified from Levi & Aguirre 1981). Deposits younger than Palaeogene are not included. The only normal faults shown are those delimiting the Quaternary Central Valley graben. The longitudinal extent of the synclinorium (in black) and the location of the cross-section are given in the accompanying map.

mechanism explains the symmetrical geological structure in which flat-lying, young volcanic flows and their feeder dykes are located in the central part of the synclinorium, and increasingly steeper, older volcanics and associated feeders occur towards its flanks. A bilateral symmetry is also revealed by a preliminary geochemical study (Levi & Nyström 1982; see also Charrier 1981) which shows that coeval volcanics from both flanks of the synclinorium have similar K_2O contents and $Zr/Ti/Y$ ratios. The structural and geochemical symmetry suggests that intra-continental spreading took place during periods of volcanic activity.

It is proposed that the ensialic troughs generated during periods of spreading-subsidence were aborted marginal basins in the sense that rifting, spreading and subsidence occurred in an ensialic environment without generation of oceanic crust, although voluminous mantle-derived flood basalts were extruded. One of these aborted basins is described. Recent metamorphic, geochronological and geochemical studies in central Chile suggest that such a basin existed during the early (to middle) Cretaceous. Relations between this and other basins along the W margin of South America during the Cretaceous are also discussed.

The early Cretaceous cycle

Outline of the geological record

The following summary is based on Carter & Aguirre (1965); Corvalán (1965); Ruiz et al. (1965); Muñoz Cristi (1968); Levi (1970); Vergara (1972); Aguirre et al. (1974); Zeil (1979); Charrier (1981); Drake et al. (1982); Aguirre (1983) and references therein.

By the end of the Jurassic, central Chile had gone through a cycle of volcanic arc and basin formation, deposition of volcanic and sedimentary material, burial metamorphism, intrusion of granitoids, and regional uplift which established continental conditions. The early Cretaceous cycle began with the development of a volcanic arc and an ensialic trough. Subsidence was initiated probably by rifting processes which were controlled by a regional block-fault pattern imposed by the late Jurassic uplift. The arc was formed within the continental margin, on eroded Jurassic rocks and the pre-Andean basement. Material from the volcanic arc was deposited in the trough ('geosynclinal' basin) and interfingered with sediments from the E.

The environment changed with time from shallow-marine to alternating shallow-marine and continental. Turbidites, limestones and continental sediments with considerable volumes of dacitic ignimbrites and subordinate basaltic flows, totalling 2–4 km, were laid down on the western ('eugeosynclinal') side of the basin during most of the Neocomian (Fig. 2). On the eastern ('miogeosynclinal') side, a sequence, c. 2 km thick, of marine clastic sediments, limestones and gypsum layers, was deposited with intercalations of dacitic tuff and basaltic lava.

During the Hauterivian the volcanism in the basin changed markedly. Eruption of basic to intermediate lavas, along deep N-trending faults, built a thick volcanic pile on its western side. The lower and middle parts of the pile are composed of flood basalts, whereas the upper part is dominated by andesitic flow breccias (Fig. 2). The volcanism continued to Albian times. It was less intense at the northern and southern ends of the basin where the sequence is dominated by marine sediments. The pile is thickest (5–8 km) in the central–S section (between 30 and 34°S), where the intercalated

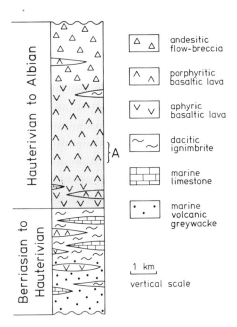

FIG. 2. A schematic stratigraphic column of the lower Cretaceous in the western flank of the synclinorium, central Chile at c. 33°S (based on Levi 1970, Table 1 and references therein). The shaded part of the column represents the main volume of lavas, the flood basalts. REE and Rb-Sr isotope compositions have been studied in samples from the part of the column indicated by A.

sedimentary rocks are largely continental. The coeval sequence on the eastern side of the basin is 1–2 km thick and composed of continental sedimentary rocks with intercalations of basaltic lava.

The subsiding volcanics and sediments suffered burial metamorphism, folding without penetrative deformation, and intrusion by granitoids. On the western side of the basin, mineral assemblages of zeolite to greenschist facies were produced in the *c.* 10 km thick sequence, and in the *c.* 4 km thick sequence in the E the assemblages range from zeolite to prehnite-pumpellyite facies. Radiometric data suggest that plutonic activity was fairly continuous and overlapped the volcanism, although it appears to have reached a maximum during the middle Cretaceous, immediately after folding. Whether the bulk of granitoids emplaced during the early (to middle) Cretaceous cycle, as well as in subsequent cycles, is related to extensional or folding events is, however, a subject for debate.

After considerable uplift along N-trending lineaments, and erosion, a new cycle of basin formation, volcanism and sedimentation was initiated in the late Cretaceous.

Closeness in time between extrusion and burial metamorphism of the flood basalts

The lower Cretaceous rocks underwent burial metamorphism before folding, since isograds are parallel to the overall fold structure. Also, late Cretaceous conglomerates, unconformably overlying the lower Cretaceous volcanic pile, contain pebbles representative of the entire spectrum of metamorphic facies observed in the pile (Levi 1970).

According to palaeontological data (Aguirre & Egert 1965), the porphyritic basalts in the middle part of the flood basalt sequence (Fig. 2) were extruded at some time during the Hauterivian to Aptian, *i.e.* between 126 and 108 Ma (timescale after van Hinte (1976)). Rb-Sr data for six 'unaltered' samples of these basalts plot (Fig. 3a) close to a reference line for 117 Ma, the mean of the palaeontological time span. The average initial $^{87}Sr/^{86}Sr$ ratio, Sr(i), is 0.7038 ± 0.0004. The similar Sr(i) for strongly altered basalts (0.7040 ± 0.0002; Fig. 3b) is taken to indicate that diagenesis and burial metamorphism closely followed extrusion. However, one amygdale sample (250D; Sr(i) = 0.7040 ± 0.0001) is almost Rb-free and consequently reflects the Sr(i) of its host rock at the time of metamorphism (i.e. when the amygdale was

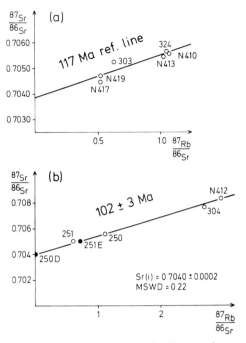

FIG. 3. Rb-Sr whole rock diagrams for early Cretaceous porphyritic flood basalts, central Chile. Analytical data are given in the Appendix. (a) Samples of 'unaltered' basalt from six flows. The scatter is probably due to incipient alteration, present in even the best-preserved samples. The reference line for 117 Ma is based on palaeontological data. (b) Samples of strongly altered basalt, i.e. spilite (circles) and amygdales (filled circles) from three of the flows plotted in (a).

formed). The Rb-Sr whole rock systems closed at 102 ± 3 Ma (Fig. 3b), which would be about 10–20 Ma after the flows were extruded.

Spreading-subsidence

The overall symmetry in structure, stratigraphy and geochemistry can be explained in terms of a simple depositional-deformation history. However, these symmetries are also consistent with a mechanism of spreading-subsidence, and this is preferred for the following reasons:

(i) The lower Cretaceous rocks in both flanks of the synclinorium constitute a belt, several hundred km long, predominantly of volcanic rocks. Faults associated with block movements, and numerous feeder dykes (locally dyke swarms) parallel to the trend of the belt, indicate extension at 90°

to deep, persistent fractures along the belt. Also, rifting is interpreted as important in the development of a belt of apatite-rich iron deposits associated with the flood basalts and their comagmatic intrusions (Oyarzún & Frutos 1982).

(ii) Rapid subsidence must have occurred because, in spite of a high rate of accumulation of volcanic material in the basin, the depositional surface remained close to sea-level, and because burial metamorphism closely followed extrusion. Furthermore, the subsiding volcanics did not experience long periods at high temperatures and pressures, probably because they were removed laterally by intracontinental spreading. This interpretation is consistent with the prevalence of metastable mineral assemblages in the metamorphosed volcanics (Levi *et al*. 1982), and the fact that greenschist-facies assemblages occur at a stratigraphical depth of several km without the development of pumpellyite-actinolite-facies assemblages.

(iii) High-level emplacement of granitoids in a linear belt is consistent with crustal spreading. Drake *et al*. (1982) suggest that the emplacement of granitoids rifted and extended the continental margin in a way analogous to mid-ocean ridge spreading.

Mantle-derived flood basalts

The early Cretaceous flood basalts exceed $60,000 \, km^3$ and show a rather limited spatial and temporal variation in primary mineralogy and chemistry (Chávez & Nisterenko 1974; Levi *et al*. 1982; Oyarzún & Frutos 1982). The predominant type is highly porphyritic with unzoned phenocrysts of labradorite (up to 3 cm), clinopyroxene, magnetite and olivine (altered), in a groundmass of the same minerals with minor amounts of K-feldspar-quartz intergrowths. The lavas are K-rich calc-alkaline basalts according to the classification of Irvine & Baragar (1971), though transitional to andesites (Table 1). A calc-alkaline affinity is also shown by their $Zr/Ti/Y$ ratio, although a tholeiitic character might be inferred from their high Fe content and low Mg/Fe ratio (cf. Oyarzún & Frutos 1982). The rare-earth element (REE) pattern of the porphyritic lavas (Fig. 4) indicates an evolved basaltic type. The initial Sr ratio is low (0.7038 ± 0.0004; Fig. 3a).

The early Cretaceous flood basalts have a composition resembling that of lavas of corresponding SiO_2 content occurring in thin continental margins (e.g. central-southern Chile:

TABLE 1. *Compositional range in porphyritic flood basalts of early Cretaceous age*

	Wt%		a	b
SiO_2	48.5–53.5	La/Yb	6.5	8.9
TiO_2	0.8–1.1	La/Y	0.5	0.9
Al_2O_3	15.9–17.7	K/La	1020	814
FeO^{total}	8.1–11.9	P/La	74	49
MnO	0.2–0.3	Th/U	3.6	3.6
MgO	2.0–4.6	Zr/Y	3.8	5.4
CaO	6.7–9.6	Hf/Yb	1.4	1.7
Na_2O	1.8–3.6	Ni/Co	1.1	1.0
K_2O	1.4–3.4	Sc/Cr	0.6	0.6
P_2O_5	0.1–0.4	Sc/Ni	0.8	1.1
	ppm			
Ba	450–680			
Rb	90–180			
Sr	400–530			
Y	31–39			
Zr	125–160			

The ranges are based on analyses of eight representative samples from two profiles, at *c*. 33°S (Levi 1969) and 30°S (Chávez & Nisterenko 1974), central Chile. Also given are some elemental ratios considered by Bailey (1981) to be sensitive discriminators of tectonic setting. a = this paper; b = Bailey's median values for continental island-arc andesites. Trace elements and ratios are based on four to six samples from the profile at 33°S.

Chávez & Nisterenko 1974; López-Escobar *et al*. 1977), and the inner part of continental island arcs (e.g. NW Japan: Katsui *et al*. 1978; see also the compilation by Bailey 1981). The

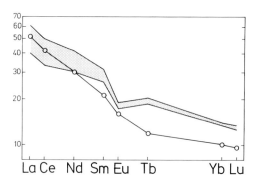

FIG. 4. Chondrite-normalized REE pattern for porphyritic flood basalts of early Cretaceous age, central Chile (shading covers range in samples 324, N410, N417 and N419). Instrumental neutron activation analytical procedures are given in Nyström (in press). The REE pattern of continental island-arc andesites, compiled by Bailey (1981; circles = median values) is given for comparison.

low initial Sr ratio of the flood basalts, their uniformity and large volume indicate a mantle source. Their evolved character might be due to contamination with mantle-derived crustal material. In fact, granitoids and volcanics with low initial Sr ratios (James *et al*. 1974; Zentilli 1974; McNutt *et al*. 1975) constitute the Jurassic 'basement' to the northern continuation of the early Cretaceous flood basalts in Chile.

An aborted marginal basin

It is generally considerd that the formation of rifts and marginal basins is preceded by doming (e.g. Bott 1981; Hsui & Toksöz 1981). The uplift and establishment of continental conditions in central Chile at the end of the Jurassic might be interpreted as due to an upwelling phase prior to rifting. Rifting, block movements and volcanism subsequently led to the development of a volcanic arc within the continental margin, and a basin to the E of the arc.

Upwelling mantle material, whether the result of induced convection (e.g. Hsui & Toksöz 1981), mantle plumes (e.g. Sengor & Burke 1978; Anderson 1981) or other processes, initiated a spreading phase. Partial melting of crustal rocks by heat from the rising mantle material (Hildreth 1981) might account for the predominant dacitic ignimbrites of the Neocomian arc volcanism (Fig. 2). The mantle material, which probably started its ascent several million years earlier (cf. Zorin 1981), eventually yielded magmas which were erupted in the basin as flood basalts, from the beginning of the Hauterivian to Albian time. Despite their large volume, no oceanic lithosphere was formed and the basin failed to evolve into a marginal basin *sensu stricto*. This failure could have been due to *slow* upwelling of the mantle material, which was hot enough to permit considerable assimilation of crustal rocks at depth, thus producing a magma of evolved character. Although this character might alternatively reflect chemical heterogeneity in the upper mantle, it is considered that assimilation and slow spreading below the basin could explain the composition of the flood basalts and the lack of oceanic crust. In addition, a slow rate of cooling would explain why most of the basalts are highly porphyritic with unzoned phenocrysts. The less common aphyric flood basalts (Fig. 2) could represent pulses of more rapidly rising magma and the change to intermediate lavas during Aptian–Albian times might be a consequence of more extensive assimilation, possibly coupled with still slower spreading (cf. Uyeda & Kanamori 1979).

In addition to the back-arc rifting, the emplacement of granitoids in the volcanic arc and along its border with the basin contributed to the extension of the continental margin (Drake *et al*. 1982; Oyarzún & Frutos 1982).

Intracontinental spreading-subsidence and plate subduction

Intracontinental events can be correlated with spreading rates in the ocean. An episode in the SE Pacific, with a spreading rate of 5 cm yr^{-1} during the interval 125–110 Ma (Larson & Pitman 1972), can be correlated with extension in the western margin of South America (Frutos 1981). This interval corresponds to the time of extrusion of the flood basalts (cf. Oyarzún & Frutos 1982). The subsequent (110–85 Ma) folding, intrusion of granitoids and uplift occurred during an episode of rapid spreading in the ocean (18 cm yr^{-1}: Larson & Pitman 1972). The plutonic activity culminated after a folding event (Aguirre *et al*. 1974; Aguirre 1983) which can be correlated with the Oregonian (mid-Cretaceous) orogeny around 100 Ma. However, the studies of Ramberg (1967) demonstrate that the kinds of broad folds seen in the central Andes could be a consequence of magmatic activity and block movements.

Cyclic repetition of alternating intracontinental spreading and rapid sea-floor spreading and plate subduction can account for the geological evolution of central Chile during the Mesozoic and Palaeogene, wherein volcanism, burial metamorphism, folding, intrusion of granitoids and uplift occurred cyclically. For example, basic lavas, similar in mineralogy and chemistry to the flood basalts of the early Cretaceous cycle (though quantitatively less important), and silicic ignimbrites, were erupted in basins formed during the late Cretaceous and Tertiary. However, the duration of each cycle was short (*c*. 40 Ma) and the paucity of good radiometric and palaeontological dates is a serious limitation in precise correlation with oceanic spreading.

Cretaceous marginal basins in western South America

Basin formation with eruption of mantle-derived basalts was an outstanding feature of the western margin of South America during the Cretaceous. Some of the basins developed into marginal basins proper, with the generation of oceanic crust, although most were aborted.

In the southernmost Andes (Magallanes) an ensialic basin opened at the end of the Jurassic and lasted until uplifted and destroyed by mid-Cretaceous times (Dalziel 1981). A pillow lava–sheeted dyke–gabbro complex, probably representing the upper part of an ophiolite, indicates that oceanic crust was generated in this basin. Large volumes of basaltic magmas intruded the continental crust and led to the development of an oceanic spreading centre towards the S of the basin (De Wit & Stern 1981 and references therein; Dalziel 1981). An abortive ensialic basin of similar age was formed in the Southern Andes of Aysén, approximately 500 km N of the Magellan basin (Bartholomew & Tarney, this volume). Here the lavas resemble normal arc magmas with REE patterns showing enrichment in LREE in comparison with the magmas from the oceanic spreading centre at the southernmost end of the Magellan basin (see also Baker et al. 1981).

In the central Andes of W Peru, large volumes of basaltic lava were extruded during the middle Cretaceous. Crustal extension and basin formation thinned the continental crust without generating oceanic lithosphere (Mégard 1978; Cobbing et al. 1981). In northern Peru, pillow lavas and related dyke swarms emplaced along fractures of Andean trend have been reported by Bussell (1983 and personal communication), who interprets this activity as characterizing the early stage of an episode of crustal extension which was followed by an episode of compression. Atherton et al. (1983) conclude that basaltic flows and dykes belonging to the same volcanic group, in N-central Peru, represent primitive mantle-derived material extruded in an ensialic, aborted, marginal basin. From the pattern of burial metamorphism of these volcanics, Aguirre et al. (1978) and Offler et al. (1980) inferred the presence of a steep geothermal gradient during the metamorphism which was due to abnormally large fluxes of heat associated with mantle-derived basic magmas.

In the northern Andes of W Ecuador, a volcanic arc was active from the Aptian to the Eocene. A major geological discontinuity, the Guayaquil–Bucay line, divides the Cretaceous arc rocks of southern Ecuador, extruded in an ensialic environment, from those of northern Ecuador which were built upon a basement at least partly of oceanic character (Henderson 1979). Further N, in the Western Cordillera of Colombia, a sequence of pillow lavas, diabases and associated volcaniclastic material was deposited during the late Cretaceous. According to Barrero (1979), the basaltic rocks of the sequence (low-K tholeiites) were extruded during the initial stage of an evolving island arc, probably formed upon pre-existing oceanic crust.

The various basic volcanic sequences of Cretaceous age outlined above show significant geochemical differences. Lavas of a primitive character and close affinities with ocean-ridge basalts, or island-arc tholeiites, are present in the Northern and Southern Andes, and more evolved basalts with calc-alkaline features are located in the central Andes (central Chile, this paper). Northwards from central Chile a transition to more primitive types is indicated by the early Cretaceous lavas of northern Chile (cf. Zentilli 1974; McNutt et al. 1975; Dostal et al. 1977) and the middle Cretaceous volcanics of Peru (Atherton et al. 1983). Towards the S, the lavas from the Aysén region (Bartholomew & Tarney, this volume) represent types transitional to the primitive Magellan basalts. The possible transition from the 'eugeosynclinal' basins of the central Andes to truly oceanic marginal basins was pointed out by Dalziel (1981).

The metamorphism of the Cretaceous volcanics also varies with the type of basin. It is of burial type in the ensialic basins (cf. Levi 1970; Aguirre et al. 1978; Offler et al. 1980; Levi et al. 1982), and of ocean-floor type in the marginal basins towards the N and S (cf. Dalziel et al. 1974; Stern et al. 1976; Espinosa-Baquero 1980). In central Chile, the passage from rocks of zeolite to greenschist facies requires a stratigraphic depth of the order of several km. The metamorphic gradient is steeper in Peru, and towards the Northern and Southern Andes. This consistent relationship, between the character of the basin, its metamorphic gradient and its associated basalts, suggests a common cause.

The rate and volume of upwelling mantle, and the thickness of the continental crust, where present, are likely to be controlling factors in basin generation processes. A rapid upwelling would lead to back-arc spreading and the formation of a marginal basin with primitive basalts and a steep thermal metamorphic gradient. Slow upwelling would produce basins with more evolved basalts bearing the imprint of a shallow thermal gradient of burial metamorphism. The various Cretaceous basins in western South America, thus, could be produced by variations in the rate of upwelling, which controlled the rate of spreading in relation to subsidence. It is tempting to interpret the development of the Cretaceous basins as the result of a process of lithospheric splitting initiated during that period. The progressive younging of the basins from Magallanes to the Northern Andean volcanic arcs suggests that the process proceeded from S to N.

Appendix

Rb-Sr analytical data: the superscripts I, II, and III indicate samples from the same flows. Mineralogical and chemical compositions of 'unaltered' and altered parts of the flood basalts are given in Levi *et al.* (1982). The Rb-Sr analytical procedure followed conventional routines. The samples were run on AVCO 901 and Finnegan MAT 261 mass spectrometers at the Laboratory for Isotope Geology, Stockholm. The $^{87}Sr/^{86}Sr$ ratios were normalized to $^{86}Sr/^{88}Sr = 0.1194$. The isochron was calculated according to the method of Williamson (1968), with a ^{87}Rb decay constant of $1.42 \times 10^{-11}a^{-1}$ (Steiger & Jäger 1977). Age and initial ratio errors are given at the 1σ level. Error in $^{87}Rb/^{86}Sr = 0.6\%$ (XRF); in $^{87}Sr/^{86}Sr = 0.05\%$ (AVCO 901) and $= 0.01\%$ (MAT 261).

Sample No.	Material	$^{87}Rb/^{86}Sr$	$^{87}Sr/^{86}Sr$	$^{87}Sr/^{86}Sr_0$ (117 Ma)
N417	'Unaltered' basalt	0.5187	0.70446	0.70360
N419	'Unaltered' basalt	0.5198	0.70470	0.70384
303[III]	'Unaltered' basalt	0.6258	0.70523	0.70419
N413	'Unaltered' basalt	1.0246	0.70544	0.70374
324[I]	'Unaltered' basalt	1.0507	0.70565	0.70390
N410[II]	'Unaltered' basalt	1.0680	0.70555	0.70377
				mean = 0.70384
251[I]	Strongly altered basalt	0.5985	0.70501	
250[I]	Strongly altered basalt	1.0979	0.70558	
304[III]	Strongly altered basalt	2.7133	0.70763	
N412[II]	Strongly altered basalt	2.9752	0.70830	
250D[I]	Amygdales	0.0157	0.70399	
251E[I]	Amygdales	0.7072	0.70500	

ACKNOWLEDGMENTS: We are indebted to M. P Atherton, D. S. Bartholomew, F. Hervé, G. Mason, D. Rickard, J. Tarney and F. E. Wickman who kindly reviewed the manuscript and contributed with valuable suggestions. B. P. Kokelaar is thanked for several improvements to the text. The authors greatly benefited from discussions with M. P. Atherton, A. Bussell and V. Warden about different aspects of Andean volcanism, and with J. Oyarzún concerning the genesis of the porphyritic flood basalts of central Chile. A. O. Brunfelt is thanked for providing advice and facilities for neutron activation analysis at the Mineralogical-Geological Museum, Oslo. E. Welin kindly put at our disposal the facilities at the Laboratory for Isotope Geology, Stockholm. This research was made possible thanks to grant No. 82–184 from the Swedish Agency for Research Co-operation with Developing Countries (B.L. and J.O.N.) and to a University of Liverpool Research grant and a Guggenheim Fellowship (L.A.)

References

AGUIRRE, L. 1983. Granitoids in Chile. *In*: RODDICK, J. (ed.) *Granitoids of Circum-Pacific Terrains. Geol. Soc. Am. Memoir*, **159**, 293–316.

—— & EGERT, E. 1965. Cuadrángulo Quebrada Marquesa, provincia de Coquimbo. *Carta geol. Chile*, **15**, 92 pp. Inst. Invest. geol. Santiago.

——, CHARRIER, R., DAVIDSON, J., MPODOZIS, A., RIVANO, S., THIELE, R., TIDY, E., VERGARA, M. & VICENTE, J. C. 1974. Andean magmatism: its paleogeographic and structural setting in the central part (30° to 35°S) of the Southern Andes. *Pacific Geol.* **8**, 1–38.

——, LEVI, B. & OFFLER, R. 1978. Unconformities as mineralogical breaks in the burial metamorphism of the Andes. *Contrib. Mineral. Petrol.* **66**, 361–6.

ANDERSON, D. L. 1981. Hotspots, basalts, and the evolution of the mantle. *Science* **213**, 82–9.

ATHERTON, M. P., PITCHER, W. S. & WARDEN, V. 1983. The Mesozoic marginal basin of central Peru. *Nature, Lond.* **305**, 303–6.

BAILEY, J. C. 1981. Geochemical criteria for a refined tectonic discrimination of orogenic andesites. *Chem. Geol.* **32**, 139–54.

BAKER, P. E., REA, W. J., SKARMETA, J., CAMINOS, R. & REX, D. C. 1981. Igneous history of the Andean cordillera and Patagonia plateau around latitude 46°S. *Phil. Trans. R. Soc. London*, **A303**, 105–49.

BARRERO, D. 1979. Geology of the Central Western Cordillera, west of Buga and Roldanillo, Colombia. *Pub. Geol. Esp. Ingeominas*, **4**, 1–75. Bogotá.

BODVARSON, G. & WALKER, G. P. L. 1964. Crustal drift in Iceland. *Geophys. J. R. astr. Soc.* **8**, 285–300.

BOTT, M. H. P. 1981. Crustal doming and the mechanism of continental rifting. *Tectonophysics*, **73**, 1–8.

BUSSELL, M. A. 1983. Timing of tectonic and magmatic events in the central Andes of Peru. *J. geol. Soc. London*, **140**, 279–86.

CARTER, W. D. & AGUIRRE, L. 1965. Structural geology of Aconcagua province and its relationship to the Central Valley Graben, Chile. *Bull. geol. Soc. Am.* **76**, 651–64.

CHARRIER, R. 1981. Geologie der chilenischen Hauptkordillere Zwischen 34° und 34°30′ südlicher Breite und ihre tektonische, magmatische und paläogeographische Entwicklung. *Berliner geowiss. Abh.*(A) **36**, 270 pp. Berlin.

CHÁVEZ, L. & NISTERENKO, G. 1974. Algunos aspectos de la geoquímica de las andesitas de Chile. *Pub. Dept. Geol. Univ. Chile*, **41**, 97–127.

COBBING, E. J., PITCHER, W. S., WILSON J. J., BALDOCK, J. W., TAYLOR, W. P., McCOURT, W. M. & SNELLING, N. J. 1981. The geology of the Western Cordillera of northern Peru. *Overseas Mem. 5, Inst. geol. Sci.* London, 143 pp.

CORVALÁN, J. 1965. Geologia General. *In: Geografía Económica de Chile*. Corp. Fomento Producción. Santiago, Chile, 35–82.

DALZIEL, I. W. D. 1981. Back-arc extension in the Southern Andes: a review and critical reappraisal. *Phil. Trans. R. Soc. London*, **A300**, 319–35.

——, DE WIT, M. J. & PALMER, K. F. 1974. Fossil marginal basin in the southern Andes. *Nature, Lond.* **250**, 291–4.

DE WIT, M. J. & STERN, C. R. 1981. Variation in the degree of crustal extension during formation of a back-arc basin. *Tectonophysics*, **72**, 229–60.

DOSTAL, J., ZENTILLI, M., CAELLES, J. C. & CLARK, A. H. 1977. Geochemistry and origin of volcanic rocks of the Andes (26°–28°S). *Contrib. Miner. Petrol.* **63**, 113–28.

DRAKE, R., VERGARA, M., MUNIZAGA, F. & VICENTE, J.-C. 1982. Geochronology of Mesozoic–Cenozoic magmatism in central Chile, Lat. 31°–36°S. *Earth Sci. Rev.* **18**, 353–63.

ESPINOSA-BAQUERO, A. 1980. Sur les roches basiques et ultrabasiques du bassin du Patia (Cordillère Occidentale des Andes colombiennes): Etude Géologique et Pétrographique. Thèse, Docteur ès sciences, Université de Genève.

FRUTOS, J. 1981. Andean tectonics as a consequence of sea-floor spreading. *Tectonophysics*, **72**, T21–32.

HENDERSON, W. G. 1979. Cretaceous to Eocene volcanic arc activity in the Andes of northern Ecuador. *J. geol. Soc. London*, **136**, 341–5.

HILDRETH, W. 1981. Gradients in silicic magma chambers: Implications for lithospheric magmatism. *J. geophys. Res.* **86**, 10153–92.

HSUI, A. T. & TOKSÖZ, M. N. 1981. Back-arc spreading: Trench migration, continental pull or induced convection? *Tectonophysics*, **74**, 89–98.

IRVINE, T. N. & BARAGAR, W. R. A. 1971. A guide to the chemical classification of common volcanic rocks. *Can. J. Earth Sci.* **8**, 523–48.

JAMES, D. E., BROOKS, C. & CUYUBAMBA, A. 1974. Strontium isotopic composition and K, Rb, Sr geochemistry of Mesozoic volcanic rocks of the central Andes. *Yearb. Carnegie Inst.* **73**, 970–83.

KATSUI, Y., OBA, Y., ANDO, S., NISHIMURA, S., MASUDA, Y., KURASAWA, H. & FUJIMAKI, H. 1978. Petrochemistry of the Quaternary volcanic rocks of Hokkaido, north Japan. *J. Fac. Sci. Hokkaido Univ.* Ser. IV, **18**, 243–82.

LARSON, R. L. & PITMAN, W. C. 1972. World-wide correlation of Mesozoic magnetic anomalies, and its implications. *Bull. geol. Soc. Am.* **83**, 3645–62. 3645–62.

LEVI, B. 1969. Burial metamorphism of a Cretaceous volcanic sequence west from Santiago, Chile. *Contrib. Miner. Petrol.* **24**, 30–49.

—— 1970. Burial metamorphic episodes in the Andean geosyncline, central Chile. *Geol. Rdsch.* **59**, 994–1013.

—— & AGUIRRE, L. 1981. Ensialic spreading-subsidence in the Mesozoic and Paleogene Andes of central Chile. *J. geol. Soc. London*, **138**, 75–81.

—— & NYSTRÖM, J. O. 1982. Spreading-subsidence and subduction in central Chile: a preliminary geochemical test in Mesozoic–Paleogene volcanic rocks. *III Congreso Geológico Chileno, Concepción*, B28–36.

——, —— & NYSTRÖM, J. O. 1982. Metamorphic gradients in burial metamorphosed vesicular lavas: comparison of basalt and spilite in Cretaceous basic flows from central Chile. *Contrib. Miner. Petrol.* **80**, 49–58.

LÓPEZ-ESCOBAR, L., FREY, F. A. & VERGARA, M. 1977. Andesites and high-alumina basalts from the central-south Chile High Andes: Geochemical evidence bearing on their petrogenesis. *Contrib. Miner. Petrol.* **63**, 199–228.

McNUTT, R. H., CROCKET, J. H., CLARK, A. H., CAELLES, J. C., FARRAR, E., HAYNES, S. J. & ZENTILLI, M. 1975. Initial $^{87}Sr/^{86}Sr$ ratios of plutonic and volcanic rocks of the central Andes between latitudes 26° and 29° south. *Earth planet. Sci. Lett.* **27**, 305–13.

MÉGARD, F. 1978. Étude géologique des Andes du Pérou Central. *Mémoires ORSTOM*, **86**, Paris, 310 pp.

MUÑOZ CRISTI, J. 1968. Evolución geológica del territorio chileno. *Acad. Ciencias Bol.* **1**, 18–26. Santiago, Chile.

NYSTRÖM, J. O. in press. Mobility of rare earth elements in burial metamorphosed vesicular lavas.

OFFLER, R., AGUIRRE, L., LEVI, B. & CHILD, S. 1980. Burial metamorphism in rocks of the Western Andes of Peru. *Lithos*, **13**, 31–42.

OYARZÚN, J. & FRUTOS, J. 1982. Proposición de un modelo para los depósitos Cretácicos de magnetita del Norte de Chile. Discusión de un esquema general para las mineralizaciones ferríferas asociadas al magmatismo calcoalcalino. *V Congreso Latinoamericano de Geología, Buenos Aires*, **3**, 25–39.

PÁLMASON, G. 1973. Kinematics and heat flow in a volcanic rift zone, with application to Iceland. *Geophys. J. R. astr. Soc.* **33**, 451–81.

RAMBERG, H. 1967. *Gravity, Deformation and the Earth's Crust as Studied by Centrifuge Models*. Academic Press, London, 241 pp.

RUIZ, C., AGUIRRE, L., CORVALÁN, J., KLOHN, C., KLOHN, E. & LEVI, B. 1965. *Geología y yacimientos metalíferos de Chile*. Santiago, Chile, Inst. Invest. geol. 305 pp.

SENGOR, A. M. C. & BURKE, K. 1978. Relative timing of rifting and volcanism on earth and its tectonic implications. *Geophys. Res. Lett. Washington*. **5**, 419–21.

STEIGER, R. H. & JÄGER, E. 1977. Convention on the use of decay constants in geo- and cosmochronology. *Earth planet. Sci. Lett.* **36**, 359–62.

STERN, C. R., DE WIT, M. J. & LAWRENCE, J. R. 1976. Igneous and metamorphic processes associated with the formation of Chilean ophiolites and their implications for ocean floor metamorphism, seismic layering, and magnetism. *J. geophys. Res.* **81**, 4370–80.

UYEDA, S. & KANAMORI, H. 1979. Back-arc opening and the mode of subduction. *J. geophys. Res.* **84**, 1049–61.

VAN HINTE, J. E. 1976. A Cretaceous time scale. *Bull. Am. Ass. Petrol. Geol.* **60**, 498–516.

VERGARA, M. 1972. Note on the paleovolcanism in the Andean Geosyncline from the central part of Chile. *24th Int. geol. Congr.* **2**, 222–30.

WILLIAMSON, J. H. 1968. Least-squares fitting of a straight line. *Can. J. Phys.* **46**, 1845–7.

ZEIL, W. 1979. *The Andes. A Geological Review*. Gebrüder Borntraeger, Berlin, 260 pp.

ZENTILLI, M. 1974. Geological evolution and metallogenetic relationships in the Andes of northern Chile between 26° and 29° South. Ph.D. Thesis, Queen's Univ. Kingston, Ontario.

ZORIN, Y. A. 1981. The Baikal rift: an example of the intrusion of asthenospheric material into the lithosphere as the cause of disruption of lithospheric plates. *Tectonophysics*, **73**, 91–104.

G. ÅBERG, B. LEVI & J. O. NYSTRÖM, Geologiska Institutionen, Stockholms Universitet, S-10691, Stockholm, Sweden.

L. AGUIRRE, Department of Geology, University of Liverpool, P.O. Box 147, Liverpool L69 3BX, U.K. Present address: Laboratoire de Pétrologie, Faculté des Sciences et Techniques de Saint Jérome, Université d'Aix Marseille III, 13397 Marseille Cedex 13, France.

Crustal extension in the Southern Andes (45–46°S)

D. S. Bartholomew & J. Tarney

SUMMARY: As a result of back-arc extension along the Andean margin of southern Chile, marginal basins formed during both Mesozoic and Tertiary times. Crustal extension was limited but was accompanied by tholeiitic basalt or bimodal (acid-basic) volcanism rather than the calc-alkaline magmatism normally characteristic of the Andean margin.

The basement in southern Chile is composed of highly deformed metasediments, part of a major accretionary wedge built out from the Pacific margin in late Palaeozoic to early Mesozoic times in front of the late Palaeozoic magmatic arc now situated in Argentina. In the mid-Jurassic the locus of calc-alkaline magmatism moved W to near the present continental margin, and a narrow marine basin developed behind the magmatic arc from the late Jurassic to the early Cretaceous. The formation of the basin was marked by extensive silicic volcanism, probably a product of melting of the metasedimentary basement, followed by bimodal acid-basic volcanism in the early to mid-Cretaceous and the emplacement of a major batholith. Extension was less than in the contemporaneous 'rocas verdes' back-arc basin in southernmost Chile in which new mafic crust was formed. The basin was closed and uplifted during compression in the mid-Cretaceous. However, in the late Cretaceous a new basin, floored by mafic crust, opened within the calc-alkaline arc. Tholeiitic volcanism and plutonism continued in this basin during the Tertiary while alkaline volcanism occurred in the back-arc regions. Compression in the Miocene, accompanied by further calc-alkaline plutonism, closed the basin and resulted in deep levels of the Patagonian batholith being juxtaposed with the volcanic sequences of the basin.

Convergent plate boundaries, though normally regarded as zones of compression with associated high-magnitude seismic activity, can also be zones of extension. This is most frequently manifest in intra-oceanic areas, the best examples being in the western Pacific where a series of marginal basins (Mariana, Parece Vela, Shikoku, Fiji and Lau Basins) have opened up behind relatively young, primitive island arcs by the process of back-arc spreading (see Uyeda & Kanamori 1979; Weissel 1981 and references therein). The primary cause of back-arc extension may depend on several factors, amongst which are the age and density of the subducting plate controlling the dip of the Benioff zone (Molnar & Atwater 1978), and the relative vectors of the subducting and overriding plates permitting 'roll-back' of the hinge of the subducting plate (Dewey 1980). The ultimate expression of back-arc extension is mantle diapirism which splits the volcanic arc, initiates spreading, and is associated with a change from calc-alkaline to tholeiitic magmatism (Tarney et al. 1981). However, Toksöz & Bird (1977) have pointed out that the rifting of mature continental margin lithosphere may be much more difficult than with oceanic lithosphere, due to its greater thickness and strength. Hence marginal basins resulting from back-arc spreading are more common in intra-oceanic environments than at continental margins. Examples of marginal basins formed through rifting of continental margins do exist, such as Bransfield Strait (Barker 1976; Weaver et al. 1979) at the northern end of the Antarctic Peninsula, and the Gulf of California (Dickinson & Snyder 1979; Saunders et al. 1982).

The eastern margin of the Pacific, along the Andes of S America, is very different from that in the western Pacific (see Uyeda & Kanamori 1979; Dewey 1980), and has been regarded as an essentially compressive plate boundary dominated by large calc-alkaline batholiths, intermediate to acid volcanism, and generally shallower-dipping Benioff zones. Although the record of such subduction-related magmatism and tectonism in the Andes extends back to the late Palaeozoic, it is becoming increasingly apparent that extensional phases have also been important in the evolution of the Andean margin, and can perhaps be recognized from the character of the associated volcanism and sedimentation. In this paper we examine the evidence for extensional regimes in a section across the Southern Andes between 45 and 46°S, but first we outline the evidence in the regions to the S and N.

Southernmost Chile

South of 49°S remnants of a small ensialic marginal basin, which opened in the late Jurassic to early Cretaceous, have been recognized by Katz

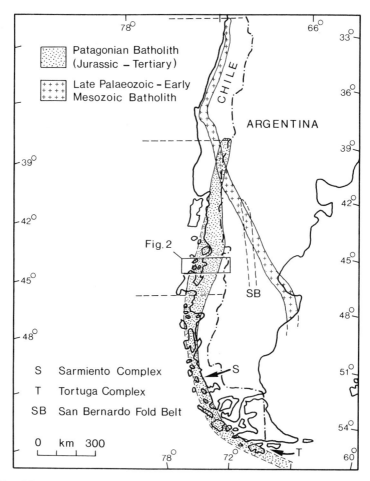

FIG. 1. Simplified geological map of southern South America. The proposed segmentation of the margin is shown by horizontal dashed lines. From N to S the segments are central, south and southernmost Chile. Only the main plutonic belts are indicated; flanking plutons are not shown.

(1973) and Dalziel *et al*. (1974a). Ophiolitic fragments (the 'rocas verdes'), particularly well displayed at Sarmiento and Tortuga (Fig. 1), indicate that crustal rifting took place parallel to the Andean margin with the formation of new oceanic lithosphere within the established volcanic arc. Because of the basin fill of early Cretaceous sedimentary and volcanogenic debris, and the deformation associated with closure of the basin in the mid-Cretaceous, it is difficult to estimate the exact amount of extension that occurred. Dalziel (1981) estimates the original width of the basin as *c*. 50 km at Tortuga in the far S, narrowing northwards to a minimum of 5 km near Sarmiento at 49°S; but how much of the basin floor in the former is new mafic crust and how much is thinned continental crust is unknown. Bruhn *et al*. (1978)

and de Wit & Stern (1981) suggest that mantle diapirism within the volcanic arc thinned and locally ruptured the crust to form the marginal basin. The extensive silicic volcanism which immediately preceded basin opening is interpreted as reflecting widespread melting of the deep crust by the uprising mantle diapir. The mafic rocks of the basin floor (Saunders *et al*. 1979; Stern 1980) are tholeiitic, with trace-element characteristics typical of ocean-floor basalts where the basin was wide, but transitional calc-alkaline characteristics where the basin was narrower. This variation compares closely to that in the lavas of Bransfield Strait (Weaver *et al*. 1979). Clearly, extension occurred at the Andean margin S of 49°S in the late Jurassic, and it is a point of interest whether this was manifest further N.

Central Chile

Levi & Aguirre (1981) and Åberg *et al*. (this volume) have demonstrated the importance of extension in the evolution of the Andean margin in central Chile, and have noted a marked cyclicity of events during the Mesozoic and early Tertiary. Each cycle began with a shift in the locus of volcanic activity, accompanied by accumulation, subsidence and burial metamorphism of thick volcanic sequences, and was terminated by folding, closely followed by emplacement of granitoid magmas, uplift and erosion. The cycle was repeated with further extension, block-faulting and subsidence. Levi & Aguirre (1981) proposed a mechanism of ensialic spreading-subsidence broadly similar to that operating in Iceland.

Crustal evolution in south Chile (45–46°S)

Metasedimentary basement

In the late Palaeozoic a broad accretionary wedge built out from the continental margin in southernmost South America to form the basement into which Mesozoic–Recent magmas were emplaced (Dalziel 1982; Forsythe 1982). Between 45 and 46°S the basement is dominantly composed of metasedimentary rocks which, prior to their burial and deformation, consisted mainly of greywackes and shales. However, the intense deformation associated with the accretionary process has largely destroyed the original bedding. No large-scale folds have been recognized although several late minor fold phases occurred. Deeper levels of the accretionary wedge are characterized by a strong new pervasive foliation, of quartz and mica-rich bands, which dips consistently towards the continent at low to moderate angles.

In the Permian and Triassic the large fore-arc basin, lying between the late Palaeozoic magmatic arc and the trench-slope break of the accretionary wedge, filled with sediments and stabilized to a continental area of low relief (Lesta & Ferello 1972).

Middle Jurassic rifting and volcanism

Intense silicic volcanism began around the Middle Jurassic, its products accumulating throughout South America, as far N as latitude 40°S, to an estimated average thickness of 1 km. The volcanics, generally known as the Tobifera, consist of thick piles of poorly-bedded silicic tuffs and breccias, interbedded with subordinate ignimbrites and lavas of dacitic to rhyolitic composition (Dalziel *et al*. 1974b). There is considerable evidence that regional extensional faulting occurred contemporaneously with eruption, and that the marked variations in thickness resulted from deposition in an area undergoing active subsidence (Bruhn *et al*. 1978). Locally the volcanics are interbedded with shallow-marine sediments of Kimmeridgian–Tithonian age (Natland *et al*. 1974), and available radiometric dates suggest that activity was roughly restricted to the interval 170–150 Ma (e.g. Bruhn *et al*. 1978; Ramos *et al*. 1982).

A pronounced unconformity exists between the Tobifera and the underlying metasedimentary basement in the region roughly coinciding with the present outcrop of the Patagonian batholith (Fig. 1). Close to the E side of the batholith, the silicic volcanics are subordinate to subaerial Middle and Upper Jurassic basaltic to andesitic lavas and tuffs (Ramos 1976; Baker *et al*. 1981). A little further to the E of the batholith, silicic Tobifera volcanics lie conformably on Toarcian sediments (Haller & Lapido 1982). The earliest super-unit of the Patagonian batholith was intruded in the Middle–Upper Jurassic and lies on the W side of the batholith (zone B, Fig. 2). It is of tonalitic to granodioritic composition, and marks the position of the new Mesozoic calc-alkaline magmatic arc. Therefore in the Jurassic at this latitude there was an abrupt westwards shift in the locus of magmatism, from its position in the Triassic within the late Palaeozoic to early Mesozoic batholith (Fig. 1). In Oxfordian times the zone of bimodal Tobifera volcanism lying E of the Jurassic calc-alkaline arc (zone B) subsided to form a back-arc basin (Ramos 1976).

Late Jurassic–Tertiary evolution (45–46°S)

The strata along the E side of the batholith (zone F, Fig. 2) contain the most complete record of the succeeding evolution of the continental margin. The western volcano-sedimentary basin (zone C) and the plutonic complex (zone D) are described in separate sections.

Along the E flank of the Patagonian batholith, volcanism continued intermittently in the late Jurassic (Ramos *et al*. 1982) and marine conditions prevailed until mid-Cretaceous times. Ramos (1976) investigated the portion just to the N of 45°S. There, owing to subsidence, the largely subaerial Middle Jurassic andesitic volcanics became submerged and the

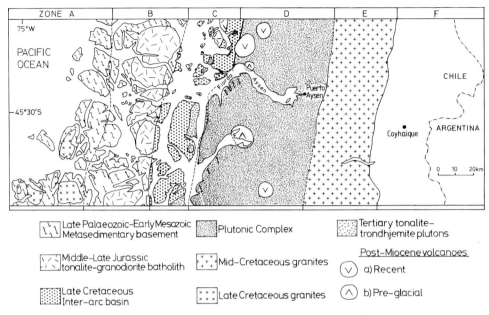

FIG. 2. Geological map of Chile between 45 and 46°S.

upper lavas are intercalated with limestones. These are succeeded by coral reef and bioclastic limestones. The base of the overlying Tithonian succession is marked by conglomerates up to *c.* 16 m thick, composed of small sub-rounded clasts of chert in a calcareous or ferruginous matrix (Ramos 1976; Ramos & Blasco 1978). The abundant chert fragments suggest that chert had been accumulating in a marine environment to the W and that a nearby subaerial volcanic arc had not by this stage become established. The conglomerates are overlain by bioclastic limestones and minor deltaic sandstone-shale-limestone units.

This sequence is conformably overlain by thick black shales, dated on the basis of their microfaunas as Valanginian–Hauterivian (Masiuk & Nakayama 1978). Similar black shales are present in many areas along the E flank of the Andes. Around Coyhaique (Fig. 2), where they are over 400 m thick, they rest unconformably on Jurassic acid volcanics (Skarmeta 1976a). To the W of Coyhaique, close to the batholith, the shales are intercalated with lavas and tuffs of intermediate composition, and to the E, near the border with Argentina, sandstones become abundant in parts of this sequence passing on to the low-lying continental platform of central Patagonia. In the early Cretaceous there was therefore a narrow marine basin with restricted circulation, between a volcanic arc and the continent. This basin was a northward extension of the Magallanes basin of southernmost South America. Ramos *et al.* (1982) described early Cretaceous dyke and sill complexes of basic alkaline rocks intruding the Neocomian black shales between 47 and 49°S. They concluded that this magmatism was the northern equivalent of the marginal basin magmatism of the 'rocas verdes' complex, but that extension had only caused incipient rifting. The lavas preserved along the E margin of the batholith between 45 and 46°S, although silica-saturated, are not calc-alkaline in character. They clearly display enrichment in Ti, Zr and Nb (Fig. 3). This suggests that a limited degree of back-arc extension and magmatism also took place in the early Cretaceous between 45 and 46°S.

In Barremian times, a short-lived andesite-dacite volcanic arc developed in the E part of the marine basin (Ramos 1978). The growth of the arc contributed to the infilling of the marine basin. However the marine regression has also been attributed to uplift caused by the mid-Cretaceous deformation (Skarmeta & Charrier 1976). This deformation, which also affected the continental margin in central and southernmost Chile, caused appreciable uplift in the region where the main body of the Patagonian batholith is now exposed (zone D and E, Fig. 2) and was coincident with the emplacement of the E zone of the batholith. The continental deposits which accumulated along the E side of

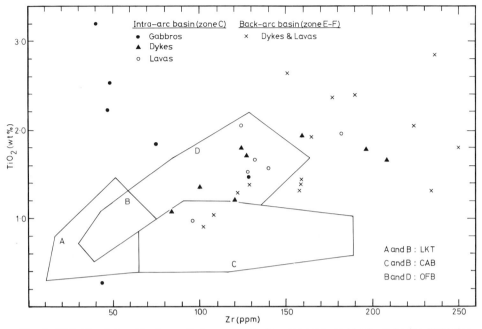

FIG. 3. TiO$_2$-Zr relationships for mafic igneous rocks from the intra- and back-arc basins. Fields for various basalt types after Pearce & Cann (1973). LKT—low-K tholeiite; CAB—calc-alkaline basalt; OFB—ocean-floor basalt.

the batholith at this time are up to 1000 m thick and include lavas of dominantly silicic composition, voluminous ash-flow tuffs and coarse detritus derived from the erosion of volcanic rocks (Baker *et al.* 1981). The entire E part of the batholith (zone E, Fig. 2) is mid-Cretaceous in age (105–80 Ma: e.g. Halpern & Fuenzalida 1978) and consists of intrusive granophyres and granites. The presence of abundant vesicles is consistent with their emplacement as magma chambers at shallow depths, from which thick sequences of ash-flow tuffs were erupted over the terrane to the E. The granites are cut by basaltic dykes of quartz tholeiite and olivine tholeiite composition, similar in chemistry to the Lower Cretaceous lavas along their E flank (Fig. 3). The coarse detritus consists largely of basic and intermediate volcanic material derived from the uplifted terrane to the W. Granite clasts are present in some of the upper beds, indicating that granites (*sensu stricto*) were already being unroofed in the source regions by this stage (Baker *et al.* 1981).

Following the emplacement of the granites, olivine tholeiite plateau basalts were erupted in the late Cretaceous (77 Ma: Baker *et al.* 1981), along the present border with Argentina (zone F, Fig. 2). This essentially back-arc activity was a precursor to the extrusion in this region of greater volumes of tholeiitic and alkali basalts,

from 57 to 43 Ma and 16 to 4 Ma (Charrier *et al.* 1979; Baker *et al.* 1981), which form part of the extensive Patagonian plateau basalt province. There was also a widespread marine incursion into the back-arc regions during the Eocene and early Oligocene, suggesting a period when extensional processes prevailed. Metasediments in the W (zone A) are intruded by a cluster of discordant plutons of late Cretaceous age (*c.* 70 Ma: Halpern & Fuenzalida 1978). The plutons are associated with undeformed granitic sheets and a suite of basic dykes.

In the Miocene considerable uplift of the Cordillera must have occurred, as coarse continental clastic material was deposited along its E flank (zone F) (Skarmeta 1976b). These sediments consist mainly of coarse, cross-bedded sandstones and conglomerates with imbricated clast fabrics, pointing to deposition in a high-energy fluvial environment. The conglomerates contain clasts of metamorphic and granitoid rocks, as well as volcanics, indicating that erosion had reached deeper levels of the Patagonian batholith in the W.

Western basin

The western basin (zone C, Fig. 2) can be regarded as an intra-arc basin, about 25 km wide, between the Jurassic belt of calc-alkaline plutons (zone B) and the plutonic complex (zone D). The history of the basin is poorly known and much of it is presently below sea level. Lavas, tuffs and shales accumulated within it until at least the Miocene and the older strata are mostly buried. However, along the E margin a major eastward-dipping Miocene shear zone marks the boundary between the basinal rocks and higher-grade, foliated rocks of the batholith, and here older rocks within the basin are also exposed.

South of Fiordo Aysen (Fig. 2) sheared basalt and chert sequences, and lesser rhyolites, sediments and gabbros are in tectonic contact with foliated, often gneissic, rocks of the batholith. The volcanics and sediments pass laterally into less deformed sequences and are part of the Mesozoic basin infill, and not of the underlying highly deformed metasediments of the basement accretionary wedge. At the N entrance to Fiordo Aysen there is an 8 km wide slice of gabbros and mafic volcanics, metamorphosed in the amphibolite facies. Furthermore, just N of 45°S, numerous mafic dykes cut rocks forming the E part of the basin. Dyke injection was most intense over an E–W coastal section more than 10 km long, where the dykes intrude medium-grained gabbros and locally constitute up to 40% of the outcrop, giving the appearance of a sheeted-dyke complex.

Fractionation trends of these basic rocks are essentially tholeiitic rather than calc-alkaline. A plot of TiO_2 against Zr (Fig. 3) shows that most dykes and lavas plot in the ocean-floor basalt field of Pearce & Cann (1973). The gabbros, however, show a trend of strong TiO_2 enrichment, indicating that titanomagnetite was a cumulus phase. The magmas must have had an extensive earlier fractionation history, judging by their predominantly quartz-normative nature, low Ni and Cr and relatively high levels of incompatible elements, including REEs (Fig. 4). The rocks are chemically similar to the gabbros, dykes and pillow lavas of the Sarmiento Complex of the marginal basin to the S of 49°S (Fig. 1) (Saunders *et al.* 1979).

On Isla Magdalena (44°30′S), just to the N of the area, Fuenzalida & Etchart (1976) noted a 4000 m thick sequence of mafic pillow lavas interbedded with greywackes, shales and chert of possible late Jurassic to early Cretaceous age. The volcanics most probably belong to the same episode of submarine volcanism as that described here, and the evidence indicates that the basin formed by rifting in the late Cretaceous or early Tertiary. During late Jurassic to early Cretaceous times, when the 'rocas verdes' marginal basin formed to the S, volcanism occurred in a marine basin in the back-arc regions (zones E–F) of the calc-alkaline arc between 45 and 46°S. It is therefore unlikely that a second marginal basin formed on the oceanward side of the arc at this time. However, after the mid-Cretaceous compression, magmatism shifted to the W and, in the Tertiary, arc mag-

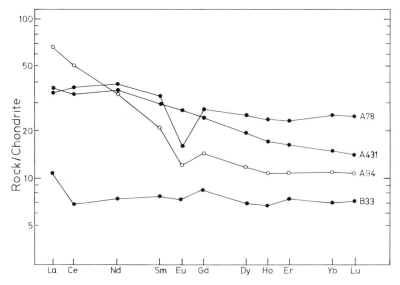

FIG. 4. Chondrite-normalized REE patterns for a dyke (A431), a Tertiary low-K tholeiite lava (B33) and a tonalite (A78) from the intra-arc basin (zone C), and for a typical tonalite (A94) from the plutonic complex (zone D).

matism occurred within, or close to, zone C. It is therefore concluded that the W basin opened at some time in the late Cretaceous. In central Chile it is known that extension occurred in the late Cretaceous, accompanied by fissure eruptions of large volumes of mafic lavas (Levi & Aguirre 1981).

Between the Eocene and Miocene a suite of tonalite-trondhjemite plutons was emplaced within the W basin, in places intruding coeval low-K tholeiitic lavas. Their geochemical characteristics are strikingly different from those of the calc-alkaline plutons of the adjacent batholith. There is a trend towards high Fe/Mg ratios, the concentrations of Zr, Ba and the REEs are high, but the REEs are unfractionated and have marked negative Eu anomalies (Fig. 4). These geochemical features are closely similar to those of Deception Island dacites from the Bransfield Strait marginal basin (Weaver et al. 1979).

Plutonic complex (Patagonian batholith)

The plutonic complex (zone D, Fig. 2), to the E of the Miocene shear zone, represents a deep section of the Patagonian batholith which has been uplifted and exposed by erosion. The most abundant rocks are variably foliated diorites, quartz diorites and tonalites, and the average composition of the complex is rather more basic than that of most circum-Pacific batholiths. The western portion consists in part of basement metasediments containing numerous small intrusions of basic to intermediate composition. These intrusions are commonly garnet-bearing and have melted, and hybridized with, the surrounding metasediments. Also there are several larger gabbro plutons, which have been amphibolitized and disrupted by later tonalite intrusion. However, most of the plutonic complex is composed of numerous small intrusions of tonalite and diorite, commonly intermingled on a scale of tens or hundreds of metres. Distinctive within the complex are the large number of mafic inclusions in the granitoids and numerous amphibolite dykes in all stages of deformation and disruption. Field relationships suggest that the mafic inclusions resulted from repeated injection of sheets of mafic magma into hot, ductile tonalite magma at depth, with the mafic material being disrupted and drawn out by later flow to give a gneissic appearance to the complex. The magmas in the plutonic complex have a clear calc-alkaline signature.

Strain imposed later within the plutonic complex is heterogeneous and is greatest in several shear zones which probably developed contemporaneously with the main bounding Miocene shear zone. Late Miocene K-Ar hornblende dates have been obtained from rocks close to the main shear zone between 42 and 43°S (Herve et al. 1979) and from both foliated and unfoliated tonalite in the 10 km to the E of the shear zone at 45°40'S (R. M. MacIntyre, unpublished data). Some of the complex, therefore, was emplaced in Miocene times. Halpern & Fuenzalida (1978) obtained late Miocene single biotite Rb-Sr ages across the entire width of the complex between 45 and 46°S. However, since rocks from the E of the plutonic complex, near Puerto Aysen, gave a poor Rb-Sr isochron of 107 ± 18 Ma and a cross-cutting pegmatite vein gave a model age of 92 Ma (D.S.B. unpublished data), the biotite ages cannot be ages of granitoid emplacement. Instead it would seem that the E part of the plutonic complex was emplaced at about the same time as the eastern granite (zone E, Fig. 2), and the Rb-Sr biotite systems were either reset in the Miocene compression, or did not close until uplift and cooling at that time.

A line of recent stratovolcanoes and associated smaller volcanic centres, of calc-alkaline basalt to basaltic andesite composition, has developed in the W part of the plutonic complex (Fig. 2). Within the area studied two pre-glacial volcanoes of this group have been deeply dissected to reveal underlying tonalite-diorite plutons.

Crustal evolution (45–46°S): summary and synthesis

The evidence indicates that during the late Palaeozoic and early Mesozoic a broad accretionary wedge developed at the continental margin in southernmost South America, to the W of a calc-alkaline magmatic arc. In the Middle Jurassic rifting of this metasedimentary basement occurred and volcanism was initiated to the W of previous activity. The volcanism was restricted to a linear belt of uplift which broadly coincides with the zone of later Mesozoic and Cainozoic magmatism. Large volumes of silicic volcanics, possibly generated by anatexis of the metasedimentary basement (Bruhn et al. 1978), were erupted along this developing rift and accumulated within subsiding basins to the E. With continuing extension and extraction of crustal melts the rift axis progressively subsided and voluminous basic magmas were extruded. A subduction-related magmatic arc developed on the western side of the rift zone (zone B) in the late Jurassic, and around Oxfordian times the rift zone subsided below sea level and most volcanism probably ceased in the back-arc regions.

In the early Cretaceous calc-alkaline magmatism shifted eastwards to the eastern part of zone D. Extension and subsidence allowed thick shales to accumulate in a narrow back-arc basin but, in contrast to southernmost Chile, generation of new mafic crust never took place. However, bimodal volcanism did occur in the western part of this basin (mainly zone E) during the early to mid-Cretaceous. In the early Cretaceous, basic and intermediate volcanics were erupted in zone E and the extreme western part of zone F. Then mid-Cretaceous compression caused uplift and the destruction of the basin, with considerable silicic volcanism and erosion of earlier volcanics in zone E, accompanied by the emplacement of the eastern, granitic portion of the batholith (Fig. 5). The granite is intruded by late basic dykes. The basic to intermediate magmas are similar in chemistry to the mildly alkaline basalt-trachyandesite suites of continental rift zones. In the Barremian, at about the onset of mid-Cretaceous compression, a short-lived andesite-dacite arc formed in the eastern part of the back-arc basin (zone F).

During the late Cretaceous, extension was re-established at the continental margin. It is considered that the intra-arc marginal basin (zone C) opened at this time, and plateau basalts were erupted in the back-arc regions (far east of zone F). The marginal basin was floored in part by new mafic crust. In the far W (zone A) the local emplacement of a bimodal olivine tholeiite-granite suite took place in the late Cretaceous. Trenchward migration of magmatic activity in the late Cretaceous suggests that a new, steeper subduction zone was established after the mid-Cretaceous compressive event.

Magmas with tholeiitic characteristics, some resembling island-arc tholeiites, were emplaced through the young crust of the marginal basin (zone C) during the Tertiary. In addition, calc-alkaline magmas of Tertiary age were injected as a suite of basic dykes in the eastern part of zone B, and calc-alkaline magmas of basic to intermediate composition were emplaced both in the western part of the plutonic complex (zone D) and within the marginal basin. The eruption of alkali basalts, and a temporary restoration of marine conditions, suggest a return to extensional tectonics in the back-arc regions (zone F) during the Tertiary.

In the Miocene a major compressive event led to thrusting within the calc-alkaline arc (zone D) and underthrusting of the marginal basin (zone C) beneath the crust to the E, causing appreciable uplift in the region of the plutonic complex (zone D), and intense erosion towards the E (Fig. 5). The unroofed complex of foliated calc-alkaline intrusive rocks, with numerous amphibolite dykes, abundant mafic inclusions and relict metasedimentary screens, is interpreted to represent deep levels of the main continental margin magmatic arc.

Since the Miocene, calc-alkaline magmatism has occurred in the western part of the plutonic complex and alkaline activity has continued intermittently in the back-arc regions.

Extension and compression at the Andean continental margin

The Andean continental margin between 45 and 46°S has been subjected to alternating periods of extension and compression, resulting

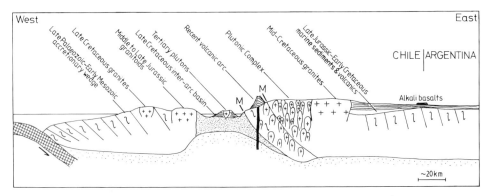

FIG. 5. Schematic E–W section through the continental margin around 45–46°S. (M—Miocene thrusts.)

not only in a complex migration of the loci of magmatism and sedimentation, but also in changes in the character of the associated magmatic activity. The normally dominant calc-alkaline activity of compressional periods was replaced by basaltic or bimodal silicic-basic volcanism as basins developed during extension. The erupted basalts, and the coeval plutons emplaced within the intra-arc basin, have tholeiitic fractionation trends but transitional calc-alkaline trace-element characteristics. These features are similar to those of intra-oceanic arc–marginal basin systems, and in particular to those of ensialic basins, such as Bransfield Strait, Sarmiento and Gulf of California, which have developed at continental margins by rifting and 'back-arc' spreading. To what extent the migration of the locus of igneous activity has been a function of the changing dip of the Benioff zone is not possible to determine, but the subduction geometry would be expected to change after any phase of back-arc mantle diapirism leading to the formation of a back-arc basin. Moreover, closure of such basins, involving a return to normal crustal thickness, could explain the observed juxtaposition of deep-level, high-grade, foliated zones of the batholith and the high-level volcanic rocks of the marginal basin.

The Andean margin of southernmost South America can be divided into three main segments (Ramos *et al.* 1982) that responded differently to the periods of extension and compression: central Chile (33–39°S), S Chile (39–47°S) and southernmost Chile (47–56°S) (Fig. 1). In late Jurassic to early Cretaceous times, while new oceanic lithosphere was formed in a back-arc basin in southernmost Chile, emplacement of oceanic lithosphere in the rifted continental crust was never achieved in S Chile (39–47°S), and instead large volumes of silicic magma were generated beneath the back-arc basin and emplaced as granites along the E side of the batholith. This could account for the much broader outcrop of the Patagonian batholith between 39 and 47°S.

The effects of compression upon the three segments were also different. In southernmost Chile thrusting occurred mainly along continent-verging structures in the mid-Cretaceous and resulted in the uplift of the marginal basin relative to the crust to the E (Nelson *et al.* 1980). Later eastward propagation of similarly oriented thrusts in the Tertiary gave rise to a foreland fold and thrust belt (Winslow 1982). In central Chile (33–39°S), where the Mesozoic to Cainozoic igneous rocks lie to the E of the main late Palaeozoic batholith, the mid-Cretaceous compression first folded the western volcanics and was then transmitted eastwards to produce décollement along Oxfordian evaporite horizons in the eastern basin (Aguirre *et al.* 1974). Miocene compression produced movement between basement blocks in the E, with overthrusting directed towards the continental interior. However, in S Chile (39–47°S) the volcanics and sediments immediately to the E of the batholith, although uplifted, are essentially undeformed. The back-arc basin was destroyed by uplift and magmatism in the mid-Cretaceous, without deformation of the strata to the E and, in the Miocene, thrusting along ocean-verging shear zones uplifted the main calc-alkaline magmatic arc (zone D) relative to the late Cretaceous marginal basin. However, further E, in Argentina, the San Bernardo fold belt (Fig. 1) reflects contemporaneous tectonism (Zambrano & Urien 1970), which suggests cover deformation as a result of movement along ocean-verging thrust faults (Gonzalez 1971).

ACKNOWLEDGMENTS: We are particularly grateful to the Instituto de Investigaciones Geologicas of Chile for help during fieldwork, especially Dr M. Suarez, J. Skarmeta and O. Perez. We thank Dr R. M. MacIntyre for K-Ar dates and Dr A. N. Halliday for aid with Rb-Sr isotopic analysis. These studies were supported by the Natural Environment Research Council. DSB acknowledges receipt of a NERC studentship.

References

AGUIRRE, L., CHARRIER, R., DAVIDSON, J., MPODOZIS, A., RIVANO, S., THIELE, R., TIDY, E., VERGARA, M. & VICENTE, J.-C. 1974. Andean magmatism: its paleogeographic and structural setting in the central part (30° to 35°S) of the Southern Andes. *Pacific Geol.* **8**, 1–38.

BAKER, P. E., REA, W. J., SKARMETA, J., CAMINOS, R. & REX, D. C. 1981. Igneous history of the Andean Cordillera and Patagonian Plateau around latitude 46°S. *Philos. Trans. R. Soc. London*, **A303**, 105–49.

BARKER, P. F. 1976. The tectonic framework of Cenozoic volcanism in the Scotia Sea region. *In*: GONZÁLEZ, O. (ed.) *Proc. Symposium on Andean and Antarctic Volcanology Problems.* IAVCEI Spec. Series, 330–46.

BRUHN, R. L., STERN, C. R. & DE WIT, M. J. 1978. The bearing of new field and geochemical data on the origin and development of a Mesozoic volcano-tectonic rift zone and back-arc basin in southernmost South America. *Earth planet. Sci. Lett.* **41**, 32–46.

CHARRIER, R., LINARES, E., NIEMEYER, H. & SKARMETA, J. 1979. K-Ar ages of basalt flows of the Meseta Buenos Aires in southern Chile and their relation to the southeast Pacific triple junction. *Geology*, **7**, 436–9.

DALZIEL, I. W. D. 1981. Back-arc extension in the southern Andes: a review and critical reappraisal. *Phil. Trans. R. Soc. London*, **A300**, 319–35.

—— 1982. The early (pre-Middle Jurassic) history of the Scotia Arc Region: A review and progress report. *In*: CRADDOCK, C. (ed.) *Antarctic Geoscience*, 111–26. University of Wisconsin Press, Madison.

——, DE WIT, M. J. & PALMER, K. F. 1974a. Fossil marginal basin in the southern Andes. *Nature, Lond.* **250**, 291–4.

——, CAMINOS, R., PALMER, K. F., NULLO, F. & CASANOVA, R. 1974b. South extremity of Andes: Geology of Isla de los Estados, Argentine Tierra del Fuego. *Bull. Am. Ass. Petrol. Geol.* **58**, 2502–12.

DEWEY, J. F. 1980. Episodicity, sequence and style at convergent plate boundaries. *In*: STRANGEWAY, D. W. (ed.) *The Continental Crust and Its Mineral Deposits. Spec. Pap. Geol. Assoc. Can*. **20**, 553–76.

DE WIT, M. J. & STERN, C. R. 1981. Variations in the degree of crustal extension during formation of a back-arc basin. *Tectonophysics*, **72**, 229–60.

DICKINSON, W. R. & SNYDER, W. S. 1979. Geometry of subducted slabs related to San Andreas transform. *J. Geol.* **87**, 609–27.

FORSYTHE, R. D. 1982. The Late Palaeozoic to Early Mesozoic evolution of southern South America: a plate tectonic interpretation. *J. geol. Soc. London*, **139**, 671–82.

FUENZALIDA, R. & ETCHART, H. 1976. Evidencias de migracion volcanica reciente desde la linea de volcanes de la Patagonia Chilena. *In*: GONZALES, O. (ed.) *Proc. Symposium on Andean and Antarctic Volcanology Problems*. IAVCEI Spec. Series, 392–7.

GONZALEZ, R. 1971. Descripcion geologica de la Hoja 49c, 'Sierra San Bernardo', Provincia de Chubut, Republica Argentina. *Dir. Nac. Geol. Mineria*. Bol. No. 112.

HALLER, M. J. & LAPIDO, O. R. 1982. The Jurassic–Cretaceous volcanism in the Septentrional Patagonian Andes. *Earth Sci. Rev.* **18**, 395–410.

HALPERN, M. & FUENZALIDA, R. 1978. Rb-Sr geochronology of a transect of the Chilean Andes between latitudes 45° and 46°S. *Earth planet. Sci. Lett.* **41**, 60–6.

HERVE, F., ARAYA, E., FUENZALIDA, J. & SOLANO, A. 1979. Edades radiometricas y tectonica neogena en el sector costero de Chiloe Continental, Xa region. *Actas Segundo Congreso Geologico Chileno*, F1–18.

KATZ, H. R. 1973. Contrasts in tectonic evolution of orogenic belts in the southwest Pacific. *J. R. Soc. N.Z.* **3**, 333–62.

LESTA, P. & FERELLO, R. 1972. Region Extraandina del Chubut, norte de Santa Cruz. *In*: LEANZA,

A. F. (ed.) *Geologia Regional Argentina*, 601–54. Academia Nacional de Ciencias, Cordoba.

LEVI, B. & AGUIRRE, L. 1981. Ensialic spreading-subsidence in the Mesozoic and Palaeogene Andes of central Chile. *J. geol. Soc. London*, **138**, 75–81.

MASIUK, V. & NAKAYAMA, C. 1978. Sedimentitas marinas mesozoicas del Lago Fontana—Su importancia. *VII Congreso Geologico Argentino*, **2**, 361–78.

MOLNAR, P. & ATWATER, T. 1978. Interarc spreading and cordilleran tectonics as alternates related to the age of subducted oceanic lithosphere. *Earth planet. Sci. Lett.* **41**, 330–40.

NATLAND, M. L., GONZALEZ, E., CANON, A. & ERNST, M. 1974. A system of stages for correlation of Magallanes Basin sediments. *Mem. geol. Soc. Am.* **139**, 1–126.

NELSON, E. P., DALZIEL, I. W. D. & MILNES, A. G. 1980. Structural geology of the Cordillera Darwin collisional-style orogenesis in the southernmost Chilean Andes. *Eclog. geol. Helv.* **73**, 727–51.

PEARCE, J. A. & CANN, J. R. 1973. Tectonic setting of basic volcanic rocks determined using trace element analyses. *Earth planet. Sci. Lett.* **19**, 290–300.

RAMOS, V. A. 1976. Estratigrafia de los Lagos La Plata y Fontana, provincia del Chubut, Argentina. *Actas Primer Congreso Geologico Chileno*, **1**, A43–64.

—— 1978. El vulcanismo del Cretacico inferior de la Cordillera Patagonica. *VII Congreso Geologico Argentino*, **1**, 423–35.

—— & BLASCO, G. 1978. El titoniano de arroyo Pedregoso, Lago Fontana, provincia del Chubut. *Primer Congreso Latinoam. Paleont.* (abstract).

——, NIEMEYER, H., SKARMETA, J. & MUNOZ, J. 1982. Magmatic evolution of the Austral Patagonian Andes. *Earth Sci. Rev.* **18**, 411–43.

SAUNDERS, A. D., FORNARI, D. J., JORON, J.-L., TARNEY, J. & TREUIL, M. 1982. Geochemistry of basic igneous rocks recovered from the Gulf of California: Deep Sea Drilling Project Leg 64. *In*: CURRAY, J. R., MOORE, D. G. *et al.* (eds) *Init. Rep. Deep Sea drill. Proj.* **64**, 595–642. U.S. Government Printing Office, Washington, D.C.

——, TARNEY, J., STERN, C. R. & DALZIEL, I. W. D. 1979. Geochemistry of Mesozoic marginal basin floor igneous rocks from southern Chile. *Bull. geol. Soc. Am.* **90**, 237–58.

SKARMETA, J. 1976a. Evolucion tectonica y paleogeografica de los Andes Patagonicos de Aisen (Chile) durante el neocomiano. *Actas Primer Congreso Geologico Chileno*, **1**, B1–15.

—— 1976b. Estratigrafia del Terciario sedimentario continental de la region central de la Provincia de Aysen, Chile. *Rvta. Asoc. Geol. Argent.* **31**, 73–82.

—— & CHARRIER, R. 1976. Geologia del sector fronterizo de Aysen entre los 45° y 46° de latitud sur, Chile. *VI Congreso Geologico Argentino*, **1**, 267–86.

STERN, C. R. 1980. Geochemistry of Chilean ophiolites: evidence for the compositional evolution of the mantle source of back-arc basin basalts. *J. geophys. Res.* **85**, 955–66.

TARNEY, J., SAUNDERS, A. D., MATTEY, D. P., WOOD, D. A. & MARSH, N. G. 1981. Geochemical aspects of back-arc spreading in the Scotia Sea and Western Pacific. *Phil. Trans. R. Soc. London*, **A300**, 263–85.

TOKSÖZ, M. N. & BIRD, P. 1977. Formation and evolution of marginal basins and continental plateaus. *In*: TALWANI, M. & PITMAN, W. C. (eds) *Island Arcs, Deep Sea Trenches, and Back-Arc Basins*, 379–93. Maurice Ewing Series, Am. Geophys. Union, Washington, D.C.

UYEDA, S. & KANAMORI, H. 1979. Back-arc opening and the mode of subduction. *J. geophys. Res.* **84**, 1049–61.

WEAVER, S. D., SAUNDERS, A. D., PANKHURST, R. J. & TARNEY, J. 1979. A geochemical study of

magmatism associated with the initial stages of back-arc spreading: the Quaternary volcanics of Bransfield Strait, from South Shetland Islands. *Contrib. Miner. Petrol.* **68**, 151–69.

WEISSEL, J. K. 1981. Magnetic lineations in marginal basins of the western Pacific. *Phil. Trans. R. Soc. London*, **A300**, 223–47.

WINSLOW, M. A. 1982. Neotectonics and Cenozoic sedimentation in southernmost South America: Late Tertiary transcurrent faulting and the foreland fold and thrust belt of southern Chile. *In*: CRADDOCK, C. (ed.) *Antarctic Geoscience*, 143–54. University of Wisconsin Press, Madison.

ZAMBRANO, J. J. & URIEN, C. M. 1970. Geological outline of the basins in southern Argentina and their continuation off the Atlantic shore. *J. geophys. Res.* **75**, 1363–96.

D. S. BARTHOLOMEW & J. TARNEY, Deparment of Geology, University of Leicester, Leicester, LE1 7RH, U.K.

Processes of formation and filling of a Mesozoic back-arc basin on the island of South Georgia

B. C. Storey & D. I. M. Macdonald

SUMMARY: South Georgia contains a diverse suite of rocks representing different stages in the development of a back-arc basin on the Gondwana margin during the Mesozoic. Basin formation initially involved intrusion of large gabbroic plutons into deformed paragneisses and calc-alkaline rocks of a magmatic arc, and then extension of this crust by dyke emplacement. Partial melting of the metasediments formed a migmatitic aureole around the basic rocks. Rifting of this thinned continental crust led to the emplacement of mafic crust composed of lavas, dykes and plutonic gabbros with oceanic crustal characteristics. On an upper slope marginal to the arc, the mafic crust is overlain by a thinly bedded sequence of tuffs and mudstones deposited by turbidity currents. Overlying andesitic volcaniclastic rudites, deposited by debris flows, represent a major channel-fill sequence. The basin was filled by thick sequences of both arc-derived volcaniclastic sandstones and silicic detritus with a continental provenance. The volcaniclastic sediments form a fault-controlled aggrading system, deposited by high- and low-density turbidity currents in an elongate basin.

The sub-Antarctic island of South Georgia, located on a microcontinental block 2000 km E of Cape Horn (Fig. 1), is thought to be a continuation of the Andean Cordillera (Dalziel *et al.* 1975; Tanner 1982) moved eastwards by major transform movement during Tertiary times (Barker & Griffiths 1972). South of 50°S an accretionary fore-arc terrane, active in the Triassic and early Jurassic, was replaced by an arc–back-arc basin system in the late Jurassic and Cretaceous (de Wit 1977; Dalziel 1981; Tanner *et al.* 1981).

Geology of South Georgia

Summaries of the geology and stratigraphy of South Georgia and references to earlier work are given by Tanner (1982) and Thomson *et al.* (1982). Four major units record the development of the arc-basin system on the South Georgia block (Fig. 1). These are: (i) extended continental basement rocks of the basin floor (Drygalski Fjord Complex: Storey *et al.* 1977; Storey & Mair 1982; Storey 1983); (ii) mafic rocks of the basin floor (Larsen Harbour Formation: Mair, in press; Storey & Mair 1982); (iii) calc-alkaline volcanic, sedimentary (Annenkov Island Formation: Pettigrew 1981) and intrusive rocks (Tanner *et al.* 1981) of the magmatic arc; and (iv) sedimentary fill of the basin (Cumberland Bay, Ducloz Head, Sandebugten and Cooper Bay Formations: Tanner & Macdonald 1982; Macdonald & Tanner 1983; Storey, in press; Macdonald 1982a).

The island is split by the Cooper Bay disloca-tion zone, a 1 km wide ductile shear zone (Storey, in press) which traverses the South Georgia block (Simpson & Griffiths 1982). Rocks of the basin floor and island arc lie to the SW of this dislocation, and all the basin-fill rocks lie to the NE.

A geochemical and petrographical comparison between the mafic rocks of the Drygalski Fjord Complex and the Larsen Harbour Formation suggest a common origin that can be related to formation of a back-arc basin. The mafic rocks of the Larsen Harbour Formation are conformably overlain by andesitic tuffs of the volcanic-arc sequence. The petrography of the clasts within the basinal sedimentary rocks can be directly correlated with the volcanic arc and with continental rocks which occurred on the opposite side of the basin. This is supported by palaeocurrent evidence (Macdonald & Tanner 1983). The age relationships of the magmatic and sedimentary rocks (Tanner & Rex 1979; Thomson *et al.* 1982) are consistent with the proposed back-arc basin tectonic model. The Annenkov Island Formation and Cumberland Bay Formation are lateral equivalents.

Drygalski Fjord Complex

The complex comprises deformed paragneisses and metamorphosed clastic sediments, intruded by small calc-alkaline granitic plutons of a subduction-related magmatic arc (Storey 1983). A large volume of tholeiitic magma, with mid-ocean ridge basalt (MORB) affinities, was intruded into this sialic crust and represents

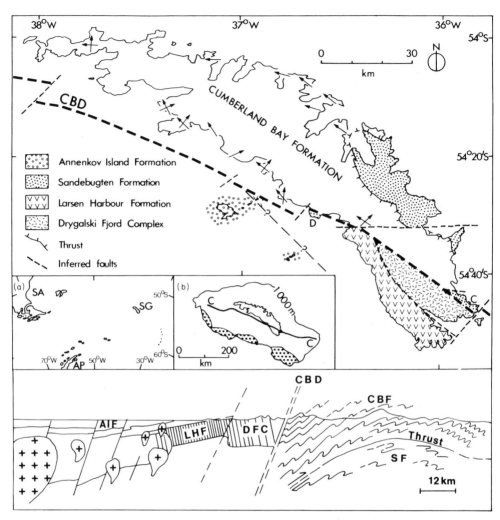

FIG. 1. Location and geology of South Georgia: (a) the position of South Georgia (SG) relative to the Antarctic Peninsula (AP) and South America (SA); (b) the South Georgia microcontinent with the batholith (stipple) and the offshore extension of the Cooper Bay dislocation zone (C–C) (after Simpson & Griffiths 1982). The main figure is a geological map of South Georgia with palaeocurrent directions in the Cumberland Bay Formation indicated by arrows (from Macdonald & Tanner 1983); note the bimodal directions along the SW coast. Correlated with the Sandebugten Formation are the Ducloz Head (D) and Cooper Bay (C) Formations. The diagrammatic cross-section is of the South Georgia microcontinent (after Tanner *et al.* 1981). CBD, Cooper Bay dislocation zone; CBF, Cumberland Bay Formation; SF, Sandebugten Formation; DFC, Drygalski Fjord Complex; LHF, Larsen Harbour Formation; AIF, Annenkov Island Formation.

the early stage of formation of the back-arc basin, in the middle–late Jurassic. Crystallization and fractionation of this magma produced massive and layered gabbros, diorites and plagiogranites within the basement rocks. Continued emplacement of basaltic magma within the gabbros and basement resulted in the development of sheeted-dyke complexes, with up to 80% extension of the continental crust. Emplacement of the basaltic rocks and their

acidic differentiates caused migmatization of the surrounding country rocks. Anatectic melts intruded, brecciated and net-veined the magmatic rocks. Some basaltic dykes, with chilled margins, cut the migmatitic aureole.

Larsen Harbour Formation

The Larsen Harbour Formation is separated from the Drygalski Fjord Complex to the E by a marked topographic break, inferred to be a fault (Storey *et al*. 1977). It comprises a westerly dipping sequence, nearly 2 km thick, of amygdaloidal, massive and pillowed basaltic lavas and lava breccias, thin andesitic and basaltic tuffs and numerous basaltic dyke complexes. The formation is intruded by a large, multiphase, plagiogranite-diorite-layered gabbro pluton, minor gabbroic plutons and dacitic and rhyolitic stocks (Mair, in press). No basement rocks are exposed.

Volcanic-arc terrane

The Annenkov Island Formation consists of a 2–3 km thick gently inclined series of thinly inter-bedded andesite crystal-lithic turbidite tuffs and mudstones (Lower Tuff Member), overlain by poorly bedded volcaniclastic breccia with subordinate sandstone, at least 1 km thick (Upper Breccia Member). From the map of Annenkov Island (Pettigrew 1981), the base of the upper member diverges by 15–20° from the bedding in the lower member; the upper member was probably deposited in a large-scale erosion feature. The sequence is intruded by hornblende and biotite andesites, sparse metabasic sills, and gabbro, monzodiorite and

granodiorite stocks and plutons ranging in age from 100 to 80 Ma (Tanner & Rex 1979). Geophysical evidence suggests that the island-arc terrane stretches to the SW edge of the continental block where a large magnetic batholith (Fig. 1) has been correlated with the Patagonian Batholith (Simpson & Griffiths 1982).

Within the Lower Tuff Member, the turbiditic tuff beds are generally 1–5 cm (max. 40 cm) thick (Fig. 2). Almost all are graded and laterally persistent, and many contain mudstone intraclasts. Sole marks are rare. Parallel lamination is most common, either in T_{be} or T_{abe} types. Cross-lamination is uncommon, occurring as single tabular sets of straight-crested ripples. The beds show no systematic coarsening or thickening trends. The mudstones commonly show faint parallel or cross-lamination and include concentrations of recrystallized radiolaria which may be lag deposits. Slumping, synsedimentary faults, small sedimentary dykes, and load structures at the base of tuff beds are all fairly common.

The Lower Tuff Member is sharply overlain by very coarse sandstone and fine granulestone beds, mostly 0.5–2 m thick (Fig. 2), of the Upper Breccia Member. These basal beds have sharp, slightly undulating bases and are either

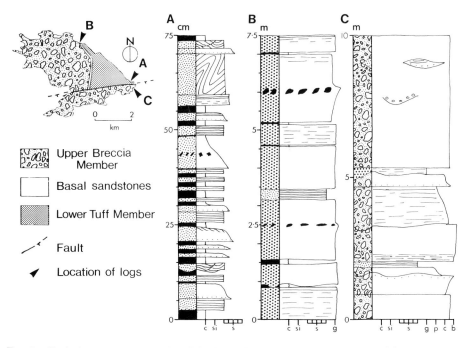

FIG. 2. Geological map of Annenkov Island showing location of graphic logs: (A) Lower Tuff Member (B) Basal sandstones (C) Upper Breccia Member. Sediment grades: c, clay; si, silt; s, sand; g, granule; p, pebble; c, cobble; b, boulder; mudstone, black; sandstone, fine stipple; granulestone, coarse stipple; rudites, large ornament.

amalgamated or separated by thin beds of structureless mudstone. The sandstones are either structureless or have faint parallel lamination, although one bed shows inverse grading near its base. All contain mudstone intraclasts up to 30 cm across and have sharp, flat tops.

The sandstones are overlain by andesitic volcanic breccias (Fig. 2) in beds generally 1–5 m thick. Bed bases are highly irregular, and locally downcut 0.5–1.5 m. Some beds exhibit crude coarse-tail grading. Clasts are angular to subrounded, vary in size from a few cm to 1.5 m and are supported by a matrix of andesitic sandstone or granulestone. In places, beds similar to those of the Lower Tuff Member and basal sandstones are intercalated with the breccias, commonly draping clasts which protrude from the top of the breccia beds. Pettigrew (1981, Fig. 15) noted extensive erosion and dragfolding of mudstones underlying the breccia beds. In a number of localities the breccias include large intraclasts of laminated mudstone and tuff.

Basin-fill formations

Four sedimentary rock formations represent various types of basin fill. All of these were deformed and metamorphosed during closure of the basin in the mid-late Cretaceous (Thomson *et al.* 1982).

Sandebugten and Cooper Bay Formations

These formations consist of similar silicic detritus and are probably equivalent. The Sandebugten Formation has been interpreted as reflecting turbidite accumulation (Dalziel *et al.* 1975; Tanner 1982). Although the complex deformation prevents reliable measurement of palaeocurrent directions, petrographical evidence strongly indicates a NE continental provenance for both formations (Winn 1978; Stone 1980).

Ducloz Head Formation

This formation consists of two members, both unfossiliferous and, possibly due to the proximity of the major ductile shear zone (Cooper Bay dislocation zone) (Storey, in press), both highly deformed. The fault-bound Inland Member is equivalent to the Annenkov Island Formation of the volcanic-arc terrane. The Coastal Member consists of massive epiclastic sub-quartzose sandstones and sandstone breccias, interbedded with thin-bedded fine sandstones, siltstones and black shales and massive, felsitic volcaniclastic breccias. The epiclastic rocks are composed of igneous and metamorphic fragments with intraformational shale clasts. Beds are structureless or crudely graded, up to 10 m thick, with angular to subangular clasts and very poor sorting. The sedimentary rocks indicate derivation from a nearby area of continental basement and silicic volcanic rocks; their association with basic pillow lavas is interpreted as indicating deposition in an active submarine rift. The absence of andesitic detritus suggests that deposition was either before commencement of calc-alkaline arc volcanism or that the site was topographically isolated from the arc. The former is more likely, and these sediments possibly represent the initial fill of the basin.

Cumberland Bay Formation

This formation is a sequence of andesitic volcaniclastic greywackes and shales. Minimum estimates of the thickness of the formation vary from 5 to 8 km (Tanner & Macdonald 1982). The greywacke beds, < 1 cm to 7 m thick, most commonly show T_a, T_{ab} and T_{abc} turbidite sequences, with mixed detritus of up to ten clast types which can be matched petrographically with the rocks of the island-arc terrane (Tanner *et al.* 1981). Such provenance is supported by palaeocurrent evidence (Fig. 1) which shows that although most transport was to the NW along the basin axis, the components were derived from the SW, from many points along the arc (Macdonald & Tanner 1983). Sparse body fossils indicate deposition from (?) uppermost Jurassic or Neocomian to the Albian (Stone & Willey 1973; Thomson *et al.* 1982). In contrast, trace fossils are numerous but belong to a restricted number of ichnogenera, suggesting deposition in a restricted, partially anoxic basin (Macdonald 1982b).

Tanner & Macdonald (1982) recognized three major facies; shale-dominated, transitional and sandstone-dominated. The first two (fine grained) facies are constant in character across the island, however the sandstone facies varies between the SW and NE coasts. In the SW the sandstone facies forms more than 50% of the succession, mean sandstone bed thickness is > 50 cm and beds are commonly thicker than 2 m. On the NE coast, sandstone facies form 30–40% of the successions, mean bed thickness is *c.* 35 cm and few beds exceed 2 m (Macdonald & Tanner 1983). The facies form thickening- and coarsening-upward cycles, as a result of migration of depositional lobes. These data, together with palaeocurrent and palaeoslope indicators, demonstrate that along the SW margin of the basin there were areas of high submarine relief, which were also major sites of coarse clastic deposition (Tanner & Macdonald 1982). Variations from the common turbidite deposits are described below:

Thin, fine beds. Within the two fine sediment facies, 30% of the beds consist solely of cross-laminated sandstone. In some shale-dominated facies, up to 50% of these are clearly not turbidites. They are generally 0.5–5 cm thick, very fine sandstones with undulating bases and tops, in places occurring as discrete lenses with festoon cross-lamination. These sandstones are better-sorted than the others and are ungraded.

Thick, coarse beds. These beds, on the SW coast, vary from 1 to 7 m in thickness and are typically of very coarse, very poorly graded or structureless sandstone. Variation in grading can be related to bed thickness. This ranges from totally structureless beds (up to 7 m) to beds showing distribution grading and a full range of Bouma structures (< 1 m). Between these types, otherwise structureless beds may have a concentration of crudely parallel-laminated coarse to medium sandstone towards the top, or a thin (< 5 cm) graded layer (normal or inverse) at the base. Thick, coarse beds may have scour and toolmarks at the base, although they are rare; the tops of these beds are sharp and flat.

Tuff beds. Tuffs composed of white, coarse prehnite after glass, in planar parallel structureless beds, 5–20 cm thick, occur randomly in the sequence.

Processes of basin formation

Four major processes were involved in the formation of the back-arc basin on South Georgia, which was initiated in the early Jurassic. They are: (i) emplacement of basic plutons; (ii) extension of continental crust by basic dykes; (iii) partial melting of continental crust; and (iv) rifting and formation of mafic crust.

Some of these processes overlapped in space and time. For example, the early gabbroic plutons are cut by extensional dykes, and the anatectic melts both invade pre-existing dykes and gabbros and are cut by basic dykes with chilled margins.

Emplacement of basic plutons

In the southern part of the Drygalski Fjord Complex, large gabbroic plutons were emplaced within deformed paragneisses, metamorphosed clastic sediments and small calc-alkaline intrusions of the magmatic arc. The gabbros are medium and coarse grained, pale-green to dark-grey and black, massive and of variable composition (troctolites, olivine gabbros, hornblende-pyroxene gabbros, 2-pyroxene gabbros, norites, leucogabbros and melagabbros). Many of the larger plutons display rhythmic layering (Fig. 3a) and cumulus textures; many are composite. In one area, ultramafic (lherzolite, peridotite and picrite) inclusions, up to 40 m across, are enclosed within the gabbros. There is marked variation in the strike of the layering, and dips are mostly steeply inclined. Disrupted layering, folding, minor unconformities (Fig. 3a), cross-layering and large (3 m) blocks of layered gabbro in massive gabbro are also present.

The layering can be ascribed to repeated intrusion of magma into a chamber where crystal fractionation takes place (see O'Hara 1977). The irregularities, mentioned above, may indicate the site of a zone of repeated intrusion. The steep dips of the cumulate layers could be due to deformation; disharmonic folding and slumping of the cumulates indicate that deformation and magmatic activity were synchronous. Intrusion was facilitated by magmatic stoping, with blocks of paragneiss floating in the roof zones of the magma chambers.

Extension of continental crust (Drygalski Fjord Complex)

Extensional dykes (Fig. 3b) intrude metasediments, granites and gabbroic rocks. The amount of extension varies markedly throughout the crustal block, from at least 80% at the western margin to isolated basic dykes within the metasediments on the E, over a distance of 6 km. Within smaller areas there are also abrupt changes in the amount of extension.

The dykes have chilled margins, and occur both singly and in multiple units up to 10 m wide. Up to five separate intrusions occur in some multiple units with successive dykes intruding along the centre of the preceding dyke. Dyke trends are mostly bimodal (Fig. 4), with NW and NE orientations. The NE dykes have azimuths ranging from 47 to 73°, and the NW dykes vary from 283 to 329°. The angle between dyke sets varies from 34 to 90°, from subarea to subarea (for details of subareas see Storey 1983). Only in a few cases is there a departure from this pattern. For example, in the northern part of the complex a third (dominant) mode is oriented N–S. The cross-cutting relationships of the dykes are complex and although the NE set is mostly younger than the NW set, variations do occur.

FIG. 3. (a) Irregular layering within gabbros of the southern part of the Drygalski Fjord Complex.
(b) Extension dykes within basement rocks of the Drygalski Fjord Complex. (c) Migmatites within
metasediments on the margin of the basic plutons of the Drygalski Fjord Complex. (d) Dyke
net-veined by anatectic granite within the Drygalski Fjord Complex.

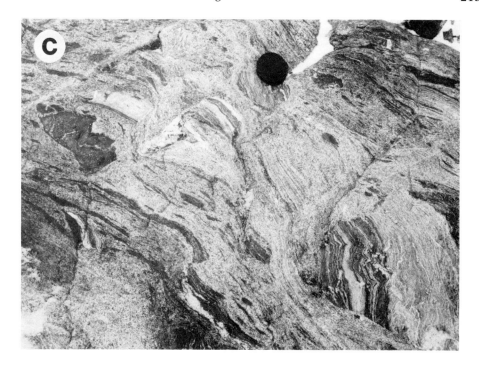

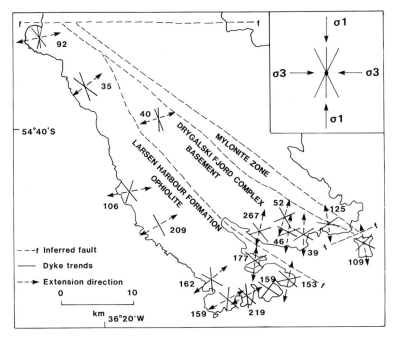

FIG. 4. Dyke trends and extension directions within the basin-floor rocks of the southern part of South Georgia, with number of readings at each locality.

If a dyke trend is related to a given stress system, the Drygalski Fjord Complex, with more than one trend, records changes in the stress field during formation of the back-arc basin, perhaps with early extension in a NE–SW direction and later NW–SE extension. However, as the cross-cutting relationships are not consistent with a single change, it is possible that the trends reflect a conjugate fracture system related to a single stress field (see inset Fig. 4), with the obtuse bisectrix of the dyke modes as the extension direction. If this is the case, the Drygalski Fjord Complex records a change from N–S extension in the SE part to E–W in the NW part. The relative timing of the dyke emplacement in different parts of the complex is unknown and so this change in orientation could be due to a spatial or temporal shift in the stress field, or due to rotation of the block during formation of the basin.

Partial melting of sialic basement

The gabbros and basic dykes suites of the Drygalski Fjord Complex are surrounded by migmatitic aureoles (Fig. 3c). Considerable disruption of the basic bodies occurs, with irregular areas of breccia in which angular to subrounded, partially assimilated basic fragments are set in a granitic matrix. Commonly the basic

dykes have acted as conduits for the migrating fluids, with the fine grained granitic neosome preferentially invading the dykes (Fig. 3d).

There is a marked bimodality in the composition of the acidic rocks within the complex. The migmatites and granitic veins were produced both by partial melting of the metasediments and by migration of acidic fluids formed by differentiation of the basic magma (Storey & Mair 1982).

Rifting and formation of mafic crust (Larsen Harbour Formation)

The main phase in the formation of the back-arc basin is characterized by complete rifting and emplacement of lavas, dykes and gabbros of the Larsen Harbour Formation. In contrast to the Drygalski Fjord Complex, this formation contains no direct evidence of continental crust and has been considered to be part of an ophiolite sequence (Mair, in press) that formed the floor of the basin. The dacites, rhyolites and plagiogranites which are associated with the mafic rocks have trace-element concentrations typical of oceanic plagiogranites (Storey & Mair 1982; Mair, in press). There is some enrichment in the alkali content of the rhyolites, which may indicate that the dykes and lavas were emplaced through thinned continental crust, but this is not conclusive as alkali elements are known to be mobile in hydrothermal sea-floor settings. It is thus suggested that the formation represents a phase of complete rifting of the extended continental crust.

As in the Drygalski Fjord Complex, there is a bi- or tri-modal dyke orientation in each subarea (Fig. 4; see Mair, in press). In the SE there are generally two trends, 070° and 130°, with 130° and 175° trends elsewhere. The age relationships are again conflicting, although the 130° trend is most commonly the earliest. As already discussed, the dyke trends may record major spatial or temporal changes in the extensional direction within the basin, irrespective of whether they are related to a single stress field or to a change in the stress system with time. The inconsistent age relationships of the dykes favour the former hypothesis.

Processes of infilling

Turbidity currents

The most important agents of transport and deposition, both within the basin and on the arc shelf, were low-density turbidity currents (usage of Nardin *et al.* 1979). However, the structureless and poorly graded beds, which are volumetrically significant within the Cumberland Bay Formation on the SW coast, and the sandstones at the base of the Upper Breccia Member of the Annenkov Island Formation, are similar to those described by Walker (1978) and Hein (1982) and ascribed by Lowe (1982) to deposition from high-density turbidity currents. Most of the coarse beds of the Cumberland Bay Formation can be assigned to suspension sedimentation from a sandy high-density turbulent flow (S_3 unit of Lowe 1982). The few inverse-graded bases resemble traction-carpet sedimentation (S_2). The range from ungraded to distribution-graded turbidites (S_3/T_a to T_{a-e}) represent a change from high- to low-density turbulent flows.

Debris flows

Debris flow deposits, although not found within the main basin-fill formations, are very important in the (?) early basin-fill Coastal Member of the Ducloz Head Formation. On the arc terrane debris flows were responsible for rapidly depositing the voluminous Upper Breccia Member, possibly within a large submarine canyon or channel.

Standing ocean currents

Standing ocean currents produced the well-sorted festoon cross-laminated sandstones in the Cumberland Bay Formation and the cross-lamination and radiolarian lag deposits in the mudstones of the Annenkov Island Formation. This reworking was a very minor basin process and the sediments are not large-scale contourite deposits in the sense of Stow & Lovell (1979).

Direct volcanic input

The prehnitized tuff beds represent minor direct volcanic input, either by airfall or pyroclastic flows entering the sea (Sparks *et al.* 1980). These deposits may have been redistributed by mass flows or reworked by standing ocean currents (Winn 1978).

Discussion

During the Jurassic, four processes interacted to extend the continental crust of South Georgia and produce a back-arc basin floored by mixed mafic and continental crust. The emplacement of large volumes of basic magma produced a crustal block transitional between oceanic and continental, and a block of mafic crust with no obvious continental characteristics.

The initial stage in the formation of the basin involved massive upwelling of mantle-derived tholeiitic magma, which is similar to the model proposed by McKenzie (1978) for sedimentary basin development involving upwelling of hot asthenosphere with rapid stretching and thinning of continental lithosphere. Haxby *et al.* (1976) produced a similar model, suggesting that mantle diapirs intrude and replace the lower part of the lithosphere.

As the basin is floored by extended continental crust, the model of McKenzie (1978, Fig. 4) is applicable. Taking extreme estimates for the subsidence time of 25 and 60 Ma, and minimum sediment thicknesses 6–8 km, an extension factor (β) between 4 and ∞ is required, with the most likely value $\beta \geqslant 10$. As these values exceed those of the Drygalski Fjord Complex, where emplacement of basic magma resulted in an overall extension factor of *c*. 1.8 (locally 5), the stretching of the continental crust must have reached a limit after which extension was accommodated by emplacement of new material to form the mafic crust.

Much of the bedding variation in the basin-fill reflects the variety of back-arc environ-

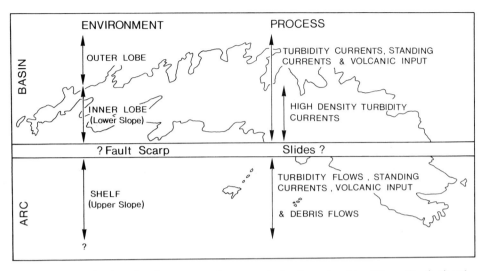

| ENVIRONMENT | PROCESS |

FIG. 5. The distribution of environments and processes in the Annenkov Island Formation (arc) and Cumberland Bay Formation (basin) during deposition.

ments. The Cumberland Bay and Annenkov Island Formations together make up a large volcaniclastic depositional system (Fig. 5) which corresponds to those developed in elongate basins (see Walker 1978).

This depositional system can be divided into the products of four distinct submarine environments (Table 1). These form an aggrading system with essentially vertical accumulation, possibly controlled by a fault at the basin margin.

The abrupt change in facies and thickness between the Cumberland Bay and Annenkov Island Formations, across the Cooper Bay dislocation zone, suggests that the dislocation lies along an earlier fault which controlled sedimentation in the basin. Contemporaneous fault movement would have prevented the upper slope and channel facies (Annenkov Island Formation) from prograding across the mid and lower fan facies (Cumberland Bay Formation). The clasts within the sandstones and tuffs of the Cumberland Bay and Annenkov Island Formations reflect a source which was an area of erosion and mixing of different volcanic rock types, possibly the shelf around volcanic islands on the arc. Here the high erosion rates and rapid palaeogeographical changes (Mitchell 1970; Mitchell & Reading 1971) would have created a supply of mixed volcanic sand on the shelf and randomly shifted major sites of deposition along the arc with time. Sigurdsson *et al.*

TABLE 1. *Summary of lithological units of the volcaniclastic depositional system*

Lithological unit	Environmental interpretation
A. Upper Breccia Member of Annenkov Island Formation	Major channel or canyon fill
B. Lower Tuff Member of Annenkov Island Formation	Upper slope deposits, bypassed by most coarse material
C. SW coastal exposure of Cumberland Bay Formation	Inner part of large coalescing lobes
D. NE coastal exposure of Cumberland Bay Formation	Outer part of large coalescing lobes

(1980) have shown, in the Lesser Antilles, how strong currents between islands can give rise to fans of volcaniclastic sand in a back-arc basin. It could be that due to shifting sources, true fan systems cannot develop in back-arc basins.

ACKNOWLEDGMENTS: We would like to thank our colleagues within the British Antarctic Survey and in particular Drs B. F. Mair and P. W. G. Tanner for discussions during this work. D.I.M.M. carried out this work while employed by the British Antarctic Survey 1975–80, and during tenure of a demonstratorship at the University of Keele.

References

BARKER, P. F. & GRIFFITHS, D. H. 1972. The evolution of the Scotia Ridge and Scotia Sea. *Phil. Trans. R. Soc. London*, **A271**, 151–83.

DALZIEL, I. W. D. 1981. Back-arc extension in the southern Andes: a review and critical reappraisal. *Phil. Trans. R. Soc. London*, **A300**, 319–35.

——, DOTT, R. H., WINN, R. D. & BRUHN, R. L. 1975. Tectonic relations of South Georgia Island to the southernmost Andes. *Bull. geol. Soc. Am.* **86**, 1034–40.

HAXBY, W. F., TURCOTTE, D. L. & BIRD, J. B. 1976. Thermal and mechanical evolution of the Michigan Basin. *Tectonophysics*, **36**, 57–75.

HEIN, F. J. 1982. Depositional mechanisms of deep-sea coarse clastic sediments, Cap Enrage Formation, Quebec. *Can. J. Earth Sci.* **19**, 267–87.

LOWE, D. R. 1982. Sediment gravity flows. II. Depositional models with special reference to the deposits of high density turbidity currents. *J. sediment. Petrol.* **52**, 279–97.

MACDONALD, D. I. M. 1982a. The sedimentology, structure and palaeogeography of the Cumberland Bay Formation, South Georgia. Ph.D. Thesis, University of Cambridge, 155 pp. (unpublished.)

—— 1982b. The palaeontology and ichnology of the Cumberland Bay Formation, South Georgia. *Bull. Br. Antarct. Surv.* **57**, 1–14.

—— & TANNER, P. W. G. 1983. Sediment dispersal patterns in part of a deformed Mesozoic back-arc basin on South Georgia, South Atlantic. *J. sediment. Petrol.* **53**, 83–104.

MAIR, B. F. In press. The geology of South Georgia. VI. Larsen Harbour Formation. *Sci. Rep. Br. Antarct. Surv.*

MCKENZIE, D. P. 1978. Some remarks on the development of sedimentary basins. *Earth planet. Sci. Lett.* **40**, 25–32.

MITCHELL, A. H. G. 1970. Facies of an early Miocene volcanic arc, Malekula Island. *Sedimentology*, **14**, 201–43.

—— & READING, H. G. 1971. Evolution of island arcs. *J. Geol.* **79**, 253–84.

NARDIN, T. R., HEIN, F. J., GORSLINE, D. S. & EDWARDS, B. D. 1979. A review of mass-movement processes, sediment and acoustic characteristics, and contrasts in slope and base-of-slope versus canyon-fan-basin floor systems. *In*: DOYLE, L. J. & PILKEY, O. H. (eds) *Geology of Continental Slopes. Soc. econ. Paleont. Miner. Spec. Pub.* **27**, 61–74.

O'HARA, M. J. 1977. Geochemical evolution during fractional crystallisation of a periodically refilled magma chamber. *Nature, Lond.* **266**, 503–7.

PETTIGREW, T. H. 1981. The geology of Annenkov Island. *Bull. Br. Antarct. Surv.* **53**, 213–54.

SIGURDSSON, H., SPARKS, R. S. J., CAREY, S. N. &

HUANG, T. C. 1980. Volcanogenic sedimentation in the Lesser Antilles arc. *J. Geol.* **88**, 523–40.

SIMPSON, P. & GRIFFITHS, D. H. 1982. The structure of the South Georgia continental block. *In*: CRADDOCK, C. (ed.) *Antarctic Geoscience*, 185–93. University of Wisconsin Press, Madison.

SPARKS, R. S. J., SIGURDSSON, H. & CAREY, S. N. 1980. The entry of pyroclastic flows into the sea. I. Oceanographic and geologic evidence from Dominica, Lesser Antilles. *J. Volcan. geoth. Res. Amsterdam*, **7**, 87–96.

STONE, P. 1980. The geology of South Georgia. IV. Barff Peninsula and Royal Bay areas. *Sci. Rep. Br. Antarct. Surv.* **96**, 45 pp.

—— & WILLEY, L. E. 1973. Belemnite fragments from the Cumberland Bay type sediments of South Georgia. *Bull. Br. Antarct. Surv.* **36**, 129–32.

STOREY, B. C. 1983. The geology of South Georgia. V. Drygalski Fjord Complex. *Sci. Rep. Br. Antarct. Surv.* **107**, 88 pp.

——. In press. The geology of the Ducloz Head area, South Georgia. *Bull. Br. Antarct. Surv.*

—— & MAIR, B. F. 1982. The composite floor of the Cretaceous back-arc basin of South Georgia. *J. geol. Soc. London*, **139**, 729–37.

——, —— & BELL, C. M. 1977. The occurrence of Mesozoic oceanic floor and ancient continental crust on South Georgia. *Geol. Mag.* **114**, 203–8.

STOW, D. A. V. & LOVELL, J. P. B. 1979. Contourites. Their recognition in modern and ancient sediments. *Earth Sci. Rev.* **14**, 251–91.

TANNER, P. W. G. 1982. Geologic evolution of South Georgia. *In*: CRADDOCK, C. (ed.) *Antarctic Geoscience*, 167–76. University of Wisconsin Press, Madison.

—— & MACDONALD, D. I. M. 1982. Models for the deposition and simple shear deformation of a turbidite sequence in the South Georgia portion of the southern Andes back-arc basin. *J. geol. Soc. London*, **139**, 739–54.

—— & REX, D. C. 1979. Timing of events in an early Cretaceous island arc-marginal basin system on South Georgia. *Geol. Mag.* **116**, 167–79.

——, STOREY, B. C. & MACDONALD, D. I. M. 1981. Geology of a Late Jurassic–Early Cretaceous island arc assemblage in Hauge Reef, the Pickersgill Islands and adjoining areas of South Georgia. *Bull. Br. Antarct. Surv.* **53**, 77–117.

THOMSON, M. R. A., TANNER, P. W. G. & REX, D. C. 1982. Fossil and radiometric evidence for ages of deposition of sedimentary sequences on South Georgia. *In*: CRADDOCK, C. (ed.) *Antarctic Geoscience*, 177–84, University of Wisconsin Press, Madison.

WALKER, R. G. 1978. Deep-water sandstone facies and ancient submarine fans: models for exploration for stratigraphic traps. *Bull. Am. Ass. Petrol. Geol.* **62**, 932–66.

WINN, R. D. 1978. Upper Mesozoic flysch of Tierra del Fuego and South Georgia island: a sedimen-tological approach to lithosphere plate recon-struction. *Bull. geol. Soc. Am.* **89**, 533–47.

DE WIT, M. J. 1977. The evolution of the Scotia Sea as a key to the reconstruction of southwestern Gondwanaland. *Tectonophysics*, **37**, 53–81.

B. C. STOREY, British Antarctic Survey, Natural Environment Research Council, Madingley Road, Cambridge CB3 0ET, U.K.

D. I. M. MACDONALD, Sedimentology Branch, BP Exploration Co. Ltd, Britannic House West, London EC2Y 9BU, U.K.

Proximal volcaniclastic sedimentation in a Cretaceous back-arc basin, northern Antarctic Peninsula

G. W. Farquharson, R. D. Hamer & J. R. Ineson

SUMMARY: During Cretaceous times, the northern Antarctic Peninsula was the site of an active ensialic magmatic arc. Volcanism was dominated by pyroclastic eruptions with rare lava flows. Marine conglomerates and sandstones formed a volcaniclastic apron along the eastern margin of the arc and represent the proximal deposits of an extensive back-arc basin. Volcanogenic material, redeposited by turbidity currents and other sediment gravity flows, forms an important part of the proximal basin fill. Air-fall tuffs and eruption-induced sediment flows form a small but significant part of the succession and large exotic slide-blocks of Jurassic sediment are a distinctive feature of the Lower Cretaceous strata on James Ross Island. Aeromagnetic data and regional geology indicate that the arc-back-arc basin boundary was fault-controlled. Sedimentation within the basin was strongly influenced by both the steep, unstable nature of the faulted arc flanks and the coeval volcanism.

From at least early Mesozoic through to late Tertiary times, the Antarctic Peninsula lay to the E of an active trench related to easterly subduction of the Pacific plate beneath Gondwana. An associated magmatic arc, typical of convergent continent–ocean plate boundaries and bearing a close similarity to the western cordillera of southern South America (Saunders & Tarney 1982), developed over the area now occupied by the Antarctic Peninsula. Volcaniclastic strata (Jurassic–Tertiary) occur to the E of this arc, on the margins of the Weddell Sea (Fig. 1), and represent an apron of arc-derived debris that accumulated along the edge of an ensialic back-arc basin. This paper considers the Lower Cretaceous rocks of the

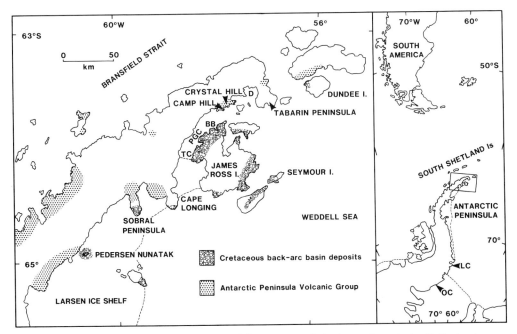

FIG. 1. Location map showing the outcrop of Cretaceous rocks in the northern Antarctic Peninsula. BB: Brandy Bay, D: Duse Bay, PGC: Prince Gustav Channel, TC: Tumbledown Cliffs. Inset shows position of study area relative to South America and the Antarctic Peninsula. LC: Lassiter Coast, OC: Orville Coast.

219

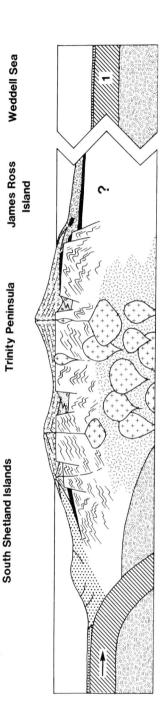

South Shetland Islands

Trinity Peninsula

James Ross Island

Weddell Sea

Back-arc sedimentary rocks, including large slide blocks of Nordenskjöld Formation

Alluvial sedimentary rocks (Farquharson 1982b)

Late Jurassic mudstone–tuff sequences of Trinity Peninsula (Nordenskjöld Formation, Farquharson 1982a) and South Shetland Islands (mudstone member, Smellie et al. 1980)

Subduction zone complex (Scotia metamorphic complex, Tanner et al. 1982) and ?fore-arc terrane

?Carboniferous–Triassic deformed metagreywackes of Trinity Peninsula (Trinity Peninsula Group, Hyden & Tanner 1981) and South Shetland Islands (Miers Bluff Formation, Smellie 1981)

?Pre-Mesozoic basement (Hamer & Moyes 1982, Pankhurst 1983)

Plutonic rocks (Saunders & Tarney 1982)

Volcanic rocks (Hamer 1983)

Oceanic crust with a thin capping of sedimentary rocks

Mantle

1   At present limited marine geophysical observations indicate oceanic crust (pre–late Jurassic to early Cretaceous) in the eastern Weddell Sea (LaBreque & Barker 1981)

Fig. 2. Interpretive section across the northern Antarctic Peninsula in early Cretaceous times. Bransfield Strait did not open until the late Cenozoic.

back-arc basin succession exposed in the northern Antarctic Peninsula. The deposits are described and the tectonic and volcanic influences on their accumulation are discussed.

Stratigraphy and tectonic setting

The Mesozoic and Tertiary geology of the northern Antarctic Peninsula and South Shetland Islands is dominated by calc-alkaline plutonic and volcanic rocks. They intrude or rest unconformably on metagreywackes of the (?) Carboniferous–Triassic Trinity Peninsula Group (Hyden & Tanner 1981) and Miers Bluff Formation (Smellie 1981) (Fig. 2) that were deformed during the early Mesozoic. Along the E coast of the Antarctic Peninsula, however, scattered outcrops of migmatitic gneiss constitute an older crystalline basement representing the leading edge of the Gondwana massif (Pankhurst 1983).

Evolution of the northern Antarctic Peninsula during the late Mesozoic can be divided into two distinct phases of volcanic arc development. In late Jurassic times much of the area was covered by an epeiric sea in which radiolaria-rich mudstones accumulated (Nordenskjöld Formation: Farquharson 1982a). Volcanism regularly contributed air-fall tuffs to the marine succession, although the absence of epiclastic detritus suggests that at this time the magmatic arc lacked appreciable subaerial relief. Uplift, possibly in response to increased magmatic activity (Pankhurst 1982), led to emergence of the arc during the earliest Cretaceous (Fig. 2). This phase is represented by the development of local fault-bounded alluvial basins (Farquharson 1982b) and the onset of widespread terrestrial calc-alkaline volcanism (Hamer 1983). Contemporary with emergence was the initiation of an ensialic back-arc basin within which an extensive and thick (≈ 5 km) pile of arc-derived debris accumulated. To the rear of the arc volcaniclastic deposition continued from early Cretaceous (Hauterivian/Barremian) through to Eocene/(?)Oligocene times (Farquharson 1982a). The widespread arc volcanism was dominated by explosive eruptions of intermediate to acidic pyroclastic material, except in the South Shetland Islands (Fig. 1), closer to the trench, where pyroclastic rocks are subordinate to basic lavas (Hamer 1983).

An extensional tectonic regime, associated with the gradual cessation of subduction, characterized the final stages in the arc's development (Barker 1982). Expressions of this occur in the small marginal basin of Bransfield Strait (Fig. 1), which opened 1.3 Ma ago (Barker 1982), and the late Cenozoic alkali basalts (James Ross Island Volcanic Group: Nelson 1975) that unconformably overlie the volcaniclastic sedimentary rocks in the back-arc basin.

Basin morphology

Aeromagnetic data together with outcrop geology delineate the boundary between the late Mesozoic arc and back-arc basin. The majority of magnetic profiles perpendicular to the axis of the peninsula (Fig. 3) can be divided into two distinct sections (British Antarctic Survey, unpublished aeromagnetic data). Over the peninsula the magnetic signature comprises numerous high-amplitude short-wavelength anomalies, reflecting the basic igneous component of the arc terrane. In contrast, to the E of the peninsula the magnetic profiles are either flat or have very low gradients (Fig. 3), indicative of substantial thicknesses of sediment (Renner *et al.* 1982). Localized short-wavelength, high-amplitude anomalies to the E of the peninsula coincide with occurrences of the James Ross Island Volcanic Group (see profiles marked with a star, Fig. 3). The change from 'quiet' to 'noisy' magnetic signature generally coincides with a line drawn between outcrops of the Cretaceous marine sedimentary rocks and arc terrane rocks (plutons, volcanic rocks and associated sedimentary rocks); it defines the western edge of the preserved back-arc basin (Fig. 3).

Several lines of evidence suggest that continental crust underlies at least the western edge of the basin. Firstly, the chemistry of garnet-bearing xenoliths in (?) early Cretaceous arc terrane volcanic rocks indicates crustal thicknesses of approximately 25 km adjacent to the basin margin (Hamer & Moyes 1982). Furthermore, the aeromagnetic profiles militate against an abrupt termination of the basement at the basin margin (R. G. B. Renner, personal communication). In addition, the James Ross Island Volcanic Group, which was formed by extrusion through the basin sequence, has chemical affinities with intra-plate alkali basalts generated in sub-continental mantle (Saunders & Tarney 1982).

The basin extends southward beneath the Larsen Ice Shelf (Renner *et al.* 1982) and can be extrapolated as far S as the mainly Upper Jurassic marine sedimentary rocks of the Orville and Lassiter Coasts (Laudon *et al.*

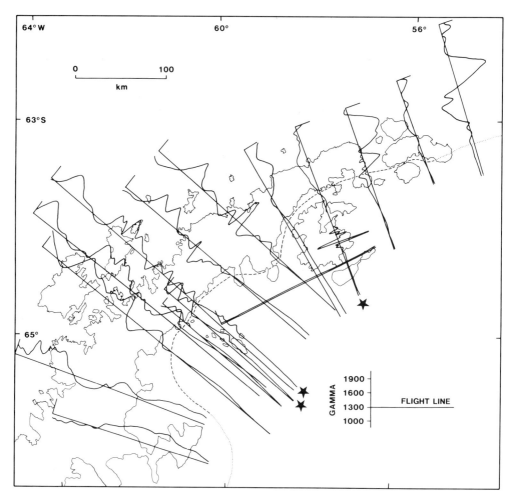

FIG. 3. Outline of the northern Antarctic Peninsula with aeromagnetic flight lines and profiles superimposed (British Antarctic Survey, unpublished data). Dashed/dotted line marks the inferred boundary between the Cretaceous arc and back-arc basin based on aeromagnetic data and outcrop geology (Fig. 1). Stars denote profiles with anomalies due to the James Ross Island Volcanic Group.

1983) (Fig. 1). The 5 km of presumed Cretaceous and Tertiary deposits, detected by seismic refraction and magnetic surveys to the S of the South Orkney Islands (Harrington *et al.* 1972), probably are a northward continuation of the basin. Thus the exposed basin in the northern Antarctic Peninsula may represent an embayment of a more extensive, possibly continous, back-arc system, the lateral extent of which is undefined. Its eastern margin may either lie further out on the continental shelf, or more likely was disrupted by the opening of the Weddell Sea and the concomitant dislocation of the Antarctic Peninsula from the Gondwana massif.

Basin processes

Although the basin was a site of deposition from early Cretaceous times through to at least the Eocene, the facies examples cited here are mainly drawn from Lower Cretaceous strata. Thus, they are penecontemporaneous with preserved arc-terrane deposits (Antarctic Peninsula Volcanic Group) and were deposited when the basin margin was actively fault-controlled.

Sobral Peninsula

On the Sobral Peninsula (Fig. 1), 750–1000 m of predominantly conglomeratic strata are exposed. Some beds have yielded dinoflagellates and calcareous nannofossils which indicate a late Hauterivian/Barremian age and a marine depositional setting (Farquharson 1982a). The succession comprises monotonous thick-bedded red-brown conglomerates with sparse units, up to 20 m thick, of well-bedded, graded, feldspar-rich conglomerates (Fig. 4a). These two rock associations are the result of contrasting sedimentary processes.

Thick-bedded conglomerates

The thick-bedded conglomerates form > 95% of the succession and are invariably clast-supported. The clasts are poorly sorted, very well rounded, up to 3 m across although they are generally 5–30 cm across, and are derived from the Trinity Peninsula Group and Antarctic Peninsula Volcanic Group. Bed thickness ranges from 0.5 to 17 m. Many of the conglomerates possess erosive basal contacts, commonly having considerable relief (up to 5 m) and inclination (up to 40°). The conglomerates typically show large-scale cross-stratification (Fig. 4b) with planar foresets 2–3 m high defined by slight variations in clast size and in places by drapes of sandstone.

Both thinning- and fining-upward, and thickening- and coarsening-upward sequences occur. The former (10–20 m thick) show erosional basal contacts overlain by conglomerate, up to 15 m thick, which is in turn overlain by progressively thinner and finer grained conglomerates and finally capped by alternating lenses of conglomerate and coarse-grained sandstone. These sequences are the result of channel aggradation and abandonment. The thickening- and coarsening-upward sequences, between 20 and 40 m thick, comprise planar stratified coarse-grained sandstone with trails of isolated clasts, overlain by pebble conglomerates with thin very coarse-grained sandstone interbeds. Cross-bedded cobble conglomerates with erosional basal contacts cap the sequence and become thicker bedded and coarser grained upwards. These sequences reflect the gradual approach of a channel, and the progressive thickening and coarsening of the channel-fill is ascribed to the plugging of a channel elsewhere. Vertical sequences within the thick-bedded conglomerates therefore indicate deposition in a system dominated by braided channels. Such channels have been described from the mid-fan region of modern submarine fans (Haner 1971) and interpreted from ancient submarine fan sequences (Walker 1978). However, there appears to be no account of a modern submarine gravelly braided system, although other workers on ancient marine conglomerates envisage their existence (Davies & Walker 1974; Johnson & Walker 1979; Hein & Walker 1982).

The type of flow responsible for the thick-bedded, cross-stratified conglomerates is problematical. Winn & Dott (1977) described large-scale traction-produced structures from Cretaceous conglomerates of the southern Andes and proposed deposition by turbidity currents. Lowe (1982) also concluded that they reflect traction sedimentation beneath gravelly high-density turbidity currents. However, the beds described here lack features associated with such sediment gravity flow deposits, which are a basal, inversely graded, traction carpet layer overlain by a normally graded suspension sedimentation layer (R_2 and R_3 of Lowe 1982). Possibly deposition occurred beneath a low-density fluid gravity flow of sufficient competence to entrain a basal bed load of pebbles and cobbles that progressed by rolling and saltation.

Graded, feldspathic conglomerates

Sequences of graded, feldspathic conglomerates form distinct pale bands within the thick-bedded conglomerates (Fig. 4a). Bed thickness is generally 80–120 cm and maximum clast size is 14 cm. Bases are planar, sharp and lack sole marks. The clasts are sub-rounded to rounded and consist largely of meta-siltstones and meta-sandstones derived from the Trinity Peninsula Group, but also include lithologies from the Antarctic Peninsula Volcanic Group. These conglomerates can be classified into four main types:

(i) *Inverse-to-normally graded, stratified* (Fig. 4c). Such beds are characterized by a basal inversely graded layer overlain by a normally graded layer that passes up into medium- to very coarse-grained sandstone. The sandstone displays diffuse planar stratification and is commonly normally graded (distribution type). Clast imbrication, with long axes parallel to flow, is a ubiquitous feature. Some beds are overlain by sharp-based planar laminated fine- to coarse-grained sandstone, up to 20 cm thick (Fig. 4c). Isolated clasts, up to 5 cm across, thin inverse-to-normally and normally graded layers, and ripple cross-lamination occur in places within these overlying beds.

These beds comprise R_2, R_3 and S_1 divisions of Lowe (1982). The conglomeratic parts closely resemble density-modified grain flows, comprising gravel supported predominantly by intergranular dispersive pressure, beneath a high-density turbidity current (Sanders 1965; Davies & Walker 1974; Lowe 1976; Mullins & Van Buren 1979; Nardin *et al.* 1979). The basal gravel-rich part of several beds contain two

FIG. 4. Field photographs of Lower Cretaceous back-arc basin deposits. Scale interval in (B), (C) and (D) is 10 cm. (A) Clast-supported, red-brown conglomerates overlying a sequence of graded, feldspathic conglomerates, Sobral Peninsula. Scale provided by figure lower centre. (B) Cross-stratified, red-brown conglomerates with erosive base, Sobral Peninsula. (C) Inverse-to-normally graded and stratified conglomerate (type i) bed, Sobral Peninsula. (D) Breccias interbedded with silty clays (dark) and thin graded sandstone beds (light), NW James Ross Island.

inverse-to-normally graded cycles (R_2, R_3, R_2, R_3 and S_1 sequences). Surging flows (Lowe 1982) or closely related pulses of sediment input (Hendry 1973) could produce such features.

(ii) *Inverse-to-normally graded.* These beds are essentially the same as those described above but lack stratification and distinct planar-laminated sandstone caps. They comprise R_2, R_3 and S_3 divisions (Lowe 1982) and are the product of high-density turbidity currents with a basal layer of pebbles and cobbles supported by dispersive pressure.

(iii) *Normally graded and stratified.* In this group a basal, normally graded layer of pebbles (up to 6 cm across) fines upwards into very coarse- to medium-grained sandstone. The pebble-rich units may be clast- or matrix-supported and imbrication is rare. The sandy upper parts of the beds are planar-stratified and in a couple of beds contain sparse ripple cross-lamination. Bed thickness is generally between 10 and 50 cm. Sharp-based planar-stratified sandstones, similar to those above type i beds, occasionally cap type iii units. The conglomerate beds comprise R_3 and S_1 divisions, typical of deposition by gravelly high-density turbidity currents (Lowe 1982).

(iv) *Non-graded.* These conglomerates lack grading, stratification and imbrication. Largest clast size ranges from 1 to 12 cm and matrix percentage varies between 5 and 95%. Particle support by matrix strength, particularly in the matrix-supported beds, was probably a major cause of the lack of an organized fabric and grading. In the clast-rich beds dispersive pressure contributed towards the support of particles. These disorganized conglomerates are interpreted as deposits from sandy debris flows transitional to modified grain flows (cf. Surlyk 1978).

Normally graded sandstones, 10 cm thick, with stratified upper parts are locally interbedded with the conglomerates. These S_3 T_b sequences are the products of low density turbidity currents (Lowe 1982).

The matrix and sandstone partings of the graded conglomerates are characterized by a large (up to 67%) modal content of euhedral plagioclase feldspar with lesser embayed, clear quartz crystals and clinopyroxene. This petrography implies an origin either as resedimented unconsolidated tuffs or as direct eruption-related flows. Thick sequences of (?)early Cretaceous welded ash-flows tuffs (Pankhurst 1982) accumulated on the arc, including areas adjacent to the back-arc basin margin, and it is possible that some ignimbrites entered the basin. Sparks *et al.* (1980) have shown that

dense pyroclastic flows can move underwater but no ignimbrites have yet been recognized in the Cretaceous marine succession of the northern Antarctic Peninsula. However, the graded, feldspathic conglomerates (particularly types i and iii) are remarkably similar to the deposits of subaerial sediment flows associated with the 19 March 1982 eruption of Mount St Helens (Harrison & Fritz 1982). These deposits result from single catastrophic events and are restricted to certain river valleys. They are divisible into three zones: (i) a basal zone up to 1 m thick, of massive or normally graded, clast-supported conglomerate; (ii) a granule sandstone layer with ubiquitous horizontal stratification, generally the same thickness as the lower zone; and (iii) a layer of boulders and logs on the top surface of the flow. It is conceivable that should such flows reach the sea, incorporation of water would transform them into gravelly high-density turbidity currents capable of generating the range of deposits represented by the graded feldspathic conglomerates. The cyclical arrangement of these conglomerates on the Sobral Peninsula may thus be the result of repeated eruption-induced sediment flow events.

Subaqueous pyroclastic-flow deposits have been reported from ancient sequences in Washington (Fiske 1963), Japan (Fiske & Matsuda 1964) and Quebec (Tassé *et al.* 1978). In these occurrences inversely and/or normally graded conglomerates are overlain gradationally by horizontally stratified sands. The flow deposits described by Fiske (1963) and Fiske & Matsuda (1964) are sharply overlain by laminated or thin-bedded finer grained tuffs which they interpret as the deposits of turbidity currents fed from ash settling slowly from the eruption cloud. A similar mechanism is envisaged to account for the sharply based, distinctly finer-grained sandstones that overlie types i and iii graded conglomerates. This is supported by a positive correlation between bed thicknesses of the sandstone and conglomeratic units which implies that they are genetically related.

James Ross Island

The Lower Cretaceous (Aptian/Albian) sequence of James Ross Island (Fig. 1), approximately 1500 m thick, comprises an alternation of units dominated by conglomerates and pebbly sandstones with units of mudstones, siltstones and thin sandstones. The conglomeratic units comprise clast- and matrix-supported, poorly sorted fine pebble to boulder conglomerates, graded pebbly sandstones and medium- to coarse-grained sandstones with minor interbedded siltstones and mudstones. Inverse-to-normal and normal grading, and imbrication, are common features of the conglomerates, which were probably deposited by gravity flows similar to those interpreted for the graded feldspathic conglomerates of the Sobral Peninsula. These coarse units alternate with, and grade into, nodular silty clays, weakly lithified bioturbated mudstones and siltstones, thin, graded, coarse- to fine-grained sandstones and rare conglomerates.

Breccia beds

Breccia beds (Fig. 4d) characterize one mudstone-dominated interval of early Cretaceous age. These occur sporadically throughout the 200–250 m thick interval. Beds vary in maximum thickness between 0.2 and 7 m and amalgamation is common. They are sheet-like or broadly lenticular in form, generally with planar bounding surfaces, although some beds have hummocky upper surfaces, particularly where large clasts protrude above the top of the bed.

Internally the breccias are unstratified (Fig. 4d) and are composed mainly of tabular and equant, angular or subangular clasts of Upper Jurassic Nordenskjöld Formation (Farquharson 1982a). Clasts range in size from < 1 cm to blocks and slabs with dimensions up to 5 by 6 m. The breccias are typically very poorly sorted and clast-supported, with a variable sandy or silty clay matrix that forms between 5 and 20% of the beds. Generally the breccias are ungraded though about 30% of beds display abrupt normal grading at their top into a bed, 3–10 cm thick, comprising pebbles supported in a silty clay or sand matrix. Inverse grading, restricted to a matrix-supported basal layer, 5–20 cm thick, was observed in less than 10% of the breccia beds.

A positive correlation between maximum clast size and bed thickness of the breccias suggests that the beds were deposited from single mass flows. Poor sorting, pervasive fine-grained matrix, lack of stratification and chaotic, disorganized fabric are features characteristic of debris flow deposits (Johnson 1970). Using the terminology of Lowe (1979), the breccia beds described here are the result of flows transitional between cohesive debris flows and grain flows. Their clast-rich nature and the sparse development of a basal inversely graded layer suggest that dispersive pressure played a significant role supporting clasts during flow.

Slide-blocks

Coarse debris derived from the Upper Jurassic Nordenskjöld Formation forms a significant proportion of the Lower Cretaceous strata exposed in NW James Ross Island and includes distinctive large slide-blocks embedded within conglomerates. One block, 3 km SW of Brandy Bay (Fig. 1), is approximately 200 m thick and at least 800 m long. Graded tuff beds indicate that the block is the right way up. Bedding in the block is only locally distorted and is approximately concordant with the surrounding Lower Cretaceous strata. Two similar blocks occur in the Tumbledown Cliffs area (Fig. 1). The larger, 220 m long and 26 m thick, displays brecciation and brittle folds at its base indicating a transport direction towards the SSE. These isolated exotic blocks of Nordenskjöld Formation are interpreted as having been transported to their present position by submarine sliding.

Volcanogenic deposits

Air-fall tuffs form less than 5% of the exposed Cretaceous succession in the ensialic back-arc basin. Shardic tuffs occur in units up to 2 m thick, both at Cape Longing (Farquharson 1982a) and on James Ross Island. Accretionary lapilli, 0.5–1.5 cm in diameter, occur in discrete beds up to 20 cm thick on James Ross Island, and typically show normal grading which probably resulted from differential settling through the water column. However, substantial quantities of dispersed volcanogenic material, particularly plagioclase and quartz, occur throughout the Cretaceous strata of the James Ross Island group and Cape Longing. Some of this material may be the product of direct air-fall, either of insufficient volume to be preserved as a discrete bed or involved in a secondary sedimentary process. Most is probably the result of penecontemporaneous erosion and resedimentation of unconsolidated pyroclastic material originally deposited on, or immediately adjacent to, the volcanic arc.

Discussion

The abrupt change in aeromagnetic signature between the arc terrane and back-arc basin (Fig. 3) suggests that there is a substantial thickness of sedimentary rock at the basin margin and implies a faulted boundary. The arcuate Prince Gustav Channel between James Ross Island and the mainland (Fig. 1) has long been thought to be fault-controlled (Bibby 1966). This fault, or fault zone, has a long history of movement reflected in (?)early Cretaceous alluvial fan deposition at Camp Hill (Farquharson 1982b), flexuring of strata along the NW coast of James Ross Island (Bibby 1966), the present scarp topography on Tabarin Peninsula and the submarine trough in Duse Bay (Renner 1980). A period of mid-Cretaceous flexuring is indicated by an unconformity within the succession on NW James Ross Island which may be the result of further motion along the Prince Gustav Channel fault(s). It is probable that these faults limited the back-arc basin in this area during the Cretaceous. The proximal deposits of the basin, particularly the large-scale slide-blocks of Nordenskjöld Formation, are consistent with a fault-scarp basin margin.

By the Tertiary the western margin of the basin was no longer fault-controlled. Palaeocene–(?)Oligocene arc-derived deltaic sandstones and siltstones occur on Seymour Island (Fig. 1) (Elliot & Trautman 1982). These rocks are the youngest preserved strata of the basin and represent a regressive phase involving eastward migration of the coastline and probably cessation of the controlling nature of the original boundary fault(s). The overall succession is analogous to the basinward progradation of a volcaniclastic apron as recognized in modern marginal basins around the Pacific Ocean (Karig & Moore 1975; Carey & Sigurdsson, this volume).

Palaeocurrent directions from the Cretaceous back-arc basin deposits, and the transport direction of the Nordenskjöld Formation slide-blocks, have an approximate mean orientation towards the SE, consistent with derivation from an arc terrane on the site of Trinity Peninsula. It is as yet uncertain whether the deposits accumulated on a laterally uniform apron or on coalescing fan-shaped bodies. Preliminary results from the James Ross Island strata, however, suggest an overall radial distribution pattern away from the peninsula.

These deposits illustrate the complex nature of marine proximal sedimentation in an ensialic back-arc basin and the influence of volcanic and non-volcanic processes. Recognizable primary volcanogenic deposits, in the form of air-fall tuffs and eruption-induced sediment flow units, form only a minor part of the basin fill. Dispersed volcanogenic material, however, is an important component of the sedimentary accumulation, much of it reaching the basin following remobilization of near-shore volcaniclastic deposits. The coarse-grained volcaniclastic turbidites of mid-Cretaceous age from Cape Longing (Farquharson 1982a) are an example of resedimented volcanic sands. Such remobilization may be the result of oversteepened volcanic debris or instability caused by contemporaneous fault movement on the flanks of the arc.

Although the thick-bedded conglomerates of the Sobral Peninsula contain clasts derived predominantly from the Trinity Peninsula Group, it is possible that they represent resedimentation of subaerial and/or shallow-subaqueous volcanogenic material. At several localities on the arc terrane (e.g. Camp Hill: Fig. 1) the initial pyroclastic rocks were erupted through, and incorporated, the conglomerate infill of alluvial basins (Bibby 1966; Farquharson 1982b). These conglomerates were derived solely from the Trinity Peninsula Group. Elsewhere (e.g. Crystal Hill: Fig. 1) the pyroclastics were erupted directly through the Trinity Peninsula Group, producing agglomerates composed of clasts of the group set in a tuffaceous matrix. Erosion and redeposition of these agglomerates could produce conglomerates with an apparent Trinity Peninsula Group provenance. Similarly, the blocks of Upper Jurassic Nordenskjöld Formation from the Lower Cretaceous breccia beds of James Ross Island are embedded in a matrix containing scattered angular plagioclase and clear quartz crystals indicating the incorporation of unconsolidated tuffaceous material. Indeed it is likely that all the Cretaceous back-arc basin deposits contain at least some primary and/or resedimented pyroclastic components.

The ratio of pyroclastic to resedimented volcanogenic deposits preserved within modern arc-related basins varies greatly and may be significantly different on one side of an arc compared to the other (Bouma 1975; Sigurdsson *et al*. 1980). The relative paucity of primary volcanogenic deposits and abundance of coarse resedimented material, within the Cretaceous back-arc basin of the northern Antarctic Peninsula, indicate the strong influence of the steep and unstable faulted arc flanks on proximal sedimentation in the basin. The character of the back-arc basin deposits also reflects the continental foundation of the arc; 'basement' rock sequences (Trinity Peninsula Group) contributing significant quantities of detritus to the otherwise volcaniclastic apron.

ACKNOWLEDGMENTS: We wish to thank our colleagues at the British Antarctic Survey, and S. Garrett in particular, for valuable criticism and discussion. M. P. D. Lewis, M. C. Sharp and N. St J. Young assisted in the field.

References

BARKER, P. F. 1982. The Cenozoic subduction history of the Pacific margin of the Antarctic Peninsula: ridge crest—trench interactions. *J. geol. Soc. London*, **139**, 787–801.

BIBBY, J. S. 1966. The stratigraphy of part of north east Graham Land and the James Ross Island group. *Sci. Rep. Br. Antarct. Surv.* **53**, 37 pp.

BOUMA, A. H. 1975. Sedimentary structures of Philippines Sea and Sea of Japan sediments, DSDP Leg 31. *In*: KARIG, D. E., INGLE, J. C. *et al.* (eds) *Init. Rep. Deep Sea drill. Proj.* **31**, 471–88. U.S. Government Printing Office, Washington, D.C.

DAVIES, I. C. & WALKER, R. G. 1974. Transport and deposition of resedimented conglomerates: the Cap Enragé Formation, Cambro–Ordovician, Gaspé, Quebec. *J. sediment. Petrol.* **44**, 1200–16.

ELLIOT, D. H. & TRAUTMAN, T. A. 1982. Lower Tertiary strata on Seymour Island, Antarctic Peninsula. *In*: CRADDOCK, C. (ed.) *Antarctic Geoscience*, 287–97. University of Wisconsin Press, Madison.

FARQUHARSON, G. W. 1982a. Late Mesozoic sedimentation in the northern Antarctic Peninsula and its relationship to the southern Andes. *J. geol. Soc. London*, **139**, 721–7.

—— 1982b. Lacustrine deltas in a Mesozoic alluvial sequence from Camp Hill, Antarctica. *Sedimentology*, **29**, 717–25.

FISKE, R. S. 1963. Subaqueous pyroclastic flows in the Ohanapecosh Formation, Washington. *Bull. geol. Soc. Am.* **74**, 391–405.

—— & MATSUDA, T. 1964. Submarine equivalents of ash flows in the Tokiwa Formation, Japan. *Am. J. Sci.* **262**, 76–106.

HAMER, R. D. 1983. Petrogenetic aspects of the Jurassic–early Cretaceous volcanism, northernmost Antarctic Peninsula. *In*: OLIVER, R. L., JAMES, P. R. & JAGO, J. B. (eds) *Antarctic Earth Science*, 338–42. Australian Academy of Sciences, Canberra/Cambridge University Press, Cambridge.

—— & MOYES, A. B. 1982. Composition and origin of garnet from the Antarctic Peninsula Volcanic Group of Trinity Peninsula. *J. geol. Soc. London*, **139**, 713–20.

HANER, B. E. 1971. Morphology and sediments of Redondo Submarine Fan, southern California. *Bull. geol. Soc. Am.* **82**, 2413–32.

HARRINGTON, P. K., BARKER, P. F. & GRIFFITHS, D. H. 1972. Crustal structure of the South Orkney Islands area from seismic refraction and magnetic measurements. *In*: ADIE, R. J. (ed.) *Antarctic Geology and Geophysics*, 27–32. Universitetsforlaget, Oslo.

HARRISON, S. H. & FRITZ, W. J. 1982. Depositional features of March 1982 Mount St Helens sediment flows. *Nature, Lond.* **299**, 720–2.

HEIN, F. J. & WALKER, R. G. 1982. The Cambro-Ordovician Cap Enragé Formation, Quebec, Canada: conglomeratic deposits of a braided submarine channel with terraces. *Sedimentology*, **29**, 309–29.

HENDRY, H. E. 1973. Sedimentation of deep water conglomerates in Lower Ordovician rocks of Quebec—composite bedding produced by progressive liquefaction of sediment? *J. sediment. Petrol.* **43**, 125–36.

HYDEN, G. & TANNER, P. W. G. 1981. Late Palaeozoic—early Mesozoic fore-arc basin sedimentary rocks at the Pacific margin in Western Antarctica. *Geol. Rdsch.* **70**, 529–41.

JOHNSON, A. M. 1970. *Physical Processes in Geology*. Freeman, Cooper & Co., San Francisco, 577 pp.

JOHNSON, B. A. & WALKER, R. G. 1979. Paleocurrents and depositional environments of deep water conglomerates in the Cambro-Ordovician Cap Enragé Formation, Quebec Appalachians. *Can. J. Earth Sci.* **16**, 1375–87.

KARIG, D. E. & MOORE, G. F. 1975. Tectonically controlled sedimentation in marginal basins. *Earth planet. Sci. Lett.* **26**, 233–8.

LA BRECQUE, J. L. & BARKER, P. 1981. The age of the Weddell Basin. *Nature, Lond.* **290** 489–92.

LAUDON, T. S., THOMSON, M. R. A., WILLIAMS, P. L. & MILLIKEN, K. L. 1983. The Jurassic Latady Formation, southwestern Antarctic Peninsula. *In*: OLIVER, R. L., JAMES, P. R. & JAGO, J. B. (eds) *Antarctic Earth Science*, 308–14. Australian Academy of Sciences, Canberra/Cambridge University Press, Cambridge.

LOWE, D. R. 1976. Grain flow and grain flow deposits. *J. sediment. Petrol.* **46**, 188–99.

—— 1979. Sediment gravity flows: their classification and some problems of application to natural flows and deposits. *In*: DOYLE, L. J. & PILKEY, O. H. (eds) *Geology of Continental Slopes. Spec. Publs Soc. econ. Paleontol. Miner.*, *Tulsa*, **27**, 75–82.

—— 1982. Sediment gravity flows. II. Depositional models with special reference to the deposits of high density turbidity currents. *J. sediment. Petrol.* **52**, 279–97.

MULLINS, H. T. & VAN BUREN, H. M. 1979. Modern modified carbonate grain flow deposit. *J. sediment. Petrol.* **49**, 747–52.

NARDIN, T. R., HEIN, F. J., GORSLINE, D. S. & EDWARDS, B. D. 1979. A review of mass movement processes, sediment and acoustic characteristics, and contrasts in slope and base-of-slope systems versus canyon-fan-basin floor systems. *In*: DOYLE, L. J. & PILKEY, O. H. (eds) *Geology of Continental Slopes. Spec. Publ. Soc. econ. Paleontol. Miner.*, *Tulsa*, **27**, 61–73.

NELSON, P. H. H. 1975. The James Ross Island Volcanic Group of north-east Graham Land. *Sci. Rep. Br. Antarct. Surv.* **54**, 62 pp.

PANKHURST, R. J. 1982. Rb-Sr geochronology of Graham Land, Antarctica. *J. geol. Soc. London*, **139**, 701–11.

—— 1983. Rb-Sr constraints on the ages of basement rocks of the Antarctic Peninsula. *In*: OLIVER, R. L., JAMES, P. R. & JAGO, J. B. (eds) *Antarctic Earth Science*, 367–71. Australian

Academy of Sciences, Canberra/Cambridge University Press, Cambridge.

RENNER, R. G. B. 1980. Gravity and magnetic surveys in Graham Land. *Sci. Rep. Br. Antarct. Surv.* **77**, 99 pp.

——, DIKSTRA, B. J. & MARTIN, J. L. 1982. Aeromagnetic surveys over the Antarctic Peninsula. *In*: CRADDOCK, C. (ed.) *Antarctic Geoscience*, 363–70. University of Wisconsin Press, Madison.

SANDERS, J. E. 1965. Primary sedimentary structures formed by turbidity currents and related resedimentation mechanisms. *In*: MIDDLETON, G. V. (ed.) *Primary Sedimentary Structures and their Hydrodynamic Interpretation. Spec. Publ. Soc. econ. Paleontol. Miner., Tulsa,* **12**, 192–219.

SAUNDERS, A. D. & TARNEY, J. 1982. Igneous activity in the southern Andes and northern Antarctic Peninsula : a review. *J. geol. Soc. London,* **139**, 691–700.

SIGURDSSON, H., SPARKS, R. S. J., CAREY, S. N. & HUANG, T. C. 1980. Volcanogenic sedimentation in the Lesser Antilles Arc. *J. Geol.* **88**, 523–40.

SMELLIE, J. L. 1981. A complete arc-trench system recognized in Gondwana sequences of the Antarctic Peninsula region. *Geol. Mag.* **118**, 139–59.

——, DAVIES, R. E. S. & THOMSON, M. R. A. 1980. Geology of a Mesozoic intra-arc sequence on Byers Peninsula, Livingston Island, South Shetland Islands. *Bull. Br. Antarct. Surv.* **50**, 55–76.

SPARKS, R. S. J., SIGURDSSON, H. & CAREY, S. N. 1980. The entrance of pyroclastic flows into the sea, I. Oceanographic and geologic evidence from Dominica, Lesser Antilles. *J. Volcan. geoth. Res., Amsterdam,* **7**, 87–96.

SURLYK, F. 1978. Submarine fan sedimentation along fault scarps on tilted fault blocks (Jurassic-Cretaceous boundary, East Greenland). *Bull. Grønlands geol. Unders.* **128**, 108 pp.

TANNER, P. W. G., PANKHURST, R. J. & HYDEN, G. 1982. Radiometric evidence for the age of the subduction complex in the South Orkney and South Shetland Islands, West Antarctica. *J. geol. Soc. London,* **139**, 683–690.

TASSÉ, N., LAJOIE, J. & DIMROTH, E. 1978. The anatomy and interpretation of an Archean volcaniclastic sequence, Noranda region, Quebec. *Can. J. Earth Sci.* **15**, 874–88.

WALKER, R. G. 1978. Deep water sandstone facies and ancient submarine fans: models for exploration for stratigraphic traps. *Bull. Am. Ass. Petrol. Geol.* **62**, 932–66.

WINN, R. D. & DOTT, R. H. 1977. Large-scale traction produced structures in deep-water fan-channel conglomerates in southern Chile. *Geology,* **5**, 41–4.

G. W. FARQUHARSON, R. D. HAMER & J. R. INESON, British Antarctic Survey, Natural Environment Research Council, Madingley Road, Cambridge CB2 0ET, U.K.

LOWER PALAEOZOIC

Ordovician marginal basin development in the central Norwegian Caledonides

D. Roberts, T. Grenne & P. D. Ryan

SUMMARY: In the Caledonides of central Norway, volcano-sedimentary successions, of the Lower and Upper Hovin and Horg Groups (Ordovician to ?Lower Silurian), unconformably overlie a variable substrate which includes fragmented oceanic crust and ensimatic immature arc assemblages initially deformed and metamorphosed in earliest Ordovician times. In western areas, remnants of a mid-Ordovician, evolved island-arc complex comprise a volcanic-plutonic suite of clear calc-alkaline affinity. To the E, in the Trondheim district, continental margin-type basic to andesitic volcanics, and sediments of comparable age, denote accumulation in a fault-dissected, back-arc marginal basin characterized by marked facies variations and bipolar sediment dispersal patterns. Sediment composition reflects the nature of the contemporaneous volcanism, although an input of metamorphic detritus from a continental margin source can also be determined. Thick, massive to pillowed tholeiitic greenstones, mainly of OFB chemistry, occur in several areas and some contain thick sheeted-dyke units, gabbro complexes and subordinate plagiogranites. These ophiolitic assemblages appear to occur at two stratigraphic levels within the Lower Hovin Group, reflecting separate phases of crustal thinning and oceanic lithosphere accretion. Mega-breccias and an olistostrome may relate to abortive ophiolite displacement or intra-basinal movements in late Arenig times. During the Caradoc and Ashgill, and possibly in earliest Silurian times, continental margin-type acidic volcanism developed widely, particularly in the S, while in central and northern parts of the basin, turbidites and resedimented conglomerates, derived from the basin margins, show SW transport in the deeper-water axial trough. Although there is strong evidence that the marginal basin and arc lay at the edge of the continental plate (Baltoscandia) on the SE side of the Iapetus, with SE subduction, this is not supported by the faunas in the lowest parts of the Ordovician successions, which are largely of North American affinity. A compromise model involving an elongate microcontinental plate within the Iapetus is invoked.

Metamorphosed volcanic and sedimentary rock sequences of early Ordovician to early Silurian age constitute a significant part of the allochthon in the Caledonides of central Norway. The structural position, degree of allochthoneity and the depositional environment of these thick sequences have been subjects of debate since the late 19th century (Törnebohm 1896). For much of this time the Norwegian Caledonian lithostratigraphic successions were divided into facies types (Bailey & Holtedahl 1938; Strand 1960; Størmer 1967), related to the shelf, miogeosyncline and eugeosyncline of Stillean terminology. In this scheme the rocks of the Trondheim region of central Norway were considered to be eugeosynclinal. During the past decade the sequences have been reinterpreted to relate to a plate tectonic reconstruction. Early models delineated magmatic arc complexes, back-arc marginal basin environments and probable major (Iapetus) ocean regimes (Gale & Roberts 1972; 1974). Subsequently these models have been refined with the recognition of almost complete ophiolite assemblages in some parts of the region (Prestvik 1974,

1980; Grenne *et al.* 1980; Grenne & Roberts 1981). Also, it has been recognized that the thick basaltic greenstone units, and in some cases ophiolitic complexes, lie at different stratigraphic levels.

This paper considers the post-mid-Arenig sequences, comprising mixed sedimentary, volcanic and associated plutonic rocks, in the western part of the central Norwegian Caledonides (Fig. 1A). It is these sequences which reflect accumulation in a marginal basin.

Regional setting

The metamorphic allochthon comprises Proterozoic–Silurian rocks in an array of SE-transported nappes and thrust sheets which rest on a thin cover of autochthonous Vendian–Cambrian sediments deposited upon crystalline Precambrian basement (Roberts & Wolff 1981). In western areas, however, this basement is probably partly or largely allochthonous.

The sequences in the Trondheim area are

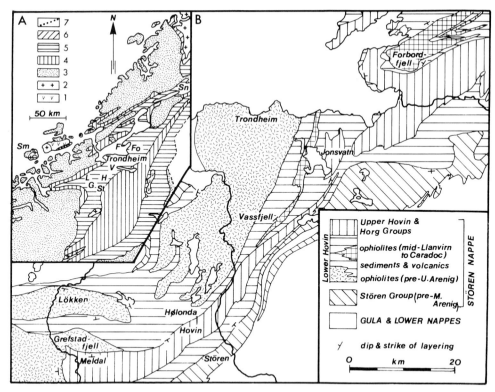

FIG. 1. (a) Location map showing the relevant principal geological units of the central Norwegian
Caledonides. 1. Gabbro and ultramafite. 2. Granite, trondhjemite, quartz diorite. 3. Precambrian
crystalline basement, lower nappes and autochthonous sediment. 4. Gula Complex, Skøtingen Nappe
and equivalents. 5. Trondheim Supergroup (minus Gula). 6. Helgeland Nappe Complex. 7. Old Red
Sandstone deposits. Fo = Forbordfjell; F = Frosta; G = Grefstadfjell; H = Hølonda; L = Løkken;
Sm = Smøla; Sn = Snåsavatn; St = Støren; V = Vassfjell. (b) Simplified geological map of the
Trondheim–Løkken–Støren district, modified from Grenne & Roberts (1981) which is based on
Wolff's (1976) map-sheet 'Trondheim'.

situated in the Støren Nappe (Gale & Roberts
1974) of the Upper Allochthon (Roberts &
Gee 1981). This nappe is a composite tectonic
sheet comprising two main units: (i) a tectono-
stratigraphically lower unit of MORB-type
tholeiitic basalts with subordinate cherts and
hemipelagic sediments (Støren Group), rare
mafic dykes and gabbros; and (ii) an unconformably overlying volcano-sedimentary succession of Ordovician to possible early Silurian
age (the Lower and Upper Hovin Groups and
Horg Group of Vogt (1945)). The Støren
Group unit, together with its substrate, the
Gula Complex, upon which it was tectonically
emplaced, is considered to have been initially
deformed and metamorphosed in earliest
Ordovician times (Furnes *et al.* 1980; Hardenby
et al. 1981). The Støren Group is interpreted as
a slice of a dismembered ophiolite (Furnes *et al.*
1980; Grenne & Roberts 1981). The deformed

sequences are cut by trondhjemite massifs and
dykes of early to middle Ordovician age (Klingspor & Gee 1981). As the Gula Complex includes
Tremadoc slates (Vogt 1940), the deformation
cycle is broadly equivalent to the later stages of
the Finnmarkian orogenic event (Sturt 1978;
Roberts & Sturt 1980), which corresponds to
the Grampian deformation of the British
Caledonides. The younger volcano-sedimentary
sequence (unit ii, above) was deposited unconformably upon the trondhjemite-intruded
Støren Group and was initially deformed in
approximately mid-Silurian times. Both these
units were transported as one nappe complex
during this later (Silurian), major orogenic
event.

It is the volcano-sedimentary sequences of
the Hovin and Horg Groups, and coeval rocks
on the island of Smøla (Fig. 1a), which furnish
the characteristic features of marginal basin and

adjacent magmatic arc volcanism and sedimentation. In some areas, well-preserved ophiolitic sequences, from gabbros, through sheeted-dyke complexes, to pillow-lava piles with associated marine sediments, have been described (Grenne *et al.* 1980; Grenne & Roberts 1981). Several of the separate sedimentary successions contain a rich fauna of graptolites, brachiopods, trilobites, molluscs, corals and conodonts (Neuman & Bruton 1974; Berry 1968; Bergström 1979; Bruton & Bockelie 1980; Ryan *et al.* 1980) which provides fairly reliable stratigraphical control. In the Arenig–Llanvirn, North American faunas dominate over European, though there are variations between areas. However, uncritical evaluation of faunas may promote misleading provenance interpretations if viewed in isolation (cf. Roberts & Gale 1978).

The W Trondheim region is an important mineralization province, noted primarily for its massive stratiform sulphide deposits, which tend to be associated with some of the thicker volcanic sequences (Grenne *et al.* 1980). Ore types differ between volcanic settings, e.g. those in mature island-arc rocks and those occurring in ophiolitic assemblages.

Magmatic and sedimentary assemblages

Ophiolite complexes

The pervasive Silurian deformation dismembered and fragmented the Norwegian Caledonide ophiolite complexes (Furnes *et al.* 1980, 1984; Roberts *et al.* 1984). In addition, erosion of these complexes is such that in some areas only parts are preserved. In the interpretation of ophiolite assemblages, the importance of assessing the geological context as well as the magmatic associations and geochemistry has been stressed by Grenne & Roberts (1980), and by Moores (1982) who introduced the concept of the *ophiolitic association*. In this 10-fold subdivision of an ideal ophiolite assemblage, the sedimentary association is a key factor in the interpretation of genesis and palaeoenvironment. In the case of the ophiolite fragments of the Trondheim region, from 4 to 8 of Moores' 10 units are exposed in any one area.

Løkken

The Løkken basaltic greenstone and gabbro complex occurs in a late, major, E–W trending synform, a few km N of the Grefstadfjell area

(Fig. 1b). Here, the Lower Hovin Group is inverted, with the plutonic member of the ophiolite occurring in the core of the synform. The gabbros, more than 1 km thick, are mainly high-level leucogabbros that locally contain small bodies of plagiogranite. Dykes increase in frequency upwards into a dolerite dyke complex of uncertain thickness. The dykes are feeders to the overlying MOR tholeiitic basalt sequence, approximately 1 km thick, and through the basalts the frequency of dykes diminishes. The basalt sequence comprises lower, non-vesicular, variolitic pillow lavas which grade upwards into generally vesicular, pillowed and massive flows. Massive stratiform sulphide deposits are common, including the Løkken Cu-Zn pyrite orebody ($c.\ 25 \times 10^6$t) with associated jaspers, various cherts and thin sulphide-oxide Fe formations, locally known as 'vasskis'. The widespread alteration and sulphide disseminations within the basalts are interpreted to result from sea-floor hydrothermal activity. Hydrothermal activity was facilitated by faults with surface expressions marked by fault-scarp breccias (Grenne 1981; Grenne & Roberts 1981).

Lower Hovin Group sediments overlying the greenstones vary from coarse breccias to siltstones with ophiolite-derived detritus and, locally, limestone fragments. Sediments also occur intercalated within the uppermost lavas W of Løkken (M. Heim, personal communication 1982). No fossils have yet been recovered from the sediments directly above the Løkken ophiolite. Quartz-feldspar porphyrites, of tonalite composition (Table 1), intrude both the ophiolite complex and the overlying breccias. They are, however, also found locally as clasts within the breccias.

Vassfjell

The Vassfjell ophiolite is situated the right-way-up in the wide greenstone belt that lies between Trondheim and Hølonda (Fig. 1b). The lowest preserved part of the ophiolite comprises *c.* 1 km of gabbro, including several small plagiogranites similar to those at Løkken. The uppermost 300–400 m of the gabbro is a transitional zone, with an increasing frequency of dolerite dykes, which passes upwards into a 100% sheeted-dyke complex *c.* 1 km thick. The dyke complex is overlain by 800–1000 m of basaltic pillow lavas with dolerite dykes intruding the lower portion of the lava pile. As at Løkken, small massive sulphide deposits, jasper horizons and the characteristic cherty Fe formations (vasskis) occur (Grenne *et al.* 1980), although here they are not as common.

TABLE 1. *Geochemical data for the quartz porphyrites cutting the pre-uppermost Arenig ophiolites; the Hølonda basic porphyrites; and the dacites and rhyolites intercalated within the Lower and Upper Hovin Groups*

	Lower Hovin				Lower and Upper Hovin			
	Quartz porphyrites		Hølonda (basic) porphyrites		Dacites		Rhyolites	
	$\bar{x}$	SD	$\bar{x}$	SD	$\bar{x}$	SD	$\bar{x}$	SD
SiO_2	70.41*	2.97	50.12	1.90	64.51	1.28	74.00	2.40
TiO_2	0.28	0.04	1.10	0.09	0.70	0.15	0.34	0.09
Al_2O_3	15.30	1.97	16.23	1.54	16.19	0.88	13.54	0.81
Fe_2O_3	2.26	0.56	9.23	1.44	4.76	1.00	1.98	0.81
MgO	0.78	0.32	4.73	1.48	2.09	0.41	0.54	0.23
CaO	2.96	1.75	7.81	1.45	3.32	0.39	0.87	0.43
Na_2O	5.61	1.26	3.70	0.90	3.08	0.86	4.09	0.80
K_2O	0.79	0.64	1.56	0.70	2.67	0.53	3.56	0.91
MnO	0.03	0.02	0.12	0.40	0.07	0.02	0.05	0.02
P_2O_5	0.07	0.03	0.39	0.05	0.13	0.03	0.04	0.02
Nb	<5[†]		14	3	13	2	15	2
Zr	131	49	159	22	250	34	224	28
Y	18	27	26	2	26	2	35	4
Sr	323	210	614	200	467	78	195	76
Rb	14	11	59	34	100	15	118	27
Ni	<5		33	17	11	10	<5	
Cr	5		90	81	47	45	10	9
V	30	14	271	43	106	44	27	18
Ba	141	113	488	238	616	274	667	137
(*n*)	8		13		5		6	

*Weight %.
[†] Ppm.

Structurally below and to the E of the complex is a thick breccia, comprising fragments derived from various parts of the ophiolite together with a high proportion of recrystallized limestone and exotic basic and acidic igneous rocks. This breccia is structurally underlain by Lower Hovin Group metagreywackes and phyllites of unknown age. In places, the breccia, which is interpreted as an olistostrome (Grenne & Roberts 1981), includes clasts from 10 to > 100 m across (Grenne 1980). As yet there is no direct evidence of the age of the Vassfjell ophiolite fragment. However, the ophiolite forms part of the Bymark Group greenstones of Carstens (1919) which have generally, though probably mistakenly, been correlated with the Støren Group (e.g. Gale & Roberts 1974; Bruton & Bockelie 1980, 1982). Some 20 km along strike SW of Vassfjell, the ophiolite assemblage is overlain by the late Arenig–Llanvirn Hølonda Limestone (Bruton & Bockelie 1980).

Grefstadfjell

The Grefstadfjell ophiolite occurs to the S and SW of the Løkken and Vassfjell complexes, respectively (Fig. 1b). At its present base are gabbros which grade upwards into a 100% sheeted-dolerite-dyke complex, which in turn is overlain by 1.1 km of pillow lavas, the lower and middle members of which contain no intercalated sediment. Small sulphide bodies are located at the dykes–lavas contact. The upper lavas (maximum 350 m) are interlayered with fossiliferous Lower Hovin Group sediments of uppermost Arenig (Ya 2) age. As the lower lavas show evidence of rapid eruption, and as there is no break evident in the accumulation of the lava pile, this age is considered to be that of much of the ophiolitic complex. The Grefstadfjell ophiolite is underlain by a fossiliferous Arenig (Ca 2–Ya 1) sequence, of non-volcanic calcisiltites and graphitic quartz siltstones of the Lo Formation (Ryan et al. 1980), thus defining a lower limit to the age of this complex. The mainly graptolitic faunas of the sediments associated with the ophiolitic complex are currently being re-examined by Dr B. Erdtmann with a view to determining a more precise biostratigraphy for this part of the Lower Hovin Group.

Greenstone units of probable ophiolitic affinity

Forbordfjell and Jonsvatn

Although geographically separated (Fig. 1b), these neighbouring volcanitic rock sequences are of comparable stratigraphic position. Also, the stratigraphic subdivisions of the sequences—based on major- and trace-element chemistry (see below; Grenne & Roberts 1980)—are similar. The successions, 0.5–2 km thick, are characterized by both massive and pillowed basaltic lavas, with local pillow breccias. A cyclic sequence, of massive lava flows succeeded by pillow basalts and then pillow breccias, tuffs and thin tuffaceous sediments, is common. Gabbro, transected by basic dykes, occurs in one part of the Forbordfjell area.

The Lower Hovin Group in these areas varies from generally shallow-marine sandstones, pelites and limestones below the lavas, with dark pelites and cherts immediately below, to mainly greywackes, fanglomerates and pelites of a deeper-water environment above. Dacitic to rhyolitic lavas and tuffs occur in these overlying sequences. Higher in the stratigraphy, fairly deep-marine turbidites have been recorded (Roberts 1972; Pedersen 1981).

The Forbordfjell and Jonsvatn basalts occur within the upper part of the Lower Hovin Group, apparently above the level of the Hølonda Limestone but below the late Caradoc phyllitic shales at the top of the group (Vogt 1945; Bruton & Bockelie 1982). They thus lie within the Llanvirn–early Caradoc.

Frosta

The greenstone lavas in the Lower Hovin Group on the Frosta peninsula (Fig. 1a) are stratigraphically and geochemically similar to those at Forbordfjell. Fossils in the nearby Tautra Limestone indicate a mid-Ordovician age or younger (N. Spjeldnæs, written communication 1974; Roberts 1975), but the precise relationship of the carbonate to the Frosta greenstones is uncertain. Cobbles of lava in an adjacent conglomerate are chemically similar to the Frosta lavas, suggesting that the clasts were derived from these lavas. The turbidites of the Upper Hovin Group, above the level of the Tautra Limestone, reflect a deepening marine environment.

Age relationships of the ophiolitic complexes

It has been traditional to assume that all greenstone complexes in the Trondheim region form part of the Støren Group, which Vogt (1945) defined as being older than the Lower Hovin Group and separated from it by a tectonometamorphic break. Earlier, however, Carstens (1919) had demonstrated that thick greenstone-gabbro units also occur within the Lower Hovin Group, which was later corroborated by Roberts (1975). One such unit is the Grefstadfjell complex which is considered not to belong to the Støren Group for the following reasons:

(i) Pillow lavas at the top of the lava pile are intercalated with cherts, manganiferous sediments and black shales. Graptolites from these black shales are of a similar age to those within the Bogo Formation sediments of the Lower Hovin Group.

(ii) These shale and chert sequences are cut by channels containing volcanic breccias which also cut shales of the Bogo Formation.

(iii) At the extreme E end of the Grefstadfjell complex, vesicular pillow lavas, occurring at high levels in the ophiolite, interdigitate with greenstone breccias and conglomerates that pass laterally into sediments of the Lower Hovin Group, thus showing that these breccias formed contemporaneously with lava eruption and are not the product of a subsequent tectonic disturbance as suggested by Vogt (1945).

The precise stratigraphic relationship between the Grefstadfjell, Løkken and Vassfjell ophiolites is not known. However, all show similar geochemical traits (see below) and have been subjected to the same tectonothermal history as the sediments of the Lower Hovin Group. None of these three complexes contains pre-Lower Hovin Group medium-grade fabrics as would be expected had they been subjected to the earliest Ordovician metamorphic event recorded in the Trondheim area (Hardenby *et al.* 1981; Klingspor & Gee 1981), which is generally taken as being equivalent to the post-Støren Group (Trondheim) disturbance. Although the bases of the Løkken and Vassfjell complexes are nowhere exposed, these ophiolites, like the Grefstadfjell complex, are overlain by sediments of the Lower Hovin Group to which they have contributed detritus. The balance of evidence at the present time therefore suggests that these complexes are more properly assigned to the Lower Hovin Group, rather that to the Støren Group. The interpretation that these complexes are broad correlatives, representing fragments of an originally extensive pre-uppermost Arenig (Ya 2) ophiolite, requires that they now occupy a major, SE-facing F_1 isoclinal fold-nappe, with Løkken on the inverted limb and Grefstadfjell–Vassfjell on the normal limb.

The Jonsvatn and Forbordsfjell ophiolitic fragments, the Snåsavatn upper lavas (see below) and the Frosta lavas, on the other hand, all appear to be of a slightly younger age. If this is confirmed then they would represent the products of a later spreading event.

Subduction-related igneous activity and basin fill

Smøla

Low-grade volcanic rocks on the island of Smøla (Fig. 1a) are, in part, intercalated with fossiliferous limestones of Arenig–Llanvirn age (Bruton & Bockelie 1979), with faunas comparable to those found in the Hølonda Limestone (Neuman & Bruton 1974). The volcanic rocks range from high-Al basalt through andesite to subordinate rhyolite, with the chemistry defining a distinct calc-alkaline trend (Roberts 1980). Following a minor episode of folding and faulting, the sequence was intruded by quartz diorite and granodiorite in late Ordovician times. The complete sedimentary-magmatic association is indicative of development in a mature, evolved, magmatic arc. Several magnetite-chalcopyrite-pyrite occurrences are associated with the volcanic rocks and skarn, and these supported a local mining industry in the 18th century. This mineralization can be traced for some 200 km along strike northeastwards to the Snåsavatn area. Comparisons with the chemical and sedimentary characteristics of the contemporaneous ophiolitic and basinal sequences in the Løkken–Hølonda district indicate that the Smøla arc lay to the 'W' of the main basin; details are given by Roberts (1980).

Snåsavatn

The basalts in the vicinity of Snåsavatn (Fig. 1a), 1.5–1.8 km thick, are unusual for the region in that no clear pillow structures have yet been recognized. The basalts, which are locally interbedded with basic to intermediate tuffs, overlie the Snåsa Limestone of mid-Ordovician or younger age (N. Spjeldnæs, written communication 1974). Whereas the lower basalts are of calc-alkaline affinity, the bulk of the lavas are tholeiitic (Roberts 1982b) and comparable to the Forbordfjell and Jonsvatn basalts of similar age. Thus, the sequence is interpreted to represent a transition from arc to basin-spreading volcanism.

In the NE Snåsavatn area, rocks above the basalts are not exposed, but to the SW the Upper Hovin Group sequence (Carstens 1960; Springer Peacey 1964) lies in this position. The Upper Hovin Group here comprises a turbiditic greywacke-shale-conglomerate association.

Central-southern areas

Here, a variety of hypabyssal, volcanic and volcaniclastic rocks, apparently unrelated to the ophiolitic assemblages, occurs in the Lower Hovin Group and partly in the Upper Hovin Group. In the Meldal–Hølonda district, for example, volcanic breccias and tuffs are common at lower levels in the Lower Hovin Group, and in places the lowest part of the sequence is cut by late Arenig–early Llanvirn quartz porphyrite dykes of tonalitic composition. The Hølonda porphyrites, of early Llanvirn age, are basaltic to andesitic in composition and occur both as flows and as subvolcanic intrusions. At one locality, Hølonda porphyrite dykes transect strongly sheared Løkken basalts (Grenne & Roberts 1981).

Higher in the sequence, in the Jonsvatn–Stjørdal district, dacitic to rhyolitic lavas and tuffs occur near the top of the Lower Hovin Group and in the Upper Hovin Group. Further SW, towards Meldal, chemically similar acidic volcanic rocks first occur in the Upper Hovin Group (Table 1). These acidic rocks were mostly reworked in a marine environment, although rare ignimbrites are still preserved.

The sediments of the Lower Hovin Group include limestones, black and green shales, sandstones, volcanic conglomerates and megabreccias, with marked changes in facies. In the SW the Upper Hovin Group comprises calcareous sandstones and acidic volcaniclastics containing abundant polycrystalline quartz fragments. To the NE, in the Forbordfjell–Frosta district, comparatively deep-marine turbidites predominate (Pedersen 1981). Shale geochemistry in the Lower Hovin Group (Ryan & Williams 1984) can be related to a basic and intermediate volcanic source at lower levels, whereas an acidic component is dominant higher up.

Geochemistry

Pre-late Arenig ophiolite assemblages

The basaltic lavas of the older, Løkken–Vassfjell–Grefstadfjell ophiolitic complexes show clear Fe- and Ti-enrichment trends (Grenne *et al.* 1980; P. D. Ryan, unpublished data) typical

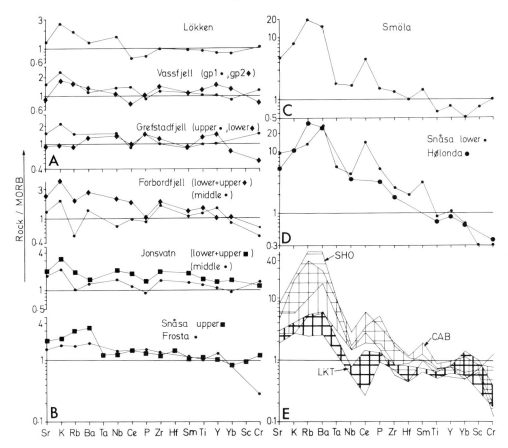

FIG. 2. Rock/MORB diagrams (Pearce 1980) for: (a) the pre-uppermost Arenig ophiolites; (b) the Llanvirn–Caradoc ophiolitic greenstones; (c) Smøla metabasites; (d) Hølonda basic porphyrites and Snåsa upper basic lava unit; and (e) a compilation of data for comparison purposes from Cenozoic and Mesozoic active continental margins and ensimatic island-arc complexes, showing the fields from both environments of island-arc tholeiites (IAT), calc-alkaline basalts (CAB) and shoshonitic basalts (SHO). Data from: Atherton *et al.* (1979); Brown *et al.* (1977); Dostal *et al.* (1977); Jakes & White (1972); Jakes & Gill (1974); Lopez-Escobar *et al.* (1977); Mackenzie & Chappel (1972); Meijer & Reagan (1981); Morrison (1980); Myashiro (1974); Saunders *et al.* (1980); Siegers *et al.* (1969).

of spreading-ridge ophiolites. MORB-normalized trace-element diagrams show nearly flat trends (Fig. 2a). The slight enrichment of the large ion lithophile (LIL) elements, Sr, K, Rb and Ba, may not be diagnostic in view of the mobility of these elements (Humphris & Thompson 1978), but for all three complexes average contents of high field strength (HFS) elements are close to or only slightly above those typical of tholeiitic rocks of mid-ocean ridges. Also, tectonic environment discrimination based on stable elements only, such as in Ti-Zr-Y and Ti-Cr diagrams (Grenne *et al.* 1980; P. D. Ryan, unpublished data), clearly demonstrates the OFB affinity of these complexes. Although the composition of the three complexes is generally fairly uniform, the uppermost lavas of the Grefstadfjell area are somewhat anomalous with a more primitive character, in places approaching komatiitic compositions (P. D. Ryan, unpublished data). However, the incompatible-element ratios are comparable to those of the lower, MORB-like, major part of the complex.

Younger greenstone units

The thick basaltic sequences of Forbordfjell, Jonsvatn, Frosta and Snåsavatn (upper lavas) show tholeiitic fractionation trends similar to those of the older ophiolites. The earliest lavas in these units are characterized by a hint of within-plate basalt geochemistry (Grenne & Roberts 1980; Roberts 1982a,b), reflected in Fig. 2b by a slight enrichment of most incompatible elements (including the LIL elements) and a decreasing trend towards Y and Yb. Higher in the sequence the composition changes towards more normal MORB, although the youngest basalts at Forbordfjell and Jonsvatn revert to slight within-plate tholeiitic characteristics. Details of the geochemistry are given by Grenne & Roberts (1980).

Subduction-related igneous activity

The Smøla suite of high-Al basalts, andesites, dacites and minor rhyolites defines a typical calc-alkaline fractionation trend (Roberts 1980). Trace-element ratios and abundances in the basic members (Fig. 2c) conform to those of island-arc calc-alkaline basalts, with characteristically high LIL/HFS element ratios, low Ta and Nb, and a low, generally decreasing trend from Zr towards Yb. An ensimatic arc origin for these rocks is suggested by the predominance of basalt and basaltic andesite.

Elsewhere in the region, intrusive quartz porphyrites of tonalitic composition (Table 1) are the first non-ophiolite magmatic products. The intrusions were contemporaneous with the late Arenig–early Llanvirn uplift of the early ophiolites. By comparison with similar low-K acidic rocks elsewhere (Payne & Strong 1979; Size 1981), and as shown experimentally by Helz (1976), it is suggested that these magmas formed by small degrees of melting of a basaltic source, possibly the amphibolitic lower parts of the ophiolites which may have suffered partial fusion in connection with uplift and displacement of this oceanic crust.

The Hølonda porphyrites, of basaltic to andesitic composition, show calc-alkaline fractionation trends (unpublished data). However, the incompatible elements towards Ti (Fig. 2d) are distinctly enriched compared to normal island-arc calc-alkaline basalts. The high contents of LIL elements in the basic porphyrites conform to this 'enriched' nature and suggest that, although these elements are susceptible to alteration during sea-floor weathering and metamorphism, abundances of elements such as Sr, K, Rb and Ba are not greatly altered from the original values in these thick massive flows and subvolcanic intrusions. Such high K_2O/Na_2O ratios and high alkali and HFS element contents (Table 1) are more typical of calc-alkaline basalts of active continental margins than of island arcs (e.g. Jakes & White 1972).

The Snåsavatn lower lavas are chemically very similar to the Hølonda porphyrites (Fig. 2d); clear calc-alkaline trends, together with levels of HFS elements that are far too high for island-arc lavas, again indicate an origin close to a continental margin. The comparatively low Zr contents of the rhyolites and dacites from the central-southern areas (Table 1) are also diagnostic of subduction-related acidic volcanism (Pearce 1980). Furthermore, the clear predominance of acidic effusive rocks in this upper part of the sequence is also strongly suggestive of an active continental margin environment, rather than an island-arc origin (Jakes & White 1972).

Discussion

The Ordovician magmatic and sedimentary sequences in the W Trondheim region are interpreted as having accumulated in a back-arc marginal basin. The interpretation that an arc-basin couplet is represented is based on volcanic stratigraphy and geochemistry, as well as on sedimentological studies. The occurrence of ophiolite complexes, in some cases with intercalated fossiliferous dark shales, indicates that, prior to Silurian tectonic disruption, the basin was sited above a zone of crustal dilation associated with accretion of oceanic lithosphere at a spreading ridge. The nature of the sedimentary infill argues against a major ocean environment, whereas several of the features outlined above accord with the marginal basin interpretation. The late Cambrian–early Ordovician Finnmarkian orogeny (Sturt 1978; Roberts & Sturt 1980; Ryan & Sturt 1984) pre-dated the basin development. In the Trondheim region this orogeny included eastward obduction of pre-mid-Arenig ocean floor and immature arc products on to a continental margin complex.

In the Grefstadfjell area, the oldest recorded Lower Hovin Group sediments of the Lo Formation (mid-Arenig, Ca 2: Ryan et al. 1980) define a minimum age for the Finnmarkian event. The formation comprises limestones and graphitic quartz siltstones of continental derivation (Ryan & Williams 1984). At this time there was rapid generation of oceanic crust, now represented by the ophiolite sequences at Løkken, Grefstadfjell and Vassfjell. The petrochemistry of these basalts is that of normal MORB; interbedded sediments are Mn-shales, cherts, jasper and 'vasskis', as well as cupriferous massive sulphide orebodies. Such an association suggests the existence of thin oceanic crust forming the floor to a basin dilating by accretion of mantle-derived magmas.

Cessation of spreading and basalt generation was marked by development of locally thick accumulations of breccias which, in the Vassfjell area, are of olistostrome character, comprising mega-blocks derived mainly from the adjacent ophiolite. This development, in late Arenig times, may relate to local abortive obduction or intra-basinal disruption of the oceanic crust, coeval with a temporary change to a compressive or transpressive regime, with the breccias accumulating both ahead of, and along faults within, the rising ophiolite. In some areas the intrusion of tonalitic quartz porphyrites was contemporaneous with and immediately succeeded breccia formation. Late Arenig–early Llanvirn basaltic to andesitic porphyrites that intrude sheared Løkken basaltic greenstones suggest that, locally, the uplift and dissection of the ophiolites was accompanied by penetrative deformation.

The non-ophiolitic component of the breccias varies between complexes. Fragments of the tonalitic rocks are abundant in the breccias above the Løkken basalts, the Vassfjell olistostrome contains blocks of island-arc calc-alkaline rocks (T. Grenne, unpublished data), and the Grefstadfjell breccias contain material that is similar to the Hølonda porphyrites. These breccias may thus record the passage of the ophiolite slabs through different segments of the marginal basin.

The tonalitic intrusions are believed to be derived from the melting of a basaltic source, as there is no indication in their composition of continental crust derivatives. However, the late Arenig–early Llanvirn Hølonda porphyrites, lavas and subvolcanic intrusions are compositionally closer to continental margin rocks than to island arc. These rocks are temporally closely associated with the richly fossiliferous Hølonda Limestone (Neuman & Bruton 1974; Bergström 1979). This suggests that the ophiolite slabs, with their associated oceanic sediments, were either uplifted or perhaps subject to abortive obduction on to the continental margin where sediments of the Lo and Bogo Formations accumulated and the Hølonda porphyrites were emplaced during latest Arenig–early Llanvirn times. This event was broadly coeval with the construction of a magmatic arc of distinctive calc-alkaline chemical character, represented on Smøla and possibly at Snåsavatn, where associated limestones have some faunal affinity with the Hølonda Limestone. While the Smøla rocks are ensimatic, though they matured during gradual thickening towards

intermediate-type crust, the calc-alkaline basalts at Snåsavatn are of a more continental margin character. This maturing of the Smøla rocks, contemporaneous with a migration of continental margin-type volcanism east- or southeastwards (to the Hølonda and Snåsavatn areas), is thought to relate to a constriction of the back-arc marginal basin.

The thick, pillowed to massive, variolitic lavas of the Forbordfjell and Jonsvatn complexes are evidence that in mid-Ordovician times in the main basin there was a second spreading stage. Geochemically, these lavas differ from the earlier ophiolitic basalts in that both the early and the late lavas are enriched in incompatible elements and are of within-plate type (Grenne & Roberts 1980), reflecting a thickening crust to the marginal basin. However, the bulk of the lavas are of MORB type, with no hint of subduction-zone influence; the Frosta basalts are similar. The upper lavas of the Snåsavatn area are also of ocean-floor type, though with a trace of within-plate characteristics, and are of broadly equivalent age. These basalts, succeeding the calc-alkaline lavas, are thought to reflect an oceanward migration of the subduction zone and arc, with marginal basin accretion thus taking over up-sequence. Alternatively, they may have formed in a narrower ensialic basin where adjacent subduction had ceased (cf. Saunders & Tarney, this volume).

Following the second period of tholeiitic basalt generation, immediate uplift is recorded in locally thick conglomerates. The overlying sediments, at least in central and northern districts, reflect a deepening of the depositional environment into Caradoc–Ashgill times. Over this period the volcanism changed from basic and intermediate to acidic with the incoming of dacite and rhyolite lavas (Carstens 1960; Loeschke 1976). These are of continental margin character and indicate that subduction was still active in a gradually more evolved continental margin environment. The acidic volcanics tend to thin northwards and tuffs are dominant in the deeper parts of the basin. Lavas are well developed in the shallower-water SW areas where ignimbrites occur locally. There, the boundary between the Lower and Upper Hovin Groups is an unconformity, marked by uplift and gentle folding (Ryan *et al.* 1980). In the Forbordfjell–Frosta district, to the N, the contact is conformable within a sequence of fairly deep-marine turbidites and debris-flow conglomerates. These turbidites and debris flows were deposited in distal to proximal fan systems and submarine canyons, respectively (Pedersen l981). Palaeocurrent directions indicate transport from both NW and SE into a longitudinal or axial dispersal system with current flow from NE to SW. The depositional basin thus narrowed and became shallower towards the NE (Pedersen 1981), a feature previously developed in Llanvirn times.

In the N, the flysch-type deposition with abundant conglomerates and pebbly mudstones continued into the Caradoc to ?lowermost Llandovery Upper Hovin Group, although shallower-water sandstones appear higher in the sequence. In the S, coarse-grained volcanogenic sediments and acidic volcanic rocks overlie limestones. The volcanogenic sediments, which are the youngest pre-Devonian sediments known in the district, contain abundant polycrystalline quartz in sandstones and quartzite cobbles in conglomerates derived from an easterly source.

Thus the late Arenig–Caradoc evolutionary history of the basin records a varied sedimentary and magmatic development, including oceanic lithosphere accretion and related pelagic sedimentation through to turbiditic infill, with a coeval magmatic arc to the NW of the main basin. Sediment dispersal patterns in the S mainly indicate a southeasterly continental source, and further N palaeocurrents indicate submarine dispersal from both NW and SE into a marine trough deepening to the SW. Clast material from the NW was predominantly derived from the arc and adjacent shelf limestones, and that from the SE was largely of continental derivation (Pedersen 1981). The bulk of the evidence thus suggests the presence of a continental block along the E or SE margin of the basin, with the arc constructed to the W of the basin, above easterly dipping and subduct-ing oceanic lithosphere. The magmatic activity represented on Smøla and at Hølonda supports this model, the Smøla rocks being of evolved island-arc character whereas most of the volcanics at Hølonda to the E are more like a continental margin type. The spatial and temporal variations of spreading ridge, island-arc and continental margin-type volcanism suggest a model involving at least two phases of widening and subsequent rapid narrowing of the marginal basin with displacement of the trench and subduction zone southeastwards close to the continental margin.

With reference to the major plate tectonic configuration, it is suggested that the marginal basin and arc were located in the eastern part of the Iapetus, adjacent to either the Baltoscandian plate or an elongate microcontinental plate separated from the Baltoscandian plate by a marine gulf (Roberts 1980). However, the Arenig–mid-Llanvirn faunas in the sediments deposited within the marginal basin are largely of North American affinity. Clearly, any model which is based solely on these faunas belonging strictly to either the North American or the Baltoscandian plate is too simplistic. The compromise model involving an elongate microcontinental plate within the Iapetus, with arc and basin volcanism and sedimentation related to a destructive plate margin along its NW side, offers the most satisfactory explanation of the current palaeontological, geochemical and sedimentological data.

ACKNOWLEDGMENTS: The authors are grateful to Drs D. L. Bruton and H. Furnes for written and oral discussion on points arising from this manuscript, and to two anonymous reviewers for their helpful comments before the final revision.

References

ATHERTON, M. P., McCOURT, W. J., SANDERSON, L. M. & TAYLOR, W. P. 1979. The geochemical character of the segmented Permian coastal batholith and associated volcanics. *In*: ATHERTON, M. P. & TARNEY, J. (eds) *Origin of Granite Batholiths*, 45–64. Shiva, Kent.

BAILEY, E. B. & HOLTEDAHL, O. 1938. Northwestern European Caledonides. *Reg. Geologie der Erde*. **2**(II), 1–76. Akademie-Verlag, Leipzig.

BERGSTRÖM, S. 1979. Whiterockian (Ordovician) conodonts from the Hølonda Limestone of the Trondheim region, Norwegian Caledonides. *Nor. geol. Tiddskr*. **59**, 295–307.

BERRY, W. B. N. 1968. Age of Bogo shale and western Ireland graptolite faunas and their bearing on dating early Ordovician deformation and metamorphism in Norway and Britain. *Nor. geol. Tiddskr*. **48**, 217–30.

BROWN, G. M., HOLLAND, J. G., SIGURDSSON, H., TOMBLIN, J. F. & ARCULUS, R. F. 1977. Geochemistry of the lesser Antilles volcanic island arc. *Geochim. cosmochim. Acta*, **41**, 785–801.

BRUTON, D. L. & BOCKELIE, J. F. 1979. The Ordovician sedimentary sequence on Smøla, west central Norway. *Nor. geol. Unders*. **348**, 21–31.

—— & —— 1980. Geology and Palaeontology of the Hølonda area, western Norway—A fragment of North America? *Proc. IGCP Cal. Orogen Symp. Virginia Poly. Inst. Memoir* **2**, 41–7.

—— & —— 1982. The Løkken–Hølonda–Støren areas. *Paleontol. Contrib. Oslo Univ*. **279**, 77–86.

CARSTENS, C. W. 1919. Oversigt over Trondhjemsfeltets bergbygning. *Kgl. Norske Videnskp. Selsk. Skr.* **1**, 152 pp.

—— 1960. Stratigraphy and volcanism of the Trondheimsfjord area Norway. *Nor. geol. Unders.* **212a**, 27–39.

DOSTAL, J., ZENTILLI, M., CAELLES, J. C. & CLARK, A. M. 1977. Geochemistry and origin of volcanic rocks of the Andes (26°–28°S). *Contrib. Miner. Petrol.* **63**, 113–28.

FURNES, H., ROBERTS, D., STURT, B. A., THON, A. & GALE, G. H. 1980. Ophiolite fragments in the Scandinavian Caledonides. *Proc. Int. Ophiolite Symp. Cyprus*, 582–600.

——, RYAN, P. D., GRENNE, T., ROBERTS, D., STURT, B. A. & PRESTVIK, T. 1984. Geological and geochemical classification of the ophiolite fragments in the Scandinavian Caledonides. *Proc. Uppsala Caled. Symp.* Wiley, London.

GALE, G. H. & ROBERTS, D. 1972. Palaeogeographical implications of greenstone petrochemistry in the Southern Norwegian Caledonides. *Nature (Phys. Sci.)* **238**, 60–1.

—— & —— 1974. Trace element geochemistry of Norwegian lower Palaeozoic basic volcanics and its tectonic implications. *Earth planet. Sci. Lett.* **22**, 380–90.

GRENNE, T. 1980. Vassfjellet area. In: WOLFF, F. C. (ed.) *Guide to Excursions. Excursions across part of the Trondheim Region, Central Norwegian Caledonides. Nor. geol. Unders.* **356**, 159–64.

—— 1981. The Høydal deposit, Løkken—an ophiolite-hosted massive sulphide deposit in the central Norwegian Caledonides. (Abstract). *Trans. Inst. Min. Metall. B* **90**, 58–9.

—— & ROBERTS, D. 1980. Geochemistry and volcanic setting of the Ordovician Forbordfjell and Jonsvatn greenstones, Trondheim region, central Norwegian Caledonides. *Contrib. Miner. Petrol.* **74**, 374–86.

—— & —— 1981. Fragmented ophiolite sequences in Trondelag, central Norway. *Excursion Guide, Uppsala Symposium.* B. 12.

——, GRAMMELTVEDT, G. & VOKES, F. M. 1980. Cyprus-type sulphide deposits in the western Trondheim district, central Norwegian Caledonides. *Proc. Int. Ophiolite, Symp. Cyprus*, 727–43.

HARDENBY, C., LAGERBLAD, B. & ANDREASSON, P. G. 1981. Structural development of the Northern Trondheim nappe complex, central Scandinavian Caledonides. (Abstract). *Terra Cognita*, **1**, 50.

HELZ, R. T. 1976. Phase relations of basalts in their melting ranges at $P_{H_2O} = 5Kb$. II Melt compositions. *J. Petrol.* **17**, 139–93.

HUMPHRIS, S. E. & THOMPSON, G. 1978. Trace-element mobility during hydrothermal alteration of oceanic basalts. *Geochim. cosmochim. Acta*, **42**, 127–36.

JAKES, P. & GILL, J. 1974. Rare earth elements and the island arc tholeiitic series. *Earth planet. Sci. Lett.* **9**, 17–28.

—— & WHITE, A. J. R. 1972. Major and trace element abundances in volcanic rocks of Orogenic areas. *Bull. geol. Soc. Am.* **83**, 29–40.

KLINGSPOR, I. & GEE, D. G. 1981. Isotopic age-determination studies of the Trøndelag trondjemites. (Abstract). *Terra Cognita*, **1**, 55.

LOESCHKE, J. 1976. Petrochemistry of eugeosynclinal magmatic rocks of the area around Trondheim (central Norwegian Caledonides). *Neues Jahrb. Mineral Abh.* **128**, 1–44.

LOPEZ-ESCOBAR, L., FREY, F. A. & VERGARA, M. 1977. Andesites and high-alumina basalts from the Central-South Chile High Andes: geochemical evidence bearing on their petrogenesis. *Contrib. Miner. Petrol.* **63**, 199–228.

MACKENZIE, D. E. & CHAPPEL, B. W. 1972. Shoshonitic and calc-alkaline lavas from the Highlands of Papua, New Guinea. *Contrib. Miner. Petrol.* **35**, 50–62.

MEIJER, A. & REAGAN, M. 1981. Petrology and geochemistry of the island of Sarigan in the Mariana Arc; calc-alkaline volcanism in an oceanic setting. *Contrib. Miner. Petrol.* **77**, 337–54.

MOORES, E. M. 1982. Origin and emplacement of ophiolites. *Rev. Geophys. Space Phys.* **20**, 735–60.

MORRISON, G. W. 1980. Characteristics and tectonic setting of the shoshonite rock association. *Lithos*, **13**, 97–108.

MYASHIRO, A. 1974. Volcanic rock series in island arcs and active continental margins. *Am. J. Sci.* **274**, 321–55.

NEUMAN, B. & BRUTON, D. 1974. Early Ordovician fossils from Hølonda, Trondheim region, Norway. *Nor. geol. Tidsskr.* **56**, 69–115.

PAYNE, J. & STRONG, D. 1979. Origin of the Twillingate Trondjemite, north-central Newfoundland: partial melting in the roots of an island arc. *In*: BARKER, F. (ed.) *Trondhjemites, Dacites and Related Rocks*, Elsevier, Amsterdam. 489–516.

PEARCE, J. A. 1980. Geochemical evidence for the genesis and eruptive setting of lavas from the Tethyan ophiolites. *Proc. Int. Ophiolite Symp. Cyprus*, 261–72.

PEDERSEN, P. Å. 1981. Resedimenterte konglomerater og turbiditer på overgangen mellom undre/Øvre Hovingruppe (Llandeilo-Caradoc) i Åsen-Området, Nord Trøndelag. Cand. real. Thesis, Univ. of Bergen, 127 pp.

PRESTVIK, T. 1974. Supracrustal rocks of Leka, Nord Trøndelag. *Nor. geol. Unders.* **311**, 65–87.

—— 1980. The Caledonian ophiolite complex of Leka, north central Norway. *Proc. Int. Ophiolite Symp. Cyprus*, 555–66.

ROBERTS, D. 1972. Penecontemporaneous folding from the Lower Palaeozoic of the Trondheimsfjord region, Norway. *Geol. Mag.* **109**, 235–42.

—— 1975. The Stokkvola conglomerate—a revised stratigraphical position. *Nor. geol. Tidsskr.* **55**, 361–71.

—— 1980. Petrochemistry and palaeogeographic setting of the Ordovician volcanic rocks of Smøla, central Norway. *Nor. geol. Unders.* **359**, 43–60.

—— 1982a. *En foreløpig rapport om geokjemien au Frostagrønnsteinene*. NGU report (unpubl.), 4 pp.

—— 1982b. Disparate geochemical patterns from the

Snåsavatn greenstone, Nord Trøndelag, central Norway. *Nor. geol. Unders.* **373**, 63–73.

—— & GALE, G. H. 1978. The Caledonian–Appalachian Iapetus Ocean. *In*: TARLING, D. H. (ed.) *Evolution of the Earth's Crust*, 255–342. Academic Press, London and New York.

—— & GEE, D. 1981. Caledonian tectonics in Scandinavia. (Abstract). *Terra Cognita*, **1**, 69–70.

—— & STURT, B. A. 1980. Caledonian deformation in Norway. *J. geol. Soc. London*, **137**, 241–50.

—— & WOLFF, F. C. 1981. Tectonostratigraphic development of the Trondheim region Caledonides, Central Norway. *J. Struct. Geol.* **3**, 487–94.

——, STURT, B. A. & FURNESS, H. 1984. Volcanite assemblages and environments in the Scandinavian Caledonides and the sequential development history of the mountain belt. *Proc. Uppsala Caled. Symp.* Wiley, London.

RYAN, P. D. & STURT, B. A. 1984. Early Caledonian orogenesis in north-western Europe. *In*: STURT, B. A. & GEE, D. (eds) *The Caledonide Orogen—Scandinavia and Related Areas*. Wiley, London.

——- & WILLIAMS, D. M. 1984. The shale geochemistry of the Hovin group, Meldal, Sør Trøndelag, Norway. *In*: STURT, B. A. & GEE, D. (eds) *The Caledonide Orogen–Scandinavia and Related Areas*. Wiley, London.

——, SKEVINGTON, D. & WILLIAMS, D. M. 1980. A revised interpretation of the Ordovician stratigraphy of Sør Trøndelag and its implications for the evolution of the Scandinavian Caledonides. *Proc. IGCP Cal. Orogen. Symp. Virginia, Poly. Inst. Memoir*, **2**, 99–103.

SAUNDERS, A. D., TARNEY, J. & WEAVER, S. 1980. Transverse geochemical variations across the Antarctic peninsula: Implications for the genesis of calc-alkaline magmas. *Earth planet. Sci. Lett.* **46**, 344–60.

SIEGERS, A., PICKLER, H. & ZIEL, W. 1969. Trace element abundances in the 'Andesite' Formation of Northern Chile. *Geochim. cosmochim. Acta*, **33**, 882–87.

SIZE, W. 1981. Origin of Trondhjemite in relation to Appalachian–Caledonide palaeotectonic settings. (Abstract). *Terra cognita*, **1**, 73.

SPRINGER PEACEY, J. 1964. Reconnaisance of the Tømmerås Anticline *Nor. geol. Unders*. **227**, 13–84.

STØRMER, L. 1967. Some aspects of the Caledonian geosyncline and foreland west of the Baltic Shield. *J. geol. Soc. London*, **123**, 185–214.

STRAND, T. 1960. Cambro-Silurian deposits outside the Oslo region. *Nor. geol. Unders*. **208**, 151–69.

STURT, B. A. 1978. The Norwegian Caledonides—introduction. *Geol. Surv. Canada Paper*. **78-13**, 13–15.

TÖRNEBOHM, A. E. 1896. Grunddragen af det centrala Skandinaviens bergbyggrad. *Kgl. Sv. vet. akad. Handl.* **28**, 1–212.

VOGT, TH. 1940. Geological notes of the *Dictyonema* locality and the upper Gauldal district in the Trondheim area. *Nor. geol. Tidsskr.* **20**, 171–92.

—— 1945. The geology of part of the Hølonda-Horg district, a type area in the Trondheim region. *Nor. geol. Tidsskr.* **25**, 449–528.

WOLFF, F. C. 1976. Geologisk kart over Norge, berggrunnskart Trondheim 1:250,000. *Nor. geol. Unders*.

D. ROBERTS, Norges Geologiske Undersøkelse, Postboks 3006, 7001 Trondheim, Norway.

T. GRENNE, Geologisk Institutt, Norges Tekniske Høgskole, 7034 Trondheim-NTH, Norway.

P. D. RYAN, Department of Geology, University College, Galway, Ireland.

The Ordovician marginal basin of Wales

B. P. Kokelaar, M. F. Howells, R. E. Bevins, R. A. Roach & P. N. Dunkley

SUMMARY: The Lower Palaeozoic Welsh Basin was founded on immature continental crust. During late Precambrian–early Cambrian times, volcanism and sedimentation were influenced by NE–SW-trending faults which defined the NW and SE margins of the basin. During the Cambrian, marine sediments infilled a graben and at the end of the Tremadoc widespread tectonism was associated with an island-arc volcanic episode. In the Ordovician this subduction-related activity was succeeded by mainly tholeiitic volcanism related to back-arc extension, with the locus of arc volcanism sited further N, in the Lake District—Leinster Zone of the Caledonides. In Wales, the Ordovician volcanic activity shifted in time and space. In S Wales volcanism persisted from the middle Arenig through the Llanvirn. In N Wales the volcanism can be broadly divided into dominantly pre-Caradoc activity in southern Snowdonia and an intra-Caradoc episode in central and northern Snowdonia. In eastern Wales, including the Welsh Borderland, and in Llŷn, both episodes are represented. In all areas faults greatly influenced both volcanism and sedimentation. Intrusive activity was dominated by high-level emplacement of sills. Granite (*s.l.*) stocks are restricted to central and northern Snowdonia and Llŷn and many were coeval with extrusive volcanism.

Volcanism in the basin was essentially bimodal with voluminous eruptions of tholeiitic basalts with ocean-floor affinities, and of rhyolites. Minor volumes of andesite to rhyodacite resulted from low-pressure fractional crystallization of the tholeiitic basalts. Available evidence suggests that the rhyolites resulted mainly from crustal fusion, although in some instances evolution by crystal fractionation from intermediate magma has been proposed. Calc-alkaline assemblages are petrographically distinct, of minor occurence and, contrary to previous conclusions, are relatively insignificant in the characterization of the tectonic environment of the basin.

Throughout the basin, volcanism was generally succeeded by deposition of black muds and then turbidite-dominated sequences.

The major outcrops of Ordovician igneous rocks in Wales and the Welsh Borderland are shown in Fig. 1. In N Wales these rocks were first described in detail by Ramsay (1881). In a distinguished essay, Harker (1889) described the petrography of the volcanic rocks, defined possible centres and even postulated the extent of related extrusions. Later work concentrated on establishing the lithostratigraphical detail in selected areas, and the broad biostratigraphical divisions (for references see Bassett 1969). Apart from a few expressions of doubt, notably by Dakyns & Greenly (1905) and Williams (1927), the early consensus was that, because the extrusive rocks were interbedded with marine sedimentary rocks, they had been emplaced in a submarine environment. However, with the independent recognition by Oliver (1954) and by N. Rast, R. V. Beavon and F. J. Fitch (summarized by Fitch 1967) that many of the rhyolites are welded ash-flow tuffs, this assumption was revised in the belief that such rocks could not form after subaqueous emplacement. Consequently the palaeogeographic setting was modified to account for widespread and repeated emergence of land. Recently, theoretical considerations (Sparks *et al.* 1980) and field evidence (see below) have shown that

subaqueous emplacement and welding of silicic tuffs is likely and does occur, and it is now argued (e.g. Howells & Leveridge 1980) that the environment was dominantly marine and that the few determinable volcanic islands were small and short-lived.

In SW Wales notable early investigations are recorded in accounts of the volcanic rocks on Ramsey Island (Kidd 1814) and around Fishguard (Reed 1895). The first detailed studies of the volcanic successions were made by Cox (1915, 1930), Cox & Jones (1914), Cox *et al.* (1930a,b), Pringle (1930) and Thomas & Cox (1924), and later by Evans (1945) and Thomas & Thomas (1956). Recently, the principal aim of investigations (Bevins 1979; Bevins & Roach 1979a,b; Kokelaar 1982; Kokelaar *et al.*, this volume; Lowman & Bloxam 1981) has been a modern interpretation of the sequences in terms of environments and emplacement processes.

To the E, near Llandeilo, the lithostratigraphy and biostratigraphy of Llanvirn strata, which include volcanic rocks, has been described by Williams (1953), Lockley & Williams (1981) and Williams *et al.* (1981), and the nature of Caradoc igneous rocks around Llanwrtyd Wells was reported by Stamp & Wooldrige (1923). The igneous rocks of the Builth–Llan-

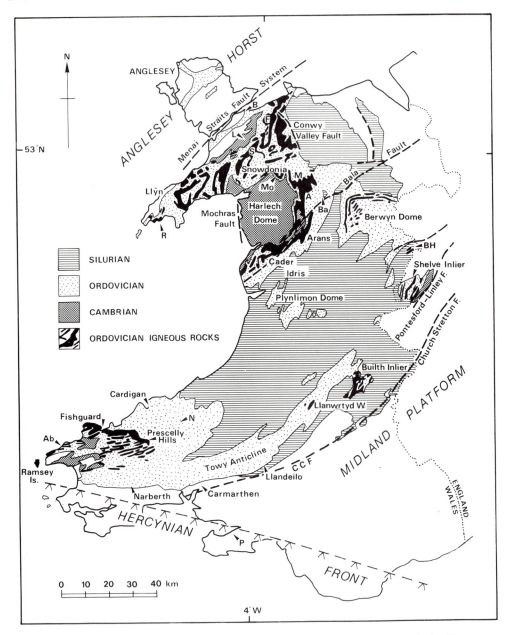

FIG. 1. Simplified geology of the Welsh Basin. (A—Arenigs; Ab—Abereiddi; B—Bangor; Ba—Bala; BH—Breidden Hills; CCF—Carreg Cennen Fault; F—Foel Fras Volcanic Complex; L—Llanberis; M—Migneint; Mo—Moelwyns and Manods; N—Newcastle Emlyn; P—Penclawdd; R—Rhiw; S—Snowdon).

drindod Inlier (hereafter abbreviated to the Builth Inlier) were described in detail following the mapping of Jones & Pugh (1941, 1946, 1948a,b, 1949). These authors described an Ordovician shoreline with reworking of predominantly subaerial Llanvirn volcanic rocks, and gave an account of the complex form of nearby basic intrusions of Caradoc age.

The igneous rocks of the Shelve and Breidden Hills areas have been described by Blyth (1938, 1944) and Watts (1885, 1925), and the palaeogeography of tuffs in the Berwyn Dome has been discussed in detail by Brenchley (1972, 1978) and Brenchley & Pickerill (1980).

Recently the Welsh Ordovician igneous rocks have figured significantly in numerous models of Caledonide plate tectonics (e.g. Fitton & Hughes 1970; Phillips *et al.* 1976; Fitton *et al.* 1982), and there is a consensus that they were emplaced in an ensialic basinal environment, which developed along a destructive plate margin associated with closure of the Iapetus Ocean to the NW.

In this paper recent detailed field and petrochemical studies of the major outcrops of Ordovician igneous rocks in the Welsh Basin are summarized, to demonstrate the volcanic and associated sedimentary and tectonic processes that occurred, and to revise existing views concerning the petrochemistry and hence the tectonic affinities of the basin. Detailed descriptions, maps and interpretations of several of the volcanic terranes are given separately in a field guide (Kokelaar *et al.*, this volume). Correlations of Ordovician strata across the basin are imprecise and there is a wealth of parochial lithostratigraphical nomenclature (see Allen 1982). In order to emphasize processes, this paper treats the major outcrops separately with minimal use of formal terminology.

The base of the Arenig is taken as the base of the Ordovician. A revised Ordovician timescale (McKerrow *et al.* in press) indicates a 57 Ma duration, with the beginning of the Arenig at 492 Ma, the Llanvirn at 470 Ma, the Llandeilo at 461 Ma, the Caradoc at 454 Ma and the Ashgill at 442 Ma.

Basement and general tectonic setting

The Lower Palaeozoic rocks of Wales accumulated on continental crust which extended in a belt from the Midland Platform in Britain, through SE Newfoundland and New Brunswick, into the NE Appalachians (Kennedy 1979). Radiometric evidence suggests that in Britain this basement is mostly younger than *c.* 900 Ma, although slivers of older crust may be present (Thorpe *et al.* 1984; see also Hampton & Taylor 1983). It comprises gneisses, schists and igneous rocks with calc-alkaline affinities, overlain by low-grade, commonly flyschoid sedimentary and various igneous rocks which locally range into the Lower Cambrian. The latter are mostly calc-alkaline products of ensialic volcanic arcs, although some show affinities with ocean-floor basalts and others with continental tholeiites. The rocks represent craton development, mainly by accretion of volcanic arcs (Thorpe 1974, 1979; Rast *et al.* 1976). A pronounced structural grain in

the Precambrian basement is indicated by repeated activation, during the Palaeozoic, of certain fault zones. A strong NE–SW grain is evident, both at the margins and within the basin, and a lesser N–S grain is also determinable (see below). The NE–SW lineaments in particular probably reflect major tectonic discontinuities in the basement.

Faunal evidence (Cocks & Fortey 1982) indicates that, during the Arenig and lower Llanvirn, New Brunswick, E Newfoundland, S Britain, France and Bohemia constituted the northern margin of Gondwanaland, located at southern hemisphere cold temperate to subarctic latitudes. Studies of palaeomagnetism, however, indicate southern tropical latitudes (Piper 1978; Smith *et al.* 1973, 1981). Disparate faunas in the Baltic region are attributed (Cocks & Fortey 1982) to separation from Gondwanaland by an intervening true ocean, Tornquist's Sea. At this time both Gondwanaland and Baltica were separated by the Iapetus Ocean from the equatorially situated North American continent. Closure of Tornquist's Sea during the Caradoc is indicated by faunal similarity between S Britain and Baltica from the Ashgill through the Silurian. Also, from Caradoc times earlier faunal distinctions between opposing sides of the Iapetus Ocean began to diminish, heralding its late-Silurian–early-Devonian closure.

The development of the Ordovician basin of Wales was broadly contemporaneous with subaerial and submarine transitional tholeiitic and calc-alkaline arc volcanism in the Lake District–Leinster Zone, which presently lies to the NW, beyond the Precambrian of the Anglesey (Irish Sea) Horst (Stillman & Francis 1979). Most plate tectonic models (e.g. Phillips *et al.* 1976; Fitton *et al.* 1982) indicate southeasterly subduction of Iapetus oceanic lithosphere from a trench NW of the Lake District–Leinster arc, even though no associated fore-arc terrane has been positively identified and the original relative positions of the Anglesey Horst and the Lake District, and hence the Welsh Basin and the arc to the NW, are uncertain. The presence of a volcanic arc to the NW places the Welsh Basin in a back-arc setting with a subjacent mantle wedge and subduction zone. Whilst suspecting that future work will show this model to be a simplification, it is considered the best available at present.

Cambrian developments

Many major structures which influenced the development of the Ordovician basin in Wales were established before or during the Cambrian.

In NW Wales the Anglesey Horst and the fault zone along its SE side exerted a profound influence on early Cambrian volcanism and sedimentation. In the Llanberis and Bangor regions more than 2 km of acidic ash-flow tuff of the Arfon Group (Arvonian of Greenly (1944); Reedman *et al.* 1984) was ponded in a narrow graben, bound on the NW by the Dinorwic Fault. Only minor volumes of tuff escaped from the confines of the graben during eruption, suggesting extremely rapid volcano-tectonic subsidence. Similarly, the succeeding sediments, including coarse alluvial and shallow-water sandstones, vary dramatically in thickness and lithology and are related to continued movement along the Dinorwic Fault, and along several other closely spaced NE-trending faults, including the Aber Dinlle Fault (see Institute of Geological Sciences 1984). The NW limit of the Arfon Group probably lay along the scarp-foot of the NE-trending Berw Fault on Anglesey (Reedman *et al.* 1984). Clear correlations in the Arfon Group across the faults on to Anglesey, and considerations of provenance, show that there have been no substantial transcurrent movements on the faults since deposition of the Arfon Group (Reedman *et al.* 1984; cf. Nutt & Smith 1981). The faults are the first indication of a profound crustal weakness which was to be repeatedly activated and would define the NW margin of the Ordovician marginal basin. Historic earthquakes along the Dinorwic Fault (Wood 1974) testify to its continued movement. The faults are here collectively referred to as the Menai Straits Fault System (Fig. 1), although no fault actually lies in the straits.

Faults, with probable basement control, influenced sedimentation in the succeeding Llanberis Slates Formation (Howells *et al.*, in press), with coarse turbiditic sandstones interbedded with silty mudstones infilling rifted depressions (Webb 1983). Correlation between the thick Cambrian succession of greywackes, sandstones, siltstones and mudstones in the Harlech Dome (Institute of Geological Sciences 1982) and the succession about Llanberis is extremely difficult, but from S to N there is a marked thinning due both to active faulting, which restricted accumulation, and to late Cambrian overstep (George 1963).

In SW Wales the Cambrian succession rests unconformably on late Precambrian andesite lavas and tuffs, granites, granophyres and diorites. Above a thin basal conglomerate, the lower Cambrian sequence (Caerfai Group) records a transgression with deepening of the depositional environment from intertidal,

through shallow marine, to below wave base (Crimes 1970a). These strata are unconformably overlain by further shallow-marine sediments (Solva Group), which contain clasts derived from the Precambrian (Williams & Stead 1982) and are succeeded by black mudstones with turbidites in the upper part (Menevian Group: Rushton 1974). The overlying *Lingula* Flags, once thought to have resulted from turbiditic input of coarse terrigenous detritus (George 1970; Stead & Williams 1971), are now interpreted as having accumulated in an intertidal to subtidal environment (Turner 1977).

The general derivation of detritus from the S and SE (Crimes 1970a) suggests reactivation along a major basement fault or fault zone which lay in this direction and defined the basin margin.

To the ENE the Towy (Tywi) Anticline and the Carreg Cennen Fault strike into the Pontesford–Linley and Church Stretton Faults (Fig. 1). This fault system defined the boundary between the Midland Platform to the SE and the Cambrian and Ordovician basins to the NW. East of the fault system, a sequence of Cambrian shallow-water sandstones and carbonates, 700 m thick, oversteps the platform (basement) and numerous tectonically induced breaks with either erosion or non-deposition can be determined. These sediments contrast markedly with thicker, deeper-water, clastic-dominated sequences to the W, although by late Cambrian times extensive open-sea conditions were established across most of the area (Bassett 1980).

In the late Tremadoc a major episode of tectonism and volcanism terminated the Cambrian basinal sedimentation (Kokelaar 1977, 1979; Kokelaar *et al.* 1982). In N Wales most basal Arenig strata show either disconformable or low-angle unconformable relationships with mainly Upper Cambrian strata (Mawddach Group: Allen *et al.* 1981), but to the N of Llanberis there is major overstep of the Cambrian, with the Arenig resting on the Arfon Group in the Bangor area (Wood 1969; Reedman *et al.* 1983) and on the Precambrian on Anglesey. Clearly, the Menai Straits Fault System had been reactivated.

Along the E side of the Harlech Dome, Cambrian sediments were folded, faulted and locally eroded prior to subaerial eruption and shallow subvolcanic intrusion of magmas with distinct island-arc characteristics transitional between low-K tholeiitic and calc-alkaline (Kokelaar 1977, 1979). The volcano, partly preserved as the Rhobell Volcanic Complex, was centred over a N–S basement weakness known as the Rhobell Fracture. As well as influencing the location of the volcano, this defined the E margin of a newly emergent horst of Cambrian strata, the proto-Harlech Dome. The Mochras Fault (Fig. 1) may represent the complementary basement weakness (Kokelaar 1977, 1979). Tectonism along the Rhobell Fracture persisted through and outlived the volcanic episode such that the volcanic complex was folded, faulted and deeply eroded before being unconformably overlain by the basal Arenig. This volcanic-arc episode, also represented in S Wales by the Trefgarn Volcanic Group of Tremadoc or lowermost Arenig age, is considered to mark the beginning of the southeasterly subduction of Iapetus lithosphere that eventually led to the closure of the Iapetus Ocean. Clasts of igneous rock from this volcanic episode are common in the lowermost Arenig strata in N Wales and indicate that the arc volcanoes were extensive. However, the widespread Arenig transgression reflected a distinct change in tectonic conditions and heralded the development of the Ordovician marginal basin (Kokelaar 1979; Dunkley 1979).

The Ordovician marginal basin

The most extensive outcrop of Ordovician strata in N Wales (Fig. 1) forms a belt around the S, E and N margins of the Harlech Dome (southern Snowdonia), extending to a broad synclinorium further N (central and northern Snowdonia). On the western edge of the latter outcrop Cambrian rocks are exposed but further W an Ordovician sequence occurs in Llŷn (Lleyn). To the E of Snowdonia the Ordovician is overlain by the Silurian but reappears further E in the Berwyn Dome. Other major outcrops occur in SW Wales, in E Wales in the Builth Inlier and near Llanwrtyd Wells, and along the Welsh Borderland, most notably in the Shelve Inlier and the Breidden Hills area.

Southern Snowdonia

Around the Cambrian rocks of the Harlech Dome, the Ordovician rocks crop out from the Cader Idris escarpment in the S, through the Aran and Arenig Mountains (referred to as the Arans and Arenigs respectively) on the E, to the Manod and Moelwyn Mountains (Manods and Moelwyns) in the N. The sequence is dominated by volcanic rocks of Arenig–early Caradoc age.

Sedimentation

The Arenig sediments are mostly composed of a thin, impersistent basal conglomeratic sandstone with phosphatic oncoliths (the so-called *Bolopora undosa*: Hofmann 1975), overlain by quartzose and feldspathic sandstones, laminated sandstones and silty mudstones, and tuffites. The sediment constituents indicate two main sources; the quartz grains from uplift and erosion of thick greywackes of Cambrian age, and feldspars and lithic clasts from igneous rocks of the late Tremadoc volcanic-arc episode. In the Migneint area, sparse metamorphic clasts indicate a lesser source, possibly in the Precambrian rocks of the Anglesey Horst (Lynas 1973).

SE of the Bala Fault, in the Arans, thick and coarse alluvial conglomerates, composed of clasts of Tremadoc volcanic-arc igneous rock, accumulated along and just NW of a fault-scarp with uplift on the SE (cf. Dunkley 1978). The source of this material must have lain to the E or SE, indicating the former existence of the igneous rocks far beyond their present outcrops (see Institute of Geological Sciences 1982). Northwest of the Bala Fault, along the E side of the proto-Harlech Dome (horst), block movements along N–S faults dominated patterns of erosion and sedimentation and reflected reactivation of the Rhobell Fracture. Here the Arenig rocks over most of the outcrop have been interpreted as reflecting delta progradation over the erosion surface recently submerged between horsts, later giving way to more general marine conditions (Lynas 1973; Ridgeway 1976; Kokelaar 1979). Flaggy feldspathic sandstones, interbedded with mudstones containing a restricted graptolite fauna, occur near the top of the Arenig sequence in the Arenigs and indicate a shallow subtidal environment (Zalasiewcz 1982).

The overlying sequence, through the Llanvirn and probably into the early Caradoc, is dominated by volcanic rocks which are interbedded with marine siltstones and mudstones, without widespread influxes of coarse terrigenes. Coarse sandstones do occur locally and their high content of volcanic clasts and restricted occurrence suggest that they are directly related to local uplift and reworking. Desiccation cracks in mudstones associated with the volcanics about the Arenigs (Zalasiewicz 1982) reflect temporary emergence. Chamositic algal concretions in mudstones in the Arans indicate a low rate of detrital sedimentation in an environment interpreted as shallow marine (Dunkley 1978).

Throughout southern Snowdonia the volcanic rocks are overlain by mudstones and silt-stones of early Caradoc age with little evidence of reworking. Most of the succeeding Caradoc and Ashgill strata are argillaceous. The upper-most beds of Caradoc age are pelagic black graptolitic mudstones which were deposited at this time across most of the basin in N Wales. These are overlain by mudstones, siltstones and, locally, sandstones of Ashgill age. These sediments generally become more turbiditic upwards and hence similar to the succeeding Silurian strata.

Volcanism

The volcanic rocks and their associated sedi-ments in the Ordovician of southern Snowdonia have recently been termed the Aran Volcanic Group (Ridgway 1975; Dunkley 1978; 1979; Institute of Geological Sciences 1982). The age of the group ranges through the Arenig to poss-ibly early Caradoc. In the S, on Cader Idris, there is an alternating sequence, up to 2 km thick, of acid and basic extrusive rocks with interbedded sediments (Cox 1925; Davies 1959). The acidic horizons are mainly of rhyoli-tic ash-flow tuffs, both welded and non-welded, and the basic rocks comprise two horizons of basaltic pillow lavas and thin tuffs interbedded with graptolitic mudstones. By their sedimen-tary association and lack of extensive rework-ing, the volcanic rocks are interpreted as having been subaqueously emplaced. Several major and minor dolerite sills, and semi-concordant and transgressive bodies of granophyre, also occur. SW of Cader Idris a N–S oriented fault was active, to the W of which several volcanic and sedimentary horizons are missing and are replaced by a condensed sequence of nodular and oolitic mudstones and ironstones with non-sequences (Jones 1933; Dunkley 1979).

To the NE, in the Arans, a similar bimodal alternation of extrusive rocks is recognized (Dunkley 1978, 1979), although in this direc-tion the group shows overstep and thins consid-erably. The acidic rocks are represented by thick, welded and non-welded, multiple ash-flow tuff units, locally underlain by rhyolite lavas and domes with carapaces of autobreccia and surrounding aprons of poorly sorted, poorly bedded breccias resulting from their col-lapse. The basic component is represented by pillowed basalts, hyaloclastites and tuffs, the latter indicating eruption above the Pressure Compensation Level (Fisher, this volume). Common slumping of these basic extrusions, together with the muds with which they were interbedded, is indicated by the widespread and repeated occurrence of debris flow and high-density turbidity current deposits. The thinning of the sequence is mainly attributable to uplift on the SE side of the Bala Fault, which is also reflected in the sediments (see above; Dunkley 1978). Dunkley (1979) considered that, in the Arans in contrast to elsewhere, each of the three lower acidic ash-flow tuffs were emplaced and eroded subaerially prior to rapid subsi-dence and re-establishment of marine condi-tions. However, with no evidence of reworking at the top of the uppermost multiple ash-flow tuff unit, which is intercalated with and overlain by mudstones, subaqueous emplacement was interpreted.

To the N of the Arans, the Aran Volcanic Group can be traced from the Bala Fault northwards into the Arenigs (Zalasiewcz 1982) and the Migneint area (Lynas 1973). In this direction successive formations overlap and the sequence thins. The lower acidic rocks comprise two subaqueously emplaced non-welded ash-flow tuffs, massive at the base and thinly bed-ded at the top due to secondary sloughing of tuff which was subsequently redeposited from sediment gravity flows or suspension. Subaque-ous emplacement is readily interpreted in the succeeding sequences of basaltic pillow lavas, hyaloclastites, slumped thick agglomeratic and crystal tuffs, and tuffites. These vary consider-ably in thickness and lithology along strike, and wedge-out laterally into marine siltstones and mudstones. The tuffites range from coarse peb-ble to fine silt in grade and display bedding from massive to thin and flaggy. Although they are predominantly of basaltic composition, an intermediate and acid component is present in places. As in the Arans, these are interpreted as having accumulated from debris and turbiditic flows which redistributed previously emplaced, or stored (Carey & Sigurdsson, this volume), pyroclastic debris, and locally incorporated sediment from the basin floor.

In the Arenigs, high-level penecontem-poraneous dacite domes intrude a sequence of crystal tuffs, debris flows and mudstones locally with desiccation cracks (see above; Zalasiewcz 1982). Lynas (1973) recognized similar intru-sions further N in the Migneint area, and associ-ated debris-flow breccias and volcanogenic conglomerates were interpreted as resulting directly from slumping of unlithified deposits caused by magma emplacement. However, this association was restricted to the E of a penecon-temporaneous N–S fault which profoundly affected the basin-floor topography and associ-ated sedimentation pattern, and is attributable to the continued influence of the Rhobell Frac-ture (Kokelaar 1979).

The uppermost, multiple, acidic ash-flow tuff unit of the Aran Volcanic Group extends across the total outcrop on the E side of the Harlech Dome. Here the tuffs are welded and non-welded, generally massive, although locally with ill-defined bedding, and typically comprise shards, feldspar and quartz crystals, and few lithic clasts, in a matrix originally of vitric dust. Impersistent dust-tuffs and breccias occur in places. The tuffs overlie, and are themselves overlain by, marine siltstones and the lack of extensive reworking of the flow tops or incision by drainage channels suggests that they were emplaced subaqueously, as in the Arans and on Cader Idris, beneath wave base and away from the influence of strong currents. The absense of intercalated sedimentary rocks suggests that accumulation must have been rapid, although mudstone is intercalated at one horizon in this sequence in the NE Arans.

In the Manods and Moelwyns area, on the N side of the Harlech Dome, the Aran Volcanic Group is dominated by rhyolite lavas with coeval high-level rhyolite sills and intrusive domes (Bromley 1965). A deeper, fault-bound, boss-like intrusion indicates the site of a possible source, and the disposition of the rocks about pronounced N–S faults suggests fault influence in magma ascent and eventual distribution. To the SW the group becomes dominated by debris-flow breccias, tuffs and tuffites which thin to a feather-edge further SW.

Throughout the outcrops of the Aran Volcanic Group the sources of the explosive acidic eruptions are problematical. Davies (1959) showed that a major transgressive granophyre intrusion on Cader Idris could be traced to the uppermost tuffs and suggested that it might represent their source. Within the group in the Arans, Dunkley (1979) recognized the remnants of an extrusive rhyolite dome, *c.* 600 m thick, which, with its subjacent intrusion, might mark an eruptive centre. In the SW Moelwyns, a volcanic neck, 260 m in diameter and containing rhyolite lapilli tuff and breccia with some accidental fragments, has been correlated with tuffs which lie stratigraphically some 460–600 m above (Bromley 1968). Such structures are uncommon and are unlikely to represent the vents of major eruptions.

The basic volcanic rocks are more local in distribution than the rhyolitic tuffs and there is no expression of a major centre. Rather, the magma reached the surface step-wise via the extensive basic sills and transgressive sills which are virtually ubiquitous beneath the highest basic extrusive rocks. At depth in the sequence these intrusions are thickest and of coarse grained dolerite, but at successively shallower levels they are thinner, finer grained and show pillowed and peperitic margins where they interacted with wet host sediments and tuffs during emplacement.

Central and northern Snowdonia

From the Moelwyns in the S, the folded Ordovician strata form a broad outcrop about the Snowdon massif and through northern Snowdonia to W of the Conwy Valley Fault, where they lie in juxtaposition with Silurian rocks to the E. The outcrop is dominated by volcanic rocks that lie almost entirely within the Caradoc.

Sedimentation

The basal Arenig is locally marked by bioturbated sandstones which lie with slight angular unconformity on Cambrian strata. To the NW this unconformity is more pronounced (see above and Reedman *et al.* 1983). These sandstones have been interpreted (Crimes 1970b) as having been deposited within or near the intertidal zone to below wave-base, with deposition much affected by fault movements along the SE side of the Anglesey Horst.

The overlying Arenig to early Caradoc strata comprise a monotonous sequence of marine mudstones, siltstones and impersistent sandstones (Howells *et al.* 1983), and there is no indication here of the volcanism which dominates the lower Ordovician sequence in southern Snowdonia. The strata reflect offshore sedimentation broadly keeping pace with subsidence, although temporarily reduced rates of sedimentation are indicated by impersistent beds of pisolitic ironstone. Towards its top, the sequence is dominantly of mudstones, which possibly reflect a eustatic transgression (Leggett 1980). Locally developed slumped mudstones and mudstone breccias, at the top of the sequence, reflect disruption of the basin floor by faulting and tilting immediately prior to the onset of volcanism.

Two major volcanic groups have been determined, a lower, Llewelyn Volcanic Group and an upper, Snowdon Volcanic Group (Howells *et al.* 1983). Sedimentation was little affected by the early Llewelyn group volcanism. Marine mudstones and siltstones are intimately associated with the thick accumulations of extrusive rocks, which themselves show little evidence of reworking, suggesting that the volcanic piles subsided rapidly and, or, were emplaced in deep water. However, towards the top of the Llewelyn Volcanic Group (in the Capel Curig Volcanic Formation) the associated sediments show a shallowing of the depositional environment, with incursions of coarse terrigenes, especially in the N of the district. These coarse sandstones have been interpreted (Howells & Leveridge 1980) as the deposits of subaerial alluvial fans encroaching into the basin due to rapid fault-block uplift to the N and W. From the NW to the SE a low-profile palaeoslope developed, through a complex coastal zone, into a marine environment. The position of the coastline is interpreted as having been controlled by an active NE–SW trending flexure, probably marking a concealed basement fracture, with subsidence to the SE.

This flexure continued to affect sedimentation in the interval between the emplacement of the Llewelyn and Snowdon Volcanic Groups and similar facies variations can be determined, with alluvial and fluvial sandstones in NE Snowdonia, thick-bedded, coarse-grained quartzose sandstones, locally cross-bedded and intercalated with siltstone in central Snowdonia, and offshore mudstones and siltstones in E Snowdonia.

Within the Snowdon Volcanic Group in central Snowdonia there is no clear indication of background sedimentation, though all the faunas in the volcaniclastic deposits are shallow marine. In NE Snowdonia siltstones and black graptolitic mudstones are intercalated with the thick sequences of volcanic rocks which show no signs of reworking, although some thin turbiditic sandstones can be distinguished. A relatively deep-water environment here is envisaged.

Widespread accumulation of pelagic black graptolitic mudstones after Snowdon group volcanism suggests general volcanic and tectonic quiescence by this time. The mudstones are equivalent to those seen in the uppermost part of the Caradoc in southern Snowdonia and indeed throughout much of N Wales. From strata preserved in NE Snowdonia, as in the S, sediments of Ashgill age compare more closely with the tubiditic accumulations of the lower Silurian than with the preceding Ordovician strata.

Volcanism

The Caradoc volcanic rocks are predominantly of acid or basic composition. They mainly belong to the Llewelyn Volcanic Group or Snowdon Volcanic Group, although there are distinctive expressions of volcanism in the intervening sedimentary rocks throughout the area, and also in the closely overlying strata in the NE.

The Llewelyn Volcanic Group (Howells *et al.* 1983) is largely confined to northern Snowdonia. Early eruptions from a number of penecontemporaneously active centres produced petrographically distinct, local volcanic piles. In NE Snowdonia an acidic centre is marked by high-level rhyolite intrusions with flow-folded and autobrecciated rhyolite lavas. To the SW these lavas abut and interdigitate with porphyritic trachyandesite lavas and tuffs erupted from a centre which in part is marked by intrusions of similar composition. These rocks constitute the Foel Fras Volcanic Complex (Fig. 1), a centre which is unique in this part of the basin in being dominated by magmas of an intermediate composition. To the S the lavas and tuffs from this centre interdigitate with basalts, pillow breccias and hyaloclastites, and also with welded rhyolitic ash-flow tuffs and lavas, the eruptive centre of which is marked by a rhyolite plug.

As noted above, the volcanism producing these lower piles in the Llewelyn Volcanic Group had little effect on the associated marine sediments. However, it is evident from variations in the thickness of the volcanics and associated sediments that locally both were restricted during accumulation by active faults. Further to the S these volcanic rocks thin through a zone of mixed tuffitic debris, isolated tuffs and thin lavas, to a feather-edge in marine siltstones on the W of the Snowdon massif (Howells *et al.*, in press).

The upper part of the Llewelyn Volcanic Group (the Capel Curig Volcanic Formation) is restricted to N and E Snowdonia and represents the first major phase of acidic ash-flow volcanism. It accumulated by eruptions from three centres, two of which were established in the subaerial environment in the N of the district (see above), with the third at the edge of the offshore marine environment to the SE (Howells & Leveridge 1980). From the subaerial sources, ash-flows transgressed from land into the sea where they retained sufficient heat to develop lobate bases by fluidization of subjacent unlithified sediments, and welded fabrics (Francis & Howells 1973; Kokelaar 1982). The accumulation about the subaqueous centre is characterized by mass- and debris-flows from previously emplaced pyroclastic debris, with few primary ash-flows. Accretionary lapilli tuffs around this third centre indicate that the eruptive column temporarily had a subaerial expression.

In the predominantly sedimentary sequence which separates the Llewelyn and Snowdon Volcanic Groups, volcanic activity is reflected in an extensive horizon of welded, rhyolitic ash-flow tuffs in central and western Snowdonia (Pitt's Head Tuff and Llwyd Mawr Ignimbrite: Roberts 1969), local basalt lavas and hyaloclastites, and repeated thin beds of distal ash-fall tuff (Howells *et al.* 1978, 1981). In the marine sediments which occur immediately below and above the ash-flow tuffs in central Snowdonia, there is no obvious change of facies or composition and this, coupled with the lack of signs of subaerial erosion (such as incision by drainage channels), indicates subaqueous emplacement. To the S of Snowdon the top of the sequence is marked by a thick wedge of breccias (the Llyn Dinas Breccias: Beavon 1963) which were probably produced by the disruption of strata during faulting and, or, as magma moved to a higher level immediately prior to the first major eruption of the Snowdon Volcanic Group. The breccias have been interpreted as representing sector collapse of a caldera wall (Beavon 1980). Further N they are missing, and there is unconformity at the base of the Snowdon Volcanic Group (see below).

The Snowdon Volcanic Group is essentially that defined by Williams (1927). In the vicinity of Snowdon it is composed predominantly of volcanic rocks, broadly with an acid–basic–acid cycle, whereas to the N and E intercalations of volcaniclastic and sedimentary rocks occur increasingly.

On Snowdon the lowest formation (Lower Rhyolite Tuff Formation) comprises up to 600 m of unbedded, mostly non-welded, rhyolitic ash-flow tuffs with local pyroclastic breccias at the base and bedded and reworked tuffs, conglomeratic in places, at the top (Howells *et al.*, in press). The basal deposits show faulting against subjacent sediments which can only have occurred during emplacement of the tuff (see Kokelaar *et al.*, this volume). Late-stage rhyolite intrusions within the main body of the tuff can be traced into overlying autobrecciated flow-folded lavas. Littoral erosion and minor unconformity with succeeding strata are evident on Snowdon. It is considered that this central area lies close to the source, which was mostly contained in a submarine environment although a small island was eventually formed and immediately eroded (see below). To the N of Snowdon the massive rhyolite tuff passes laterally into bedded tuffs which are intercalated with sedimentary rocks reflecting a continuously submarine environment proximal to the volcanic centre (Fig. 2). To the E a more distal, deeper-water environment is defined where two subaqueously emplaced, non-welded, rhyolitic ash-flow tuffs are interbedded with marine siltstones and mudstones (Howells *et al.* 1973). These two ash-flows probably changed from being hot gas-charged to cool water-saturated, during transport from their source near Snowdon.

In central Snowdonia the lowest formation of the Snowdon Volcanic Group is overlain by

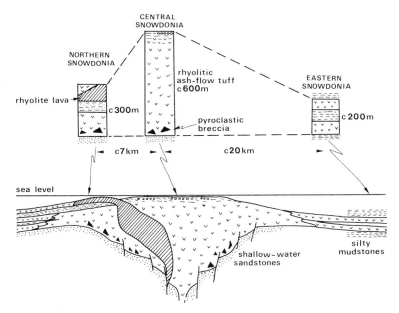

FIG. 2. Proximal to distal variations in accumulation of the lowest formation of the Snowdon Volcanic Group.

basic volcaniclastic deposits with basalt lavas, commonly pillowed (Bedded Pyroclastic Formation). The volcaniclastic deposits include local piles of hyaloclastite and bedded lithic breccias, but are mostly very well bedded, very coarse to very fine sandstones which commonly show cross-bedding, wave and current ripples, slumps, dewatering structures and bioturbation. These overall characters and associated faunas indicate a shallow, intertidal to subtidal environment.

In contrast, in E and NE Snowdonia the equivalent strata are rhyolitic tuffs interbedded and admixed with black graptolitic mudstones. Here a subaqueous volcanic centre has been determined in a thick sequence of rhyolitic ash-flow tuffs, tuffite debris flows and mudstones, which passes laterally into a thinner sequence of thin, in places blocky, ash-flow tuffs and fine grained ash-fall tuffs striped with laminae of black mudstone (Howells *et al.* 1978, 1981).

Thus two contrasting volcanic expressions and sedimentary environments have been defined within the Snowdon Volcanic Group; in central Snowdonia, of basic composition erupted in a shallow-marine environment, and in the E, of acid composition emplaced in a deeper-marine setting (Fig. 3). It has been proposed (Howells 1977) that these environments were separated by an active ridge, possibly controlled by a deep-seated fracture.

The upper part of the Snowdon Volcanic Group crops out mainly in E and NE Snowdonia. Here it is characterized by a widespread accumulation of acidic ash-flow tuffs with much incorporated black mudstone. An horizon of basaltic tuffs (Howells *et al.* 1981) in the black mudstones overlying the Snowdon Volcanic Group in NE Snowdonia reflects temporarily

renewed activity and is the final expression of Caradoc volcanism in Snowdonia. Overlying black graptolitic mudstones show no evidence of reworking of the underlying volcanics.

The nature of the source of the major ash-flow tuffs of the Snowdon Volcanic Group remains problematical. The distribution of the lower acidic rocks, their variations in thickness and lithological characters indicate a source in central Snowdonia. It was here that a central subaerial volcano-tectonic structure was proposed by Rast (1969) and Bromley (1969) and later elaborated by Beavon (1980). However, current work indicates a mainly submarine vent and that the main structural control of the volcanism was NE–SW and approximately N–S oriented faults, which were active prior to the volcanism and controlled the upward movement and eventual distribution of the magma. What has to be explained is the great thickness, up to 600 m, of ash-flow tuff close to its source, and the dramatic thinning away from it (Fig. 2). Near the source the tuff is underlain by marine sediments, locally with unconformity, and faulting of the base contemporaneous with tuff emplacement is evident. Unconformity and conglomerates at the top of the tuff in the source area indicate erosion of a small island located here, and the succeeding basic volcaniclastic deposits indicate a general shallow-water environment. For such a relationship to occur, the underlying strata must have sloughed or been eroded away prior to the eruption, perhaps owing to uplift and, or, tilting, and massive subsidence must then have occurred during the eruption to accommodate locally the great thickness of tuff. Normal faulting is envisaged such that the pile of tuff, rather than forming a wholly positive topographic feature with a flat

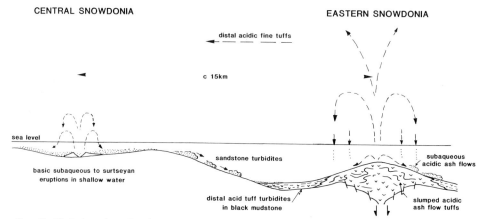

FIG. 3. Variations in volcanism and emplacement environment in the middle formation of the Snowdon Volcanic Group.

base, was mostly ponded in a deep depression. The dramatic thinning of the tuff excludes the possibility that subsidence occurred after the eruption. Except perhaps in the latest stage, the eruption must have occurred subaqueously. The two well-defined ash-flow tuffs in E Snowdonia escaped from the depression when the accumulation rate of the ash exceeded the subsidence.

As in southern Snowdonia, the dolerite sills and transgressive sills throughout this area tend to be thick and coarse grained at depth, and thinner and finer grained with wet host interactions at the shallowest levels. Proximity of the latter features to the localized piles of extrusive basalt strongly suggests that the sills represent the feeder systems.

In NE Snowdonia, at one of the basaltic centres locally active during Snowdon Volcanic Group times (Howells *et al.*, in press), distal rhyolitic ash-flow tuffs of the Snowdon group can be traced through a sequence of basaltic tuffs and hyaloclastites, the latter having a limited distribution close to their vent(s). For such a relationship to occur the basaltic tuffs could not have formed a marked positive feature on the basin floor. They either accumulated in a depression or local rapid subsidence occurred during their emplacement.

East of Snowdonia

To the E of the Conwy Valley Fault and its projection S to the Bala district, the expressions of Caradoc volcanism are less prominent. However, the distal representatives of the two major rhyolitic ash-flow eruptive cycles, at the top of the Llewelyn Volcanic Group and at the base of the Snowdon Volcanic Group, have recently been determined (Campbell 1983). The sedimentology and petrography of three other Caradoc tuff units around Bala have been described by Schiener (1970). In this area the Ashgill lies unconformably on the Caradoc, which suggests that uplift occurred to the E of the Conwy Valley Fault and its putative extension. Continued fault control over sedimentation is suggested by the presence of an intra-Ashgill unconformity, linked by Bassett *et al.* (1966) to movements along the Bala Fault.

The pre-Ashgill unconformity can be determined further to the E, in the Ordovician inlier of the Berwyn Dome (Brenchley & Pickerill 1980, and references therein). Here the Caradoc sequence includes thin acidic ash-flow tuffs and air-fall pumice tuffs that are commonly reworked. The emplacement environments have been determined as shallow-marine, subtidal and subaerial, and the sources

of the main tuff units are postulated to lie to the E (Brenchley 1972) and NW (Brenchley 1978).

Llŷn

To the W of Snowdonia a sequence of Ordovician strata lies in the core of the ENE-trending syncline that crosses the Llŷn peninsula (Institute of Geological Sciences 1979). Here, volcanic rocks of both Arenig–Llanvirn and Caradoc age occur. The earlier episode is represented by pillow basalts and a thin, welded ash-flow tuff, interbedded with black graptolitic mudstones. The Caradoc volcanic rocks comprise a thick sequence of lavas and pyroclastic rocks interbedded with marine sediments, and numerous granitoid stocks (Croudace 1982). Fitch (1967) described an upward transition from basaltic andesite to rhyolite and local changes from shallow-marine to subaerial emplacement, prior to the final inundation by marine mudstones at the end of Caradoc times. The local sequence of intrusion and extrusion is intimately associated and the tuffs do not represent distal correlatives of the volcanism in Snowdonia.

SW Wales

A broad belt of Ordovician rocks (Fig. 1) crops out from Ramsey Island in the W, through N and central Pembrokeshire, including the Prescelly Hills (Mynydd Preseli), to Newcastle Emlyn and Carmarthen in the E. To the NE these strata are overlain by turbiditic sediments of Llandovery age. There is abundant evidence of volcanism during Llanvirn times with relatively minor activity during the Arenig.

Sedimentation

The Lower Ordovician of SW Wales records a period of transgression following late or post-Tremadoc uplift. Early Arenig arenaceous sediments rest with slight unconformity on the Upper Cambrian (*Lingula Flags*) on Ramsey Island and with probable disconformity on sediments of Tremadoc age in the Carmarthen area (Cope *et al.* 1978; Cope 1979; Owens *et al.* 1982). At Carmarthen they fine upwards into mudstones and siltstones with a fauna, the '*Neseuretus* Community' of Fortey & Owens (1978), indicative of a shallow-water, inshore environment. These are succeeded in turn by deeper-water mudstones with a '*Raphiophorid* Community' fauna, followed by deep-water 'oxygen-deficient' mudstones with an '*Olenid* Community' fauna (Fortey & Owens 1978).

The sequence reflects deepening of the depositional environment with time, but common and widespread turbiditic incursions occur in this sequence and in the succeeding middle Arenig, e.g. in the '*Tetragraptus* Shales' (Owens & Fortey 1982) in the Carmarthen area. Upper Arenig and Llanvirn times were dominated by offshore marine conditions with black-mud accumulation.

On the basis of the '*Olenid* Community' fauna, Fortey & Owens (personal communication 1983) suggest that, during the later part of the lower Arenig times, the Carmarthen area was the site of a restricted basin, with no free access to the N and W. This possibly resulted from the presence of fault blocks whose orientation is likely to have been NE–SW, reflecting the Precambrian 'grain' in SW Wales. Similarly, the general development of euxinic conditions in the Lower Ordovician of the area probably resulted from the formation of fault-bound basins, rather than from a eustatic rise in sea-level (see Leggett 1980). Active faulting is also indicated in the regular turbiditic incursions (e.g. in the Carmarthen area), which contain lithic clasts similar to the Precambrian igneous rocks which crop out in the area.

Black mud accumulation persisted through to late Llandeilo and early Caradoc times when the *Dicranograptus* Shales were widely deposited during eustatic transgression (Leggett 1980). In the late Caradoc and Ashgill, turbidites with associated black graptolitic muds were deposited and are now exposed in the area around Cardigan and in the Plynlimon Dome (Fig. 1; see James 1971, 1975).

To the E, along the line of the Towy Anticline around Llandeilo, black mudstones of Llanvirn age are overlain by predominantly volcanogenic conglomerates and flaggy sandstones, rare siltstones, tuffs and local rhyolite lavas. Williams *et al.* (1981) record that the volcaniclastic beds thicken and become coarser to the NE towards the Builth Inlier, which they speculatively interpret as the source area. The sediments were deposited in a sublittoral to intertidal environment, in three regressive cycles (Williams *et al.* 1981). The sequence is overlain unconformably by a thick succession of flaggy to massive-bedded limestones, sandstones, siltstones and mudstones with a shallow-water shelly fauna. These are succeeded by lowermost Caradoc black muds of the eustatic transgression. Younger strata in the Llandeilo area include fossiliferous limestones, possibly reflecting shallowing in the early Ashgill. The absence of upper Caradoc strata may be attributed to uplift along the line of the

Towy Anticline (George 1963). This line seems to define the Ordovician shelf–basin boundary.

Volcanism

The earliest evidence of volcanism in SW Wales is provided by several thin (> 0.5 m) rhyolitic volcaniclastic turbidites of approximately middle Arenig age, exposed on Ramsey Island and the nearby mainland (Bevins & Roach 1982). These occur in association with mudstones containing a mixed trilobite and graptolite fauna suggestive of accumulation in an outer shelf-like environment. The first major volcanic episode, determined on Ramsey Island, is represented by two units of rhyolitic turbiditic tuffs of upper Arenig to lower Llanvirn age. The lower part of the lower unit comprises 20 m of laminae and thin beds of recrystallized vitric dust and shards, intercalated with mudstones. The succeeding 63 m comprise thicker bedded, fine to medium grained, rhyolitic, vitroclastic debris including minor pumice and crystals, with only rare and thin mudstone partings. These are the deposits of both high- and low-density turbidity currents (Lowe 1982), with the finest material perhaps also of ash-fall origin. They accumulated in relatively deep water from distal rhyolitic eruptions. It is uncertain whether the eruptions were subaerial or submarine or whether ash-falls or ash-flows were predominant. These beds are succeeded in turn by upper Arenig mudstones and a second, closely similar, tuffaceous sequence, 50 m thick, which records another major, distal rhyolitic eruption.

However, from what is preserved and exposed it seems that the main episode of volcanism in this area occurred during the lower Llanvirn, with major volcanic piles around Fishguard (Thomas & Thomas 1956) and on Ramsey Island (Pringle 1930), and a relatively minor centre near Abereiddi (Cox 1915). Tuffaceous sediments at several horizons in the lower Llanvirn are probably the feather-edges of other accumulations and it is clearly impossible to comment on their relative importance.

Around Fishguard, 1800 m of acid and basic lava flows and volcaniclastic deposits constitute the entirely subaqueously emplaced Fishguard Volcanic Group, which together with contemporaneous intrusions comprise the Fishguard Volcanic Complex (Bevins 1982). West of Fishguard the succession shows a general alternation of activity (acid–basic–acid–basic), but detailed field examination reveals that both compositions were simultaneously available (see Kokelaar *et al.*, this volume). To the E of Fishguard, basic extrusive rocks are scarce

(Lowman 1977; Lowman & Bloxam 1981) although this seems to be compensated by abundant hypabyssal dolerites. Fluid basic magmas were emplaced mainly as complexly interdigitating pillowed and sheet flows with high-level sills, and accordingly the lava/clastic ratio is high. However, thin hyalotuff horizons, particularly towards the top of the pile, show that eruptions took place above the Pressure Compensation Level (Fisher, this volume). A minor volume of basaltic hyaloclastites is also present. Rhyolitic activity produced much lower lava/clastic ratios with voluminous sub-aqueously emplaced welded and non-welded ash-flow tuffs (Lowman & Bloxam 1981). Relatively local, thick rhyolite lava flows and domes comprise flow-banded and perlitic cores mantled by autobreccia carapaces (Bevins 1979). Collapse of the unstable marginal breccias produced aprons of poorly sorted, poorly bedded breccias. Heterolithic debris-flow deposits, with both basalt and rhyolite clasts, pumice clasts, crystals and a vitric component, represent material derived from earlier explosive and quiet eruptions which were mobilized by intrusion and, or, contemporaneous tectonism. Small volumes of intermediate magma were emplaced as tonalite intrusions and an exceptional pillowed rhyodacite lava flow (Bevins & Roach 1979a). Laterally continuous, thinly bedded and laminated tuffs and volcaniclastic intercalations are a minor but important component, deposited from sediment gravity flows and suspension. Silicic turbidites occurring within and near the top of the basic lava pile have no known local source and their presence indicates that the pile cannot have formed a marked positive topographic feature on the sea floor. It is envisaged that the basic eruptions took place within or along the margin of a graben whose subsidence mostly kept pace with growth of the pile.

Eastwards, the lateral equivalents of the Fishguard Volcanic Group, comprising local rhyolite lavas, subaqueous ash-flow tuffs and distal, silicic, mostly turbiditic, volcanogenic sediments, can be traced for *c*. 20 km (Bevins & Roach 1979b; Lowman & Bloxam 1981). It is unclear whether the thin sequences of distal, silicic sediment gravity flow deposits exposed NW and NE of Narberth (Institute of Geological Sciences 1963, 1967) are related to the Fishguard Volcanic Group or whether they belong to another centre.

A second major volcanic pile of lower Llanvirn age, exposed on Ramsey Island (Kokelaar *et al.*, this volume), differs from that near Fishguard in that only rhyolitic magmas were

erupted. Sedimentological and palaeontological evidence suggests that the volcanics were emplaced in an offshore environment starved of terrigenous detritus and dominated by deposition of black muds. Uplift to the W of a contemporary N–S fault resulted in major sliding of the entire pre-existing Ordovician succession (up to 2 km thick), and part of the Cambrian, and also generated major debris flows. Emergence of the block on the W led to littoral development of conglomerates which were subsequently redeposited in deeper water.

The lowermost volcanic member, conformably succeeding the conglomerates, comprises 165 m of massive rhyolitic lapilli tuffs with thinly bedded and laminated fine tuffs, interpreted respectively as the deposits of high- and low-density turbidity currents, perhaps with some distal ash-fall. The middle member, approximately 250 m thick, is composed of three rhyolitic ash-flow tuffs, each overlain by laminated very fine turbiditic tuffs. The ash-flow tuffs comprise ragged tube-pumice clasts, rhyolite lithics, crystals and shards, in a recrystallized vitroclastic matrix. Moulding of pumice clasts around lithics and crystals is common, although a well-developed eutaxitic foliation is not apparent. The lowest tuff comprises a lower part, 161 m thick, which is massive with flattening fabrics and columnar jointing, overlain by a fining upwards sequence, 25 m thick, which contains flattened small pumice clasts. The tuff is interpreted as a closely proximal, hot ash-flow with the graded deposit representing associated ash-fall. The eruption and emplacement are considered to have proceeded under water (with explosive activity suppressed by hydrostatic pressure), within a cupola of steam and magmatic volatiles (see Kokelaar *et al.*, this volume).

The upper member comprises graded and laminated turbiditic tuffs with two beds of poorly sorted coarse lapilli tuff. These interdigitate with, and wedge-out against, autobrecciated rhyolite lavas, and the two coarse tuffs are considered to be the marginal facies of slumped ash-flows. At the thin edge they show normal grading of lithics and inverse grading of pumice clasts, but away from the edge they are chaotic, containing convoluted rafts of fine tuffs up to 10 m long. This is interpreted as due to slumping and further flow of primary deposits with the feather-edges remaining undisturbed.

Elsewhere on Ramsey Island, intrusion of autobrecciated rhyolite caused overlying volcaniclastic sediments to slide and slump. High-level porphyritic intrusions occur about the contemporaneous N–S fault, as small rounded bosses, semi-concordant sills or thin irregular sheets. Their emplacement into unconsolidated and wet host material is evident by the peperitic and pillowed margins attended by fluidization (Kokelaar 1982). In places, subsequent slumping exposed the intrusive rocks at the sea floor.

The youngest Ordovician volcanic episode recorded in SW Wales was basaltic and occurred during upper Llanvirn times. The tuffs, the *Didymograptus murchisoni* Ash of Cox (1915), constitute two fining-upwards sequences, each grading both vertically and laterally from massive, coarse lapilli tuff to well-bedded, medium and fine tuffs. The tuffs, together up to 100 m thick but thinning laterally to the finer facies, are composed of moderately to highly vesicular basalt lava fragments, some angular, some bomb-like, and clearly the eruption occurred above the Pressure Compensation Level (Fisher, this volume). The coarser, more proximal deposits are of debris flows but, more distally and upwards in each cycle, these are superceded by deposits of high- and low-density turbidity currents. They represent slumping of the unstable flanks of a tephra pile during eruption, with waning activity in each cycle generating successively more minor sediment gravity flows. Overlying deposits of mixed tephra and mudstone reflect late slumping of the pile when it was partly mantled by pelagic sediments.

Builth–Llandrindod Inlier and the Welsh Borderland

Sedimentation

In the Builth Inlier, black graptolitic mudstones of lower Llanvirn age are overlain by a thick volcanic sequence which accumulated under subaerial and shallow-marine conditions (Jones & Pugh 1949). A return to deeper-water conditions occurred during early Llandeilo times and black graptolitic mud accumulation persisted through to the Caradoc.

The strata of the Welsh Borderland show a facies change across, and controlled by, the NE–SW-trending Pontesford–Linley Fault. To the W, 4 km of Ordovician strata are exposed in the Shelve Inlier. These comprise transgressive shallow-water sandstones at the base, overlain by a mixed offshore (probably outer-shelf) sequence. To the E, the oldest strata are of lower Caradoc age, deposited during eustatic transgression (Leggett 1980). The strata are about 200 m thick and generally of shallow-water facies although lateral facies changes and diachronism of certain units indicate contemporaneous faulting (Bassett 1980) or a sustained transgression over a varied relief (P. J. Brenchley, personal communication 1983).

Volcanism

In the Builth Inlier, Llanvirn lavas and pyroclastics were erupted into shallow-water and subaerial environments (Jones & Pugh 1949) at the basin margin. Extensive shallow-marine reworking produced a range of volcaniclastic deposits, commonly with a derived shelly fauna. Wide dispersal to the SW (Llandeilo area) has been postulated (see above).

Hypabyssal intrusions of complex form (Jones & Pugh 1946, 1948a,b) intrude these and younger rocks and by their interactions with wet unlithified Caradoc sediments they must belong to a later volcanic episode. Thin tuffs of Caradoc age are exposed at the NE end of the inlier, and a minor Caradoc episode is recorded in the Llanwrtyd Wells area, where basic and acid volcanics were erupted.

To the N, in the Shelve Inlier, nearby Llanvirn volcanism is reflected in volcaniclastic sediment gravity flow deposits and subaqueous ash-flow tuffs (B. D. T. Lynas, personal communication 1983). In a Caradoc volcanic episode, magmas of basic through intermediate to acid compositions were emplaced here and in the Breidden Hills, and conglomerates and volcaniclastic turbidites indicate reworking (R. J. Dixon, personal communication 1983).

Petrochemistry

Alteration

All of the igneous rocks in the foregoing account have suffered low- or very low-grade regional metamorphism (Roberts 1981; Bevins et al. 1981; Bevins & Rowbotham 1983). In addition, many have suffered deuteric alteration or have interacted chemically with ambient seawater and, or, connate brines, during or soon after emplacement. Accordingly the determination of original rock compositions is problematical and analyses must be interpreted with caution.

Most of the pyroclastic rocks were initially composed mainly of glass. The margins and the mesostasis of many basic lavas and some high-level intrusions were originally sideromelane, and most high-level intrusive and extrusive rhyolites were obsidian. Such originally glassy rocks are completely devitrified and recrystallized. Also, adjacent to relatively undeformed initially holocrystalline rocks, they are commonly strongly cleaved. This localization of strain is attributable to the alteration of the glass to phyllosilicate-rich rocks; sideromelane or palagonite to chlorite, and obsidian to quartzo-feldspathic intergrowths with chlorite and sericite. Extreme hydration and alteration are clearly implied. Analyses of pyroclastic rocks are interpreted with caution for the additional reason that they may have suffered fractionation by purely physical processes during eruption, transport and deposition.

Originally glassy rocks have only been analysed where holocrystalline equivalents are lacking, and rocks that obviously have been intensely deformed, badly weathered or subjected to hydrothermal activity, have been avoided. The analyses presented are of the petrographically freshest rocks available but, even so, there is a wide range of alteration states and because of the potential for element redistribution, this range has been investigated, particularly in the basic rocks, in order to determine those elements which may be assumed with reasonable confidence to approximate closely to primary concentrations. It has been concluded (Kokelaar 1977; Dunkley 1978; Bevins 1979, 1982) that, of the elements determined in most of the samples, contents of Ti, P, Zr, Y and Nb reliably approximate to primary compositions. These elements are of course renowned for their immobility (e.g. Pearce & Cann 1973; Pearce 1975). Also, in basic and intermediate rocks, contents of La, Ce, Fe, Mg and Si are regarded as acceptable indicators of original composition. Silica commonly shows strong correlations with Zr, for example (coherent secondary mobility not expected), and with petrographical classifications based on relict mineralogy and texture (Kokelaar 1977; Dunkley 1978). Na, K, Rb, Sr, Ba, H_2O^+ and Fe^{2+}/Fe^{3+} are clearly prone to considerable selective alteration.

These conclusions may be to some extent anticipated if the petrography of the rocks is considered, wherein accessory apatite and zircon remain unaltered, Fe-Ti oxides although altered are commonly associated with leucoxene and sphene so that Ti and Nb are probably fixed locally, clinopyroxene is quite commonly fresh, and plagioclase ranges from calcic labradorite to almost pure albite and is commonly associated with sericite, epidote, clinozoisite, prehnite and pumpellyite. The latter minerals indicate that Ca and Al may also be fixed locally.

For reasons implicit from the above, discussion bearing on Ordovician igneous rocks in Wales that rely heavily on whole-rock normative analyses or require immobility of the alkali elements (e.g. Fitton & Hughes 1970; Hughes 1977; Lowman & Bloxam 1981; Fitton *et al.* 1982) must be considered as weakly founded.

Southern Snowdonia

Dunkley (1979) determined that at outcrop the intrusive and extrusive rocks in the Aran Volcanic Group, in the Arans and on Cader Idris, show a marked basic–acid bimodality, with progressively diminishing volumes of intermediate rocks from basaltic andesite to rhyodacite, and voluminous (*c.* 60%), mainly pyroclastic, rhyolites. The basic and intermediate rocks together show continuous element variation (Figs 4 and 5) with marked Fe and Ti enrichment which, considered with their relative abundances, is characteristic of a tholeiitic series derived by low-pressure fractional crystallization of mantle-derived parental magma. The large volume of rhyolites, and absence of any geophysical evidence to suggest large volumes of mafic residua at depth, led Dunkley (1978, 1979) to conclude that the silicic magmas were the products of crustal fusion. Analytical data are equivocal in this respect as the intermediate to acid discontinuity in variation trends (Fig. 4), such as for Zr, Y, Nb, Ce and La, could, for example, for attributed to late fractionation of apatite and zircon from intermediate compositions, although the wide scatter of plots might be more consistent with variable crustal fusion. Discriminant analyses (Fig. 6; also in Ti *v.* Cr (after Pearce 1975), Dunkley 1978, Fig. 4: 14c) show the basic rocks to have strong affinities with ocean-floor basalt.

Around the E and N margins of the Harlech Dome, other pre-Caradoc igneous rocks are predominantly rhyolitic with some of basic and intermediate composition. The latter are moderately Fe-rich (data from Hughes 1977) and with the presently available data a petrogenetic scheme similar to that for the Arans and Cader Idris rocks is envisaged.

Central and northern Snowdonia

There is as yet only sparse geochemical data available from the Caradoc igneous rocks of N Wales.

The lithologically distinctive intrusive and extrusive rocks of the Foel Fras Volcanic Complex show calc-alkaline affinities in their lack of basic compositions and absence of Fe enrichment (Fig. 5). Rocks from the remaining, major Caradoc outcrops show marked basic–acid bimodality with moderate to marked Fe enrichment and sparse intermediate compositions. Discriminant plots (Fig. 6) of basic rocks cluster about the junction of the fields of ocean-floor, calc-alkaline and within-plate basalts (see also Floyd *et al.* 1976). The derivation of the acid rocks, including the numerous

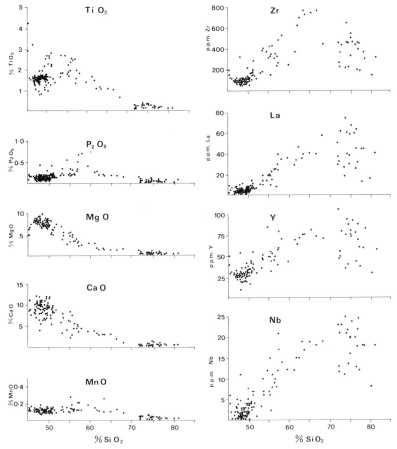

FIG. 4. Selected chemical variation diagrams of rocks from the Aran Volcanic Group in the Arans and on Cader Idris (data from Dunkley 1978).

high-level granitoid intrusions, is uncertain. As in southern Snowdonia their large volume relative to basic rocks, and geophysical considerations, tend to favour an origin by crustal fusion.

Llŷn

In Llŷn, Caradoc basic and intermediate rocks are scarce in comparison with acidic types, but do, however, show marked Fe enrichment (Fig. 5) and are here interpreted as members of a tholeiitic suite. A detailed geochemical study (Croudace 1982) indicates that the andesitic lavas probably result from approximately 70% low-pressure fractional crystallization of a transitional tholeiitic magma, and that the granitoid intrusions are related to the andesites by further low-pressure crystal fractionation. The least fractionated basic rocks in the high-level layered Rhiw Intrusion (Fig. 1) bear primary hornblende and

are mildly alkalic (Cattermole 1976), but the hydrous nature of the magma, along with the strong possibility of metasomatism and the lack of published trace-element data, preclude characterization of the parent magma.

SW Wales

Petrographically, chemically and petrogenetically the rocks of the Fishguard Volcanic Complex are closely similar to those of the Aran Volcanic Group and its associated intrusions.

At outcrop, the Fishguard Volcanic Complex shows a basic–acid bimodality with subordinate intermediate compositions. Bevins (1982) demonstrated that the basic to intermediate compositions represent a tholeiitic series derived by low-pressure fractional crystallization of a parental magma which originated as a partial melt of upper mantle. Fractionation of plagioclase, clinopyroxene and minor olivine

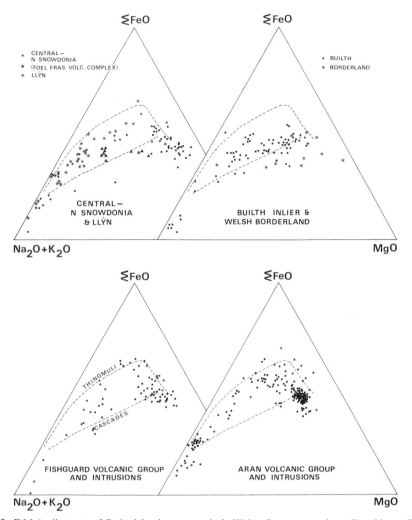

FIG. 5. F.M.A. diagrams of Ordovician igneous rocks in Wales. Some scatter is attributable to alkali metasomatism. (Data from Floyd *et al.* 1976; Hughes 1977; Lowman 1977; Dunkley 1978; Furnes 1978; Bevins 1979; Croudace 1982; J. Esson & W. J. Wadsworth (unpubl.); R. E. Bevins, G. J. Lees & R. A. Roach (unpubl.).)

resulted in whole-rock Fe- and Ti-enrichment trends (Figs 5 and 6), and Fe enrichment in clinopyroxenes from the intrusions (different in composition to disequilibrium (quench) clinopyroxenes in the basalts), which are distinctly tholeiitic in character.

The origin of the acidic rocks in the complex is uncertain. Partly on the basis of rare-earth element (REE) analyses, Bevins (1982, Fig. 8) tentatively suggested that the rhyolites may be extreme derivatives of the fractionation series. Trace-element concentrations can be interpreted to support this view, but alteration of the original element contents and an origin by

crustal fusion, cannot be ruled out with the data presently available.

The suggestion (Lowman & Bloxam 1981) that the Fishguard Volcanic Complex is calc-alkaline in nature derives mainly from a consideration of alkali element contents, which are notably susceptible to alteration (see above), and the suggestion is untenable in the light of more complete data (Bevins 1979, 1982). In discriminant trace-element diagrams (Fig. 6), the basic rocks of the complex show strong affinities with ocean-floor basalts, although they are slightly richer in Ti than these and the Aran rocks.

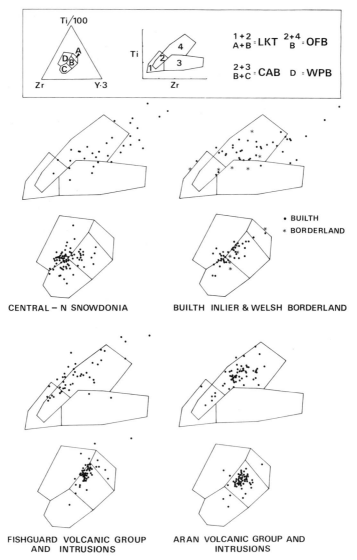

FIG. 6. Discriminant diagrams of Ti, Y and Zr in rocks of basaltic composition (after Pearce & Cann 1973). Fields: LKT—low-K tholeiite; CAB—calc-alkaline basalt; OFB—ocean-floor basalt; WPB—within-plate basalt. (Data from Floyd *et al.* 1976; Hughes 1977; Lowman 1977; Dunkley 1978; Furnes 1978; Bevins 1979; J. Esson & W. J. Wadsworth (unpubl.); R. E. Bevins, G. J. Lees & R. A. Roach (unpubl.).)

The separate suite of noritic gabbros to the SW of Fishguard (Roach 1969; Bevins & Roach 1979b, 1982) shows a tholeiitic parentage, evidence of low-pressure fractional crystallization prior to emplacement (locally attended by further fractionation, e.g. St David's Head and Carn Llidi) and, in the least evolved rocks, ocean-floor basalt affinities (P. A. Floyd & R. A. Roach, unpublished data). A large positive magnetic anomaly, along strike to the SW, indicates the possible presence offshore of a large coeval intrusion.

We have no direct evidence bearing on the origin of the voluminous rhyolitic magmas represented by the Ramsey Island volcanic centre, although speculatively the closeness of the noritic gabbros to this centre may be significant. An association through extreme fractionation or crustal fusion remains to be determined.

Builth–Llandrindod Inlier and the Welsh Borderland

The Llanvirn episode of volcanism represented in the Builth Inlier shows a fractionation trend characteristic of calc-alkaline magmatism (Fig. 5). MgO-rich compositions have not been determined and no bimodality is apparent at outcrop. Furnes (1978) considers the basic rocks, which are strongly plagioclase-phyric, to be high-alumina basalts derived by a moderate degree of partial melting of upper mantle. Discriminant plots (Fig. 6) show clustering rather similar to that of the basic rocks from central and northern Snowdonia. Llanvirn and Caradoc rocks from various localities in the Welsh Borderland show some petrochemical similarities to the Builth rocks (Figs 5 and 6).

Synthesis

The petrochemical characteristics of the major or distinct Ordovician igneous complexes in Wales are summarized in Table 1. We recognize that we have grossly 'averaged' the character of volcanism in central and northern Snowdonia, where several volcanic centres and perhaps distinct suites were formed. Also, the character of what appear as minor igneous outcrops but may be relicts or feather-edges of unexposed major complexes, is unknown. Thus our sample may be biased.

The compiled geochemical data are from many different sources and we are unable to assess the magnitude of the likely inter-laboratory variations in precision and accuracy. Because of this, and the incomplete availability of many 'useful' elements, we cannot comment in detail on possible small variations in mantle source compositions, mantle partial melting histories and controls of fractionation. However, we believe that the differences in the trends of relative Fe enrichment (Fig. 5), and in the cluster locations and trends in discriminant plots of Ti, Y and Zr (Fig. 6) are real.

At outcrop the volcanism is dominantly bimodal and the common fractionation continuum, basalt to rhyodacite, is dominantly tholeiitic. Strictly calc-alkaline series are restricted to locations near the E margin of the Welsh Basin and one minor centre in northern Snowdonia. Such dominance of tholeiitic series and basalts with ocean-floor affinities, with transitions to arc-like compositions, are characteristic of basalts emplaced in back-arc marginal basin environments (e.g. Saunders & Tarney, this volume).

These conclusions are at variance with those of Fitton & Hughes (1970), Hughes (1977) and Fitton *et al.* (1982). The latter authors claim that in the Ordovician rocks there is no evidence of low-pressure fractional crystallization from basaltic compositions, that the intermediate rocks constitute a distinct calc-alkaline suite unrelated to the basic rocks, and that the overall character is calc-alkaline. But the 'earliest tholeiitic lavas' of Fitton *et al.* (1982, p. 611) are late Tremadoc basalts and dolerites of the Rhobell Volcanic Complex, which clearly are transitional between island-arc tholeiites and calc-alkaline basalts and reflect an island-arc regime distinctively different to the Ordovician regime (Kokelaar 1977, 1979; Dunkley 1978, 1979; Kokelaar *et al.* 1982). The rest of the sample considered representative of the 'main group' (Ordovician) (Hughes 1977), is strongly biased towards intermediate compositions, with relatively numerous analyses from the Builth Inlier, the Welsh Borderland, the Foel Fras Volcanic Complex (Llwytmor andesites) and Llŷn. The major outcrops are under-represented. The sample also includes the late Tremadoc or lowermost Arenig rocks of the Trefgarn Volcanic Group which are like those of the Rhobell Volcanic Complex (Kokelaar 1977) and belong to the earlier volcanic-arc regime. The characteristic low-pressure fractionation series in the major complexes were overlooked owing to the widespread and inadequate

TABLE 1. *Summary of petrochemical characteristics of the major or distinct Ordovician igneous complexes in Wales*

Igneous complex	Age	Fe enrichment	Character	Rocks	Affinity
Fishguard	Llanvirn	Marked	Tholeiitic	BiA	OFB
Arans–Cader Idris	Pre-Caradoc	Marked	Tholeiitic	BiA	OFB
Llŷn	Caradoc	Marked	Tholeiitic	BiA	?
Central and northern Snowdonia	Caradoc	Moderate to marked	Tholeiitic (-transitional)	BiA	Mixed
Foel Fras	Caradoc	None	Calc-alkaline	IA	?
Builth Inlier	Llanvirn	None	Calc-alkaline	BIA	Mixed

B—basic, I—intermediate, A—acid; lower case signifies minor proportion.

sampling programme. The rocks with calc-alkaline affinities from the Builth Inlier, the Welsh Borderland and Foel Fras are volumetrically far less important than the tholeiitic series rocks and cannot be deemed representative of the Welsh Basin. Also, the Llŷn suite was incorrectly interpreted as calc-alkaline. Furthermore, contrary to Fitton *et al.* (1982, p. 623), there is considerable similarity between many of the basic rocks and modern ocean-floor tholeiites.

However, we do agree with other authors that the origin of the acidic rocks is enigmatic. Their relatively large volumes and compositional discontinuity with intermediate rocks, coupled with the lack of geophysical anomalies suggestive of major mafic residua, tend to favour an origin by partial fusion of deeply underlying crust. Croudace (1982) has interpreted the granitoid intrusions of Llŷn as having evolved from andesitic liquids by crystal fractionation, with removal of apatite and zircon to account for REE patterns. The solution to the problem at the major outcrops must await the availability of detailed trace-element analyses.

Conclusions

The volcanism, sedimentation and tectonic activity in the Welsh Basin during the Ordovician are consistent with the general model of Ordovician plate tectonics, which places the basin in a back-arc setting. There seems to be no single close modern analogue, although in New Zealand the Taupo Volcanic Zone (Cole, this volume), and its extension into the Bay of Plenty (Lewis & Pantin, this volume), show many conditions and processes in common with the Welsh Basin.

The marginal basin of Wales developed on immature continental crust and, although crustal tension is indicated by the tectonic and sedimentary histories, and by the character of the magmas, oceanic lithosphere was not formed. The basin was predominantly the site of marine sedimentation, partly on shallow shelves and partly in deeper water. Topographic relief within the basin resulted mainly from the contemporaneous activity of normal faults with NE–SW and N–S trends. These faults produced marked facies changes, either directly, for example where they resulted in local emergence above sealevel, or indirectly, where they restricted the deposition from various pyroclastic and sediment gravity flows. However, because the strata of the basin floor were largely unlithified, it is unlikely that steep

fault-scarps could have developed. Dip-slip movement at depth would, at shallow levels, probably have caused monoclinal flexuring initially, perhaps with minor scarps, and then sliding and slumping. Allochthonous sequences of sediments and tuffs in SW Wales (Ramsey Island) involve up to 2 km of the basin succession and show the massive scale at which such sliding could occur.

The largely unlithified and partly compacted nature of the basin-floor strata is reflected in the abundance of sills, which show that non- or weakly vesiculating basic and acid magmas, by virtue of their greater density, tended to flow laterally, interacting with wet host material, rather than reaching the sea floor (cf. Guaymas Basin, Gulf of California: Saunders *et al.* 1982). Partly because of this, the extrusive volcanism, which occurred at numerous local centres across the basin, was greatly influenced by the contemporary fault system. The major piles of basic extrusive rocks (e.g. in the Fishguard Volcanic Group and in NE Snowdonia) appear to have accumulated in contemporaneously developed depressions which were probably fault- and, or, flexure-bound and fed by transgressive sills and, or, dykes which utilized the faults to reach shallow levels. The extrusions formed little or no constructional feature on the basin floor.

The same development of an infilled volcano-tectonic depression is interpreted in some major rhyolite tuff accumulations (e.g. in the lowest part of the Snowdon Volcanic Group), which also must have had bases which were markedly convex downwards, although at times ash-flows escaped from these centres. We consider these 'inverse volcanic piles' to be a direct reflection of the extension in the Welsh Basin. Although the depressions associated with the major rhyolitic eruptions may loosely be analogous to calderas, in that major magma evacuation must have been attended by subsidence and ponding of pyroclastics, the location of the centre and form of the 'pond' were almost certainly controlled by established faults (cf. the major rhyolitic centres of the Taupo Volcanic Zone of New Zealand: Cole, this volume).

Silicic pyroclastic rocks occur in all the major volcanic horizons of the basin and, although relatively scarce, deposits of highly vesicular basic clasts are present in most sequences. Explosive eruptions of silicic magma can occur at water depths in excess of 1 km, but the limiting depth for large-scale vesiculation (Pressure Compensation Level: Fisher, this volume) of most basic magmas is less than 500 m (McBirney 1963). Probably most of the subaqueous eruptions in the Welsh Basin occurred at water depths close to or shallower than 500 m. However, within the basin, volcanic islands were uncommon and ephemeral, and even quite major eruptions commonly caused little or no change in the facies or composition of the 'normal' sediments occurring immediately beneath and above the volcanic horizon.

Subaqueously emplaced silicic ash-flow tuff, welded or non-welded, is a major constituent of the volcanic successions. Major ash-fall layers are scarce. Although some ash-flows crossed from a subaerial into a submarine environment, others were erupted subaqueously where it seems that the surrounding water, by confining the volcanic explosions and hence eruptive columns, resulted in the production of less ash-fall, with less dispersion, than would have been the case in a subaerial eruption.

Volcanogenic sediments, particularly those of mass-gravity flows resulting from sliding and slumping of unstable deposits, also are a major component of the submarine volcanic successions. Oversteepening during accumulation, shallow intrusive activity and tectonic movements were common causes of such secondary distribution processes.

ACKNOWLEDGMENTS: We are grateful to Drs M. G. Bassett, P. J. Brenchley, R. M. Owens and W. J. Wadsworth for discussion and constructive criticism of an early draft of the manuscript. Also, Drs J. Esson, P. A. Floyd, G. J. Lees and W. J. Wadsworth kindly allowed us to use their unpublished petrochemical data. Jane Bidgood is thanked for typing the numerous versions of the manuscript, and we are grateful to Pat Andrews, who assisted in the preparation of the paper in many ways. M.F.H. publishes with the permission of the Director of the British Geological Survey.

References

ALLEN, P. M. 1982. Lower Palaeozoic volcanism in Wales, the Welsh Borderland, Avon, and Somerset. *In*: SUTHERLAND, D. S. (ed.) *Igneous Rocks of the British Isles*, 65–91. Wiley, London.

——, JACKSON, A. A. & RUSHTON, A. W. A. 1981. The stratigraphy of the Mawddach Group in the Cambrian succession of North Wales. *Proc. Yorkshire geol. Soc.* **43**, 295–329.

BASSETT, D. A. 1969. Some of the major structures of early Palaeozoic age in Wales and the Welsh Borderland: An historical essay. *In*: WOOD, A. (ed.) *The Pre Cambrian and Lower Palaeozoic rocks of Wales*, 67–116. University of Wales Press, Cardiff.

——, WHITTINGTON, H. B. & WILLIAMS, A. 1966. The stratigraphy of the Bala district, Merionethshire. *Q. Jl geol. Soc. Lond.* **122**, 219–71.

BASSETT, M. G. 1980. The Caledonides of Wales, the Welsh Borderland, and Central England. *In*: OWEN, T. R. (ed.) *United Kingdom: Introduction to General Geology and Guides to Excursions 002, 055, 093 & 151*, 34–48. International Geological Congress, Paris, 1980. Institute of Geological Sciences, London.

BEAVON, R. V. 1963. The succession and structure east of the Glaslyn river, North Wales. *Q. Jl geol. Soc. Lond.* **119**, 479–512.

—— 1980. A resurgent cauldron in the early Palaeozoic of Wales, U.K. *J. Volcan. geoth. Res., Amsterdam*, **7**, 157–74.

BEVINS, R. E. 1979. The geology of the Strumble Head–Fishguard region, Dyfed, Wales. Ph.D thesis. Univ. Keele (unpubl.).

—— 1982. Petrology and geochemistry of the Fishguard Volcanic Complex, Wales. *Geol. J.* **17**, 1–21.

—— & ROACH, R. A. 1979a. Pillow lava and isolated-pillow breccia of rhyodacitic composition from the Fishguard Volcanic Group, Lower Ordovician, S. W. Wales, United Kingdom. *J. Geol.* **87**, 193–201.

—— & —— 1979b. Early Ordovician volcanism in Dyfed, S.W. Wales. *In*: HARRIS, A. L., HOLLAND, C. H. & LEAKE, B. E. (eds) *The Caledonides of the British Isles—reviewed*. *Spec. Publ. geol. Soc. London*, **8**, 603–9.

—— & —— 1982. Ordovician igneous activity in south-west Dyfed. *In*: BASSETT, M. G. (ed.) *Geological Excursions in Dyfed, South-West Wales*, 65–80. National Museum of Wales, Cardiff.

—— & ROWBOTHAM, G. 1983. Low-grade metamorphism within the Welsh sector of the paratectonic Caledonides. *Geol. J.* **18**, 141–67.

——, ROBINSON, D., ROWBOTHAM, G. & DUNKLEY, P. N. 1981. Low grade metamorphism in the Welsh Caledonides. *J. geol. Soc. London*, **138**, 634.

BLYTH, F. G. H. 1938. Pyroclastic rocks from the Stapeley Volcanic Group at Knotmoor, near Minsterly, Shropshire. *Proc. Geol. Assoc. Lond.* **49**, 392–404.

—— 1944. Intrusive rocks of the Shelve area, south Shropshire. *Q. Jl geol. Soc. Lond.* **99**, 169–204.

BRENCHLEY, P. J. 1972. The Cwm Clwyd Tuff, North Wales: A palaeogeographical interpretation of some Ordovician ash-shower deposits. *Proc. Yorkshire geol. Soc.* **39**, 199–224.

—— 1978. The Caradocian rocks of the north and west Berwyn Hills, North Wales. *Geol. J.* **13**, 137–64.

—— & PICKERILL, R. K. 1980. Shallow subtidal sediments of Soudleyan (Caradoc) age in the Berwyn Hills, North Wales, and their palaeogeographic context. *Proc. Geol. Assoc. Lond.* **91**, 177–94.

BROMLEY, A. V. 1965. Intrusive quartz latites in the

Blaenau Ffestiniog area, Merioneth. *Geol. J.* **4**, 247–56.

—— 1968. A volcanic vent in the Tremadocian rocks south-east (*sic*) of Blaenau Ffestiniog, Merionethshire, North Wales. *Geol. J.* **6**, 7–12.

—— 1969. Acid plutonic igneous activity in the Ordovician of North Wales. *In*: WOOD, A. (ed.) *The Pre-Cambrian and Lower Palaeozoic rocks of Wales.* 387–408. University of Wales Press, Cardiff.

CAMPBELL, S. D. G. 1983. The geology of an area between Bala and Betws y Coed, North Wales. Ph.D. thesis, Univ. Cambridge (unpubl.).

CATTERMOLE, P. J. 1976. The crystallisation and differentiation of a layered intrusion of hydrated alkali olivine-basalt parentage at Rhiw, North Wales. *Geol. J.* **11**, 45–70.

COCKS, L. R. M. & FORTEY, R. A. 1982. Faunal evidence for oceanic separations in the Palaeozoic of Britain. *J. geol. Soc. London*, **139**, 465–78.

COPE, J. C. W. 1979. Early history of the southern margin of the Tywi anticline in the Carmarthen area, South Wales. *In*: HARRIS, A. L., HOLLAND, C. H. & LEAKE, B. E. (eds) *The Caledonides of the British Isles—reviewed. Spec. Publ. geol. Soc. London*, **8**, 527–32.

——, FORTEY, R. A. & OWENS, R. M. 1978. Newly discovered Tremadoc rocks in the Carmarthen district, South Wales. *Geol. Mag.* **115**, 195–8.

COX, A. H. 1915. The geology of the district between Abereiddy and Abercastle. *Q. Jl geol. Soc. Lond.* **71**, 273–340.

—— 1925. The geology of the Cader Idris range (Merioneth). *Q. Jl geol. Soc. Lond.* **81**, 539–94.

—— 1930. Preliminary note on the geological structure of Pen Caer and Strumble Head, Pembrokeshire. *Proc. Geol. Assoc. Lond.* **41**, 274–89.

—— & JONES, O. T. 1914. Pillow lavas in North and South Wales. *Report of the British Association for the Advancement of Science.* 1913 (Birmingham), p. 495.

——, GREEN, J. F. N., JONES, O. T. & PRINGLE, J. 1930a. The geology of the St David's district, Pembrokeshire. *Proc. Geol. Assoc. Lond.* **41**, 241–73.

——, ——, —— & —— 1930b. The St David's district. Report of the summer field meeting, 1930. *Proc. Geol. Assoc. Lond.* **41**, 412–38.

CRIMES, T. P. 1970a. A facies analysis of the Cambrian of Wales. *Palaeogeogr. Palaeoclimatol. Palaeoecol.* **7**, 113–70.

—— 1970b. A facies analysis of the Arenig of Western Lleyn, North Wales. *Proc. Geol. Assoc. Lond.* **81**, 221–39.

CROUDACE, I. W. 1982. The geochemistry and petrogenesis of the Lower Palaeozoic granitoids of the Lleyn Peninsula, North Wales. *Geochim. cosmochim. Acta*, **46**, 609–22.

DAKYNS, J. R. & GREENLY, E. 1905. On the probable Peléan origin of the felsitic slates of Snowdon and their metamorphism. *Geol. Mag.* **42**, 541–9.

DAVIES, R. G. 1959. The Cader Idris Granophyre and its associated rocks. *Q. Jl geol. Soc. Lond.* **115**, 189–216.

DUNKLEY, P. N. 1978. The geology of the south-

western part of the Aran Range, Merionethshire, with particular reference to the igneous history. Ph.D. thesis. Univ. Wales, Aberystwyth (unpubl.).

—— 1979. Ordovician volcanicity of the SE Harlech Dome. *In*: HARRIS, A. L., HOLLAND, C. H. & LEAKE, B. E. (eds) *The Caledonides of the British Isles–reviewed. Spec. Publ. geol. Soc. London*, **8**, 597–601.

EVANS, W. D. 1945. The geology of the Prescelly Hill, north Pembrokeshire. *Q. Jl geol. Soc. Lond.* **101**, 89–110.

FITCH, F. J. 1967. Ignimbrite volcanism in North Wales. *Bull. Volcanol.* **30**, 199–219.

FITTON, J. G. & HUGHES, D. J. 1970. Volcanism and plate tectonics in the British Ordovician. *Earth planet. Sci. Lett.* **8**, 223–8.

——, THIRLWALL, M. F. & HUGHES, D. J. 1982. Volcanism in the Caledonian orogenic belt of Britain. *In*: THORPE, R. S. (ed.) *Andesites*, 611–36. Wiley, London.

FLOYD, P. A., LEES, G. J. & ROACH, R. A. 1976. Basic intrusions in the Ordovician of North Wales—geochemical data and tectonic setting. *Proc. Geol. Assoc. Lond.* **87**, 389–400.

FORTEY, R. A. & OWENS, R. M. 1978. Early Ordovician (Arenig) stratigraphy and faunas of the Carmarthen district, south-west Wales. *Bull. Br. Mus. nat. Hist. (Geol.)* **30**, 225–94.

FRANCIS, E. H. & HOWELLS, M. F. 1973. Transgressive welded ash-flow tuffs among the Ordovician sediments of N.E. Snowdonia, N. Wales. *J. geol. Soc. London*, **129**, 621–41.

FURNES, H. 1978. A comparative study of Caledonian volcanics in Wales and west Norway. D. Phil. thesis. Univ. Oxford (unpubl.).

GEORGE, T. N. 1963. The Palaeozoic growth of the British Caledonides. *In*: JOHNSON, M. R. W. & STEWART, F. H. (eds) *The British Caledonides*, 1–33. Oliver and Boyd, Edinburgh.

—— 1970. *British Regional Geology. South Wales.* 3rd edition, H.M.S.O., London. 152 pp.

GREENLY, E. 1944. The Arvonian rocks of Arvon. *Q. Jl geol. Soc. Lond.* **100**, 269–87.

HAMPTON, C. M. & TAYLOR, P. N. 1983. The age and nature of the basement of southern Britain: evidence from Sr and Pb isotopes in granites. *J. geol. Soc. London*, **140**, 499–509.

HARKER, A. 1889. *The Bala Volcanic Series of Caernarvonshire.* Cambridge, 130 pp.

HOFMANN, H. J. 1975. *Bolopora* not a bryozoan, but an Ordovician phosphatic, oncolitic accretion. *Geol. Mag.* **112**, 523–6.

HOWELLS, M. F. 1977. The varying patterns of volcanicity and sedimentation in the Bedded Pyroclastic Formation and Middle Crafnant Volcanic Formation in the Ordovician of central and northern Snowdonia *J. geol. Soc. London*, **133**, 404 (abstract).

—— & LEVERIDGE, B. E. 1980. The Capel Curig Volcanic Formation. *Rep. Inst. geol. Sci. Lond.* **80/6**, 23 pp.

——, FRANCIS, E. H., LEVERIDGE, B. E. & EVANS, C. D. R. 1978. *Capel Curig and Betws-y-Coed: Description of 1:25000 Sheet SH75. Classical*

Areas of British Geology, Institute of Geological Sciences. H.M.S.O., London, 73 pp.

——, LEVERIDGE, B. E. & EVANS, C. D. R. 1973. Ordovician ash-flow tuffs in eastern Snowdonia. *Rep. Inst. geol. Sci. Lond.* **73/3**, 33 pp.

——, ——, —— & NUTT, M. J. C. 1981. *Dolgarrog: Description of 1:25000 Sheet SH76. Classical Areas of British Geology, Institute of Geological Sciences.* H.M.S.O., London, 89 pp.

——, —— & REEDMAN, A. J. In press. *Outline of the Geology of the Bangor 1:50,000 Geological Sheet (106).* Institute of Geological Sciences.

——, ——, —— & ADDISON, R. The lithostratigraphical subdivision of the Ordovician underlying the Snowdon and Crafnant Volcanic Groups, North Wales. *Rep. Inst. geol. Sci. Lond.* **83/1**, 11–15.

HUGHES, D. J. 1977. The petrochemistry of the Ordovician rocks of the Welsh Basin. Ph.D. thesis. Univ. Manchester (unpubl.).

INSTITUTE OF GEOLOGICAL SCIENCES. 1963. *Geological Survey 1:63,360 Haverfordwest (228) Sheet.* Ordnance Survey, Southampton.

—— 1967. *Geological Survey 1:63,360 Carmarthen (229) Sheet.* Ordnance Survey, Southampton.

—— 1979. *Geological Survey 1:625,000 Sheet 1.* Ordnance Survey, Southampton.

—— 1982. *Geological Survey 1:50,000 Harlech (135 and part of 149) Sheet.* Ordnance Survey, Southampton.

—— 1984. *Geological Survey 1:50,000 Bangor (106) Sheet.* Ordnance Survey, Southampton.

JAMES, D. M. D. 1971. The Nant-y-Moch Formation, Plynlimon inlier, West Central Wales. *J. geol. Soc. London,* **127**, 177–81.

—— 1975. Caradoc turbidites at Poppit Sands (Pembrokeshire), Wales. *Geol. Mag.* **112**, 295–304.

JONES, B. 1933. The geology of the Fairbourne–Llwyngwril district, Merioneth. *Q. Jl geol. Soc. Lond.* **89**, 145–71.

JONES, O. T. & PUGH, W. J. 1941. The Ordovician rocks of the Builth district; a preliminary account. *Geol. Mag.* **78**, 185–91.

—— & —— 1946. The complex intrusion of Welfield, near Builth Wells, Radnorshire. *Q. Jl geol. Soc. Lond.* **102**, 157–88.

—— & —— 1948a. A multi-layered dolerite complex of laccolithic form, near Llandrindod Wells, Radnorshire. *Q. Jl geol. Soc. Lond.* **104**, 43–70.

—— & —— 1948b. The form and distribution of dolerite masses in the Builth–Llandrindod inlier, Radnorshire. *Q. Jl. geol. Soc. Lond.* **104**, 71–98.

—— & —— 1949. An early Ordovician shoreline in Radnorshire, near Builth Wells. *Q. Jl geol. Soc. Lond.* **105**, 65–99.

KENNEDY, M. J. 1979. The continuation of the Canadian Appalachians into the Caledonides of Britain and Ireland. *In:* HARRIS, A. L., HOLLAND, C. H. & LEAKE, B. E. (eds) *The Caledonides of the British Isles—reviewed. Spec. Publ. geol. Soc. London,* **8**, 33–64.

KIDD, J. 1814. Notes on the mineralogy of the neighbourhood of St David's Pembreshire. *Trans. geol. Soc. Lond. (old series),* **2**, 79.

KOKELAAR, B. P. 1977. The igneous history of the Rhobell Fawr area, Merionethshire, North Wales. Ph.D. thesis. Univ. Wales, Aberystwyth (unpubl.).

—— 1979. Tremadoc to Llanvirn volcanism on the southeast side of the Harlech Dome (Rhobell Fawr), N. Wales. *In:* HARRIS, A. L., HOLLAND, C. H. & LEAKE, B. E. (eds) *The Caledonides of the British Isles—reviewed. Spec. Publ. geol. Soc. London,* **8**, 591–6.

—— 1982. Fluidization of wet sediments during the emplacement and cooling of various igneous bodies. *J. geol. Soc. London,* **139**, 21–33.

——, FITCH, F. J. & HOOKER, P. J. 1982. A new K-Ar age from uppermost Tremadoc rocks of north Wales. *Geol. Mag.* **119**, 207–11.

LEGGETT, J. K. 1980. British Lower Palaeozoic black shales and their palaeo-oceanographic significance. *J. geol. Soc. London,* **137**, 139–56.

LOCKLEY, M. G. & WILLIAMS, A. 1981. Lower Ordovician Brachiopoda from mid- and southwest Wales. *Bull. Br. Mus. nat. Hist. (Geol.)* **34**, 1–75.

LOWE, D. R. 1982. Sediment gravity flows. II. Depositional models with special reference to the deposits of high-density turbidity currents. *J. sediment. Petrol.* **52**, 279–97.

LOWMAN, R. D. W. 1977. The geology, petrology and geochemistry of an area of Lower Palaeozoic rocks east of Fishguard, north Pembrokeshire. Ph.D. thesis. Univ. Wales (unpubl.).

—— & BLOXAM, T. W. 1981. The petrology of the Lower Palaeozoic Fishguard Volcanic Group and associated rocks E of Fishguard, N. Pembrokeshire (Dyfed), South Wales. *J. geol. Soc. London,* **138**, 47–68.

LYNAS, B. D. T. 1973. The Cambrian and Ordovician rocks of the Migneint area, North Wales. *J. geol. Soc. London,* **129**, 481–503.

McBIRNEY, A. R. 1963. Factors governing the nature of submarine volcanism. *Bull. Volcanol.* **26**, 455–69.

McKERROW, W. S., LAMBERT, R. ST. J. & COCKS, L. R. M. In press. The Ordovician, Silurian and Devonian Periods. *In:* SNELLING, N. J. (ed.) *Geochronology and the Geological Record. Spec. Publ. geol. Soc. London.*

NUTT, M. J. C. & SMITH, E. G. 1981. Transcurrent faulting and the anomalous position of pre-Carboniferous Anglesey. *Nature, Lond.* **290**, 492–5.

OLIVER, R. L. 1954. Welded tuffs in the Borrowdale Volcanic Series, English Lake District, with a note on similar rocks in Wales. *Geol. Mag.* **91**, 473–83.

OWENS, R. M. & FORTEY, R. A. 1982. Arenig rocks of the Carmarthen–Llanarthney District. *In:* BASSETT, M. G. (ed.) *Geological Excursions in Dyfed, South-west Wales,* 249–58. National Museum of Wales, Cardiff.

——, ——, COPE, J. C. W., RUSHTON, A. W. A. & BASSETT, M. G. 1982. Tremadoc faunas from the

Carmarthen district, South Wales. *Geol. Mag.* **119**, 1–38.

PEARCE, J. A. 1975. Basalt geochemistry used to investigate past tectonic environments on Cyprus. *Tectonophysics*, **25**, 41–67.

—— & CANN, J. R. 1973. Tectonic setting of basic volcanic rocks determined using trace element analyses. *Earth planet. Sci. Lett.* **19**, 290–300.

PHILLIPS, W. E. A., STILLMAN, C. J. & MURPHY, T. 1976. A Caledonian plate tectonic model. *J. geol. Soc. London*, **132**, 579–609.

PIPER, J. D. A. 1978. Palaeomagnetism and palaeogeography of the Southern Uplands block in Ordovician times. *Scott. J. Geol.* **14**, 93–107.

PRINGLE, J. 1930. The geology of Ramsey Island (Pembrokeshire). *Proc. Geol. Assoc. Lond.* **41**, 1–31.

RAMSAY, A. C. 1881. The geology of North Wales. *Mem. geol. Surv. G.B.* **3**. (2nd edition), 611 pp.

RAST, N. 1969. The relationship between Ordovician structure and volcanicity in Wales. *In*: WOOD, A. (ed.) *The Pre-Cambrian and Lower Palaeozoic Rocks of Wales*, 305–35. University of Wales Press, Cardiff.

——, O'BRIEN, B. H. & WARDLE, R. J. 1976. Relationships between Precambrian and Lower Palaeozoic rocks of the 'Avalon Platform' in New Brunswick, the northeast Appalachians and the British Isles. *Tectonophysics*, **30**, 315–38.

REED, F. R. C. 1895. The geology of the country around Fishguard, Pembrokeshire. *Q. Jl geol. Soc. Lond.* **51**, 149–95.

REEDMAN, A. J., LEVERIDGE, B. E. & EVANS, R. B. 1984. The early Cambrian Arfon Basin of NW Wales. *Proc. Geol. Assoc. Lond.* **95**, in press.

——, WEBB, B. C., ADDISON, R., LYNAS, B. D. T., LEVERIDGE, B. E. & HOWELLS, M. F. 1983. The Cambrian–Ordovician boundary at Aber and Betws Garmon, Gwynedd, North Wales. *Rep. Inst. geol. Sci. Lond.* **83/1**, 7–10.

RIDGWAY, J. 1975. The stratigraphy of Ordovician volcanic rocks on the southern and eastern flanks of the Harlech Dome in Merionethshire. *Geol J.* **10**, 87–106.

—— 1976. Ordovician palaeogeography of the southern and eastern flanks of the Harlech Dome, Merionethshire, North Wales. *Geol. J.* **11**, 121–36.

ROACH, R. A. 1969. The composite nature of the St David's Head and Carn Llidi Intrusions of North Pembrokeshire. *In*: WOOD, A. (ed.) *The Pre-Cambrian and Lower Palaeozoic Rocks of Wales*, 409–33. University of Wales Press, Cardiff.

ROBERTS, B. 1969. The Llwyd Mawr ignimbrite and its associated volcanic rocks. *In*: WOOD, A. (ed). *The Pre-Cambrian and Lower Palaeozoic Rocks of Wales*, 337–56. University of Wales Press, Cardiff.

—— 1981. Low grade and very low grade regional metabasic Ordovician rocks of Llŷn and Snowdonia, Gwynedd, North Wales. *Geol. Mag.* **118**, 189–200.

RUSHTON, A. W. A. 1974. The Cambrian of Wales and England. *In*: HOLLAND, C. H. (ed.) *Lower Palaeozoic rocks of the World. II. Cambrian of*

the British Isles, Norden and Spitzbergen, 43–121. Wiley, London.

SAUNDERS, A. D., FORNARI, D. J. & MORRISON, M. A. 1982. The composition and emplacement of basaltic magmas produced during the development of continental-margin basins: the Gulf of California, Mexico. *J. geol. Soc. London*, **139**, 335–46.

SCHIENER, E. J. 1970. Sedimentology and petrography of three tuff horizons in the Caradocian sequence of the Bala area (North Wales). *Geol. J.* **7**, 25–46.

SMITH, A. G., BRIDEN, J. C. & DREWRY, G. E. 1973. Phanerozoic world map. *Spec. Pap. Palaeontol. Lond.* **12**, 1–42.

——, HURLEY, A. M. & BRIDEN, J. C. 1981. *Phanerozoic Paleocontinental World Maps.* Cambridge University Press, Cambridge, 102 pp.

SPARKS, R. S. J., SIGURDSSON, H. & CAREY, S. N. 1980. The entrance of pyroclastic flows into the sea. II. Theoretical considerations on subaqueous emplacement and welding. *J. Volcan. geoth. Res., Amsterdam*, **7**, 97–105.

STAMP, L. D. & WOOLDRIGE, S. W. 1923. The igneous and associated rocks of Llanwrtyd, Brecon. *Q. Jl geol. Soc. Lond.* **79**, 16–46.

STEAD, J. T. G. & WILLIAMS, B. P. J. 1971. The Cambrian rocks of North Pembrokeshire. *In*: BASSETT, D. A. & BASSETT, M. G. (eds) *Geological Excursions in South Wales and the Forest of Dean*, 180–98. South Wales Group of the Geologists' Association, Cardiff.

STILLMAN, C. J. & FRANCIS, E. H. 1979. Caledonide volcanism in Britain and Ireland. *In*: HARRIS, A. L., HOLLAND, C. H. & LEAKE, B. E. (eds) *The Caledonides of the British Isles—reviewed. Spec. Publ. geol. Soc. London*, **8**, 556–77.

THOMAS, G. E. & THOMAS, T. M. 1956. The volcanic rocks of the area between Fishguard and Strumble Head, Pembrokeshire. *Q. Jl geol. Soc. Lond.* **112**, 291–314.

THOMAS, H. H. & COX, A. H. 1924. The Volcanic Series of Trefgarn, Roch and Ambleston (Pembrokeshire). *Q. Jl geol. Soc. Lond.* **80**, 520–48.

THORPE, R. S. 1974. Aspects of magmatism and plate tectonics in the Precambrian of England and Wales. *Geol. J.* **9**, 115–36.

—— 1979. Late Precambrian igneous activity in Southern Britain. *In*: HARRIS, A. L., HOLLAND, C. H. & LEAKE, B. E. (eds) *The Caledonides of the British Isles—reviewed. Spec. Publ. geol. Soc. London*, **8**, 579–84.

——, BECKINSALE, R. D., PATCHETT, P. J., PIPER, J. D. A. DAVIES, G. R. & EVANS, J. A. 1984. Crustal growth and late Precambrian–early Palaeozoic plate tectonic evolution of England and Wales. *J. geol. Soc. London*, **141**, 521–36.

TURNER, P. 1977. Notes on the depositional environment of the *Lingula* Flags in Dyfed, South Wales. *Proc. Yorkshire geol. Soc.* **41**, 199–202.

WATTS, W. W. 1885. On the igneous and associated rocks of the Breidden Hills in East Montgomeryshire and West Shropshire. *Q. Jl geol. Soc. Lond.* **41**, 532–46.

—— 1925. The geology of South Shropshire. *Proc. Geol. Assoc. Lond.* **36**, 321–63.

WEBB, B. C. 1983. Early Caledonian structures in the Cambrian Slate Belt, Gwynedd, North Wales. *Rep. Inst. geol. Sci. Lond.* **83/1**, 1–6.

WILLIAMS, A. 1953. The geology of the Llandeilo district, Carmarthenshire. *Q. Jl geol. Soc. Lond.* **108**, 177–208.

——, LOCKLEY, M. G. & HURST, J. M. 1981. Benthic palaeocommunities represented in the Ffairfach Group and coeval Ordovician successions of Wales. *Palaeontology*, **24**, 661–94.

WILLIAMS, B. P. J. & STEAD, J. T. G. 1982. The Cambrian rocks of the Newgale–St David's area. *In*: BASSETT, M. G. (ed.) *Geological Excursions in Dyfed, South-west Wales*, 27–49. National Museum of Wales, Cardiff.

WILLIAMS, H. 1927. The geology of Snowdon (North Wales). *Q. Jl geol. Soc. Lond.* **83**, 346–431.

WOOD, D. S. 1969. The base and correlation of the Cambrian rocks of North Wales. *In*: WOOD, A. (ed.) *The Pre-Cambrian and Lower Palaeozoic Rocks of Wales*, 47–66. University of Wales Press, Cardiff.

—— 1974. Ophiolites, mélanges, blueschists and ignimbrites: early Caledonian subduction in Wales? *In*: DOTT, R. H. & SHAVER, R. H. (eds) *Modern and Ancient Geosynclinal Sedimentation. Soc. Econ. Palaeontol. Mineral. Spec. Publ.* **19**, 334–44.

ZALASIEWCZ, J. 1982. Stratigraphy and palaeontology of the Arenig area, North Wales. Ph.D. thesis. Univ. Cambridge (unpubl.).

B. P. KOKELAAR, School of Environmental Sciences, Ulster Polytechnic. Newtownabbey, Co. Antrim BT37 0QB, U.K.

M. F. HOWELLS, Wales Geological Survey Unit, British Geological Survey, Bryn Eithyn Hall, Llanfarian, Aberystwyth, Dyfed SY23 4BY, U.K.

R. E. BEVINS, Department of Geology, National Museum of Wales, Cathay's Park, Cardiff CF1 3NP, U.K.

R. A. ROACH, Department of Geology, University of Keele, Keele, Staffordshire ST5 5BG, U.K.

P. N. DUNKLEY, British Geological Survey, Keyworth, Nottingham NG12 5GG, U.K.

Mud–magma interactions in the Dunnage Mélange, Newfoundland

B. E. Lorenz

SUMMARY: The Dunnage Mélange, part of back-arc basinal assemblages in north central Newfoundland, is host to a suite of dacitic to rhyolitic intrusions. Intrusion occurred during the late stages of mélange formation. Lobate, pillowed, corrugated and pahoehoe-like igneous contact surfaces, complex interlayering of host mudstone and dacite, and the occurrence of peperite, indicate that the magma intruded unconsolidated mud and the adjacent sediment was fluidized. This evidence for shallow emplacement of the magma is inconsistent with the existence of a tectonic overburden during mélange formation, and supports an olistostromal origin for the Dunnage Mélange.

The Cambro–Ordovician Dunnage Mélange, part of back-arc basinal assemblages in north central Newfoundland, is host to a suite of dacitic to rhyolitic intrusions, collectively known as the Coaker Porphyry (Kay 1972). The intrusions are confined to the mélange terrane, where they comprise approximately one-third of the outcrop area. The age, physical characteristics, and distribution of these intrusions suggest that they were intruded during mélange formation (Williams & Hibbard 1976). In this paper, the physical relationships between the Coaker Porphyry and the host mélange are described and their origin is explained. Implied conditions during intrusion, and the origin of the mélange, are discussed.

Regional setting of the Dunnage Mélange

The Appalachian Orogen in Newfoundland is divided into four major tectonostratigraphic zones (Williams 1979) (Fig. 1). The Dunnage Zone, including the Dunnage Mélange, consists of ophiolites, early Palaeozoic mafic volcanics and marine sediments, and represents the remains of the ancient Iapetus Ocean. The Humber Zone, to the W, represents the ancient continental margin of North America, the easternmost part of which was deformed and metamorphosed during emplacement of the Taconic allochthons on to the continental margin. To the E the Gander and Avalon Zones are enigmatic suspect terranes (Williams & Hatcher 1982) whose relationships to the Appalachian Orogen are poorly understood. They have been interpreted as continental masses from the eastern edge of the Iapetus Ocean.

The Dunnage Zone is underlain by an ophiolitic basement which is exposed along the

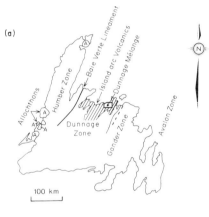

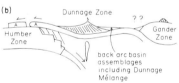

FIG. 1. (a) Generalized map of Newfoundland showing the location of major zones and geological features. Volcanic-arc and arc-flanking assemblages of the Dunnage Zone are shown in the shaded area. After Williams (1978). (b) Diagram, modified from Malpas (1977), depicting the currently accepted model of Lower Ordovician Newfoundland, showing the position of the Dunnage Mélange relative to the subduction zone and island arc. Question marks indicate the enigmatic nature of the relationship of the Gander Zone to the westerly parts of the Appalachian Orogen.

western edge of the zone, and locally in horsts in the central part of the zone (Haworth et al. 1978; Lorenz & Fountain 1982). Thick piles of Lower Ordovician volcanic and volcaniclastic

271

rocks in the western and central parts of the Dunnage Zone have been interpreted as having formed in an island-arc environment (see Hibbard & Williams 1979, and references therein).

The Dunnage Mélange is part of an assemblage of arc-derived basinal sediments that lies to the E of the island-arc volcanics. Previous workers have disagreed as to whether this basinal assemblage was located in a fore-arc or a back-arc position. Most of the advocates of a fore-arc environment based their conclusions on the presence of the Dunnage Mélange itself, which they interpreted as having formed in the trench of a W-dipping subduction zone, in a similar manner to the Franciscan mélanges (Dewey 1969; Kay 1976; McKerrow & Cocks 1978). Hibbard & Williams (1979) proposed that the Dunnage Mélange is an arc-flanking olistostrome rather than a trench-fill mélange, but they were unable to distinguish between the fore- and back-arc environments. However, most of the regional evidence favours the currently accepted tectonic model for Newfoundland in which closure of the Iapetus Ocean was

effected by an E-dipping subduction zone located along the Baie Verte Lineament (Fig. 1) (cf. Malpas & Stevens 1977; Haworth *et al.* 1978). The Baie Verte Lineament, comprising tectonized and imbricated ophiolitic slices, is interpreted as a suture zone analogous to the Dun Mountain Belt of New Zealand. Subduction of the leading edge of the continental margin (Humber Zone) resulted in the emplacement on to the continent of allochthonous slices of slope and rise clastic sediments and ophiolite. Subduction ceased by Caradocian time. This model places the Dunnage Mélange in a back-arc setting (Fig. 1b).

Local setting of the Dunnage Mélange

A summary of the previous work pertaining to the Dunnage Mélange can be found in Hibbard & Williams (1979).

The Dunnage Mélange is a chaotic unit, with inclusions mainly of sedimentary and mafic volcanic rocks in an argillaceous matrix. The inclu-

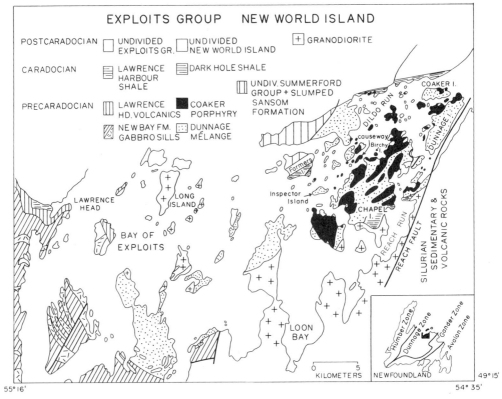

FIG. 2. Generalized geological map of the Bay of Exploits area (modified after Williams & Hibbard 1976) showing the distribution of Coaker Porphyry.

sions range in size from a few mm up to 1 km across and comprise < 50% of the mélange. The Dunnage Mélange crops out in an area 40 km long with a maximum width of 10 km (Fig. 2). It is bound to the E by the Reach Fault, to the S by the granodioritic Loon Bay Batholith and to the NW by the granodioritic Long Island Batholith. To the W, the mélange interdigitates with the sediments and volcanics of the Exploits Group and a ghost stratigraphy of this group can be determined within the mélange (Hibbard & Williams 1979). The base of the Dunnage Mélange is not exposed. To the S, on Chapel Island, the mélange grades upwards into folded, vertically dipping, S-facing, bedded siltstones, greywackes, and conglomerates. To the N, it is in contact with N-facing, vertically dipping turbidites of the Caradocian Dark Hole Formation (Horne 1969). Palaeontological evidence indicates a Cambro–Ordovician age for the components of the Dunnage Mélange (Hibbard *et al.* 1977).

The Coaker Porphyry

The igneous rocks in the Dunnage Mélange were described (Heyl 1936; Patrick 1956; Williams 1964) before the mélange was recognized (Horne 1968; Kay 1970). Although basalts were later recognized to be older blocks, rather than flows interlayered with the shales, the silicic rocks continued to be described as intrusions. Kay (1970, 1972, 1973, 1976) named and described the igneous intrusions of the NE Dunnage Mélange, the Coaker Porphyry, the Causeway Xenolith Phase of the Coaker Porphyry, and the Puncheon Diorite.

The Coaker Porphyry is a suite of peraluminous dacitic to rhyolitic hypabyssal intrusions. It is restricted to the area between Dildo and Reach Runs (Fig. 2), where the individual intrusions occur as irregular stocks that are weakly elongate in a NE–SW direction. The Coaker Porphyry is distinguished by the large size and concentration of its component intrusions, its intrusive relationships, the occurrence of ultramafic and mafic xenoliths, and its geochemistry. Williams & Hibbard (1976) suggested that the Coaker Porphyry intruded the mélange while the host sediments were unconsolidated.

Intrusive relationships of the Coaker Porphyry

The Coaker Porphyry occurs in the following forms: (i) stocks; (ii) dykes; (iii) complexly

interlayered with mudstone and folded; (iv) lobes, in isolation or as part of types i–iii; and (v) breccias.

Stocks

Stocks are the most common intrusive form and range from a few meters to about 3 km across. They are irregular, rounded and commonly have undulose, lobate, or pillowed contacts. Their surface is generally corrugated with the wavelength of the corrugations greater than the amplitude; wavelength is usually 1 cm or less but locally 6 or 7 cm. A few stocks display cracked surfaces. Chilled margins and vesicles are rare and marginal trachytic-textured dacite shows pronounced flow-banding with alignment of mica flakes and plagioclase laths parallel to contacts.

Dykes

Dykes are irregular in thickness and shape; in mudstones they split, then abruptly pinch out. Some terminate in pillow forms although they display straight, smooth, parallel contacts where they intrude brittle materials such as volcanic blocks and hornfels. They are composed of trachytic-textured dacite, and are locally rich in xenoliths which are concentrated in the centre by flow segregation. Unlike the stocks, the dykes commonly have chilled margins.

Interlayered Coaker Porphyry and mudstone

Coaker Porphyry that is interlayered and folded with mudstone is the least common and most complex form in which the intrusions occur.

On Inspector Island (Fig. 2), in an area about 500 m long and 150 m wide, an elongate, flow-banded stock of rhyolite is bordered by small rhyolite dykes that split into sheets interlayered with the host mudstone. The interlayered rhyolite and mudstone are complexly folded (Fig. 3). Commonly the limbs and hinges comprise several layers of rhyolite and mudstone that bear no systematic relationship to adjacent folds. Individual layers display complex folding within large folds, pinch and swell structures, boudinage, bifurcation, and brecciation (Fig. 3). Fold hinges in rhyolite commonly thicken into lobate structures that locally contain subparallel, discontinuous layers of mudstone. Other hinges give rise to third limbs that extend parallel to the axial surface or are folded around with the original limbs (Fig. 4). The surfaces of the small rhyolite dykes are commonly convoluted

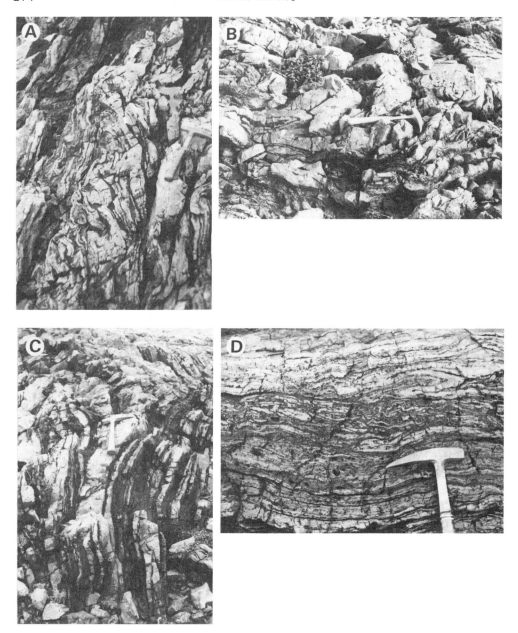

FIG. 3. (a) Complex mixing within a larger folded structure of interlayered dacite (light) and mudstone (dark), Birchy Island. (b) Contrasting behaviour of rhyolite layers within a fold, Inspector Island. (c) Bifurcation of rhyolite layers in mudstone, Inspector Island. (d) Break-up of layering in a fold limb, Birchy Island.

into ropy structures resembling pahoehoe. The layering varies from being dominantly rhyolite with narrow (1 mm) stringers of mudstone, to being dominantly mudstone with narrow (2–3 cm) stringers of rhyolite.

On Birchy Island (Fig. 2) similar relation-ships occur in an area about 200 m long and 15–20 m wide. Although here the overall struc-ture is of a recumbent isoclinal antiform (Fig. 5), the internal layering is chaotic. The inner part of the antiform consists of interlayered mudstone and trachytic dacite, and the outer

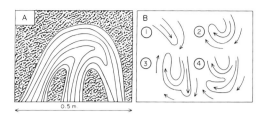

FIG. 4. (a) A three-limbed fold of Coaker Porphyry (white) in mudstone (mottled). (b) Mechanism of formation of three-limbed folds during slumping of mud and magma.

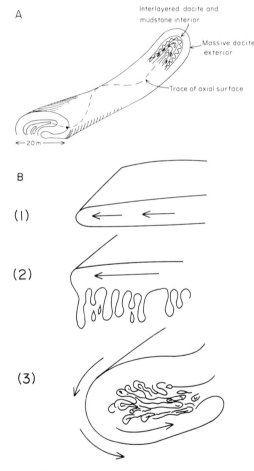

FIG. 5. (a) Schematic drawing of the body of interlayered and folded dacite and mudstone on Birchy Island. (b) A possible mechanism of formation of the body depicted in (a). (1) A sill of dacite intrudes wet mud. (2) Magma sags into the less dense underlying mud. (3) The mass folds into its final form during contemporaneous slumping.

part of massive trachytic dacite. The interlayering occurs from a fine scale with some layers only mm thick, where the rock resembles a laminated shale, to dacite layers up to 1 m thick. Individual layers display the same kinds of disruption as those on Inspector Island.

Lobes

Bulbous or lobate protuberances of trachytic dacite occur on the sides of stocks, the terminations of dykes, and the hinges of folds of dacite interlayered with mudstone. They range from *c*. 30 cm to 4 m in diameter. Mostly they occur singly, but in places as clusters with intervening mudstone. Commonly a flow lineation, defined by plagioclase laths, lies concordant with the contact.

Breccias

Breccia is scarce, occurring locally in association with the stocks in two distinct forms:
(i) As a mantle of breccia gradational into massive dacite. Dacite fragments in the breccia are set in a matrix composed primarily of coarse, elongate anhedral secondary quartz crystals with a weak subparallel alignment, with minor carbonate and wisps of mudstone. Some of the fragments display a jigsaw-fit near the contact with the massive body of the stock and elsewhere the breccia commonly contains drusy cavities. Because of the gradational contacts between the breccia and the massive stocks, and the presence of mudstone and silicification in the matrix, these breccias are interpreted as peperites (cf. Kokelaar 1982).
(ii) Dacite breccia with mudstone matrix occurs in dykes that cut some stocks. Similar features, developed during cooling of viscous rhyolite, are described by Kokelaar (1982) who ascribes them to the opening of cooling joints with simultaneous injection of hot fluidized mud and peperite formation.

Interpretation

In agreement with Williams & Hibbard (1976), the Coaker Porphyry is interpreted to have been intruded into Lower Ordovician unconsolidated wet sediments that comprised the matrix of the Dunnage Mélange. The evidence supporting this interpretation is as follows:
(i) Corrugated and pahoehoe-like surfaces on some of the intrusions suggest that both magma and host were fluid at the time of intrusion. Similar features have been described by

Kokelaar (1982) on a rhyolite sill which intruded Ordovician wet muds on Ramsey Island, SW Wales. Also, cooling cracks on Coaker stocks resemble the 'breadcrust' structures from the Hammond Sill, interpreted by Hoyt (1961) as intrusive into wet sediment.

(ii) Lobate margins of the intrusions are similar to pillows developed at the flow fronts of subaqueous lava flows.

(iii) The peperitic intrusive margins indicate intrusion into and fluidization of wet sediment (Kokelaar 1982).

The folded interlayering of Coaker Porphyry and mudstone with thickening and extension of hinges, the pinching out of mudstone layers within intrusive rock and *vice versa*, and the lack of systematic structure, also indicate deformation of unconsolidated materials. The trachytic texture of the rhyolites and dacites, and the absence of strain within euhedral phenocrysts of quartz, muscovite, and feldspar, indicate that the interlayered structures of igneous and sedimentary rocks are primary and not tectonic. For such relationships to occur the following require explanation: (i) the mechanism of interlayering of mud and magma; (ii) the origin of folding; and (iii) the persistence of very thin magma layers within the original mud.

Kokelaar (1982) has argued that the complex intermingling of magma and mud results from fluidization of the mud by steam generated at contacts. Fluidized sediments offer little resistance to intrusion, and extremely complex contacts between magma and sediment can result because magma can extend unrestricted in any direction. Fluidized sediments also insulate the magma from immediate chilling, thus allowing persistence of very thin magma layers.

The style of folding indicates that the magma and mud slumped in an unconsolidated state. This may have been triggered by the continued intrusion of the magma into muds at the edge of an escarpment or on a slope. Magma intrusion contemporaneous with slumping is envisaged such that a sheet of magma is folded into mud as the combined mass moves down-slope. The magma entering the hinge continues to flow down-slope, producing the third limb which may in turn be folded over (Fig. 4b).

Two factors support the interpretation that the folding and interlayering are slope-controlled rather than a primary function of intrusion, such as by squirting of the magma into the mud: (i) the scarcity of the phenomenon, in contrast to an abundance of dykes and other simple forms, suggests that a special condition—the local presence of a slope—is required for its formation; and (ii) there is an apparently gravity-controlled asymmetry inherent in these forms. Sheets interpreted as intrusive into wet sediments, on Ramsey Island, SW Wales, and near Waterford, SE Ireland, have smooth upper contact surfaces, and irregular bulbous protrusions extending from their lower surfaces. This asymmetry would appear to be the result of loading of the denser magma into the less-dense (fluidized) sediments. On Inspector Island, the small dykes extend from an elongate stock, not simply loading as they might in a pile of stagnant mud, but folded into complex forms. On Birchy Island, the massive outer layer of the isoclinal antiform probably originated as a thick dyke from which the thin dykes of its inner part were formed, with the whole unit slumped into its present form (Fig. 5b).

The relationships of the Coaker Porphyry support the interpretation that it intruded unconsolidated sediments. Fluidization of the associated sediments indicates depths of intrusion of less than 1.6 km (Kokelaar 1982), and the presence of dacite cobbles in conglomerates in the Dunnage Mélange indicates penecontemporaneous erosion of extrusive or very shallow intrusive Coaker units. Shallow emplacement is thus implied.

Conclusions

The Coaker Porphyry intruded the water-saturated, unlithified Lower Ordovician mud matrix of the Dunnage Mélange. Complex interlayering and folding of mud and magma occurred as a result of intrusion of the magma into mud on slopes. The evidence suggests that the Coaker Porphyry was emplaced at shallow levels, which is inconsistent with the existence of a large tectonic overburden at the time of intrusion. The relationships between the Coaker Porphyry and its host mudstones are inconsistent with a tectonic origin for the Dunnage Mélange.

ACKNOWLEDGMENTS: I wish to thank D. F. Strong, R. K. Stevens, and H. Williams for their guidance and financial support for the field work done during 1981 and 1982. I have benefitted greatly from discussions with G. Horne, D. Reusch, W. S. McKerrow, N. Rast, D. Karig, R. Flood, and P. Wonderley. Thanks are due also to D. F. Strong, R. M. Easton, R. K. Stevens, and P. Kokelaar for their constructive criticism of the manuscript. I particularly appreciate the warm hospitality of Mrs Stella Small and Dr Ann Roberts, and the expert boatmanship of Mr Elmo Small of Summerford, Newfoundland. I also wish to thank Mr Winston Howell for drafting the diagrams.

References

DEWEY, J. F. 1969. Evolution of the Appalachian/Caledonian orogen. *Nature, Lond.* **222**, 124–9.

HAWORTH, R. T., LEFORT, J. & MILLER, H. G. 1978. Geophysical evidence for an east-dipping Appalachian subduction zone beneath Newfoundland. *Geology*, **6**, 522–6.

HEYL, G. R. 1936. Geology and mineral deposits of the Bay of Exploits area, Newfoundland. *Bull. Newfoundld Dept Nat. Res. geol. Sec. 3*.

HIBBARD, J. P. & WILLIAMS, H. 1979. Regional setting of the Dunnage Mélange in the Newfoundland Appalachians. *Am. J. Sci.* **279**, 993–1021.

——, STOUGE, S. & SKEVINGTON, D. 1977. Fossils from the Dunnage Mélange, north-central Newfoundland. *Can. J. Earth Sci.* **14**, 1176–8.

HORNE, G. S. 1968. Stratigraphy and structural geology of southwestern New World Island area, Newfoundland. Unpublished Ph.D Thesis., Columbia University.

—— 1969. Early Ordovician chaotic deposits in the central volcanic belt of northeastern Newfoundland. *Bull. geol. Soc. Am.* **80**, 2451–64.

HOYT, C. L. 1961. The Hammond Sill—an intrusion in the Yakima Basalt near Wenatchee, Washington. *NW. Sci.* **35**, 58–64.

KAY, M. 1970. Flysch and bouldery mudstone in northeast Newfoundland. In: LAFOIE, J. (ed.) *Flysch Sedimentology in North America. Spec. Pap. geol. Ass. Can.* **7**, 155–64.

—— 1972. Dunnage Mélange and Lower Paleozoic deformation in north-west Newfoundland. *Int. geol. Congr. 24th*, sec. 3, 122–33.

—— 1973. Tectonic evolution of Newfoundland. *In*: DEJONG, K. A. & SCHOLTEN, R. (eds) *Gravity and Tectonics*, 313–26. Wiley, New York.

—— 1976. Dunnage Mélange and subduction of the Protacadic Ocean, northeast Newfoundland. *Spec. Pap. geol. Soc. Am.* **175**, 49 pp.

KOKELAAR, B. P. 1982. Fluidization of wet sediments during the emplacement and cooling of various igneous bodies, *J. geol. Soc. London*, **139**, 21–33.

LORENZ, B. E. & FOUNTAIN, J. C. 1982. The South Lake Igneous Complex, Newfoundland: a marginal basin—island arc association. *Can. J. Earth Sci.* **19**, 490–503.

MCKERROW, W. S. & COCKS, L. R. M. 1978. A Lower Paleozoic trench-fill sequence, New World Island, Newfoundland. *Bull. geol. Soc. Am.* **89**, 1121–32.

MALPAS, J. 1977. Petrology and tectonic significance of Newfoundland ophiolites, with examples from the Bay of Islands. *In*: COLEMAN, R. G. & IRWIN, W. P. (eds) *North American Ophiolites. Bull. Oregon Dept geol. Min. Ind.* **95**, 13–23.

—— & STEVENS, R. K. 1977. The origin and emplacement of the ophiolite suite with examples from western Newfoundland. *Geotectonics*, **11**, 453–66.

PATRICK, T. O. N. 1956. Comfort Cove, Newfoundland, map with notes. *Pap. geol. Surv. Can. 55/31*.

WILLIAMS, H. 1964. Notes on the orogenic history and isotope ages in the Botwood Map area, northeastern Newfoundland. *Pap. geol. Surv. Can. 64/17*, 22–9.

—— 1978. Geological development of the northern Appalachians: its bearing on the evolution of the British Isles. *In*: BOWES, D. R. & LEAKE, B. E. (eds) *Crustal Evolution in Northwestern Britain and Adjacent Regions. Geol. J. Spec. Issue*, **10**, 1–22.

—— 1979. Appalachian Orogen in Canada. *Can. J. Earth Sci.* **16**, 792–807.

—— & HATCHER, JR, R. D. 1982. Suspect terranes and accretionary history of the Appalachian orogen. *Geology*, **10**, 530–6.

—— & HIBBARD, J. P. 1976. The Dunnage Mélange, Newfoundland. *Pap. geol. Surv. Can. 76/1A*, 183–5.

BRENNA E. LORENZ, Department of Earth Sciences, Memorial University of Newfoundland, St John's, Newfoundland, Canada A1B 3X5

The Late Precambrian and early Palaeozoic marginal basin of South China

Pan Guoqiang

SUMMARY: An integrated study based on recent geological mapping and encompassing the evolution of sedimentary facies, volcanism, plutonism and metamorphism in the Sinian and early Palaeozoic rocks of SE China has led to a reinterpretation of the area as representing a marginal basin situated at the SE margin of the Yangtze continental plate. The marginal basin comprised an active basin or trough and an inactive or marginal sea. The two parts display different sedimentary facies and tectonic and volcanic histories. It is concluded that in the early stages of evolution magma was derived from the mantle but later magmas were generated mainly in the crust.

The Precambrian and early Palaeozoic marginal basin of southern China was an extensive back-arc depression, over 2000 km long and 700 km wide, situated between a volcanic arc to the SE and an epicontinental sea to the NW. An analysis of the sedimentary facies, facies variations and volcanism within the basin indicates that for most of its history it comprised an outer, tectonically active zone adjacent to the continental slope of the island arc, and an inner, transitional zone adjacent to the upper continental slope of the stable continental area. This configuration is similar to that proposed by Karig & Sharman (1975) who divided marginal basins into active and inactive basins separated by a remnant arc. The two parts can also be related to the classical geosyncline, the internal eugeosyncline corresponding to the active basin and the external miogeosyncline to the transitional or inactive part.

History of development of the South China Marginal Basin

The initiation and early development of the South China Marginal Basin occurred in the Sinian period which commenced at c. 800 Ma. The basin persisted until Middle Silurian times and therefore occupies the post-Jinning and pre-Guangxi stage in the stratigraphic development of China (Wang Hongzheng & Wang Ziqiang 1981).

In Early Sinian times the main tectonic elements in southern China were the overthrusting Yangtze plate to the NW and the underthrusting Nanhua plate to the SE. The plates were in contact along a NW-dipping subduction zone, along the Shaoxing–Lishui–Nanping–Haifeng line. In Late Sinian times, the Zhe-jiang–Fujian–Guangdong island arc formed on the inner side of the subduction zone. A marginal basin, wide in the SW and narrow in the NE, lay to the NW of the island arc and adjoining the epicontinental sea of the Yangtze platform (Fig. 1). Within the marginal basin a positive area, named 'Hunan–Jiangxi old island', occupied a zone from E Hunan to NE Jiangxi and separated the active marginal basin to the SE from the inactive marginal basin to the NW. Also in Late Sinian times, local uplift occurred at the margins of the Yangtze epicontinental sea (Fig. 1).

The broad configuration of the basin remained the same during the Cambrian except that the active basin extended towards W Hunan and E Guizhou (Fig. 2). In the active basin in Jiangxi, N Guangdong and SE Hunan, thick sediments accumulated, indicating continual subsidence of several thousand metres. This contrasts strongly with the thin, non-compensated sediments of the inactive basin (Fig. 2).

During Early and Middle Ordovician times subsidence in the basin was slow. At the beginning of the Late Ordovician, the interaction of the eastward subduction along the Ailaoshan subduction zone at the SW margin of the Yangtze Plate and the westward subduction of the Nanhua oceanic plate caused the regions of S Guizhou, N Guangxi and SE Yunnan to be uplifted and accreted to the Yangtze plate, with a consequent contraction of the marginal basin (Fig. 3). By Silurian times only a narrow trough existed in W Zhejiang, W Fujian and the Qinzhou–Fangchen area (Fig. 4). In Middle and Late Silurian times further uplift, accompanied by folding in the E, resulted in connection of the various islands and the disappearance of the marginal basin.

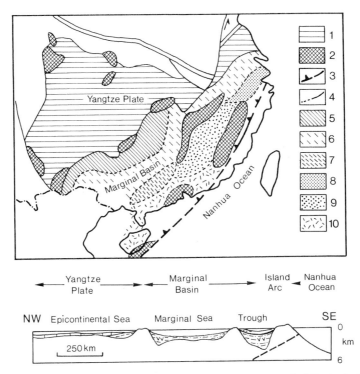

FIG. 1. Palaeogeographical map and section of the South China Marginal Basin during the Late Sinian. 1. Sedimentary area of Yangtze plate. 2. Island arc and remnant rise. 3. Subduction zone. 4. Lithofacies boundary. 5–10: Lithofacies. 5. Siliceous-carbonaceous. 6. Pelitic-siliceous. 7. Clastic-siliceous. 8. Clastic-siliceous–carbonaceous. 9. Pyroclastic and pelitic flysch. 10. Clastic flysch.

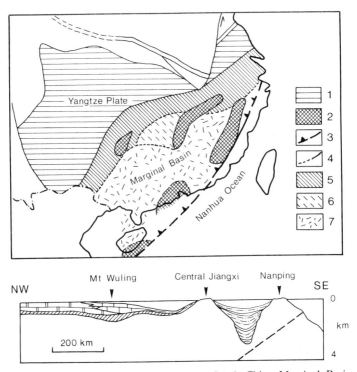

FIG. 2. Palaeogeographical map and section of the South China Marginal Basin during the Middle and Late Cambrian. 1–4 as in Fig. 1. 5–7: Lithofacies. 5. Pelitic-carbonate. 6. Carbonaceous-siliceous-pelitic-carbonate. 7. Pelitic sandy flysch-like and flysch.

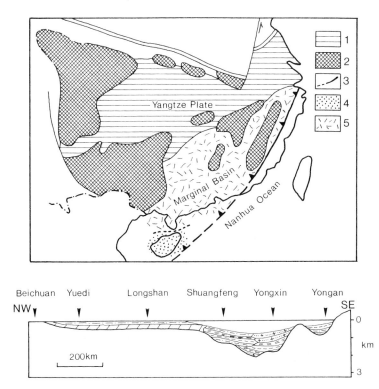

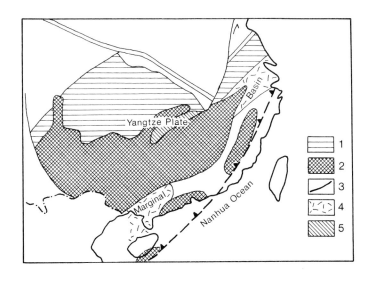

FIG. 3. Palaeogeographical map and section of the South China Marginal Basin during the Late Ordovician. 1. Sedimentary area of Yangtze plate. 2. Erosion area of Yangtze plate and island arc. 3. Lithofacies boundary. 4. Sandy pelitic flysch-like facies. 5. Pelitic flysch facies.

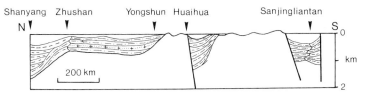

FIG. 4. Palaeogeographical map and section of the South China Marginal Basin during the Early Silurian. 1–3 as Fig. 3. 4. Clastic-pelitic flysch-like facies. 5. Pelitic-carbonaceous facies.

Sediment associations in the South China Marginal Basin

Within the marginal basin characteristic sedimentary facies can be related to variations in the amplitude and rates of relative subsidence or uplift of the basement, the intensity of penecontemporaneous fault movement and the occurrence of magmatic activity. Typically the sedimentary facies of the epicontinental sea, the inactive marginal basin and the active marginal basin and inner part of the island arc, reflect the stable, transitional and active tectonic environments respectively. The lateral and vertical migration of the various facies can be modelled to demonstrate the development of the South China Marginal Basin.

In the inactive basin siliciclastic sediments are most common. Locally, carbonaceous sediments containing sedimentary V, U and Mo deposits accumulated slowly in non-compensated deep euxinic basins, as for example in the Sinian, Cambrian and Ordovician sequences of central Hunan. Except for rare intercalations of pyroclastic material, volcanogenic sediments are absent in the inactive basin.

The active basin was dominated by arenopelitic, flysch-like and flysch facies with common siliceous interbeds. Volcanic rocks occur in sequences of Sinian–Ordovician age. The rates of subsidence and sedimentation were in approximate balance here, with the result that a succession of great thickness accumulated, the Sinian system alone being 2–5 km thick.

The number of sediment facies being deposited within the marginal basin was reduced in the later stages of its development. This can be determined in both the inactive and active parts of the basin and is associated with continuous accretion of the basin and thickening of the continental plate.

Volcanism in the South China Marginal Basin

Volcanic activity commenced in Sinian times and recurred intermittently, but with decreasing intensity, into Ordovician times. The eruptive centres were almost entirely restricted to the active part of the marginal basin. The products of large-scale eruptions of the Early Sinian period crop out extensively in Jiangxi, Fujian and Guangdong and include spilite, keratophyre and rhyolite together with pyroclastic rocks of corresponding compositions. In southern Jiangxi, some small basic and ultrabasic intrusions were emplaced, and Fe mineralization is associated with the volcanic sequences in central Jiangxi and Guangdong. In general the Cambrian period was a time of volcanic quiescence with only small quantities of rhyolite and acid tuff erupted in southern Jiangxi. Acidic volcanic rocks in northern Guangdong and southern Jiangxi are of Early and Late Ordovician age respectively.

The dominant phase of activity in early Sinian times has been studied in most detail. The volcanic assemblages are intercalated with bathyal-abyssal flysch and carbonate sequences. In the Yingyangguan area of Guangdong, three separate eruption cycles have been distinguished with each major cycle composed of two to five minor cycles with thicknesses of up to 350 m. With the exception of the rhyolite of the eastern flanks of the Wuyi Mountains, the volcanic rocks of Guangdong and Jiangxi all belong to the spilite-quartz keratophyre association. They have been modified by regional metamorphism, generally to the greenschist facies but locally into the amphibolite facies. Those in the greenschist facies exhibit distinct relict textures and are porphyritic. Spilite commonly displays pilotaxitic, intergranular and diabasic textures and where phenocrysts of albite (An_{5-10}) are present the rock is termed spilitic porphyrite. Typically the keratophyres show flattened amygdales and trachytic texture with albite (An_7), the quartz-keratophyres contain phenocrysts of quartz and albite, and occasionally of relict anorthoclase.

Although the Na_2O content of the Early Sinian volcanic rocks is generally greater than their K_2O, most samples have $Na_2O/K_2O < 2$. According to the classification of Middlemost (1972) they should therefore be assigned to the potassic type (Fig. 5). The correlation diagrams of $FeO/MgO:SiO_2$ and $FeO/MgO:FeO$ (Fig. 6) indicate that they belong to the calc-alkaline series.

The basic magmas, rich in Mg, but low in Ca and Al, display a low degree of differentiation and probably originated in the upper mantle. Ultrabasic rocks such as diopsidite and forsterite peridotite are associated with the basic rocks in southern Jiangxi. Upwelling of the magma was probably facilitated by major fractures developed during crustal extension in the back-arc region, possibly caused by convection-induced eddy currents in the asthenosphere. Deep and persistent fractures, such as those identified on both flanks of the Wuyi Mountains, in the Yunkai Mountains and between Suichuan and Wanan, probably acted

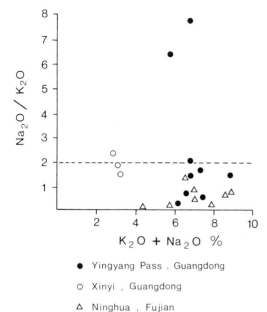

FIG. 5. Alkalinity of the Early Sinian volcanic rocks.

as major channels for the eruption of the magmas.

Post-Cambrian magmas were acidic, occurring as granitoid plutons and acidic extrusive rocks. The acidic magmas are interpreted as having originated by crustal anatexis. Guo Linzhi *et al.* (1980) noted the close spatial coincidence of prograding metamorphic zones, mig-

matite zones and zones of granitoid emplacement within the basin, and suggested that they were related to zones of deep fractures. The metamorphism, migmatisation and melting developed in the later stages of the basin evolution when continuing accretion at the margin of the Yangtze plate and compression and depression of the marginal basin restricted it to a fairly narrow trough. The three polymorphs of Al_2SiO_5, andalusite, kyanite and sillimanite, occur in the metamorphic series developed in the older parts of the basin succession of the region. The sequence of events envisaged began with the development of high compressive and shear stresses, which initiated or reactivated a major deep fracture zone, accompanied by medium-pressure metamorphism. Subsequently, heat ascending the fracture zone from deeper levels of the lithosphere produced a low-pressure, high-temperature andalusite–sillimanite type metamorphic belt. Migmatization and, eventually, anatexis with the production of granitic magmas took place in the zones of highest heat and chemical energy transfer. Intrusion and eruption of the magmas occurred during stress release. The granitoid rocks, which developed from the end of the Cambrian to the Silurian, thus display an evolutionary trend from migmatitic granitoids to hybrid and finally magmatic granitoids.

ACKNOWLEDGMENTS: The author is extremely grateful to A. J. Reedman for assistance in the preparation of the manuscript.

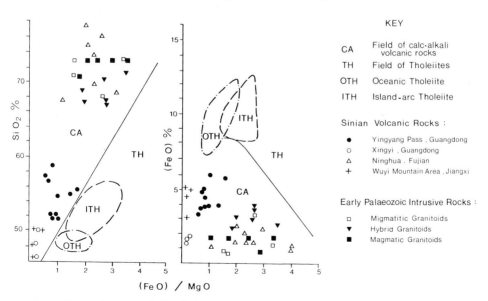

FIG. 6. Petrochemical variations of Sinian and early Palaeozoic igneous rocks.

Pan Guoqiang

References

Guo Lingzhi *et al.* 1980. *The Geotectonic Framework and Crustal Evolution of South China.* International Exchange Geological Research Papers. The Geological Publishing House, China (in Chinese).

Karig, D. E. & Sharman, G. E. 1975. Subduction and accretion in trenches. *Bull. geol. Soc. Am.* **86**, 377–89.

Middlemost, E. A. K. 1972. A simple classification of volcanic rocks. *Bull. Volcanol.* **36**, 382–97.

Wang Hongzhen & Wang Ziqiang 1981. An outline of palaeogeographical development in China since the mid-Proterozoic Era. *J. Stratigraphy*, **2** (in Chinese).

Pan Guoqiang, Geological Bureau of Anhui Province, Ministry of Geology and Mineral Resources, People's Republic of China.

Lower Palaeozoic volcanism of northern Qilianshan, NW China

Zhang Zhijin

SUMMARY: Thick volcanic sequences of mid-Cambrian–mid-Silurian age crop out in northern Qilianshan in NW China. Volcanism was concentrated in a marine basin almost 1000 km long situated at the SW margin of the N China–Korea paraplatform. The basin was subsequently deformed during Caledonian folding. Calc-alkaline, mildly alkaline and alkaline volcanic rock suites have been identified and a variety of ore deposits, genetically related to the volcanism, have been discovered.

The Qilianshan (Qilian Mountains) form the NE part of the Nanshan ranges on the borders of Tsinghai and Kansu Provinces in NW China (Fig. 1). They form a remote and rugged area with the highest summit over 6000 m. Sequences of submarine Lower Palaeozoic volcanic rocks were recognized in northern Qilianshan over 30 yr ago and since then the completion of a regional geological survey, at 1:200,000, and exploration for mineral resources, have provided considerable new data concerning the geology of the area. In particular, ultrabasic rock masses, belts of glaucophane schist and associated spilite–keratophyre volcanic suites have been studied in detail (e.g. Song Suhe *et al.* 1980; Song Zhigao 1980; Wang Quan *et al.* 1976; Xiao Xuchang *et al.* 1978). In this paper the geotectonic setting and character of the Lower Palaeozoic volcanism of northern Qilianshan are briefly described.

Geotectonic setting and history of volcanism

The Lower Palaeozoic volcanic belt of northern Qilianshan, over 900 km long and almost 80 km wide, comprises slightly metamorphosed, basic to intermediate and more rarely acidic, volcanic rocks of Middle Cambrian–Middle Silurian age. The belt forms a part of the Qilian eugeosyncline which subsequently became part of the Qinling–Qilian–Qunlun fold belt of Caledonian age.

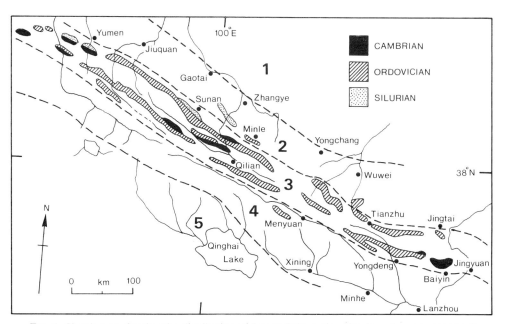

FIG. 1. Sketch map showing the distribution of Lower Palaeozoic volcanic rocks and geotectonic units at Qilianshan, NW China. 1. Alashan Block. 2. Hexi Corridor transitional zone. 3. Marginal basin assemblage at northern Qilianshan. 4. Island-arc zone. 5. Ocean basin zone.

Xiao Xuchang *et al.* (1978) interpret the volcanic belt of northern Qilianshan as representing a narrow marginal basin formed on an oceanic plate adjacent to a continent termed the N China–Korea paraplatform. To the SW of the marginal basin lay an island-arc uplift zone, which now crops out in mid-Qilianshan, and to the NE the Alashan Block, a part of the N China–Korea paraplatform, is separated from the volcanic belt by a narrow transitional zone known as the Hexi Corridor, within which volcanic rocks are poorly represented (Fig. 1). Major fault zones form the boundaries between these geotectonic units.

Early Cambrian strata are absent in northern Qilianshan. Subsidence and sedimentation commenced in the Middle Cambrian and thick sequences of marine sediments and volcanics accumulated into Early Ordovician times. Volcanism was accompanied by the intrusion of ultrabasic masses. The volcanic activity intensified to a peak in Middle Ordovician times when it was tectonically controlled with fissure–type eruptions dominant and central-type eruptions more rarely developed. Activity declined in the Late Ordovician and volcanic rocks are rare in the Silurian clastic sequences. The characteristics of the main phases of activity are outlined below.

Middle Cambrian

Isolated eruptive centres were sporadically developed within the basin during Middle Cambrian times. The volcanism was most intense in the E and W of the basin. The thickest volcanic sequences occur in Baiyin in the E, Xiangmao Mountain and Yingzui Mountain to the W of Yumen in the northern part of the belt, and in the Heihe area of the middle of the belt. In the Baiyin district, an extensively differentiated suite of spilite–quartz-keratophyre tuffs and lavas, totalling over 1700 m in thickness, was erupted (Zhang Zhijin 1976). Spilite and spilite-porphyrite were dominant in the middle and western sections.

Late Cambrian

The region was volcanically quiescent for most of the Late Cambrian period. The only evidence of volcanism is in the few beds of tuff intercalated in the sequences between the northern slopes of Xiangmao Mountain and the Yumen Petroleum River.

Early Ordovician

Volcanism of Early Ordovician age was concentrated in the W part of the belt with lesser development in the Dakeiha and Langshidanggou areas, S of Tianzhu, within the E sector. The main extrusions were of basaltic and andesitic lavas which are intercalated with mudstone, sandstone and limestone. Locally, small quantities of acidic material were erupted.

Middle Ordovician

The most widespread volcanism of the Qilianshan volcanic belt occurred in the Middle Ordovician. The eruptions were dominantly basic though in some areas, such as Dacha in Sunan, Chelengou in Wuwei and Zhongbao in Yongdeng, relatively large amounts of acidic material were also erupted. In addition, alkaline volcanics crop out in Zhongbao district. In general the eruptions are interpreted as having been of fissure-type which were distributed in a zone parallel to the axis of the basin, although in the Sunan and Yongdeng districts volcanic sequences resulting from explosive, central-type eruptions have been recognized.

The volcanic sequences exceed 3000 m in thickness in the Zhongbao district and a conspicuously rhythmic pattern of activity, quiet effusion–violent eruption–intermittent eruption–quiescence, is recognized. The sequences also demonstrate a regular evolution of the magmas from basic through intermediate to acid and finally alkaline types.

Late Ordovician

Volcanism waned in the early part of the Late Ordovician and the zones of activity migrated from the centre of the basin to marginal depressions. A thick sequence of carbonate and clastic sediments was deposited with minor intercalations of basic to intermediate tuffs in two areas, Yaomoshan in Yumen and Suyouhe in Sunan. At a later stage, volcanic activity increased but was still confined to limited areas such as Yumen, Gulang in Wuwei and Menyuan. Eruptions were of fissure-type and involved basic to intermediate magmas, the basic rocks comprising mildly alkaline spilite and the intermediate rocks normal calc-alkaline andesite.

Early and Middle Silurian

Early Silurian volcanic rocks occur in the N of Qilianshan and reflect weak and intermittent

TABLE 1. *The average chemical composition of Lower Palaeozoic volcanic rocks from Qilianshan, NW China*

Rock type	Calc-alkaline			Mildly alkaline									Alkaline		
	1	2		3		4		5		6	7	8	9	10	11
Age	MO	MO	MC	MO	MC	MC	MO	MC	MO	MC	MC	MC	MO	MO	MO
SiO_2	44.86*	54.77	48.85	45.57	47.76	48.05	47.92	53.82	52.67	71.66	75.06	73.66	54.36	52.38	58.32
TiO_2	1.32	0.70	0.81	0.81	1.18	0.88	0.92	0.38	1.39	0.43	0.15	0.20	0.75	0.84	0.56
Al_2O_3	15.73	16.21	15.65	13.75	14.71	16.52	16.97	14.70	15.52	12.85	12.73	13.91	17.38	17.61	18.02
Fe_2O_3	1.89	2.90	3.39	2.93	2.25	1.38	2.60	3.01	3.46	1.90	1.16	1.12	1.96	2.50	0.83
FeO	6.99	2.75	5.97	7.72	8.36	5.98	5.55	5.00	6.96	1.96	1.24	0.72	3.05	3.74	2.65
MnO	0.17	0.10	0.14	0.18	0.19	0.17	0.14	0.14	0.18	0.06	0.05	0.01	0.07	0.11	0.05
MgO	9.20	2.62	7.51	8.47	7.59	4.58	5.73	6.01	4.82	0.85	0.38	0.98	2.82	3.50	1.84
CaO	8.13	6.68	6.98	8.73	6.89	7.05	7.98	6.43	5.02	1.55	1.32	0.42	4.23	5.14	2.94
Na_2O	2.91	1.82	2.93	3.45	4.30	3.40	3.47	3.44	5.60	5.51	5.27	4.36	4.51	4.55	5.39
K_2O	1.03	0.33	2.21	0.80	0.59	4.51	1.10	0.38	0.20	0.58	1.29	3.35	5.63	4.76	5.67
P_2O_5	0.18	0.21	0.32	0.19	0.18	0.58	0.29	0.10	0.26	0.16	0.05	0.02	0.39	0.48	0.26
LOI	8.64	0.21	4.56	7.47	6.16	7.01	6.65	6.83	4.97	2.82	1.08	1.43	4.86	4.93	>1.54
Total	101.05	100.84	99.32	100.07	100.16	100.11	99.32	100.24	101.05	100.33	99.78	100.18	100.01	100.54	>98.07

*Percent
1. Basalt 2. Andesite 3. Spilite 4. Potash spilite 5. Spilite-porphyrite 6. Keratophyre 7. Quartz-keratophyre 8. Potash quartz-keratophyre 9. Trachyte 10. Alkali trachyte 11. Pseudoleucite-phonolite. MO—Middle Ordovician. MC—Middle Cambrian. LOI—loss on ignition.

eruptions. Intermediate to acid tuffs predominate, although small amounts of andesite and basaltic lava crop out in the Shuiguanhe and Xiaohegou districts, Sunan. In the Middle Silurian sequences, the only volcanic rocks are small amounts of basic and intermediate tuff at Quannasgou Mountain, Yumen in the W section of the belt.

Volcanic facies and chemical characteristics of the Lower Palaeozoic volcanic rocks

Five volcanic facies can be recognized in northern Qilianshan; a subvolcanic facies, volcanic-pipe facies, explosive facies, effusive facies and a volcanic-sedimentary facies. Of these, the explosive and effusive facies are most widely developed. The various eruptive facies can be related to compositionally similar high-level intrusives; e.g. spilite, spilite-porphyrite, keratophyre and quartz-keratophyre lavas and tuffs have their intrusive counterparts in bodies of diabase, albite-diorite and quartz-albite-porphyry. Amongst the spilites of the effusive facies, well-developed pillow lavas are interbedded with fossiliferous marine limestones and flysch-like greywackes which indicate that the lavas were emplaced in a submarine environment.

Compositionally, the volcanic rocks can be assigned to three major series; calc-alkaline, mildly alkaline, and alkaline (Table 1). Rocks of the calc-alkaline and mildly alkaline series are volumetrically the most important with alkaline rocks found only in a few localities. Calc-alkaline basalt and andesite, together with lesser amounts of the mildly alkaline spilite–quartz-keratophyre association, were abundant in the various Ordovician eruptive phases, for example in the Lower Ordovician of Daqingyangshan, Jiuquan and in the Middle Ordovician in Shihuigou, Yongdeng and Chelungou, Wuwei. The differences between

the two rock series are related to variations in their autometamorphic albitization. The slightly alkaline rocks are rich in sodium, with Na_2O generally greater than K_2O (Table 1), and their feldspars are mainly albite or oligoclase. In places they are relatively rich in potassium and potash spilites and trachytes occur. The mildly alkaline spilite–quartz-keratophyre association is particularly widely developed in the Middle Cambrian eruptive phase.

Reliable data of the alkaline volcanic rocks have been obtained only from Shihuigou, Yongdeng. The principal rock types include trachyte, pseudoleucite-phonolite, trachybasalt, tephrite and augite-leucitite. They occur at the top of the Middle Ordovician and may be in part of Late Ordovician age.

Mineralization associated with Lower Palaeozoic volcanism

Commercial deposits of copper, lead, zinc and iron, all closely associated with volcanic rocks, have been found in northern Qilianshan. The deposits are of two kinds; the copper and polymetallic ores are volcanic pneumato-hydatogenic fissure infills, and the iron, together with sub-economic manganese ores are of sedimentary-exhalative origin.

The copper and polymetallic ores are associated mainly with thick volcanic sequences reflecting prolonged volcanism. Most of the deposits occur either in sodium-rich rocks of the spilite–quartz-keratophyre series or in associations of normal calc-alkaline rocks. The wall rocks are most commonly those of the explosive or effusive facies and penecontemporaneous fractures associated with eruptive centres, and volcanic domes, are favourable sites for mineralization.

ACKNOWLEDGMENTS: The author is particularly grateful to A. J. Reedman for assistance in preparation of the final draft of the manuscript.

References

SONG SUHE *et al*. 1980. The major types and metallogenetic characteristics of magmatic rocks in China. *Geological Scientific and Technological Information of North West China*. **3**, 1–9 (in Chinese).

SONG ZHIGAO, 1980. Lower Palaeozoic spilite-keratophyre suite of North Qilian Mountain and

its bearing on ophilite suite. *Bull. Chin. Acad. Geol. Sci*. (edited by the Xian Institute of Geology and Mineral Resources) **1**, 14–22 (in Chinese).

WANG QUAN *et al*. 1976. Palaeo-oceanic crust of the Qilianshan Region, western China and its tectonic significance. *Sci. geol. Sin. Peking*, **1**, 42–54.

XIAO XUCHANG *et al.* 1978. A preliminary study on the tectonics of ancient ophiolites in the Qilian Mountain, north west China. *Acta geol. Sin. Peking*, **52**, 281–94.

ZHANG ZHIJIN, 1976. Some characteristic features of the Middle Cambrian marine volcanism at the eastern section of the Qilian Mountain. *Geological Scientific and Technological Information of North West China*, **2**, 54–8 (in Chinese).

ZHANG ZHIJIN, Geological Bureau of Gansu Province, Ministry of Geology and Mineral Resources, People's Republic of China.

Volcanic and associated sedimentary and tectonic processes in the Ordovician marginal basin of Wales: a field guide*

B. P. Kokelaar, M. F. Howells, R. E. Bevins & R. A. Roach

This is a modified version of the guide which was prepared for a field discussion meeting, 1–8 September 1982, prior to the Ordinary General Meeting of the Geological Society with the theme 'Volcanic Processes in Marginal Basins'. The guide describes some of the localities in the Ordovician of Wales which have recently provided much information on processes in an environment interpreted (Kokelaar *et al.*, this volume) as being a marginal basin on the SE side of the Iapetus Ocean. In particular, the volcanic and associated sedimentary and tectonic processes are emphasized.

General geological context of the Lower Palaeozoic volcanism

The geology of Wales is dominated by the Lower Palaeozoic (Fig. 1) and it was here in the 19th century that Murchison, Sedgwick and Lapworth established the Cambrian, Ordovician and Silurian Systems, with their various series. The nomenclature of Lower Palaeozoic stratigraphy (Harland *et al.* 1982) reveals the influence of these Welsh sequences.

During Lower Palaeozoic times a major sedimentary basin, the Welsh Basin (equivalent to the geosyncline of Jones 1938), was established on a basement of late Precambrian rocks. Its NE–SW orientation was largely determined by faults along its NW and SE margins (Fig. 1). Sedimentation at the edges of the basin was discontinuous and generally of shallow-water clastics or carbonates whilst towards the centre there was slow accumulation of pelagic muds with periodic incursions of turbiditic sands and silts. An estimated 13 km of sediment accumulated towards the centre of the basin and a maximum of 5 km at the margins.

Recently, the development of this basin has been interpreted in the broad context of the evolution of the major Iapetus Ocean occurring to the NW (e.g. Phillips *et al.* 1976). With closure of this ocean a period of arc volcanism occurred at the end of Tremadoc times (Kokelaar 1979; Kokelaar *et al.* 1982) and

from Arenig to late Caradoc times volcanism was widespread in a marginal basin setting (Kokelaar *et al.*, this volume). From the evidence in the present Ordovician outcrops, major volcanic centres of the marginal basin were located in SW Wales and S Snowdonia in pre-Caradoc times and in central and N Snowdonia in Caradoc times.

In SW Wales the main episode of volcanism occurred during the Llanvirn, represented by the acid and basic igneous rocks of the Fishguard Volcanic Complex and the acidic volcanic rocks on Ramsey Island. In S Snowdonia the pre-Caradoc volcanism is represented by the Aran Volcanic Group which crops out about the S, E and N margins of the Harlech Dome. In central and N Snowdonia the Caradoc volcanic rocks comprise a lower Llewelyn Volcanic Group (Howells *et al.* 1983) and an upper Snowdon Volcanic Group. Seemingly lesser accumulations have been determined elsewhere, for example, in the Builth Inlier and in Llŷn (Llanvirn and Caradoc), and in the Berwyn Dome (Caradoc). The evolution of the Ordovician marginal basin of Wales is described by Kokelaar *et al.* (this volume).

During late Silurian–early Devonian times, with closure of the Iapetus Ocean and collision between the opposing plates, the basin sequence was deformed and uplifted to form part of the Caledonian mountain chain (Caledonides). Many of the major folds, such as the Towy Anticline, Plynlimon Dome, Harlech Dome and Snowdon Synclinorium (Fig. 1) are open structures, although locally deformation resulted in small thrusts, especially along the NW and SE margins of the basin. A mainly axial planar cleavage is widely developed, which in the mudstones was sufficiently intense to produce slates, and low-grade metamorphism of prehnite-pumpellyite or lower greenschist facies is ubiquitous.

Itineraries (Fig. 2)

Excursions A and B: Ramsey Island [SM 7024] (full day each)

To examine Arenig and Llanvirn sequences which include an early distal sequence of rhyoli-

---

*Almost invariably, little can be gained by using a hammer on the rocks described here, whereas a great deal can be lost. *Please conserve the outcrops.*

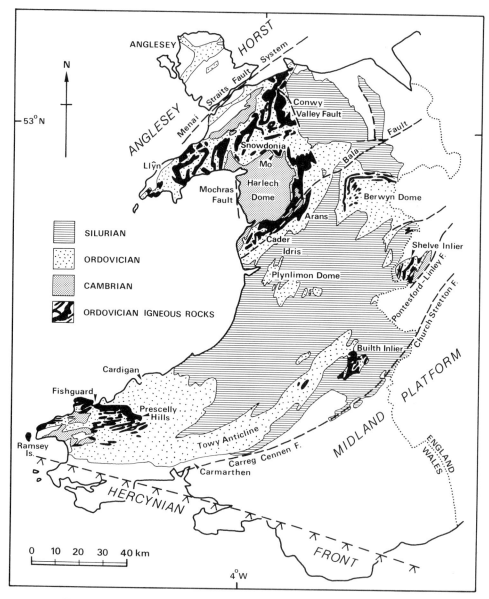

FIG. 1. Simplified geology of the Welsh Basin (Mo-Moelwyn and Manod Hills).

tic volcanogenic sediments and a later proximal volcanic pile, comprising subaqueously emplaced welded and non-welded rhyolitic ash-flow tuffs, rhyolitic domes and high-level rhyolitic intrusions with peperites. Coeval wet-sediment slides and associated debris-flow deposits are also demonstrated.

Excursion C: Abereiddi [SM 796310] (half day)

To examine proximal to distal variation in a Llanvirn sequence of basic tuffs produced by explosive submarine volcanism and deposited from debris and turbiditic flows.

Excursions D and E: Pen Caer [SM 887393 and 937404] (half and full day respectively)

To examine a thick bimodal volcanic accumulation, of Llanvirn age, comprising basaltic pillowed and sheet lavas, and tuffs, flow-banded and autobrecciated rhyolitic lavas, rhyolitic tuffs and breccias, rhyodacitic pillow lavas and

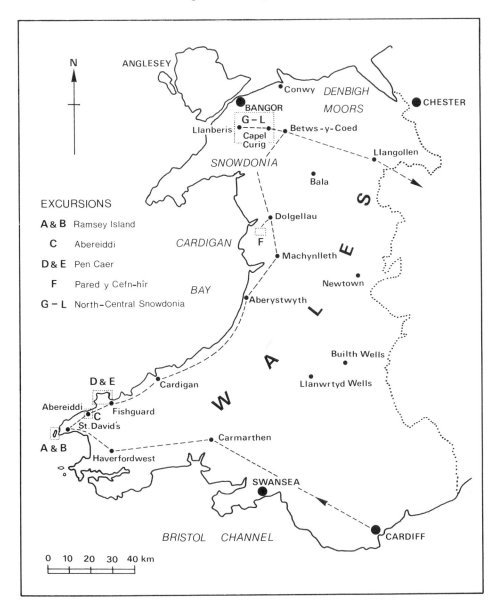

FIG. 2. Generalized route and location of excursions.

volcanogenic debris flows. Also, high-level con-temporaneous intrusions, principally of basic composition and associated with fine tuffaceous sediments which show wet-sediment sliding, are demonstrated.

Excursion F: Journey between Fishguard and Snowdonia including a brief stop at Pared y Cefn-hir [SH 662148]

At Pared y Cefn-hîr is a well-exposed sequ-ence through the pre-Caradoc Aran Volcanic Group which is restricted to S Snowdonia.

Excursion G: Llanberis Pass [SH 63005650] (half day)

To examine the Pitt's Head Tuff, basaltic lavas, rhyolitic dust tuffs and associated marine sediments of the Cwm Eigiau Formation, and an important locality which demonstrates the synvolcanic tectonism which accompanied the emplacement of the Lower Rhyolitic Tuff Formation.

Excursion H: Pen y Pass to Llyn Llydaw [SH 64705565] (half day)

To demonstrate the general sequence of the Snowdon Volcanic Group in central Snowdonia and in particular the proximal facies of the Lower Rhyolitic Tuff Formation.

Excursion I: Gallt yr Ogof [SH 69355950] (half day)

To examine the rhyolitic tuffs and associated sediments of the Capel Curig Volcanic Formation. The locality provides evidence of subaqueous emplacement of welded ash-flow tuffs with irregular transgressive basal contacts and planar tops.

Excursion J: Cwm Idwal [SH 65006030] (full day)

To examine the sequence between the Pitt's Head Tuff and the Bedded Pyroclastic Formation. The Lower Rhyolitic Tuff Formation comprises subaqueously emplaced rhyolitic lava, primary rhyolitic ash-flow tuffs and secondary, sloughed deposits of previously emplaced ash-flow material with intercalated marine sediments. The Bedded Pyroclastic Formation includes bedded basaltic tuffs deposited in shallow water and associated basaltic pillow lavas and breccias.

Excursion K: E of Capel Curig [SH 72105810] (half day)

To examine the marine sediments and associated volcanics of the upper part of the Cwm Eigiau Formation and the three rhyolitic non-welded ash-flow tuffs of the Lower Crafnant Volcanic Formation. The former include a basaltic vent entrapped in the sedimentary sequence, reworked rhyolitic tuffs and a debris-flow breccia. The ash-flow tuffs, interbedded with marine siltstones, are in part the distal equivalents of the Lower Rhyolitic Tuff Formation in central Snowdonia.

Excursion L: Sarnau [SH 77305890] (half day)

To examine the Middle Crafnant Volcanic Formation which comprises a sequence of thin, blocky rhyolitic ash-flow tuffs and water-settled dust tuffs interbedded with graptolitic black

mudstones. The localities demonstrate thixotropic deformation of tuffs and sediments as a result of loading and, or, seismic activity.

Excursions A and B: Ramsey Island[*] (Figs 3, 4 and 5)

Ramsey Island displays proximal and distal rhyolitic volcanic rocks which were in the main erupted and emplaced in a marine environment in which pelagic muds accumulated. Coeval rhyolitic intrusions show various interactions with wet and unlithified ashes and muds. Penecontemporaneous tectonism is reflected in major gaps and repetitions in the stratigraphy and in debris-flow deposits, which result from major slides and mass-gravity reworking of strata respectively.

The N–S Ramsey Fault divides the island into two blocks with contrasting stratigraphy (Fig. 3). To the E of the fault an upper Cambrian and an almost complete lower Arenig sedimentary sequence is succeeded by the Aber Mawr Formation of upper Arenig–lower Llanvirn age, comprising mudstones with distal rhyolitic tuffs. These are overlain by the lower Llanvirn Porth Llauog Formation, including black mudstones, siltstones and minor tuffs, but characterized by several major and many minor deposits of cohesive debris flows. These mostly comprise a mixture of the lithologies of the underlying strata. This formation is overlain by tuffs which are correlated with the lower part of the rhyolitic Carn Llundain Formation in the block to the W of the Ramsey Fault. On the western side, however, these tuffs grade down into a conglomerate which unconformably overlies the Trwyn Llundain Sandstones, probably of lower Middle Cambrian age (equivalent to the Solva Group on the mainland). The pre-Carn Llundain Formation sequence E of the fault is missing to the W. To the E of the Ramsey Fault and W of the Road Uchaf Slide, the strata including part of the upper Cambrian and up to at least the base of the Porth Llauog Formation are allochthonous, with numerous gaps and repetitions in the stratigraphy occurring across major, unlithified wet sediment slide planes.

Initial tectonism caused the large-scale sliding of the pre-Porth Llauog Formation succes-

[*]Ramsey Island is privately owned and permission for access *must* be sought from the owner. Access to localities A4 and B8 is steep, loose and exposed, and most other localities, around cliff edges and on steep slopes, must be approached with caution.

sion and then, more locally, uplift on the W side of the Ramsey Fault resulted in shedding of all strata down to the lower Middle Cambrian beds. The debris-flow deposits of the Porth Llauog Formation on the E side of the fault reflect this stage of tectonism which, from the presence of the several moderately thick horizons of pelagic black mudstones (with lower Llanvirn graptolite and trilobite faunas), must have been quite protracted. The extrusive volcanism of the upper part of the Carn Llundain Formation and the numerous coeval high-level intrusions are probably centred on the fault. A late Caledonian penetrative cleavage affects most argillaceous rocks, although strain is markedly inhomogeneous around and between the rhyolitic intrusions.

Excursion A (Fig. 3), confined to the E of the Ramsey Fault, examines the distal tuffs of the Aber Mawr Formation, the cohesive debris-flow deposits of the Porth Llauog Formation and striking exposures of rhyolites intruded into wet sediments. Excursion B (Figs 4 and 5) examines the products and processes of emplacement of the relatively proximal volcanism of the Carn Llundain Formation, W of the Ramsey Fault.

Locality A1: Aber Mawr

At the N end of the bay, black mudstones, presumed to be of Arenig age (*Tetragraptus* mudstone), dip steeply S and are hornfelsed and spotted up to 10 m from their discordant contact with the Carn Ysgubor microtonalite boss. To the S of a fault, faintly laminated black mudstones of lower Llanvirn age (*D. bifidus* mudstone) are contorted and dip mainly W and SW. These mustones contain several thin (up to 1.5 m) felsite sheets which are sinuous and discontinuous, semiconcordant or discordant, and display features indicative of emplacement into wet sediment.

South of the *D. bifidus* mudstone, rhyolitic tuffs (Lower Aber Mawr Tuff Member) are conformably overlain by upper Arenig mudstones (pencil slates) so that a major break must occur between the *D. bifidus* mudstone and the tuffs. No fault is evident and the boundary is interpreted as having been a major slide plane in wet sediment, similar to those widely developed across the N and NE of Ramsey Island. The laminations in the mudstones are discordant to the strata above and below, indicating deformation during movement.

The Lower Aber Mawr Tuff Member comprises *c*. 20 m of laminae and thin beds of fine rhyolitic tuff intercalated with mudstones, grad-ing up into 63 m of thicker bedded, medium- and fine-grained rhyolitic tuffs, with only minor silicified mudstone partings. The tuffs contain whole crystals and fragments of quartz and plagioclase. Pumice fragments and shards are discernible in some coarser units but thorough recrystallization, into diffuse segregations of quartzo-feldspathic aggregates within more micaceous areas, is common. Carbonate nodules are also common.

Within the tuffs three bed forms can be determined. Firstly, there are massive beds, 1.5–5 m thick with mudstone intraclasts, which are most common in the upper part of the member. Some beds show normal coarse-tail grading, and parallel laminations towards their top. Flute casts occur at some bed bases. These beds are regarded as mainly suspension deposits of high-density turbidity currents (the S_3 division of Lowe (1982) or T_a of Bouma (1962)), with deposition by traction sedimentation producing the laminations, T_b. The second bed form, up to 2 m thick but mostly much thinner, includes fine parallel laminations and no evidence of grading. The laminations are commonly convoluted and locally slump-folded, and the beds are in places associated with others dominated by current ripples or cross-lamination. Beds of this second type are traction sediments, T_{b-d}, of low-density turbidity currents which may represent fractions of the immediately underlying and, or, more proximal S_3 units. The third type is of thin beds mainly of massive, although in places laminated, very fine grained tuff. These are probably suspension deposits, T_{d-e}, associated with the turbidity currents, although some may represent distal ash-fall.

The turbiditic tuff sequence is interpreted as a moderately deep-water distal accumulation, either from a single rhyolitic centre where activity increased in frequency and intensity, or from more than one centre. There is no evidence to indicate whether the eruptions were submarine or subaqueous, ash-falls (Plinian) or ash-flows, or both.

This lower tuff member is succeeded by *c*. 50 m of distinctive sooty-black pencil slates, of upper Arenig age. The overlying Upper Aber Mawr Tuff Member comprises two tuff sequences separated by black mudstones. These tuffs are similar to the upper part of the lower member, and they are considered to represent two further episodes of distal rhyolitic volcanism.

The succeeding black mudstones, locally with faint laminations, contain a *D. bifidus* (lower Llanvirn) fauna. They generally dip steeply S

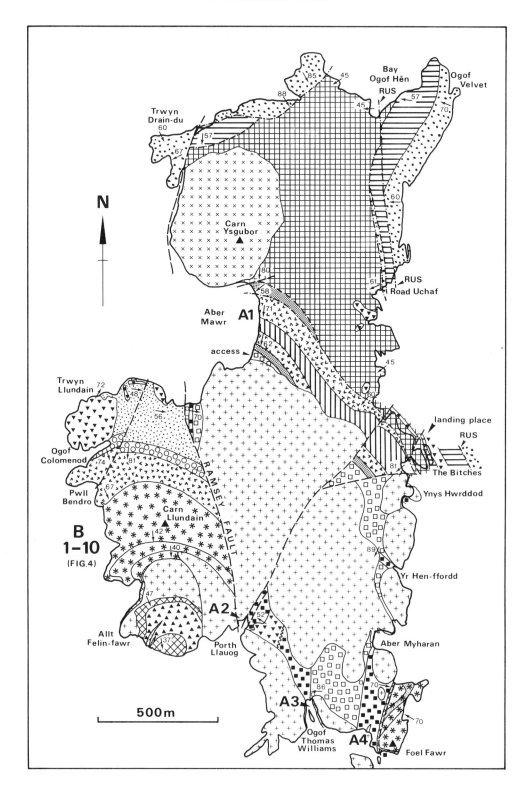

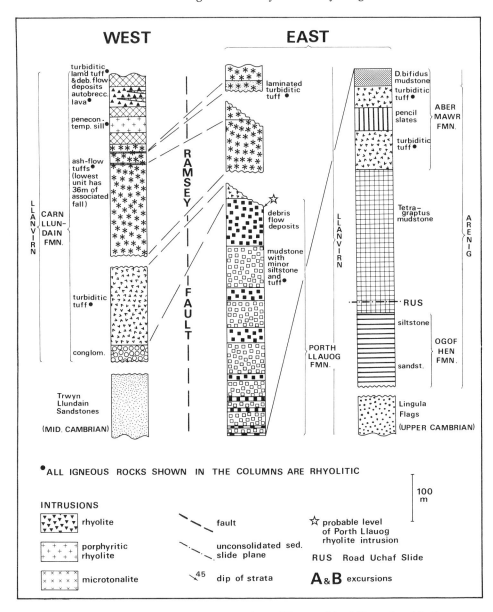

WEST   EAST

FIG. 3 (above and opposite). Simplified geology of Ramsey Island (Excursion A1–4).

but are disturbed and overturned in places. The mudstones are the youngest strata certainly involved in the major wet sediment sliding on Ramsey Island. The Aber Mawr section is only a small part of the allochthonous pile of sediments and tuffs present on the island.

The base of the succeeding Porth Llauog Formation is taken at the lowest debris-flow unit within the *D. bifidus* mudstone. This bed, up to 2 m thick, contains angular to rounded clasts of

the fine Aber Mawr tuffs and mudstone, in a mudstone matrix.

Locality A2: Porth Llauog

In the cliffs on the SE of the bay, NW-dipping beds of the Porth Llauog Formation have been disturbed by penecontemporaneous shallow intrusion of rhyolite. The brecciated rhyolite has gently domed and locally transgressed the bedded sequence, parts of which have slipped to the NW and overlie their lateral equivalents. The breccia in the rhyolite contact zone is peperitic, comprising mostly angular and irregularly shaped fragments, some with crenulated margins, in a matrix of fine angular rhyolitic fragments and silicic mudstone.

The bedded sequence on top of the rhyolite, up to 9 m thick, comprises grey silty mudstone overlain by massive, rhyolitic, pebbly and gritty sandstones intercalated with thinly bedded and laminated rhyolitic sandstones and siltstones. The massive, poorly sorted sandstones are composed of angular to sub-rounded pebbles and granules of rhyolite and mudstone lenticles, in a sandy matrix of rhyolitic fragments, and quartz and plagioclase crystals. The beds, up to 75 cm thick, mostly show normal coarse-tail grading, although some are inversely graded close to their base. Erosion and, locally, convolution of the underlying beds of laminated fine sandstone or siltstone is apparent. These sandstones include beds of current ripple cross-lamination and bedding planes with straight to sinuous asymmetic ripple patterns. These strata are interpreted as turbidites. The massive, poorly sorted beds are predominantly suspension deposits, S_3, of high-density turbidity currents with traction carpet, S_2, deposits at some bases. The fine sand and silt units show features of deposition from low-density currents and are traction T_{b-d} and suspension T_{d-e} deposits. The sequence reflects slumping of rhyolitic hyaloclastite formed during submarine extrusion of rhyolite. The shallow intrusion which disturbed the turbiditic hyaloclastites was probably coeval and emplaced shortly after the sediments.

The bedded sequence is sharply overlain by a massive cohesive debris-flow deposit (Lowe 1979), approximately 20 m thick, comprising randomly oriented ellipsoidal Fe-Mn concretions, clasts of rhyolite tuff, poorly sorted rhyolitic sandstones and breccias, pumice, crystals of quartz and plagioclase, and mudstone, supported in a black silty mudstone matrix. Also, blocks of similar deposits can be distinguished where the mudstone matrix and clast content are slightly different to the host. Both the mudstone blocks and the matrix have yielded a *D. bifidus* and older faunas.

This deposit is faulted against similar deposits which dip SE. A distinctive pale-brown and yellow debris-flow deposit with abundant rounded (although tectonically deformed) rhyolitic pebbles, angular clasts of rhyolite and pumice in a fine rhyolitic matrix is succeeded by another muddy debris-flow deposit with rhyolitic clasts.

The cohesive debris flows reflect contemporaneous fault activity, including that on the Ramsey Fault with major uplift to the W. The deposits include most of the lithologies of the underlying sequence and well-rounded pebbles, mostly of rhyolite, probably derived from an emergent supra-littoral environment.

Locality A3: Ogof Thomas Williams (viewed from the cliff tops)

Numerous steep sided, rounded and irregular domes of intrusive rhyolite crop out extensively in the S of Ramsey Island, mostly occurring with the dominantly black argillites of the Porth Llauog Formation pinched between them. The rhyolites are composed of abundant euhedral to anhedral quartz and plagioclase phenocrysts in a fine, quartzo-feldspathic groundmass which is locally spherulitic. Chilled margins cannot be distinguished although perlitic fractures occur close to some contacts. Hornfelsing of host rock is absent, but thin silicified mudstone veneers adhere to some contact surfaces.

On the E and W walls of Ogof Thomas Williams, clearly displayed intrusion surfaces are broadly undulose with distinct and extensively developed wrinkles, commonly closely spaced and parallel over tens of metres. In addition, the surfaces display larger-scale tucks and folds which locally preserve veneers of the original host. Such patterned surfaces are common and indicate slight withdrawal of magma soon after development of maximum volume, so that the still plastic margin shrivelled. A bulbous protrusion, several metres across, extends from the foot of the W wall and irregular sheets form the stacks in the centre of the cove.

Locality A4: W of Foel Fawr

Minor intrusions of porphyritic rhyolite occur 125 m W of Foel Fawr within black mudstones of cohesive debris flows. Here the contacts are distinctively peperitic, with angular and irregular clasts of rhyolite supported in mudstone, commonly close to the site of fragmentation. Also, extremely irregular fingers, bulbous protrusions and isolated pillows of rhyolite occur within the host. These features indicate emplacement of the rhyolites into the debris-flow deposits while they were still wet and unlithified.

The cohesive debris-flow deposits, *c.* 100 m thick, are lithologically similar to those at locality A2. The presence of several flow units is indicated by slight differences in clast content and large-scale slump-folding is apparent in places. To the SE, around the headland, the deposits are less argillaceous with a preponderance of rhyolite, pumice and tuffaceous mudstone clasts with quartz and plagioclase crystals in a fine tuffaceous mudstone matrix. Rounded boulders of this material occur in a matrix of similar composition.

In the cliffs S of Foel Fawr, the debris-flow deposits are succeeded by mudstone and thin-bedded tuffs which are correlated with those of the Carn Llundain Formation cropping out W

of the Ramsey Fault at Pwll Bendro (Locality B4).

Locality B1 (Fig. 4)

Here contacts are exposed between irregular rhyolite intrusions and a well-bedded sequence of moderately sorted medium to coarse micaceous sandstones and siltstones (Trwyn Llundain Sandstones, Middle Cambrian). Mostly the con-

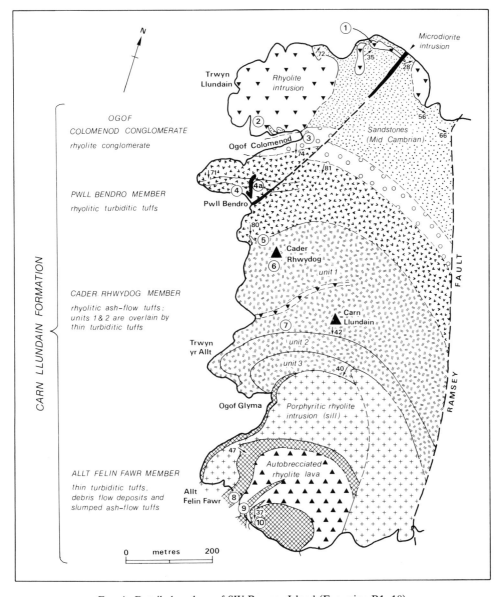

FIG. 4. Detailed geology of SW Ramsey Island (Excursion B1–10).

tacts are sharp and discordant with finger-like apophyses of rhyolite intruding the sediments. Rhyolite margins, locally flow-banded parallel to the contacts, are commonly brecciated, and in places the adjacent sediments are also brecciated up to several metres from the contact. Some rhyolite margins are peperitic, with trains of rhyolite fragments within the sediments. Such contact relationships indicate that the rhyolites were fluid when intruded into a locally unlithified, wet host.

On Ramsey Island, there are numerous minor intrusions of rhyolite. The majority intrude *D. bifidus* mudstones and so post-date the Aber Mawr Formation, and those in this vicinity pre-date the Carn Llundain Formation. Intrusion contemporaneous with tectonism (Porth Llauog Formation times) is thus indicated.

Locality B2

Here flinty and locally flow-banded rhyolite contains extensive and irregular zones of breccia. Massive rhyolite grades through rhyolite with hairline fractures and jig-saw breccias, into breccias comprising angular to subrounded rhyolitic clasts in a relatively chlorite-rich matrix. Some post-brecciation movement is indicated in zones where adjacent clasts do not fit together.

Late-consolidation exsolution of volatiles from the rhyolites probably accounts for most of the brecciation. Movement may have resulted from explosive release of volatiles, volatile streaming or, perhaps, further intrusion. Streaming volatiles, or later percolating fluids, corroded the clasts and produced the altered matrix. The explosive release of magmatic volatiles probably caused fragmentation of the host sediments, although explosive boiling of connate water trapped in lithified sediments would produce similar results.

Locality B3

The Ogof Colomenod Conglomerate is the basal member of the Carn Llundain Formation (Llanvirn). It is 35 m thick and rests with extreme angular unconformity on the Middle Cambrian sandstones and on the rhyolitic intrusions. The poorly sorted conglomerate comprises well-rounded cobbles and pebbles of rhyolite, supported in a matrix of rounded small pebbles and granules, mostly of rhyolite, but towards the base also of fine sandstone. Distinct bedding is absent, although slight variations in matrix proportion occur throughout and indi-

cate the presence of sub-units. Sorting improves slightly, and mean clast size generally decreases, upwards. In places, large clasts have been tectonically moulded against one another and small clasts squeezed into the intervening spaces. At its top the conglomerate grades over 3–4 m into grits which, through a further 2 m, are interbedded with the succeeding tuffs.

The conglomerate is interpreted as a rapid accumulation from a series of sediment-gravity flows in which the clasts were probably mainly supported by dispersive pressure due to particle collisions. Thus the deposits may reflect a type of coarse grain flow. The flows probably originated in a littoral or supralittoral environment and their deposition in at least moderately deep water is inferred from the succeeding tuffs. Such flows could have been initiated by a tsunami or directly by volcano-tectonism.

Locality B4

The Pwll Bendro Member, *c.* 165 m thick, comprises massive rhyolitic lapilli tuffs interbedded with thinly bedded and laminated finer tuffs. The lapilli tuff beds range from a few cm to 20 m thick (mapped unit, locality B4a), but are most commonly between 1 and 2 m. They are composed of abundant lapilli of rhyolite and randomly oriented tube pumice, with quartz and plagioclase crystals, in an intensely recrystallized fine vitroclastic matrix. Sorting is poor and some beds contain fewer lithics and more pumice towards their top. Normal coarse-tail grading, with rhyolite pebbles concentrated towards the bases of beds, can be distinguished near the bottom of the sequence. Most bases are flat, although locally the coarse tuff downcuts obliquely or by steps into the underlying tuffs by as much as 30 cm. With decreasing lithic content, these tuffs grade into thin beds of pumice-rich medium and fine tuff, some of which show normal grading with crystal concentration towards their bases. These massive tuffs commonly become thinly bedded or laminated towards their tops, fining rapidly into parallel laminated tuffs composed mainly of crystal fragments and recrystallized shards. Cross-laminated beds are present but uncommon. Locally the thin beds and laminae have been disturbed by dewatering or soft-sediment faulting. In a small cliff face (locality B4a) the thick lapilli tuff displays a rotational slump-with-scar, involving its base and the underlying laminated tuffs which are also disrupted by dewatering.

The massive tuffs are suspension deposits (S_3) from cold, locally erosive, high-density turbidity currents, with the thinner-bedded and laminated tuffs representing traction and suspension deposits (T_{b-e}) from less dense turbidity currents and, perhaps, distal fall of ash suspended above the vent. The absence of intercalated pelagic sediments, and the uniform, predominantly juvenile character of the deposits, suggest rapid deposition associated with a single eruption or rapid succession of eruptions.

Although this sequence appears to be more proximal than those of the Aber Mawr Formation (Locality A1), the individual thin beds and lack of very coarse material suggest that the source was not close. The absence of reworking implies deposition in at least moderately deep, tranquil water. Positive determination of the style of eruption, or of a subaerial or submarine source, does not seem possible, although the absence of accretionary lapilli suggests the latter environment. Although some massive lapilli tuff beds may have originated as discrete ash-flows derived directly from subaerial or submarine column collapse, the Pwll Bendro tuffs do not exhibit simple relations with an eruption column as modelled by Fiske & Matsuda (1964). Most beds are more likely to be the result of slumping of tuff penecontemporaneous with eruption. As in the Aber Mawr turbiditic tuffs, the low-density turbidity current deposits probably represent fractions of the immediately underlying and, or, more proximal S_3 units.

Localities B5, B6 and B7

The Cader Rhwydog Member mainly comprises three rhyolitic ash-flow tuffs: the Cader Rhwydog Tuff, the Trwyn yr Allt Tuff, and the Ogof Glyma Tuff in upward sequence (respectively units 1–3 in Fig. 4). Each is overlain by thinly bedded and laminated turbidite tuffs which above the upper ash-flow tuff are termed the Allt Felin Fawr Member. Thus, thin turbiditic tuffs constitute the background deposition in the upper members of the Carn Llundain Formation. These localities are within the Cader Rhwydog Tuff (unit 1) which is 186 m thick and comprises 161 m of uniformly coarse lapilli tuff, overlain by tuff grading into extremely fine vitric tuff at the top.

Locality B5

Here the tuff rests unconformably on a surface which slightly cuts down into the underlying Pwll Bendro tuffs to the E and infills a deeply incised gully close to the cliff top. Locally, near the base, the tuff includes rounded pebbles and granules of rhyolite, angular clasts of perlitic and flow-banded rhyolite and lenticular fragments of black mudstone. At the gully this mixture extends irregularly upwards *c.* 12 m into the overlying tuff.

Locality B6

The main body of the tuff comprises ragged clasts, up to 32 cm, of tube pumice, with variable proportions of rhyolite clasts and crystals of quartz and plagioclase in a strongly recrystallized vitroclastic matrix. Although welded shard fabrics are not evident, moderate flattening of the pumice, roughly parallel to the sheet dip, is widely developed, including places along the base. Pumices moulded around lithic clasts, and porphyritic rhyolite fiamme, are common. Such flattening fabrics are as significant as welded shard fabrics in indicating heat retention. Small pockets of crystals, winnowed by volatiles streaming through ephemeral pipes, also occur. Crude columnar joints extend from the base to *c.* 100 m in the tuff and are further evidence of heat retention.

Locality B7

Although the relative proportion and grade of the main tuff constituents are variable, no sub-units or overall grading can be distinguished in the lower 161 m of the unit. However, above there is a gradation into an unbedded fining-up sequence, through crystal pumice vitroclastic tuff with strongly flattened pumice clasts, into extremely fine grained vitroclastic tuff. Fine laminations in the lowest metre of the conformably overlying tuffs show convolution in places.

The Cader Rhwydog Tuff is interpreted as a closely proximal, subaqueously emplaced, hot ash-flow with associated ash-fall, produced by column collapse and subsequent deposition from suspension in a single, entirely submarine eruption. Emplacement under water is implied by the gradation from the main massive tuff to the finely laminated top with clear evidence of 'wet-sediment' disturbance. The absence of slumping or major incursions of remobilized tuff indicates that the depositional surface here was virtually horizontal, perhaps due to ponding. Proximity to the vent is suggested by the great thickness of the tuff and the thickness of the associated fall which grades directly from the coarse grade of the flow. This proximity, together with the apparent lack of slope, indicates that the vent must have been submarine. This is also supported by the flattening fabrics in the graded fall which could not have developed if the tuff had settled from air into water and then through a water column. Similarly, the graded fall could not represent a cloud suspended in water by the ash-flow, but must have been rapidly deposited from suspension in the eruption column. It is envisaged that, compared with a subaerial Plinian event, the explosive violence of the submarine eruption would have been suppressed and the column height reduced. Once initiated, much of the eruption would have proceeded within a cupola of steam and magmatic volatiles, resulting in less chilling than might otherwise be expected in a subaqueous eruption. As the eruption waned, much of the suspended column rapidly settled through the steam and finally the remaining finer suspension settled more slowly through water. Hot clasts enveloped by steam were readily flattened during burial. If this interpretation is correct, the Cader Rhwydog Tuff includes the first recorded subaqueously welded ash-fall deposit.

To the E of the Ramsey Fault the Cader Rhwydog Tuff rests unconformably on Pwll Bendro tuffs and on sparsely pebbly mudstones of the Porth Llauog Formation. The unconformity probably developed mainly by volcano-tectonically induced slumping, with minor gullies cut by sediment-gravity flows. Apparently the unconformity surface and gully at locality B5 were overlain by a muddy debris-flow deposit which became admixed with the ash-flow, perhaps facilitated by explosive evolution of steam from trapped water.

The laminated fine turbiditic tuffs at the top of the Cader Rhwydog Tuff are conformably succeeded by the Trwyn yr Allt Tuff, a further thin sequence of turbiditic tuffs and the Ogof Glyma Tuff. The latter is similar to the Trwyn yr Allt Tuff but also includes welded shard fabrics. The overlying Allt Felin Fawr Member, comprising thinly bedded, graded and laminated turbiditic tuffs with intercalated debris-flow deposits and rhyolitic lava, is best examined at localities B9 and B10. However, near its base (locality B8), this latter member is intruded by a slightly discordant sill of porphyritic rhyolite which, at its contacts, shows remarkable features of intrusion of magma into unconsolidated wet ashes; these features have been described and interpreted by Kokelaar (1982).

Locality B8

See Kokelaar (1982, pp. 28–9).

Locality B9 (Fig. 5)

This is to examine the rhyolitic lava and the two slumped ash-flow tuffs in the lower of the two tuffaceous wedges which occur in the lava pile.

The lava, of quartz and plagioclase-phyric flow-banded rhyolite, is entirely brecciated at outcrop. Its base has locally sunk into and disturbed laminated turbiditic tuffs. To the E the lava is apparently a single unit, *c.* 35 m thick, but the two wedges in the W show it to be compound. In places near basal contacts, the breccias form a jigsaw fabric, suggesting that they are at least in part derived by *in situ* autobrecciation.

At the base of the lower wedge (columns 5 and 6) there is a thin sequence of turbiditic fine tuffs which mantles the rhyolite in the SW but does not persist to the NE, indicating topographic restriction of the turbidity currents.

The succeeding (lower) ash-flow tuff in the NE (columns 3 and 4) consists of a basal zone of coarse lapilli tuff, which grades up into a poorly sorted gritty tuff showing normal grading of rhyolite clasts and inverse grading of pumices. Lapilli in the basal zone are of subangular to subrounded perlitic rhyolite and irregular pumices, up to 20 cm, and the matrix comprises medium- to fine-grained vitroclastic tuff with quartz and plagioclase crystals and fragments of fine tuff. The top grades sharply into fine tuff. Towards the SW (columns 5 and 6) the tuff shows a flattening fabric and contains randomly oriented, contorted slabs of fine, in places laminated, crystal- and pumice-bearing tuffs. These slabs are up to 1 m thick and 10 m long. In addition, a block—2.1 m in diameter—with lobate and peperitic protrusion of porphyritic rhyolite in laminated tuffs (column 5) lies at the base. This block is derived from the top contact zone of the closely subjacent sill (locality B8).

This lower ash-flow tuff is overlain by a debris flow, up to 0.5 m thick, composed of rhyolite clasts and crystals in a mudstone matrix. Towards the NE this bed is overlapped by the upper ash-flow tuff, which is lithologically similar to the lower. In the NE (column 2) normal grading is well developed but to the SW the grading breaks down. Coarse and fine tuffs, in ill-defined patches, and large chaotically contorted slabs of laminated fine tuff, occur throughout the unit. In places the tuff grades up into bedded sandy tuffs (columns 4 and 5) and the underlying debris-flow deposit is disturbed, with flame-like protrusions penetrating upwards to 2 m (column 5). Further SW the tuff occurs as large, rounded and irregular sack-like bodies separated by thin silicified veneers of the debris-flow deposit (column 6). A contorted slab of pelagic black mudstone is included. This remarkable deposit also contains intru-

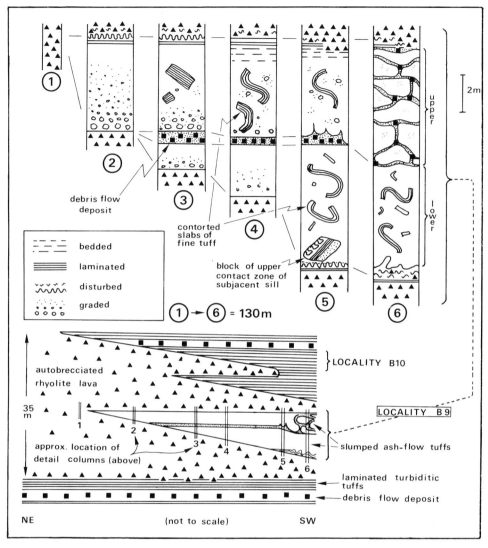

FIG. 5. Schematic cross-sections of rhyolite lava with intercalated slumped ash-flow tuffs and turbiditic tuffs, SW Ramsey Island (localities B9 and 10).

sions—up to 8 m in diameter—of porphyritic rhyolite with peperitic margins. The fine turbiditic tuffs that generally overlie this deposit are commonly contorted, soft-sediment faulted, or cut by the succeeding rhyolite breccia.

The upper and lower tuffs of the wedge are interpreted as slumped, poorly sorted ash-flow tuffs. The contorted slabs are interpreted as turbiditic and ash-fall deposits, originally overlying and genetically associated with the primary ash-flows, but subsequently incorporated during slumping and mass-gravity flow. The lateral variations of the deposits reflect telescoping of various facies. The sections are interpreted as the sides of the deposits, where the

material at the edge (NE) was deposited before that towards the centre (to the SW). It is considered that the upper slumped ash-flow ploughed into the muddy debris-flow deposit and, for reasons that presently are unclear, bodies of the tuff maintained their integrity and balled-up in the mud.

Much of the lithic rhyolite content of the slumped ash-flow tuffs was probably incorporated during eruption through and, or, flow across, lava similar to that which is presently spatially associated with the tuffs.

The block of the subjacent sill's contact zone indicates that this zone was exposed when the lower slumped tuff was deposited. As sliding along the sill's upper contact zone occurred at the time of intrusion (Kokelaar 1982), the slumping of the ash-flow tuff was at least contemporaneous with, if not a direct result of, intrusion of the sill. Thus sill emplacement was contemporaneous with the construction of the rhyolitic lava pile. The flattening fabrics in the lower slumped tuff indicate that slumping occurred soon after primary deposition, so that more or less contemporaneous emplacement of the sill, lava and primary ash-flows is indicated.

Locality B10

The upper wedge in the rhyolite pile comprises turbiditic medium- and fine-grained tuffs, and approximately 20 m of these tuffs overlie the rhyolite. A debris-flow deposit, 1–2 m thick, is also present. The turbiditic tuffs wedge-out against the rhyolite and this indicates topographic restriction of the depositing currents and precludes interpretation of the fine tuffs as being due to ash-fall. The coarser tuffs, with clasts of pumice, rhyolite, and quartz and plagioclase crystals, in a commonly well-preserved vitroclastic matrix, show normal grading, erosive bases and loading structures which are locally truncated and isolated by erosion of the parent bed. Cross, parallel, and convoluted laminations, and slumping are all locally well developed.

The angles between the upper surfaces of the rhyolitic breccia and the laminations of the turbiditic tuff are *c.* 6°. Such shallow slopes and the presence of jigsaw-fitting breccias indicate that the rhyolites are *in situ* lava-flow units and not talus deposits peripheral to a steep-sided dome.

The uppermost turbiditic tuff of the Allt Felin Fawr Member is the youngest stratum on Ramsey Island.

Excursion C: Abereiddi (Fig. 6)

Around Abereiddi Bay volcanic rocks are interbedded with strongly cleaved black pelagic mudstone of Llanvirn age. This excursion is to examine the '*Didymograptus murchisoni* Ash' (Cox 1915) on the S side of the bay.

Locality C1: NE of Aber Creigwyr (approximately 100 m NW of the ancient earthworks)

Here the '*D. murchisoni* Ash' comprises two fining-upward sequences of basaltic lapilli tuffs. The dip is *c.* 45°N and the whole 95 m-thick lower sequence crops out. The lowermost *c.* 5 m of the upper sequence is preserved as an outlier on the cliff top. The base of each sequence is sharp and plannar and that of the upper sequence rests conformably on the finer top of the lower.

The tuffs are composed mostly of angular basaltic clasts, originally glassy and porphyritic but now deformed and highly altered. Sorting is poor with scoriaceous lapilli and sparsely vesicular blocks and rare bombs, up to 40 cm in diameter, supported by a matrix of fine to coarse tuff. The lower sequence is poorly to moderately bedded, with successive massive beds being of finer grade and with more clearly defined coarse-tail grading. The upper sequence is also graded but not obviously bedded. The tuffs are interpreted as deposits from debris flows and high-density turbidity currents.

Locality C2: Melin Abereiddi

Here, the '*D. murchisoni* Ash' is *c.* 60 m thick; the lower sequence is 52 m and the upper 8 m. The base of the lower sequence (exposed

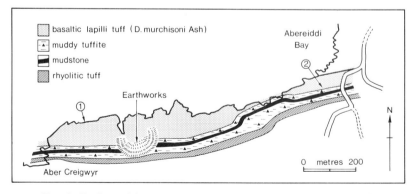

FIG. 6. Geology of the coast W of Abereiddi Bay (Excursion C1 and 2).

in the valley side just above and S of the water-fall) rests on muddy tuffites. The lapilli tuffs near the base, like those at locality C1, are poorly sorted, show crude coarse-tail grading and bedding is ill-defined. Upwards the tuffs are finer and become thinly bedded, as can be seen in the outcrops on the foreshore. The base of the upper sequence is marked by a bed of lapilli tuff and the sequence above is of well-bedded fine to coarse tuffs and lapilli tuffs. Bed thickness varies from *c*. 3 to 25 cm, with the thicker beds being generally the coarser. Most bed bases are planar although in places erosive contacts can be determined. Some beds show normal grading, but most lack regular variation in grain size. Parallel lamination and ripple drift cross-lamination also occur and the tuffs are interpreted as deposits from both high- and low-density turbidity currents.

The tuffs are overlain by an alternation of thinly bedded turbiditic tuffs with mudstone intraclasts and muddy debris-flow deposits with tuff clasts. These are succeeded by 2 m of mudstone with horizons of sparse tuff clasts, and then typical black, richly graptolitic (*D. murchisoni*) mudstones.

The tuffs become thicker, coarser and less well bedded towards the W, and the section is interpreted as the thin distal edge of a sub-marine volcanic pile. The tuffs were deposited by various sediment-gravity flows from tuffs that had previously been deposited close to the vent. Instability in the accumulating tuffs caused periodic slumping and transport of the tuffs to greater water depths. The improvement in bedding and general fining upwards in the lower unit probably reflects a decreasing rate of effusion, such that slumps became progressively smaller and dispersal of coarse material more limited. A temporary return to a relatively high effusion rate is reflected in the coarse base of the upper unit. Muds subsequently mantling the pile periodically slumped and were redeposited with, and eventually without, further turbiditic tuffs.

Excursions D and E: Pen Caer (Figs 7, 8 and 9)

On Pen Caer peninsula, NW of Fishguard, excellent coastal exposures provide a section through the lower Llanvirn Fishguard Volcanic Group. The group, up to 1800 m thick, shows a wide variety of submarine volcanic rocks and is divisible into three formations. The lowermost Porth Maen Melyn Volcanic Formation is composed principally of rhyolitic lavas, autobreccias, debris-flow breccias and bedded or massive tuffs, although rhyodacitic massive and pillowed lavas also occur (Bevins & Roach 1979). The overlying Strumble Head Volcanic Formation is composed almost entirely of basaltic pillowed and massive lavas with thin hyaloclastites and tuffs, and rare rhyolitic tuffaceous horizons. The uppermost Goodwick Volcanic Formation is predominantly rhyolitic, with further lavas, autobreccias and ash-flow tuffs although a thick sequence of bedded basaltic tuffs occurs within the formation and, at the top of the section, basalts were intruded at a high level into wet silicic tuffs and tuffites.

Excursion D (Fig. 7) is a traverse through the Porth Maen Melyn Volcanic Formation at its

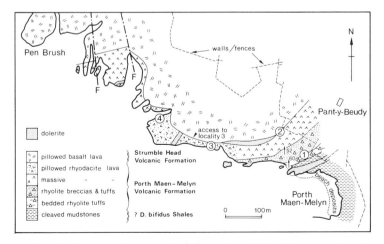

FIG. 7. Geology of the coast N of Porth Maen Melyn (Excursion D1–4).

type section. The junction with the overlying Strumble Head Volcanic Formation can also be examined.

Excursion E (Figs 8 and 9), in NE Pen Caer, illustrates the complex interdigitation of pillowed and massive basaltic and dacitic lavas with volcaniclastic silts, within the upper part of the Strumble Head Volcanic Formation, and also the predominantly rhyolitic activity of the Goodwick Volcanic Formation. Contemporaneous basaltic activity and a major penecontemporaneous slide are demonstrated.

Locality D1: lower part of the Porth Maen Melyn Volcanic Formation at Porth Maen Melyn

Porth Maen Melyn is eroded into soft cleaved mudstones, probably of the lower Llanvirn *D. bifidus* Shales (Cox 1930). No diagnostic fossils have been found. An intrusive dolerite sheet forms the promontory at the northern end of the bay and can be examined in the crags at the top of the cliff.

The lowermost unit of the Porth Maen Melyn Volcanic Formation crops out in prominent crags adjacent to the coastal path on the N side of a small valley. It comprises 10 m of bedded rhyolitic tuffs. Extensive recrystallization has resulted in the development of a peculiar spherulitic texture which is only readily observable on weathered surfaces. The spherules, up to 2 cm in diameter, and 'cigar-shaped' nodules, up to 3 cm in length, have obliterated the original vitroclastic fabric.

Overlying these tuffs is a sequence, 35 m thick, of white-weathered lithic-crystal-vitric breccias and buff crystal-vitric tuffs. Two units can be identified, both generally fining upwards although the lowermost 50 cm of the lower unit is inversely graded. The breccia clasts are predominantly of angular to sub-rounded rhyolitic lava, commonly with a perlitic texture, although in the lower part of the lower unit basalt and dolerite clasts also occur. Crystals are of plagioclase and quartz (typically bi-pyramidal) and increase in proportion in the fine tail. Shards and shard fragments are abundant and weathered-out streaky clasts probably represent pumices. This sequence of tuffs and breccias reflects sediment-gravity flow reworking of primary products of both explosive and quiet effusion of rhyolitic magmas.

Locality D2: upper part of Porth Maen Melyn Volcanic Formation and lower part of Strumble Head Volcanic Formation

Overlying, and partly loaded into, the clastic deposits of locality D1 is a massive rhyodacitic lava flow, 40 m thick. The lava is grey to green and cryptocrystalline or spherulitic with a well-developed perlitic texture. Individual perlites reach 1 cm in diameter and in the central parts of the flow flow-banding is locally developed. Rare plagioclase micro-phenocrysts are present, but otherwise the flow is thought to have been originally glassy. The top of the flow is autobrecciated, with clasts up to 5 cm, and is overlain by a thin horizon of basic hyaloclastite of the Strumble Head Volcanic Formation. This is in turn overlain by epidotized, pillowed basaltic lava. Well-developed radial joints can be observed in a number of pillows which generally also possess highly vesicular margins. Pumpellyite, epidote, albite and chlorite of metamorphic origin are abundant, although augites are pristine and pseudomorphs of igneous textures are generally preserved.

Locality D3: upper part of the Porth Maen Melyn Volcanic Formation

The massive rhyodacite lava of locality D2 passes laterally into pillows and elongate tubes and then into isolated-pillow breccias. Excellent exposures occur in the S-facing cliff section above a narrow wave-cut platform, some 300 m W of Porth Maen Melyn. Access to this platform is possible at only one place, and even here the descent is hazardous and the greatest care is necessary. The pillows, 1–3 m in diameter, are separated from each other by cleaved inter-pillow breccia generated by desquamation of the pillow margins. The rhyodacite is purple to green in colour and is petrographically similar to that at locality D2, although here distinctive spherulites are present at pillow margins and within the inter-pillow breccias. These spherulites are particularly rich in K-feldspar and are thought to have been produced during initial crystallization of the lava, in contact with seawater. The tubes and pillows result from the rapid effusion of hot magma at the steep front of the lava flow. Photographs and full descriptions of the rhyodacite pillows and pillow breccias are given in Bevins & Roach (1979).

Locality D4: 500 m WNW of Porth Maen Melyn

From the cliff top, the contact between the pillowed rhyodacitic lavas and the basaltic lavas of the Strumble Head Volcanic Formation can be recognized along the northern side of the E–W trending inlet. The lighter-coloured rhyodacite pillows are noticeably larger than the more basic ones.

Locality E1: 100 m W of Maen Jaspis (Figs 8 and 9)

Basaltic pillow lavas, dipping 40°NNE, form dip-slopes down to a few metres above the high water mark. A thin veneer of fine volcanogenic sediments is locally preserved on this pillowed surface. In the W the sediments are overlain and locally disturbed by a 1.7 m sheet of basaltic lava, with pipe vesicles at its upper and lower margins. The generally smooth top of this lava is mostly overlain by further thin, fine sediments although hollows are infilled with coarse grit and sand. Towards the E this lava sheet passes laterally into vesicular pillow lava which wedges-out up-dip, and further E the pillows become smaller and intensely cleaved. Jasper is abundant, in places forming a geopetal partial fill in drained pillows. The irregular surface at the top of the lava is partly subdued by an infill of fine silicic siltstone, which in turn is overlain by up to 1 m of interbedded acidic and basic turbiditic sandstones and further extremely fine silicic siltstone. Locally these sediments are intruded by a thin, pillowed basaltic sill.

This sequence is overlain and locally disturbed by large-pillowed plagioclase-phyric dacite lava, up to 12 m thick, which is well exposed in the cliffs. To the E the pillows are less well developed and in the W the lava wedges-out on top of a wedge of intrusive dolerite. The dolerite contains well-formed curved columnar joints and has been intruded by a small body of dacite. The pillowed dacite lava is overlain by thin porcellanous sediments and the strongly draping base of an indistinctly banded sill, 12 m thick, of porphyritic dacite. Towards its top, on a small promontory, the sill contains prehnitic spherules. To the E, across a small inlet, the dacite sill is overlain by thin sediments and a rhyolite lava with pronounced flow-folded banding. The latter is cut and partly underlain by another dolerite sill.

This locality illustrates the extremely complex interdigitation of lavas and associated intrusions at the junction between the basaltic Strumble Head formation and the predominantly rhyolitic Goodwick formation. The abundance of intercalated sediments at this horizon, including some which may in part be chemically precipitated cherts, shows that basaltic activity waned considerably before the onset of rhyolitic activity. The presence of silicic turbidites, with no known local source at this level, indicates that the basaltic Strumble Head formation did not form a marked positive topographic feature, but rather filled a depression. Fissure eruptions along faults defining a possible gra-

ben are envisaged. Weak horizons of intercalated sediment facilitated penecontemporaneous intrusions.

Locality E2: Maen Jaspis

The flow-banded rhyolite, in places spherulitic, is brecciated near its base where it has locally disturbed underlying laminated and cross-laminated fine to medium silicic volcaniclastics. Beds of pumice with normal grading also occur. To the E the thin autobrecciated top of the rhyolite is overlain by a veneer of silicic volcanogenic sediment and several massive debris flows of rhyolitic breccia which in turn are overlain by a sequence of basaltic tuffs and acidic ash-flow tuffs, although these are best examined at locality E4.

Locality E3: Penfathach

On the Penfathach headland various facies of a thick rhyolitic lava flow (or small lava dome) are well exposed. The lowest facies comprises autoclastic breccias, with angular, flow-banded clasts averaging 4–5 cm and showing minimal post-brecciation movement. The passage into the overlying massive flow-folded banded lava is sharply gradational. Large- and small-scale flow folds occur and columnar cooling joints are well developed. The white-weathered rhyolite contains sparse plagioclase phenocrysts and spherulitic quartzo-feldspathic intergrowths in a dark-grey to green groundmass of similar mineralogy, including accessory sphene, zircon and apatite. Perlitic rhyolite, with fractures up to 10 cm in diameter, overlies the massive facies and reflects the originally glassy nature of the flow. The contact with the overlying carapace of autobreccia is gradational.

During late Caledonian tectonism, strain developed inhomogeneously in the rhyolite so that intensely cleaved zones alternate with virtually undeformed zones.

Locality E4: Porth Maen

At the foot of the cliffs on the SE side of Porth Maen is a sequence of generally parallel-bedded basaltic tuffs showing both normal and inverse grading, in beds up to 2 m thick. Average clast size in the coarser beds is 2–3 mm. Maximum basaltic clast size in coarser beds is *c.* 1.5 cm, although rarer rounded to subangular rhyolitic clasts up to 6 cm also occur. There are no bomb sags and the deposits are interpreted as those of sediment-gravity flows which were contemporaneous with a nearby

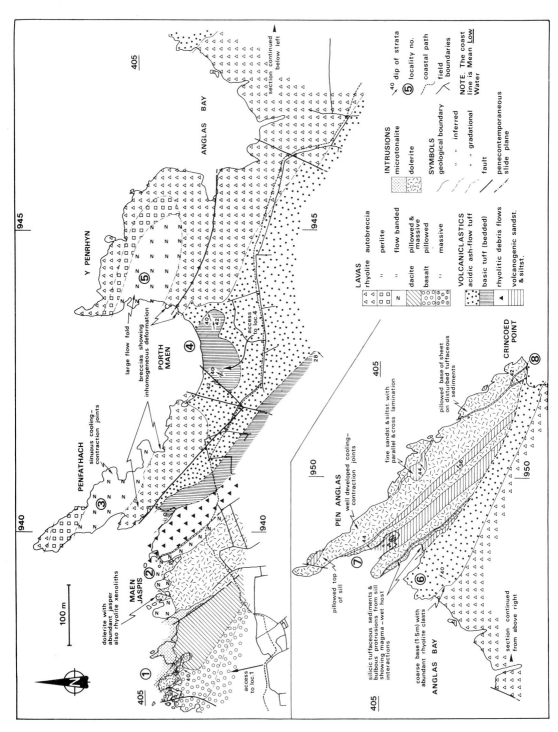

Fig. 8. Geology of NE Pen Caer (Excursion E1–8).

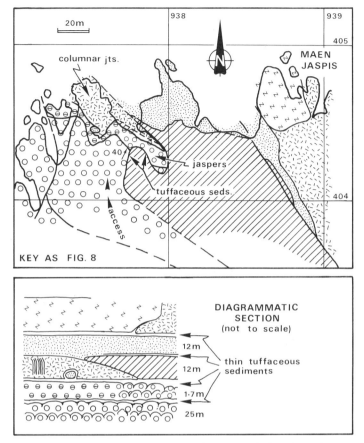

FIG. 9. Detailed geology 100 m W of Maen Jaspis (locality E1).

eruption. Soft-tuff faulting during accumulation is evident. In thin section the basic clasts are seen to be vesicular and composed of plagioclase microlites in chlorite with sphene, presumably after glass.

At the E end of the bay the basaltic tuffs are succeeded by rhyolitic ash-flow tuffs containing clasts of pumice and lava. The latter are well exposed in the cliffs on the faulted W side of the bay, and in fallen blocks.

Locality E5: Y Penrhyn

This is the eastward continuation of the rhyolitic lava described at locality E3. On the W of the headland a large-scale flow-fold is exposed and towards the SE the 'core facies', of massive and perlitic rhyolites, passes into autobrecciated rhyolite which constitutes the entire thickness for >700 m along strike.

Locality E6: E side of Anglas Bay

Approximately 1.5 m of coarse, lithic-rich, ash-flow tuffs rests on the rhyolitic autobreccia. The lithic blocks, up to 30 cm, are of the underlying rhyolitic lava proving its extrusive (subaqueous) emplacement. The coarse, basal unit is overlain by crystal-, lithic- and fiamme-rich tuffs, with fiamme mostly 10 cm across, but up to 50 cm, commonly moulded around lithic clasts of rhyolite. The tuffs fine through the succeeding 50 m and are crystal-rich at the top of the sequence. These ash-flow tuffs are succeeded by silty mudstones, fine sandstones and fine silicic tuffs which have been intruded by a series of basic sills at a shallow level. Close examination of locality E8 shows that the contact is a penecontemporaneous wet slide plane.

Locality E7: Pen Anglas

Bulbous protrusions and pillows along the contacts of the dolerite sills, and convolution, homogenization and vesiculation of the host sediments and tuffs indicate shallow-level intrusion into unlithified wet deposits. Polygonal jointing is locally well developed.

Locality E8: Crincoed Point

Here the basic sheets are pillowed with intervening fine sediments. Along a ledge just above the high-water mark the sediments underlying a pillowed body are in contact with the subjacent ash-flow tuffs. The sediments and tuffs are locally interleaved and contorted and at one point silts have been injected into the tuffs suggesting sliding of wet, unlithified materials. Clearly sedimentation, intrusion and sliding were penecontemporaneous.

Excursion F: journey between Fishguard and Snowdonia including a brief stop at Pared y Cefn-hîr (Figs 1 and 2)

NE of Fishguard the route generally follows the strike of the eastward continuation of the Fishguard Volcanic Group with the more resistant rhyolitic lavas and ash-flow tuffs forming low ridges. The tors on the northern edge of the Prescelly Hills to the S are composed of coeval dolerite. Further NE the dominating geological features are of Pleistocene age and the underlying Llandeilo mudstones and flags, and turbiditic sequence of Caradoc age (which extends beyond Cardigan), are poorly exposed. To the NE of Cardigan the route crosses turbidites of upper Llandovery age, which comprise the 1200–1500 m thick Aberystwyth Grits. The coast is roughly parallel to the axis of the sedimentary basin and the turbidites become more distal towards Aberystwyth. Major structures, a series of upright anticlines and synclines, are also aligned approximately NNE–SSW. Between Aberystwyth and Machynlleth the route crosses the western margins of several anticlinal plunge culminations of the Plynlimon Dome with Ashgill turbidites exposed in the cores.

North of Machynlleth the route passes across the conformable Ordovician—Silurian boundary, and the slate waste tips around Corris result from the extraction of both Ashgill and Llandovery cleaved silty mudstones. To the N of Corris, down sequence, the route passes on to the line of the Bala Fault which lies along the impressively straight valley of Tal y Llyn. North of this valley is the S flank of Cader Idris, composed of Ordovician volcanic rocks, mainly of pre-Caradoc age. The route to Dolgellau traverses down through the Ordovician on to the Upper Cambrian.

From Dolgellau a short detour can be made westwards along the minor road (via Gwernan Lake) to Pared y Cefn-hîr near the Gregennen lakes. Here a well-exposed sequence, of Arenig–Llanvirn age, includes conglomerates, sandstones and mudstones, volcanogenic sediments, rhyolitic lavas and tuffs, and basaltic lavas and tuffs (Cox & Wells 1921, 1927; Phillips 1966). The volcanic rocks are of the Aran Volcanic Group which extends E and N from here to the Moelwyns on the NE side of the Harlech Dome. At this locality there is much evidence of subaqueous emplacement of the volcanic rocks and redistribution of previously emplaced deposits by debris flows. The sequence is intruded by irregular sills of dolerite and granophyre. Granophyre intrudes the uppermost rhyolitic ash-flow tuffs of the Aran Volcanic Group on Cader Idris and has been considered to be coeval with these tuffs (Davies 1959).

North from Dolgellau, the rugged and spectacular scenery to the W of the road is developed in Lower Cambrian Rhinog Grits of the Harlech Dome. Turning NE towards Betws-y-Coed, the route traverses successively younger strata and passes between the major scarp of the Moelwyn Volcanic Formation to the W and the Manod Hills, of coeval intrusive rhyolites, to the E. Closer to Betws-y-Coed the route traverses the relatively thin Caradoc volcanic strata at the edge of the major accumulations of central Snowdonia.

Excursion G: Llanberis Pass (Fig. 10)

To examine the sedimentary rocks and associated volcanics of the Cwm Eigiau Formation and the base of the Lower Rhyolitic Tuff Formation.

The excursion begins at Pont y Gromlech in the Llanberis Pass. Here the Pitt's Head Tuff Member of the Cwm Eigiau Formation forms the crags on the S side of the road. This is a rhyolitic ash-flow tuff which was erupted from a centre at Llwyd Mawr in SW Snowdonia (Roberts 1969) where it was ponded in a volcano-tectonic depression and is *c.* 700 m thick. Williams (1927) considered it to be the basal member of the Snowdon Volcanic Group although Howells *et al.* (1983) have placed it in the underlying formation.

At this locality the tuff, 40–45 m thick, is typ-

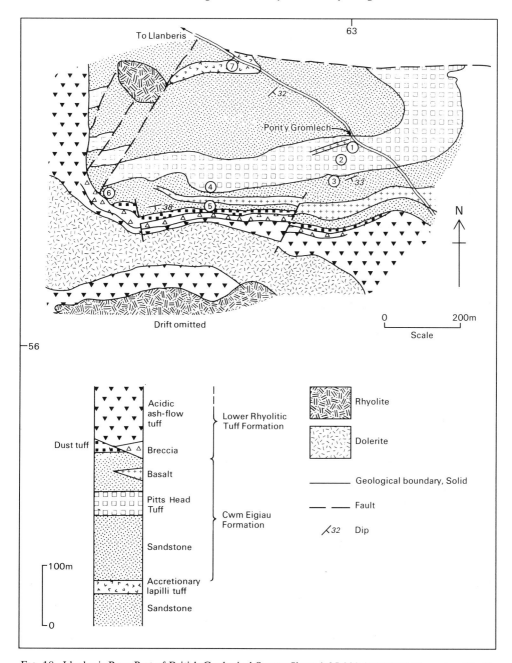

FIG. 10. Llanberis Pass. Part of British Geological Survey Sheet 1:25,000 SH65NW (Excursion G1–7).

ically bleached on its weathered surfaces where the eutaxitic foliation is accentuated by siliceous segregations. Distinctive chloritic fiamme, with indented and ragged peripheries, are scattered throughout. Here the base is not exposed; the lowest tuff is welded and is crossed by a thin basalt dyke (G1). The tuff can be examined along the W edge of the crags (G2) and within the main body the disposition of the eutaxitic planes is very irregular due to rheomorphism. Siliceous nodules are randomly distributed through the tuff although there is a marked concentration close to its top. The distribution of the nodules and the alignment of the rheomorphic fabric have been interpreted (Wright & Coward 1977) as resulting from the development of rootless vents. However, there is no indication to suggest that the escape of volatiles above the flow was concentrated about specific centres and the rheomorphism more probably reflects local slope-induced secondary movement of the tuff while in a generally plastic state.

The top of the flow (G3) is marked by 1.5 m of fine-grained, massive tuff which represents fall-out from the attendant suspended cloud. The tuff is overlain by medium- to coarse-grained, cross-bedded sandstones with shell debris of a shallow-marine environment. This sandstone infills the slightly undulose surface at the top of the tuff.

The Pitt's Head Tuff has previously been interpreted as a subaerial ash-flow tuff. However, it lies conformably in a sequence of marine sandstones which show no sedimentological change above and below the tuff, and along its extensive outcrop in SW Snowdonia there is no evidence to suggest that the tuff has been subaerially eroded. Kokelaar *et al.* (this volume) interpret the tuff as being erupted and emplaced in a subaqueous environment.

From G3 the disposition of the base of the Lower Rhyolitic Tuff Formation and the local structure about the Pass should be examined (British Geological Survey Sheet 1:25,000 SH65N/66S The Passes of Nant Ffrancon and Llanberis). The strata below the Lower Rhyolitic Tuff Formation are traversed at the edge of the screes (G4). They comprise cross-bedded sandstones, locally with dewatering structures, and a basaltic lava. The lava varies from massive, with polygonal joints, to basaltic breccia with broken pillows and pahoehoe toes. At the top of the sequence (G5), up to 6 m of well-bedded, parallel- and cross-laminated fine grained acidic tuffs underlie the Lower Rhyolitic Tuff Formation. These tuffs are widespread and include bands of fine-grained mudstone. The bedding characters indicate reworking and possibly deposition by turbidity currents. About the Pass these tuffs show a marked concentration of quartz-filled tension fractures.

The base of the Lower Rhyolitic Tuff Formation is marked by a pyroclastic breccia, up to 25 m thick, which here conformably overlies the bedded fine grained tuffs. At the base the breccia is clast supported although higher in the crags the blocks decrease in frequency and are matrix supported. The blocks, up to 0.9 m, are mainly of basalt, rhyolitic tuff and rhyolite in a matrix of finer lithic fragments, feldspar crystals and some shards. The breccia grades up into

cleaved rhyolitic tuff which higher in the crags is intruded by a thick transgressive dolerite sill. The breccia is an integral part of the Lower Rhyolitic Tuff Formation and represents the earliest eruptive phase. The abundance of basalt blocks suggests that they were incorporated from one or more of the numerous basalt centres that occur in the underlying sequence through central Snowdonia.

Westwards along the crags, the base of the pyroclastic breccia sharply downcuts the underlying sequence, eventually into juxtaposition with the Pitt's Head Tuff. At G6 the breccia abuts sandstones of the Cwm Eigiau Formation, which in this vicinity is faulted. The faults do not extend into the pyroclastic breccia. It is extremely unlikely that fault scarps would be preserved in water-saturated unlithified sediments and then covered by the pyroclastic breccia. The faults are interpreted as synvolcanic structures, resulting directly from the eruption (Kokelaar *et al.*, this volume).

At G7, on the S side of the stream, accretionary lapilli tuffs, acid tuffites and volcaniclastic sandstones are the local representatives of the Capel Curig Volcanic Formation. The accretionary lapilli, up to 1.2 cm, are concentrated in flaggy beds up to 10 cm thick, and dispersed through the associated sandstones. On fresh surfaces the lapilli are distinctive with a carapace of dark-grey, fine-grained, chloritized volcanic dust about their coarser centres. These tuffs represent the subaerial activity from a centre sited *c.* 3 km to the ENE during the late stages of accumulation of the Capel Curig Volcanic Formation (Howells & Leveridge 1980).

Excursion H: Pen y Pass to Llyn Llydaw (Fig. 11)

This short excursion is to demonstrate the general features of the Snowdon Volcanic Group in central Snowdonia; in particular the Lower Rhyolitic Tuff Formation.

The Miner's Track from Pen y Pass traverses the cleaved acidic ash-flow tuffs of the Lower Rhyolitic Tuff Formation (H1). The massive, well-cleaved tuffs comprise devitrified shards and vitric dust, with few scattered small albite crystals and ragged chloritized pumice clasts. The tuffs are non-welded and show no internal bedding.

The track crosses a small down-faulted wedge of basic tuffs and tuffites of the Bedded Pyroclastic Formation (H2). These flaggy-bedded tuffs are almost entirely composed of fragments of chloritized, vesiculated basaltic glass and

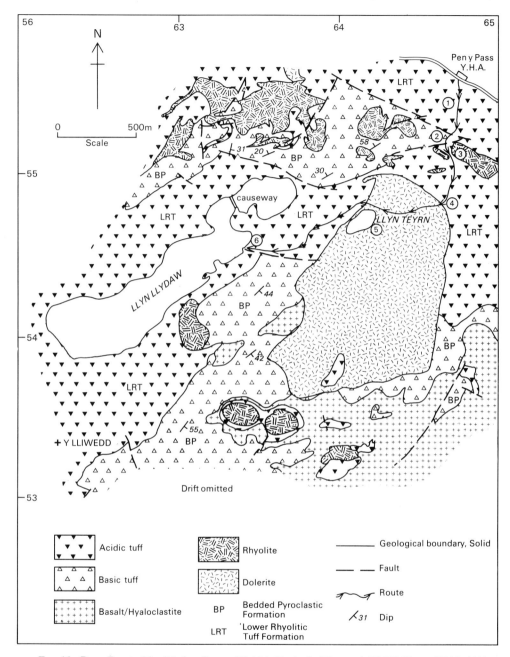

FIG. 11. Pen y Pass to Llyn Llydaw. Parts of British Geological Survey 1:25,000 Sheets SH64N/65S Snowdon, and SH65N/66S The Passes of Nant Ffrancon and Llanberis (Excursion H1–6).

basalt, of fine ash to coarse lapilli grade, with few blocks. Cross-bedding, ripple lamination, soft-sediment deformation, slumping and distinctive careous-weathered carbonate bands are common. The tuffs represent shallow-water accumulations and their source lies in the piles

of basaltic lava and hyaloclastite which are common in the sequence elsewhere in central Snowdonia.

At H3 a small rhyolite intrusion in the acidic ash-flow tuffs is well exposed. The rhyolite shows intense flow-folding on its bleached, weathered surfaces and much ochreous weathering and quartz veining at its margin, which has been locally excavated in a trial working for copper. At H4 the acidic ash-flow tuffs are indurated by a large dolerite sill; the contact is not exposed in the vicinity of the track but can be examined by following the feature of the tuffs a little distance to the N. At H5 the dolerite is medium- to coarse-grained and highly altered although a well-defined ophitic texture is easily discernible. The crags about Llyn Teyrn are spectacularly columnar-jointed.

The track up to Llyn Llydaw traverses the cleaved non-welded ash-flow tuffs of the Lower Rhyolitic Tuff Formation. This predominant lithology is remarkably uniform with no indication of flow units or internal bedding. Locally the tuffs are intensely veined with quartz. From Llyn Llydaw (H6) this thick sequence, up to 600 m, of massive tuffs is most strikingly displayed in the face of Y Lliwedd on the S side of the cwm. On the ridge to the E of Y Lliwedd, bedded acid tuffs, up to 40 m thick, at the top of the formation crop out. These tuffs are overlain by subaqueously emplaced basaltic tuffs and sediments of the Bedded Pyroclastic Formation. To the NW from this locality (H6) the S-facing flank of the cwm which is formed of a rhyolite intrusion in the Bedded Pyroclastic Formation. The contact of the rhyolite and the boundary between the Lower Rhyolitic Tuff Formation and Bedded Pyroclastic Formation can be clearly distinguished.

Thus in central Snowdonia the thickest and most uniform sequence of the Lower Rhyolitic Tuff Formation is underlain by marine sediments and overlain by marine-emplaced tuffs. The sequence is interpreted as an accumulation within a volcano-tectonic depression where emplacement was accompanied by rapid local subsidence (see Kokelaar *et al.*, this volume). Away from this centre the formation changes markedly (see Excursions J and K).

Excursion I: Gallt yr Ogof (Fig. 12)

This excursion is to examine the subaqueously emplaced and welded rhyolitic ash-flow tuffs and associated sediments of the Capel Curig Volcanic Formation in the crags at the N end of the Gallt yr Ogof ridge on the E limb of the Tryfan Anticline. Here the four tuff members of the formation are represented (Howells & Leveridge 1980).

At the W end of the crags (I1), thin flaggy and massive-bedded sandstones up to 4 m thick, interbedded with grey-green siltstones, underlie the lowest tuff of the formation. The

medium- to coarse-grained sandstones are in places pebbly, normally graded and cross-bedded. They are interpreted as deltaic deposits which encroached into a marine environment in which the evenly bedded siltstones were deposited. Temporary emergence is indicated by infilled desiccation cracks in some siltstone beds (I2).

The base of the lowest ash-flow tuff overlies sandstones at I3 and around the edge of the crags it sharply transgresses the bedding (I4). Similar irregularities were interpreted as large load casts (Francis & Howells 1973) resulting from collapse of the emplaced tuff into thixotropically disturbed sediments. More recently, Kokelaar (1982) has suggested that downcutting of the tuff into the sediments was facilitated by fluidization at the tuff–wet-sediment interface, which allowed the tuff to drill into the water-saturated sediments. The tuff is devitrified, shard-rich, with few quartz, albite and garnet crystals, and pumice clasts in a variable sericite-chlorite matrix after fine vitric dust. The tuff is welded at and above its irregular basal contact. In contrast, its top (I5) is planar with reworked crystal-rich tuff grading up into fossiliferous grey siltstones which separate it from the overlying ash-flow tuff. The tuff, underlain and overlain by marine sediments with no indication of deep erosion in its upper surface, is interpreted as being entirely subaqueously emplaced.

The base of the second ash-flow tuff is also highly irregular, with lobate extensions of the tuff occurring to 50 m below the mean base of the parent body (I6). The siltstone adjacent to the tuff is crowded with feldspar crystals which were incorporated in the silt at the time of tuff emplacement. The basal zone of the tuff, 2 m thick, is cleaved and rich in feldspar crystals. The tuff of the massive central zone is distinctively welded, with a pronounced eutaxitic foliation, and includes irregular zones of siliceous nodules. The upper zone comprises recrystallized fine vitric dust.

The third tuff member conformably overlies the fine grained top of the second member and the contact is well featured in the crags (I7). The basal few metres of the tuff are bedded and overlain by massive welded tuff, rich in albite crystals and lithic fragments. Towards the top, massive bedding features can be determined and the tuff member is overlain by blue-grey siltstones with brachiopod fragments (I8).

The fourth member comprises a basal massive non-welded ash-flow tuff, massive bedded tuffs with wispy siltstone laminations, and coarse- and fine-bedded tuffs with cross-

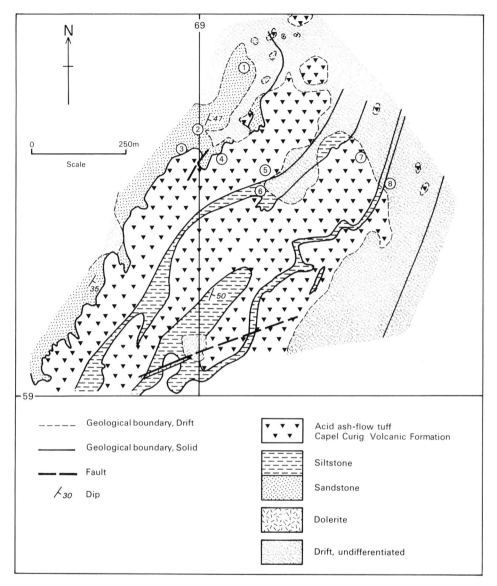

FIG. 12. Gallt yr Ogof. Part of British Geological Survey Sheet 1:25,000 SH65NE (Excursion I1–8).

bedding and scour-and-fill structures. The massive-bedded tuffs are debris-flow deposits from previously emplaced pyroclastic debris. The epiclastic element increases upwards and in places a distinctive carbonate component is indicated in the careous-weathered surfaces. The bedding in the upper zone reflects shallow-water reworking of the pyroclastic debris and, at the top, bands crowded with accretionary lapilli indicate that the column above the local eruptive centre had a subaerial expression.

The Capel Curig Volcanic Formation at Gallt yr Ogof reflects activity from three centres (Howells & Leveridge 1980). The lower two members are single ash-flow tuffs which were erupted from a centre 18 km to the N (near Conwy) and transgressed from a subaerial to a subaqueous environment just to the N of Gallt yr Ogof. In the subaerial environment the second ash-flow tuff directly overlies the first. At Gallt yr Ogof, and further S, the two members are separated by marine sedimentary rocks. To the N the bases of the tuffs are even and non-welded, but at Gallt yr Ogof the bases are irregular and welded at the contact with sediments. To the S of Gallt yr Ogof both tuffs were emplaced in a deeper-water environment.

The third member comprises an ash-flow tuff erupted subaerially from a centre *c*. 8 km to the N, which just encroached into a subaqueous environment about Gallt yr Ogof. The primary and slumped ash-flow tuffs of the fourth member were erupted from a local centre, to the SW of Gallt yr Ogof. The variably included epiclastic component with the evidence of secondary sloughing of previously emplaced pyroclastic deposits indicate the dominantly subaqueous environment of emplacement, although the occurence of accretionary lapilli towards the top of the member shows that the eruption col-

umn had at least a temporary subaerial expression.

Excursion J: Cwm Idwal (Fig. 13)

This excursion is to examine the sequence between the Pitt's Head Tuff and the Bedded Pyroclastic Formation, and in particular to examine the change in character of the Lower Rhyolitic Tuff Formation from that of the more proximal sequence described in the Snowdon massif to the S (Excursions G and H).

The route begins at Idwal Cottage. At the road cutting (J1) a section through the Pitt's Head Tuff on the NW limb of the Idwal Syncline is totally exposed. This ash-flow tuff con-

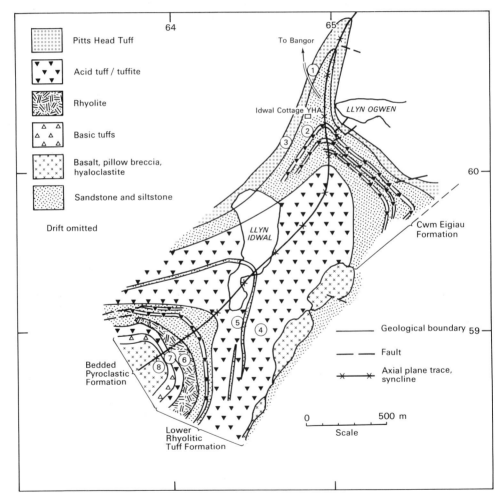

Fig. 13. Cwm Idwal. Part of British Geological Survey 1:25,000 Sheet SH65N/66S. The Passes of Nant Ffrancon and Llanberis (Excursion J1–8).

formably overlies coarse, shallow-water, cross-bedded sandstones and its basal 2 m is crowded with feldspar crystals in a non-welded, green, chlorite-rich matrix. Above, the grey-green tuff is white-weathered, silicified and comprises devitrified shards, feldspar and few rounded quartz crystals, and distinctive chloritic fiamme. The main body of the tuff, locally with well-developed columnar joints, is strongly welded with a pronounced eutaxitic fabric. Through this road section the disposition of the welded fabric is variable due to rheomorphism, probably caused by slope induced post-emplacement movement. The top of the tuff is marked by 2–3 m of fine-grained dust tuff. The tuff is overlain by shallow-water, cross-bedded sandstones, locally crowded with brachiopods. In places, these sandstones are tuffaceous although the background sedimentation of medium- to coarse-grained sands shows no influence by the underlying tuff, indicating rapid re-establishment of the supply of detritus from the area to the N where the Pitt's Head Tuff did not extend. The lack of change in the sedimentation below and above the tuff, and the absence of evidence of subaerial erosion in the upper surface of the tuff supports the interpretation (Kokelaar *et al.*, this volume) that the tuff was subaqueously emplaced. The fine-grained top of the tuff represents the fraction that settled from the water column following emplacement of the ash-flow.

Sandstones dominate the sequence between the Pitt's Head Tuff and the base of the Lower Rhyolitic Tuff Formation. In places the sequence is broken by beds of fine-grained acidic dust tuffs, such as those exposed in the quarry (J2). Here the tuffs are mixed with epiclastic debris and locally intercalated with thin beds of siltstone and fine-grained sandstones. Some beds are graded and infill scours in the tops of underlying beds and others are cross-bedded. This repeated bedded sequence and the internal structures suggest that the tuffs are the deposits of both high- and low-density turbidity currents with reworked air-fall tuffs.

At J3 the Pitt's Head Tuff on the NW limb of the syncline is well exposed in the roches moutonnées at the head of the major U-shaped valley. Here the tuff is intensely silicified with thin ribs of secondary quartz accentuating the rheomorphosed eutaxitic planes, and the development of quartz nodules. The nodules overprint the fabric and the adjacent tuff, depleted in silica, is generally dark-green with a prominent cleavage. Locally, lithophysal nodules occur in the main body of the flow.

At locality J4, close to Idwal Slabs, the pyroclastic breccia at the base of the Lower Rhyolitic Tuff Formation is well exposed. It comprises angular and subangular blocks of acidic tuff and rhyolite and common rounded blocks of vesicular basalt in a lithic, shardic tuff matrix. Locally the breccia is clast supported. In the scarps below the 'bedding planes' of Idwal Slabs the breccia grades upwards, through acidic ash-flow tuff with isolated blocks and breccia pods, into clast-free acidic ash-flow tuff. This non-welded tuff comprises recrystallized shards and vitric dust, isolated albite-oligoclase crystals and pumice fragments. Above the Idwal Slabs, bands of fine-grained tuff are developed and at J5 a bed of flinty dark blue-grey tuffaceous mudstone, 3.5 m thick, forms a prominent gully feature which extends up into the N face of Glyder Fawr. Above this mudstone the flaggy- to massive-bedded acid tuffs, locally reworked with low-angle cross-bedding, contain large carbonate nodules, up to 0.7 m. The overlying tuffs, which are crossed by the track to the back wall of the cwm, show an increasing epiclastic element, locally grading into tuffite and tuffaceous siltstones and, distinctively, the flaggy-bedded tuffs are crowded with carbonate nodules, commonly conjoined.

Above the tuffs the sequence is dominated by siltstones with thin acid tuff and sandstone beds, and a thick rhyolitic lava flow which forms the steep crags low in the back wall of the cwm (J6). The dark blue-grey rhyolite is typically flinty and its weathered surfaces commonly bleached. It is flow-folded, autobrecciated and, in hand specimen, perlitic fractures and spherulitic recrystallization can locally be distinguished. The nose of the flow is exposed in the cliff on the NW limb of the syncline.

Near the cleft of Twll Du (J7), the rhyolite is overlain by thick, massive-bedded acid tuffs and tuffites. The tuffs are generally ill-sorted, although locally graded, and are typically heterogeneous admixtures of shards, crystals, and devitrified 'glass' fragments, in a matrix with variable fine-grained quartz, feldspar, sericite, chlorite and iron oxide. A few bands of basic tuff also occur. The thicker units are interpreted as mass flows, caused by sloughing of previously emplaced pyroclastic fragments which incorporated much epiclastic debris during transport. These flows were caused by uplift in the area of the Snowdon massif following emplacement of the main body of the Lower Rhyolitic Tuff Formation.

The thick sequence of the Lower Rhyolitic Tuff Formation about the Snowdon massif (Excursion H) is dominated by unbedded, acidic, non-welded ash-flow tuff. In Cwm Idwal the thinner sequence is typically well bedded with the primary ash-flow tuffs restricted to the lower part. Above, the sedimentary influence increases markedly and is associated with secondary mass flows of pyroclastic debris. This lateral variation is interpreted as resulting from the passage from a proximal to a distal environment in relation to the eruptive centre (see Kokelaar et al., this volume).

In the core of the syncline, about Twll Du, the junction between the Lower Rhyolitic Tuff Formation and the overlying basic tuffs of the Bedded Pyroclastic Formation is clearly defined. The basic tuffs are thinly bedded, pale-green and blue coloured, fine to coarse grained, with lapilli- and block-rich bands. Normally graded, even beds are common and cross-lamination and current ripple marked surfaces occur throughout. The tuffs are dominantly of highly altered vesicular basaltic glass and basalt, few feldspar crystals and isolated single crystals and clusters of epidote. The tuffs locally grade into tuffites and tuffaceous sediments which have yielded a shelly fauna characteristic of shallow water. The common occurrence of only slightly modified scoriaceous fragments suggest that they are derived from local eruptive centres and reflect variation of erupted debris rather than a sedimentary process.

In the crags about Twll Du (J8) the bedded basic tuffs are overlain by c. 70 m of pillowed basalt and basaltic breccia. In places the pillows are well formed, up to 1.5 m in diameter but generally less than 0.5 m. The chloritized and carbonated basalt is typically pale-green, vesicular, and locally rich in albite-oligoclase phenocrysts. The groundmass, where less altered, is of feldspar laths with grains and ophitic plates of titanaugite. In altered rocks this is replaced by actinolite.

Above Twll Du the basalt is overlain by reworked basaltic tuffs with thinner basaltic flows, locally pillowed, which are exposed across the upper cwm.

Excursion K: E of Capel Curig (Fig. 14)

This excursion is to examine the Lower Crafnant Volcanic Formation, which is broadly the lateral equivalent of the Lower Rhyolitic Tuff Formation in E Snowdonia. East of the village the route traverses the upper part of the Cwm Eigiau Formation to the tuffs of the Lower Crafnant Volcanic Formation which are well featured on the ridge.

At locality K1 basaltic tuffs locally overlie a fine-grained acidic tuff at the base of a sequence of thick beds of sandstone near the middle of the Cwm Eigiau Formation. The acid tuff is fine grained, massive- to flaggy-bedded with thin siltstone intercalations. It comprises devitrified fine shards and vitric dust and represents reworked air-fall tuffs. The basaltic tuffs, containing lapilli, blocks and bombs of basalt, are well bedded and show small penecontemporaneous faults. In the overlying sequence of basaltic tuffs the bedding is centroclinal and in places slumped. Along strike to the N and S basaltic tuffs wedge-out into the sandstone sequence. The basaltic tuffs represent a small cone which was preserved by subsidence and the accumulation of overlying sandstones.

The sandstones, well featured in the ground to the N, are medium- to coarse-grained, massive-, flaggy- and cross-bedded. They contain brachiopods and generally indicate shallow-water accumulation. They also include a few persistent bands of reworked fine-grained, air-fall, water-settled dust tuffs (K2). At the top of the sandstone sequence a breccia bed, c. 3 m thick (K3), comprises rafts, up to 1 m, of sandstone and acidic dust tuffs and blocks of basalt in a coarse matrix which contains a distinctive proportion of basaltic lapilli. The breccia is interpreted as the deposit of a debris flow that was initiated by seismic activity.

The overlying siltstones, at the top of the Cwm Eigiau Formation, include some bedded, fine-grained acidic tuffs (K4). A few bands are rich in feldspar crystals. The repeated flaggy bedding suggests that the tuffs do not represent primary air-fall deposits. Normal grading and cross-laminations can be distinguished in a few bands suggesting that the beds were deposited from turbidity currents. These beds occur in approximately the position of the Pitt's Head Tuff and are possibly the sole representative here of this ash-flow tuff whose province lies in SW Snowdonia.

The base of the lowest (No. 1) acidic ash-flow tuff of the Lower Crafnant Volcanic Formation (Howells et al. 1973) crops out at K5. The lowest 2 m is dark-green, chloritic, crowded with feldspar crystals, contains few lithic clasts and is well cleaved. The chlorite is probably derived from mud incorporated into the base of the flow, which also contains well-preserved brachiopods. The basal feature of the tuff can be traced to K6 where the main body of the tuff is well exposed in sections along an eroded joint plane. The tuff shows a crude internal massive bedding, separated by impersistent, irregular, chloritized silty intercalations. The bedding is

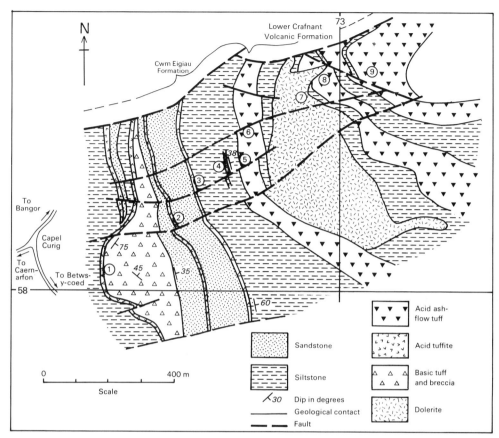

FIG. 14. East of Capel Curig. Part of British Geological Survey Sheet 1:25,000 SH75NW (Excursion K1–9).

interpreted as the early development of sub-flows which were initiated at the flow front in the more proximal part of the flow.

The non-welded tuff comprises devitrified shards, albite-oligoclase crystals and few pumice fragments, which all show a crude upward grading, in a matrix of sericite and chlorite which is interpreted as re-crystallized vitric dust. At its base the tuff contains carbonate nodules, generally weathered-out but where they are in place the internal fabric, in thin section, shows carbonate-replaced shards which are undistorted, unlike the shards outside the nodules. This indicates that the nodule development occurred soon after tuff emplacement.

The top of the ash-flow tuff is marked by a thin band, <1 m, of fine-grained tuff which represents the fine dust which settled after emplacement of the ash-flow. The source of the flow has been interpreted to be the eruptive centre of the Lower Rhyolitic Tuff Formation, in cen-

tral Snowdonia (see Excursion H). This centre was possibly temporarily subaerial during the late stages of activity although the flow here was entirely emplaced in a submarine environment. The ash-flow tuff comprises predominantly juvenile material which, considered with the wide distribution of the tuff in E Snowdonia, precludes its interpretation as the product of secondary sloughing of previously emplaced pyroclastic debris. During transport the primary ash-flow changed from a hot gas-charged to a water-charged state (Howells *et al.* 1973).

The overlying siltstones are intruded by a thick dolerite sill (K7) with locally well-developed columnar joints. In places the dolerite is gabbroic and here the ophitic texture is most apparent. Also, on weathered surfaces, the glomeroporphyritic texture characteristic of many of the bigger doleritic intrusions in N Wales is distinctive.

The No. 2 Tuff (K8) is chlorite-rich at its

base, due to incorporation of mud, and the underlying siltstones are contact-altered in proximity to the dolerite intrusion. Above, the tuff is characteristically fine grained, silicified and white-weathered. Siliceous nodules occur randomly throughout the tuff across most of its outcrop. The feldspar crystal fraction is less than in the No. 1 Tuff and the shards and crystals are generally finer. However, the source of the ash-flow is again interpreted as that of the Lower Rhyolitic Tuff Formation in central Snowdonia from which the flow escaped eastwards into deeper parts of the submarine basin (Howells *et al.* 1973).

Marine siltstones separate the No. 3 Tuff from No. 2 Tuff. The No. 3 Tuff (K9) is typically coarse grained. Its basal zone comprises large pumice blocks, which can most easily be distinguished on weathered surfaces, devitrified

glass fragments and coarse shards. A crude upward grading can be distinguished. The general characters, although of ash-flow type, are considerably less uniform than those of the lower two flows and the No. 3 Tuff here is interpreted as being proximal to its source which has been postulated to lie to the N, and not, as in the lower two flows, to the W (Howells *et al.* 1973).

Excursion L: Sarnau (Fig. 15)

To examine the Middle Crafnant Volcanic Formation at its type locality near Betws-y-Coed. Here the sequence comprises black mudstones, thin blocky rhyolitic ash-flow tuffs, and fine-grained, water-settled dust tuffs which locally grade into black mudstones. At Sarnau

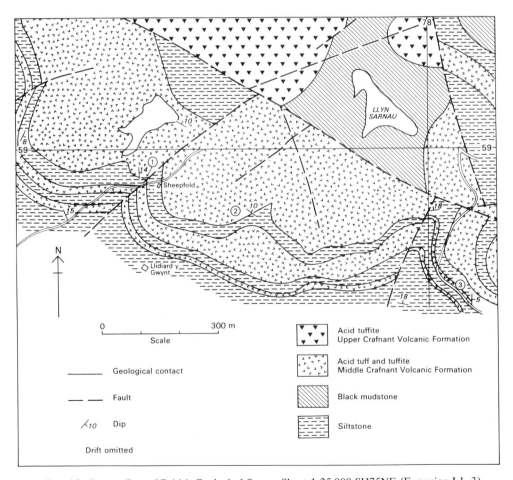

FIG. 15. Sarnau. Part of British Geological Survey Sheet 1:25,000 SH75NE (Excursion L1–3).

these lithologies form a gently dipping, evenly bedded sequence with the tuffs forming scarps and the mudstones the intervening slack features.

In the roadside exposure (L1) an acidic ash-flow tuff includes blocks, up to 0.3 m, of fine-grained silicified tuff and tuffite. The host tuff comprises fine shards and few albite crystals and is non-welded. On its weathered surface the tuff is distinctively speckled with small cream-coloured patches of segregated carbonate. Some of the included dark-grey tuffaceous blocks show indented peripheries, with 'flames' of mudstone into the adjacent tuff, which suggests that the blocks were unlithified when they were incorporated into the ash-flow tuff. From this locality the even dip-slopes and small scarps produced by the differential erosion are well displayed to the E.

Along the top of the scarp feature to the E (L2) the beds overlying the blocky ash-flow tuff are well displayed in a series of small exposures. These consist of evenly banded, fine-grained tuff, finely interbedded with, and grading into, blue-black mudstones. Where these striped beds directly overlie thin ash-flow tuff they are interpreted as representing the fine dust that settled out of the water column after the ash-flow had been emplaced. Where the beds are thicker, and not directly related to an ash-flow, they probably represent the distal fall-out, possibly subaerial, wind-transported and water-settled, from the active centre. Locally these banded fine tuffs, tuffites and mudstones, are intensely contorted by thixotropic yielding of the unlithified strata, either as a result of loading by an overlying rapidly emplaced ash-flow tuff or by seismic shock.

The locally well-developed striped sequence of the Middle Crafnant Volcanic Formation is well displayed at the edge of the forestry track (L3). Here the background sediments are clearly of euxinic black mudstones into which the tuffs were emplaced by flow or as fine dust settled from the water column. Some of the tuffs are intimately admixed with mudstone and these are interpreted as deposits resulting from secondary sloughing of unconsolidated pyroclastic and epiclastic debris. In addition, near the base of this section, there are thin ribs of turbiditic sandstones.

In NE Snowdonia, the Middle Crafnant Volcanic Formation reflects an accumulation of primary- and secondary-emplaced acidic pyroclastic debris deposited in at least moderately deep, 'stagnant' water. The Bedded Pyroclastic Formation in central Snowdonia is of basaltic volcanism in shallow water contemporaneous with this activity. These contrasting environments were separated from each other by a rising ridge, possibly along a flexure developed above a basement fracture. The turbiditic sandstones which occur in the Middle Crafnant Volcanic Formation were derived from this ridge (see Kokelaar *et al.*, this volume).

ACKNOWLEDGMENTS: The authors wish to thank BP Minerals and the National Museum of Wales for their support in production of the guide for the meeting in September 1982. M.F.H. publishes with permission of the Director, British Geological Survey, NERC.

References

BEVINS, R. E. & ROACH, R. A. 1979. Pillow lava and isolated-pillow breccia of rhyodacitic composition from the Fishguard Volcanic Group, Lower Ordovician, S.W. Wales, United Kingdom. *J. Geol.* **87**, 193–201.

BOUMA, A. H. 1962. *Sedimentology of Some Flysch Deposits: A Graphic Approach to Facies Interpretation.* Elsevier, Amsterdam. 168 pp.

COX, A. H. 1915. The geology of the district between Abereiddy and Abercastle. *Q. Jl geol. Soc. Lond.* **71**, 273–340.

—— 1930. Preliminary note on the geological structure of Pen Caer and Strumble Head, Pembrokeshire. *Proc. Geol. Assoc. Lond.* **41**, 274–89.

—— & WELLS, A. K. 1921. The Lower Palaeozoic rocks of the Arthog–Dolgelley district (Merionethshire). *Q. Jl geol. Soc. Lond.* **76**, 254–324.

—— & —— 1927. The geology of the Dolgelley District, Merionethshire. *Proc. Geol. Assoc. Lond.* **38**, 265–331.

DAVIES, R. G. 1959. The Cader Idris Granophyre and its associated rocks. *Q. Jl geol. Soc. Lond.* **115**, 189–216.

FISKE, R. S. & MATSUDA, T. 1964. Submarine equivalents of ash flows in the Tokiwa Formation, Japan. *Am. J. Sci.* **262**, 76–106.

FRANCIS, E. H. & HOWELLS, M. F. 1973. Transgressive welded ash-flow tuffs among the Ordovician sediments of N.E. Snowdonia, N. Wales. *J. geol. Soc. London,* **129**, 621–41.

HARLAND, W. B., COX, A. V., LLEWELLYN, P. G., PICKTON, C. A. G., SMITH, A. G. & WALTERS, R. 1982. *A Geologic Time Scale.* Cambridge University Press, Cambridge. 131 pp.

HOWELLS, M. F. & LEVERIDGE, B. E. 1980. The Capel Curig Volcanic Formation. *Rep. Inst. geol. Sci. Lond.* **80/6**, 23 pp.

——, —— & EVANS, C. D. R. 1973. Ordovician ash-flow tuffs in eastern Snowdonia. *Rep. Inst. geol. Sci. Lond.* **70/3**, 33 pp.

——, ——, REEDMAN, A. J. & ADDISON, R. 1983.

The lithostratigraphical subdivision of the Ordovician underlying the Snowdon and Crafnant Volcanic Groups, North Wales. *Rep. Inst. geol. Sci. Lond.* **83/1**, 11–5.

JONES, O. T. 1938. On the evolution of a geosyncline. *Q. Jl geol. Soc. Lond.* **94**, lx–cx.

KOKELAAR, B. P. 1979. Tremadoc to Llanvirn volcanism on the southeast side of the Harlech Dome (Rhobell Fawr), N. Wales. *In*: HARRIS, A. L., HOLLAND, C. H. & LEAKE, B. E. (eds) *The Caledonides of the British Isles–reviewed. Spec. Publ. geol. Soc. London*, **8**, 591–6.

—— 1982. Fluidization of wet sediments during the emplacement and cooling of various igneous bodies. *J. geol. Soc. London*, **139**, 21–33.

——, FITCH, F. J. & HOOKER, P. J. 1982. A new K-Ar age from uppermost Tremadoc rocks of north Wales. *Geol. Mag.* **119**, 207–11.

LOWE, D. R. 1979. Sediment gravity flows: their classification and some problems of application to natural flows and deposits. *Spec. Publ. Soc. econ. Paleontol. Mineral., Tulsa*, **27**, 75–82.

—— 1982. Sediment gravity flows. II. Depositional models with special reference to the deposits of high-density turbidity currents. *J. sediment. Petrol.* **52**, 279–97.

PHILLIPS, W. E. A., STILLMAN, C. J. & MURPHY, T. 1976. A Caledonian plate tectonic model. *J. geol. Soc. London*, **132**, 579–609.

PHILLIPS, W. J. 1966. The movement and consolidation of magmas—illustrated with reference to the succession of Ordovician strata and igneous rocks in the Arthog–Dolgellau district. *Welsh Geological Quarterly*, **2**, 3–9.

ROBERTS, B. 1969. The Llwyd Mawr ignimbrite and its associated volcanic rocks. *In*: WOOD, A. (ed.) *The Pre-Cambrian and Lower Palaeozoic Rocks of Wales*, 337–56. University of Wales Press, Cardiff.

WILLIAMS, H. 1927. The geology of Snowdon (North Wales). *Q. Jl geol. Soc. Lond.* **83**, 346–431.

WRIGHT, J. V. & COWARD, M. P. 1977. Rootless vents in welded ash-flow tuffs from northern Snowdonia, North Wales, indicating deposition in a shallow-water environment. *Geol. Mag.* **114**, 133–40.

B. P. KOKELAAR, School of Environmental Sciences, Ulster Polytechnic, Newtownabbey, Co. Antrim BT37 0QB, U.K.

M. F. HOWELLS, Wales Geological Survey Unit, British Geological Survey, Bryn Eithyn Hall, Llanfarian, Aberystwyth, Dyfed SY23 4BY, U.K.

R. E. BEVINS, Department of Geology, National Museum of Wales, Cathay's Park, Cardiff CF1 3NP, U.K.

R. A. ROACH, Department of Geology, University of Keele, Keele, Staffordshire ST5 5BG, U.K.